Talking to the Birds

A Compilation of Essays, Studies and Artworks by

Arto Juhani Heino

Volume I

Publisher

Ion Art - 1876406038

84 Hemphill Avenue

Mount Pritchard

2170 NSW

Talking to the Birds, Volume I

ISBN 978-1-876405-03-5

Arto J Heino 1955-

Printed by CreateSpace

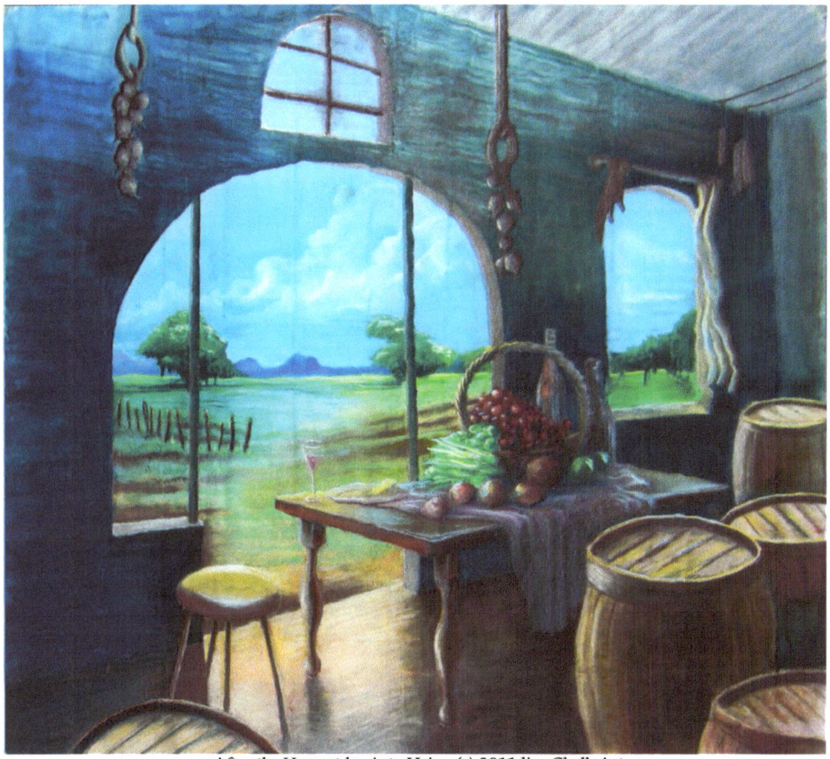

After the Harvest by Arto Heino (c) 2011 live Chalk Art

INTRODUCTION

This work was to be a series of five volumes, due to the enormous task undertaken, I have decided to compile some of my works made up of writing, essays and illustrations into one volume so that the readers can share in some of my many years of study. This volume should be read as a primer to creative intervention into the arts and sciences from all ages of mankind, not just the modern era. I would like the future of this work to become a set of detailed volumes with careful annotation and cross referencing, as I originally envisioned. The journey of this work began from where I left off in 1996, so a singular attempt to complete such a feat would require funding and staff.

As this was my sojourn away from corporate science and institutionalized methods into the esoteric study of magic squares and a new eye at what natural science and engineering had to offer the creative mind, alas, the task set forth would require another lifetime. I originally wrote about 40 pages back in 1998 involving the differing shapes and magic square connections, some of which I have included in this book. To me this is an evolving set of ideas that needed more research before I could explain some of the concepts that I had developed. As usual the pressures of a dysfunctional society and its governing bodies was to slow my progress in these matters almost to a full stop.

So to continue the creative journey after 15 years of working in the coal face of supply engineering I found many new ways of solving difficult problems in rotational machinery, these methods related to my work with magic squares and vibrational resonances. My excursions into electrical engineering was nothing more than completing the etheric picture I began when I was a child with valves and radios. The chapters with Astronomy, History and Archeology are personal rounding of an Education I never received from the school system, those teachers were more interested in indoctrinating me than teaching me any knowledge.

In 2011, I contacted Walter Cruttendon, Laird Scranton and Terry MacKinnell to to get feedback on my chapter on the historical and astronomical work. I found the input of these luminaries of great value and their kind words of encouragement added to my resolve to fulfill my artistic vision. As an artist the reader would be inclined to think that this book was to be about drawing and painting, this is a misunderstanding of the role the Artist plays in society, the Artist adds to the the ways knowledge can be interpreted and understood, this is the heart of creativity and his craft is more than pencils and paper, the Artist stands as the eyes, ears and mouth piece of how we interpret our role and our interface with the universe.

My personal background in the science of acoustics and music creation was when I first started singing in the choir at 6 or 7 years of age. Even at that age I understood the meaning of resonance, as the adults of my primary school and Sunday church noticed I had a clear and in-tune voice and that I sang to the acoustics of the room. I personally felt at one with the room acoustics, instinctively. This was very noticeable when the teacher asks me to sing solo , as the room started to vibrate with certain notes when I sang, I remember tuning my voice slightly so it would resonate with greater intensity. This was not the best way to keep in concert pitch of A=220hz , I varied between 200 to 235 at least, as I was singing from a tonic , this is what I used as the starting note, the rest of the notes were harmonically in pitch and in tune to the scale that derived from the harmonics of the natural singing not to the pianos artificial mean tone scale, which to me sounded out of tune to the room and to itself quite often. I sang solo in the church and the school choirs until I was about 13 that is when my voice changed, this ended my childhood experiments in musical tone.

I also grew up listening to bird-song communications as birds were my companions while bush walking as a very young naturalist of about 8 years of age, in the early morning light, these sounds were an education to me, spelling out the natural sound, rhythmic and tone harmonics. They also sang of their need for food, shelter and companionship, I assumed everybody understood their language and never thought anything of it until I discovered in archeology books the historical references to the symbolism of birds in all ancient cultures, I could see that the ancient Egyptians seem to understand some of the language of bird-song, I just never understood how the historians and archaeologists kept the premise that it was all about religion. I could see the Egyptians ritualistic aspects very clearly, the art of the ancients spoke to me about natural forms infused and molded with human understanding of the universe and knowledge that has been carried forward from the past generations. These experiences awoke the natural scientist in me, where observation of the natural world went hand in hand with drawing, notation and experimental verification. If there was somebody I admired most when I was this age it was Leonardo Da Vinci, I felt I knew his way of dealing with the understanding of the natural world.

At the tender young age of ten, I developed a profound insight into the symbolic meanings of shapes and colours, I augmented my curiosity with an increased reading and personal study, especially Gothic architecture and the Renaissance artists. At my local primary school I was asked to decorate one of the windows in our class room with something to celebrate Easter, I didn't skip a beat , I proceeded to paint 12 full length windows. each measuring 36 inches x 60 inches with the 12 apostles in full colour using school acrylic paint, the classroom was transformed into a church with stained glass windows. The teachers did not try to stop me and allowed me to paint during class time, this was a turning point to personal self knowledge and would set the bar line for all the other students in the school even to this day. This outlines some of my early experiences that influenced the choices of material that I explore so I can have a meaningful dialogue with the universe I live in, this I am only happy to share with all my readers, so if I can help give some insight and inspiration to others, then the task I have set myself would have been accomplished.

Currently I have at least forty notebooks comprising of one hundred pages of hand written notes that are my daily research notes along with my occasionally harried calculations and designs on loose leaf A3 and A4 pages, which seem to mount up to the roof. This out pouring is augmented by free hand sketches and cartoons that help keep my sanity from a world that is run by power mad and insane leaderships from all over the world.

When I was much younger I had wonderful dreams of a world run by an enlightened and humane government, only to be rudely awakened by their badgering of some hidden agenda that we as the people were not privy to. To my surprise I found the noise that had the people under a spell of conformity was by the popular media, simple and direct. This truth was arrived by accident when I was a copy-boy for The Australian newspaper at the age of sixteen. Part of this book has been to open the doorways to a new way of applying knowledge, a journey which began to take shape as a young man in the newsrooms when revelations of misguided leaders acquired power by the grand manipulation of information.

Please enjoy, Regards Arto

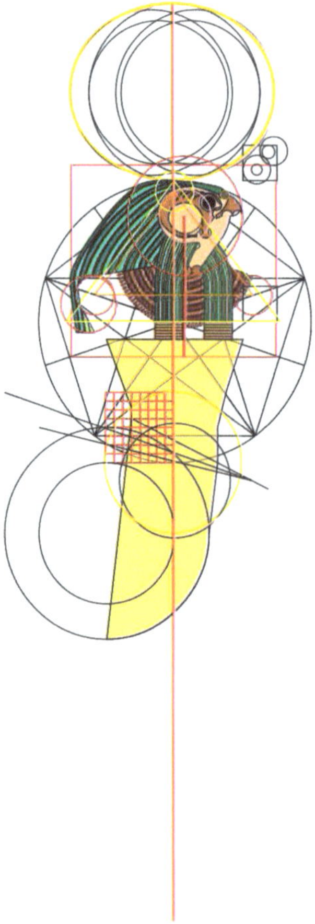

CONTENTS

Cover	*"Hats and Cane" by Arto J Heino (c) 1985*		
Title		**1**	
Introduction		**3**	
Contents		**7**	
Variables used		**11**	
Chapter 1	**Energy**	**13**	
	1.1a Movement		13
	1,1b Ether Pressure		14
1.2	Acoustics		17
	1.2a The Kings Chamber Cavity Resonator		17
	1.2b The Queens Chamber		19
	1.2c Longitudinal Sonics		20
	1.2d Home Made Keely		22
	1.2e Earth as a Half Wave Resonator		25
Chapter 2	**Machines**	**39**	
2.1	3000BC - 500 AD		39
2.2	Types of machines		40
	2.2a Motors		40
	2.2b Generators		43
	2.2c Levers and Arms		44
2.4	Interesting Machines		45
	2.4a Hydro-Automat		45
	2.4b Tesla Turbine		52
	2.4c Mechanical Oscillator		53
	2.4d Torque Converters		55
	2.4e C R M R V O		59
	2.4f Hyperbolic Impulse Drive		64
	2.4g Vortex Tube		66
Chapter 3	**Electronics**	**68**	
3.1	Electricity		68

3.1a What is Electricity? 68

3.1b Is Ether, Electricity? 72

3.2 Schematics 79

3.2a History of Symbols 79

3.2b New Schematic Ideas 80

3.3 Circuits 83

3.3a Fibonacci Circuits 83

3.4 Interesting Circuits 86

3.4a Differential Galvanometer 86

3.4b Vibrator Power Supply 87

3.4c Moray and the Nuclear Battery 90

3.4d Lester Hendershot 91

3.4e Blocking Oscillator 104

3.4f Don Smith 106

3.4 The Switch 110

3.4a Mechanical 110

3.4b Rotating Brush 110

3.5 Inductors 114

3.5a Mutual Inductance 114

3.5b Multi-filar 115

3.5c The Biflar Motor Coil Unit 119

3.5d The Magic Square Coil Technology 122

3.5e The Infinity Coil 128

3.6 Capacitors 132

3.6a Water Capacitor 132

3.6b Air Capacitor 133

3.6c Multi-Series Capacitor 136

3.7 High Voltage 137

3.7a Voltage Multipliers 137

3,7b Switched Voltage Multiplier 139

3.7c Tesla Coil Patent 568176 141

3.7d Tesla Coil Patent 568177 142

3.7e Tesla Coil Patent 568178 143

	3.7f Tesla Coil Patent 568179		144
	3.7g Tesla Coil Patent 568180		144
	3.8 Multi-Vibrator Circuits		145
	3.9 Quad Radiation Converter		147
Chapter 4	**Maths**	**152**	
	4.1 Numbers to Pi /2		153
	4.2 The Hidden Code of Bruce Cathie		154
	4.3 Decoding Carl Munck		162
	4.4 Quantum Arithmetic		169
Chapter 5	**Geometry**	**173**	
	5.1 Variety of Geometry		174
	5.1a Archimedes		174
	5.1b The Relations of Two Shapes		175
	5.1c Pythagorean Triangle Series		176
	5.2 Quantum Arithmetic Geometry		176
	5.2a Quantum Ellipse		179
	5.2b Quantum Isosceles		181
	5.3 A Formula for Shapes		185
	5.4 The Philo Line		186
	5.5 The Arto Parker Pi Series		187
	5.6 Combination Geometry		194
Chapter 6	**Art**	**196**	
	6.1 Leonardo Da Vinci		196
	6.1a The Last Supper		196
	6.2 Hand and Forearm Proportions		199
	6.3 Modern Divergence		200
Chapter 7	**Music**	**202**	
	7.1 Instruments		202
	7.1a Tutankhamen Trumpet		202
	7.1b Hexaphone		205
	7.1c The Parvatic Scale		208
	7.1d Artone 19 Guitar		211
	7.1e Quantum Scale		213

7.1f Hexartom 19 215

Chapter 8 History **220**

8.1 Modern 220

8.1a The Grand Illusion 220

Chapter 9 Archeo-Astronomy **242**

9.1 The Hidden Star Charts of the Ancients 243

9.2 Egyptian Crowns 266

9.3 Ancient Metric 267

9.4 Archeogeometry 268

9.4a Saqqara Ostrakon 268

9.4b The Kings Chamber 273

9.4c The Great Pyramid - A Quantum Solution 277

9.5 The Egyptian Star 295

Chapter 10 Magic Squares **299**

10.1 The Energy Square of Three 300

10.2 The Circular Geometry of Magic squares 312

10.3 Nested Square 15 314

10.4 Coded Square 3 316

Chapter 11 Earth **317**

11.1 Quantum Solar System 318

11.2 Earth Constants 320

Chapter 12 Inventions and Inventors **322**

12.1 Tesla Tower 323

12.2 Adams Motor 324

12.3 Sixto Ramos 331

12.4 Ucros Device 339

12.5 The Tesla Receiving Circuit 341

12.6 Flux-gate Magnetometer 344

Bibliography **347**

Index **350-356**

Back cover *by Arto J Heino (c) 2013*

Some Variables used in the text

Quantities	*Symbol*
Energy	W
Power	P
Frequency	*f*
Period	T
Time	***t***
Angular velocity	*w*
Electromotive Force	E
Potential Difference	V
Current	I
Resistance	R
Impedance	Z
Reactive Impedance	X
Magnetic Inductance	L
Dielectric Capacitance	C

Disclaimer

The author takes no responsibility for any inaccuracies or errors, that may have incurred while writing and illustrating.

This book is a compilation of many years of diverse studies, observations and creativity and thus must be construed as an continuing artistic endeavor not as a text book or story.

Regards Arto

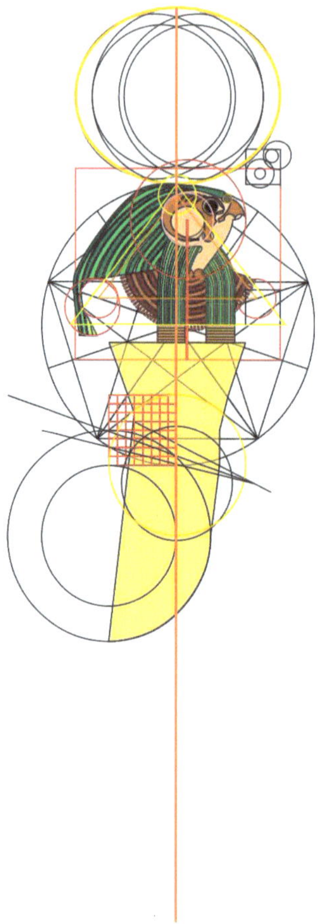

Chapter One

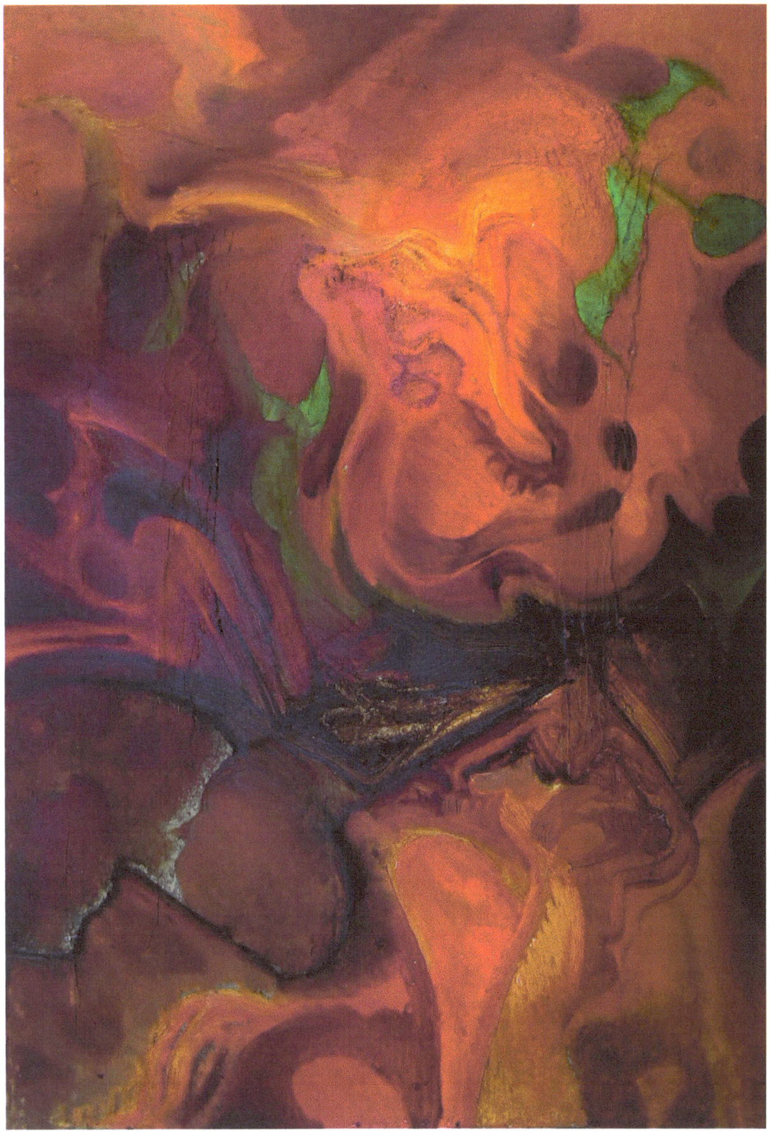

Energy

1.1a - Movement

The term energy is used as an alternative word for power which all physicists know to be incorrect, as always they never correct a term that suits their dumbing down agenda. Energy is the term for movement and thus all correlations are valid when the expertise is vacant, they do not clarify its meaning because its suits their terms, where power is energy per time and watts are energy per time per time does make it more confusing. So as we weave our understanding of modern science into practical devices, we also become disconnected from the source of understanding that energy is the real term we must deal with first before we elaborate on the names and quantities we will use to describe the workings of our devices and the description of the moving sources.

This is where our understanding of energy must be clear, it is fundamentally ***movement*** and nothing more, the elaboration is too obtuse then the picture becomes clouded. Movement entails natures laws as described by Newton and others and thus lays the simple understanding that lies at the root of the engineers mind. This description of movement, can also be represented symbolically and as the displacement of matter over distance and becoming the fundamental law. The understanding is simply stated symbolically as Energy = Mass x (Space / Time) x Time. Where Space / Time = Force. This has been stated emphatically by Le Bonn as all four items of our universe, Force, Matter, Space and Time. The fundamental reasoning of the symbolic nature is in our perceptions, Time and Space infused with matter will become Energy thus a Force of action, life results from this action and energy will have a directed path or a balanced ratio of movement. This will lead us to the nature of PHI a singular parameter as the driving ratio of all energy and the path of matter in movement and its cycloidal geometric path, because as we know the only balance we can perceive in the life ratio understanding is that Phi = 1 / PHI and Phi = PHI -1, these might seem simplistic and even naive, reassuringly these are the only principals at this most basic level that nature can work with. The conundrum is that nature does not divide itself into factional parts, broken, divided and disconnected, it ONLY uses complete whole parts, this is where our science and engineering fail to cast its principle parts as fundamental properties of this Phi ratio understanding.

1.1b – Ether Pressure

Speculation as to the nature of the universal vacuum medium has filled volumes up to the early 20th century, when the followers of Einstein began a campaign to eliminate the Ether and replace it with sterile mathematical abstraction and metaphysical probabilities.

In an attempt to visualize a different approach to how the Ether pressure establishes the gravitational force that is exerted on all bodies in the universe, I have drawn a simple 2 dimensional cross section with some preliminary mathematics to start the idea moving forward.

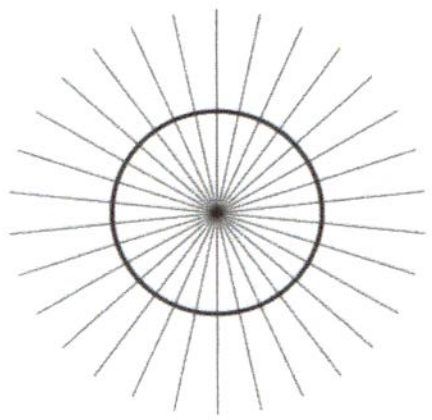

Lcircle IN = 30

Perimeter Lcircle	= 450
	= 30 x 15

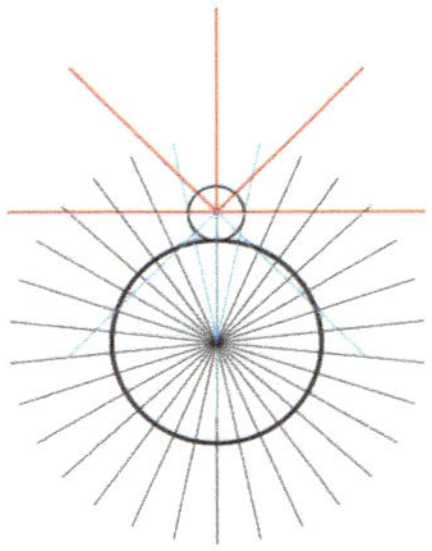

Lcircle IN = 30-3
Scircle1 IN = 8 - 3
Lc + Sc1 IN = 27 + 5
Total IN = 32

Perimeter Lcircle	= 450
	= 30 x 15
	= (30-3) x 15
	= 405
Perimeter Scircle1	= 120
	= 8 x 15
	= (8-3) x 15
	= 75
Total Perimeter IN	= 405 + 75
	= 32 x 15
	= 480

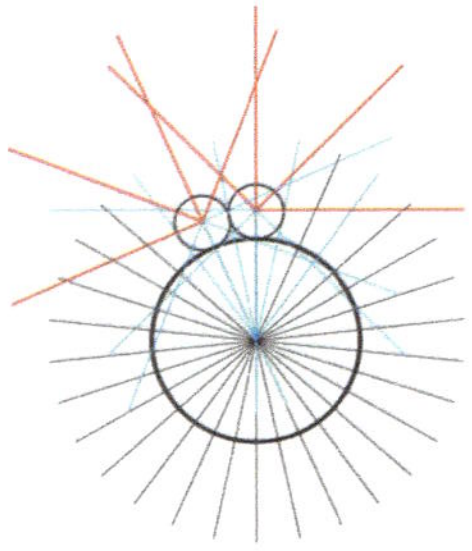

Lcircle IN = 30-5
Scircle1 IN = 8 - 4
Scircle2 IN = 8 - 4
Lc + Sc1 + Sc2 IN = 25 + 4 + 4
Total IN = 33

Perimeter Lcircle	= 25 x 15
	= 375
Perimeter Scircle1	= 4 x 15
	= 60
Perimeter Scircle2	= 4 x 15
	= 60
Total Perimeter IN	= 375 +60 +60
	= 33 x 15
	= 495

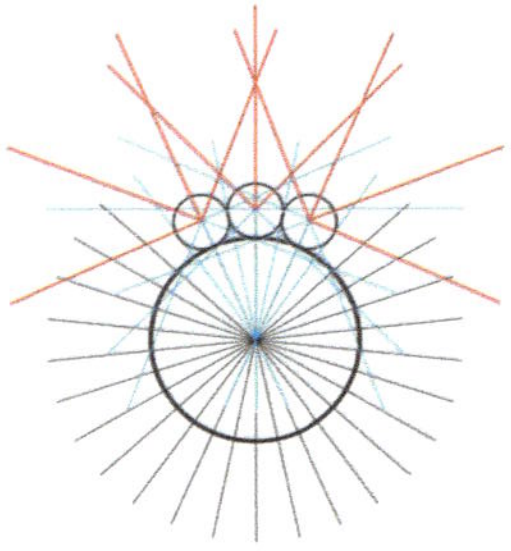

Lcircle IN = 30-7
Scircle1 IN = 8 - 5
Scircle2 IN = 8 - 4
Scircle3 IN = 8 - 4
Lc + Sc1 + Sc2 + Sc3 IN = 23 + 3 + 4 + 4
Total IN = 34

Perimeter Lcircle	= 23 x 15
	= 345
Perimeter Scircle1	= 3 x 15
	= 45
Perimeter Scircle2	= 4 x 15
	= 60
Perimeter Scircle3	= 4 x 15
	= 60
Total Perimeter IN	= 345 + 45 + 60 +60
	= 34 x 15
	= 510

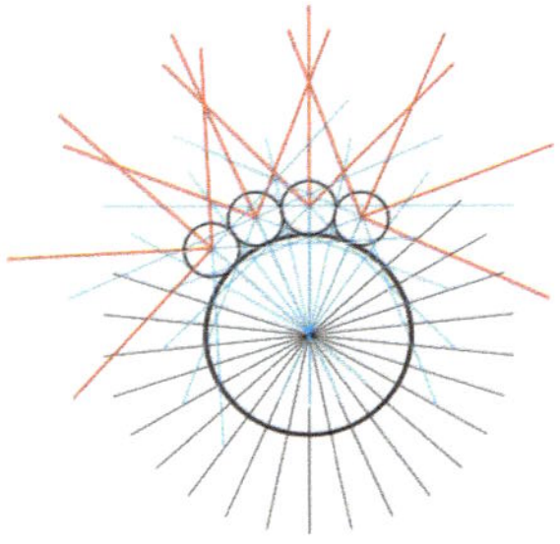

Lcircle IN = 30-9
Scircle1 IN = 8 - 5
Scircle2 IN = 8 - 5
Scircle3 IN = 8 - 4
Scircle4 IN = 8 - 4
Lc+Sc1+Sc2+Sc3+Sc4+Sc IN = 21 + 3 + 3 + 4 + 4
Total IN = 35

Perimeter Lcircle	= 21 x 15
	= 315
Perimeter Scircle1	= 3 x 15
	= 45
Perimeter Scircle2	= 3 x 15
	= 45
Perimeter Scircle3	= 4 x 15
	= 60
Perimeter Scircle4	= 4 x 15
	= 60
Total Perimeter IN	= 315 + 45 + 45 + 60 +60
	= 35 x 15
	= 525

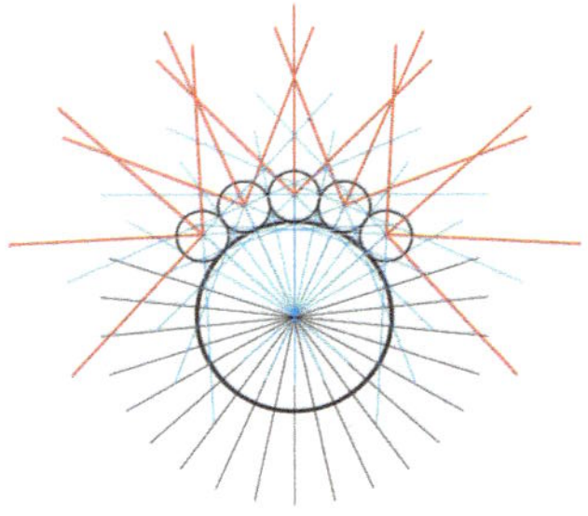

Lcircle IN = 30-11
Scircle1 IN = 8 - 5
Scircle2 IN = 8 - 5
Scircle3 IN = 8 - 5
Scircle4 IN = 8 - 4
Scircle5 IN = 8 - 4
Lc+Sc1+Sc2+Sc3+Sc4+Sc5 IN = 19 + 3 + 3 + 3 + 4 + 4
Total IN = 36

Perimeter Lcircle	= 19 x 15
	= 285
Perimeter Scircle1	= 3 x 15
	= 45
Perimeter Scircle2	= 3 x 15
	= 45
Perimeter Scircle3	= 3 x 15
	= 45
Perimeter Scircle4	= 4 x 15
	= 60
Perimeter Scircle5	= 4 x 15
	= 60
Total Perimeter IN	= 285 + 45 + 45 + 45 + 60 +60
	= 36 x 15
	= 540

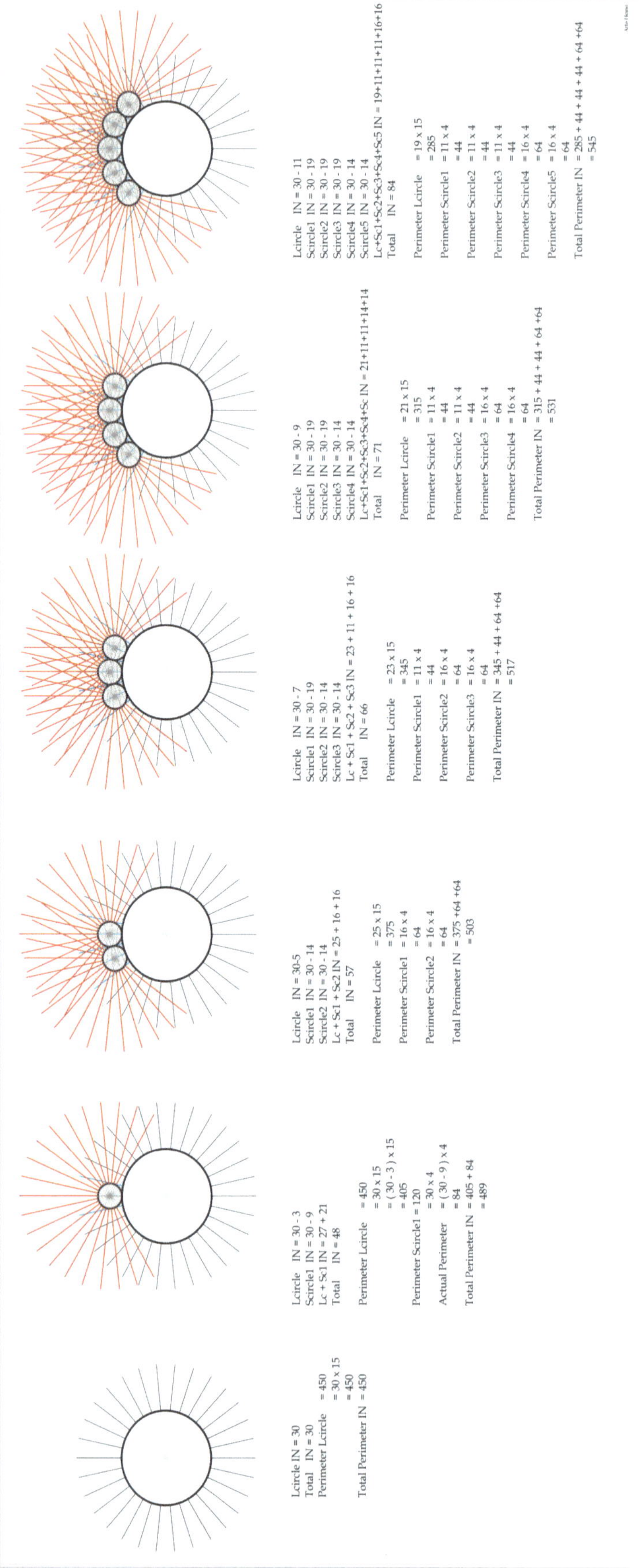
Lcircle IN = 30
Total IN = 30
Perimeter Lcircle = 450
= 30 x 15
= 450
Total Perimeter IN = 450
Lcircle IN = 30 - 3
Scircle1 IN = 30 - 9
Lc + Sc1 IN = 27 + 21
Total IN = 48
Perimeter Lcircle = 450
= 30 x 15
= (30 - 3) x 15
= 405
Perimeter Scircle1 = 120
= 30 x 4
Actual Perimeter = (30 - 9) x 4
= 84
Total Perimeter IN = 405 + 84
= 489
Lcircle IN = 30-5
Scircle1 IN = 30 - 14
Scircle2 IN = 30 - 14
Lc + Sc1 + Sc2 IN = 25 + 16 + 16
Total IN = 57
Perimeter Lcircle = 25 x 15
= 375
Perimeter Scircle1 = 16 x 4
= 64
Perimeter Scircle2 = 16 x 4
= 64
Total Perimeter IN = 375 +64 +64
= 503
Lcircle IN = 30 - 7
Scircle1 IN = 30 - 19
Scircle2 IN = 30 - 14
Scircle3 IN = 30 - 14
Lc + Sc1 + Sc2 + Sc3 IN = 23 + 11 + 16 + 16
Total IN = 66
Perimeter Lcircle = 23 x 15
= 345
Perimeter Scircle1 = 11 x 4
= 44
Perimeter Scircle2 = 16 x 4
= 64
Perimeter Scircle3 = 16 x 4
= 64
Total Perimeter IN = 345 + 44 + 64 +64
= 517
Lcircle IN = 30 - 9
Scircle1 IN = 30 - 19
Scircle2 IN = 30 - 19
Scircle3 IN = 30 - 14
Scircle4 IN = 30 - 14
Lc+Sc1+Sc2+Sc3+Sc4+Sc IN = 21+11+11+14+14
Total IN = 71
Perimeter Lcircle = 21 x 15
= 315
Perimeter Scircle1 = 11 x 4
= 44
Perimeter Scircle2 = 11 x 4
= 44
Perimeter Scircle3 = 16 x 4
= 64
Perimeter Scircle4 = 16 x 4
= 64
Total Perimeter IN = 315 + 44 + 44 + 64 +64
= 531
Lcircle IN = 30 - 11
Scircle1 IN = 30 - 19
Scircle2 IN = 30 - 19
Scircle3 IN = 30 - 19
Scircle4 IN = 30 - 14
Scircle5 IN = 30 - 14
Lc+Sc1+Sc2+Sc3+Sc4+Sc5 IN = 19+11+11+11+16+16
Total IN = 84
Perimeter Lcircle = 19 x 15
= 285
Perimeter Scircle1 = 11 x 4
= 44
Perimeter Scircle2 = 11 x 4
= 44
Perimeter Scircle3 = 11 x 4
= 44
Perimeter Scircle4 = 16 x 4
= 64
Perimeter Scircle5 = 16 x 4
= 64
Total Perimeter IN = 285 + 44 + 44 + 44 + 64 +64
= 545

V^2D	Z^2/D	Z/D	Pa/Z
ZV	Pa	V	Sq(Pa/D)
Sq(PaD)	Z	D	Z/V
Pa/V	VD	Pa/V^2	Z^2/Pa

1.2 - Acoustics

1.2a - The Kings Chamber Cavity Resonator

In chapter 9 I have worked out the size of the Kings Chamber in Quantum units, let us see if we can explore the implications of its possible functions as well, this was one of the most rewarding puzzles I had to solve of recent times. This would become more helpful as I could now scale it to something a lot smaller and useful for experimentation. I have also completed the Quantum Coffer as part of this process, armed with an understanding of quantum arithmetical methods I now could apply my work to just about any Ancient monument that was built to those exacting requirements, this information will surely help give some clarity to date these constructions so we can finally put an end to the time wasting debates that have been festering for the last 200 years.

It occurred to me a few years ago that the Ancient monuments that are of the most interest were those built pre 3000 BC(my reckoning), as the conventional dating of most monuments are not of that early era, something is distinctly wrong is the way Academia view and date the time-lines of civilizations, each new discovery pushes the dates back further than they would like to admit. To attribute many ancient buildings to a later era distorts the history in a way that denies the true origins of cultural fashions and technological beginnings. As we know writing and mathematics were fully developed at the beginning of the accepted historical time line , not at the end, the Greeks learned from the Ancient Egyptians and the Romans learned from the Greeks, this is also true in India where the Vedic texts have a fully developed society , science, art at the oldest time period. The Aztec's only rehabilitated the more ancient buildings and a culture they discovered. This brings me back to the Great Pyramid and its function, I can only make assumptions based on facts and some evidence that the Great Pyramid was some sort of machine that was designed by a people that were well versed in structural engineering that surpasses the best efforts of today. The Khemitians as I like to call them , had profound understanding of Astronomy, Geometry and Chemistry, these are evident in the materials, methods and designs of what they have left us, this is a profound and obvious fact that no real researcher can deny.

I would like to forward an idea what the possible usage was for at least part of the mechanism of the Great Pyramid, mainly the Kings Chamber acting as a Cavity Resonator . As I was working with the measurements of the Ante-chamber, I could see that the design vaguely resembled an adjustable resonant cavity mechanism. It seemed there has been a lot of conjecture about what was the original function and design layout of the Ante-chamber, the common view was a sealing mechanism of granite portcullis blocks that were strung up with ropes and made to move up and down locking the treasures of the Kings Chamber from greedy grave robbers. This idea seemed plausible if the Kings chamber really had a King with his treasures to hide, sadly there is no proof that it was ever intended for a dead Pharaoh, a simple fact that no hieroglyphs have been found in the great pyramid can attest to its lack of funerary ritual hieroglyphics that dominate the walls of real graves of the well to do Egyptians and Pharaohs. The scribbles that have been found in the sections above the Kings chamber that were attributed to the builders were the result of a rogue Archaeologists idea of getting some credentials and notification, in reality the markings were of nonsense and of the wrong period.

Christopher Dunn describes a possible Hydrogen power-plant that might be the mechanism of the Pyramid, this interested me a great deal, it started to dawn on me that this is a real possibility. The few clues I found was the sooty deposits in the air shaft in the Kings chamber and the green Dolerite ball that was found by Petrie, the unusual cavity in the South air shaft, and the 4 grooves at the end of the Ante Chamber. The early accounts reveal a few more clues, early accounts attest to finding a 6ft x 4ft slab in the antechamber, now gone, also the results that somebody had blasted their way into the kings chamber passageway, gouging a section from the top part, suggesting this small passageway was tightly closed.

Most researchers attribute the grooves in the wall of the rear of the Ante-chamber to rope clearances or such, where I think it would be ideal to create some sort of piezo vibrator or tuned half pipes to transfer a 7.84 hz tone to the inlet of the Kings Chamber, the inlet for the hydrogen gas would also have been part of the pipe mechanism as well. The Grand Gallery could have been a gas fractionation unit that was tuned both acoustically and electrically that was polarized allowing oxygen to stay in the lower portion while hydrogen could rise into the ante chamber and on through to the Kings chamber via 4 pipes and/or resonators. The action could even have water present allowing water to become part of some sluicing gas production process.

All this is just conjecture at present, the Kings Chamber has 2 shafts, one of those has a unusual cavity that looks like some sort of ball valve system, possibly to allow the valve to close or open when pressure was correct and the other shaft is connected to the outside , possibly connected to another mechanism at the outer exit.

The only way for us to discover the purpose is to a build scale model of it and test the many conjectures

I have drawn a overhead scaled version with the sonic inlet being the column of gas that is in the passageway between the Ante-chamber and the Kings Chamber . This is a typical Helmholtz Resonator design, nothing mysterious here, just well engineered for low frequency resonance. The approximate frequency response for Air was a low 2hz, this at first seemed remarkable as a lot of conjecture has been out there in the public for at least 50 years that it resonated at much higher frequencies. If you treat the chamber as a microwave resonator you will have frequencies of about 1.76Mhz, this is a suitable band for communications, but not remarkable in nature.

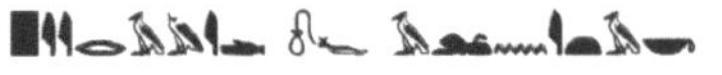

Earth Pulse Hydrogen Resonator
Kings Chamber Cavity
By Arto Juhani Heino 2013

Light	**Velocity in Vacuum C**	**299792458**	**Mtr / sec**
Earth	**Radius R**	**6370961.86815**	**Mtrs**
Earth	**Frequency C / (R x 1)**	**47.0560747662**	**Hz**
Earth	**Frequency C / (R x 2)**	**23.5280373831**	**Hz**
Earth	**Frequency C / (R x 3)**	**15.6853582554**	**Hz**
Earth	**Frequency C / (R x 4)**	**11.7640186915**	**Hz**
Earth	**Frequency C / (R x 5)**	**7.8426791277**	**Hz**
Temperature in K Chamber	**X degrees**	**21.9254182650**	**Celsius**
Hydrogen	**Velocity a 0 degrees Celsius**	**1286**	**Mtr / sec**
	0 degree Celsius in Kelvin	**273.15**	**Kelvin**
Hydrogen	**Velocity a X degrees Vel**	**1336.6166965110**	**Mtr / sec**
Name	*Dimension*	*Meters*	*Units*
Kings Chamber inlet	Width Wi	1.0600000000	**Mtrs**
(passage way)	Height Hc	1.0464800000	**Mtrs**
	Length Ln	2.5665684000	**Mtrs**
	Area Ar	**1.1092688000**	**Sqr Mtrs**
Sarcophagus	Width Ws	0.9780000000	**Mtrs**
(outside only)	Height Hs	1.0493000000	**Mtrs**
	Length Ls	2.2763000000	**Mtrs**
	Volume	**2.3359741150**	**Cubic Mtrs**
Kings Chamber	Width Wc	5.2324000000	**Mtrs**
	Height Hc	5.8500010427	**Mtrs**
	Length Lc	10.4648000000	**Mtrs**
	Volume of K Chamber	**320.3227712880**	**Cubic Mtrs**
	Volume of Sarcophagus Minus	2.3359741150	**Cubic Mtrs**
	Total Volume VL	**317.9867971730**	**Cubic Mtrs**
Helmholtz Resonator	**Frequency Vel/(2xPIxSqr(Ar/(VLxLn)))**	**7.8426791277**	**Hz**

7.8426791277 Cycles per Second

7 + 1/2 + 1/3 + 1/107 = 7.8426791277

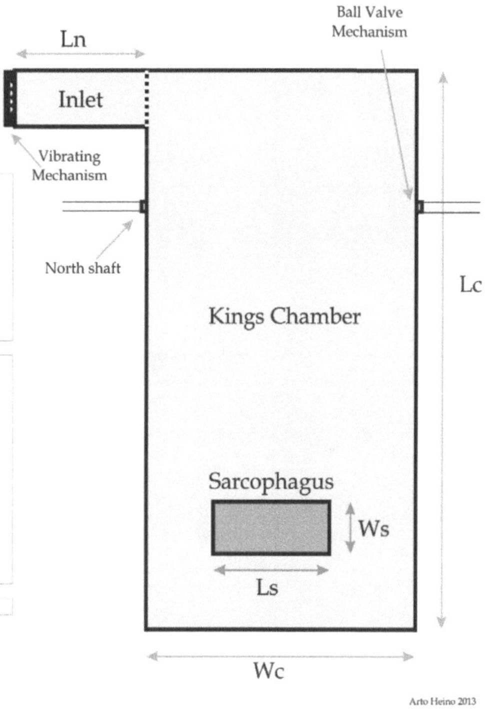

The most surprising result of looking at the resonant values for the Ante-chamber was a close match of 47Hz, ever though I did not have all the exact measurements for the missing parts. I made an approximate of what has been conjectured by many other researchers, the result was amazing to say the least. I will be working on the Grand Gallery and the Queens chamber as the next set of parameters to see if they reveal any evidence of a mufti-chamber pyramidic gasifier that breaks-down water into useful gas energy products such as Hydrogen and Oxygen.

1.2b - The Queens Chamber Resonator

The unified approach would also include the Queens Chamber as it is called by archeologists, the gas that would occupy the true status would be Oxygen, as this gas is heavier than Hydrogen. The tuning of this chamber is 2.4766 Hz, as the chart indicates. Could the subterranean chamber be the water cavity resonator that has a direct connection to the floor shaft in the Queens Chamber. This would mean the level of the water is at the entrance at the base level of the Pyramid, while the subterranean chamber is directly related to the true ground water table.

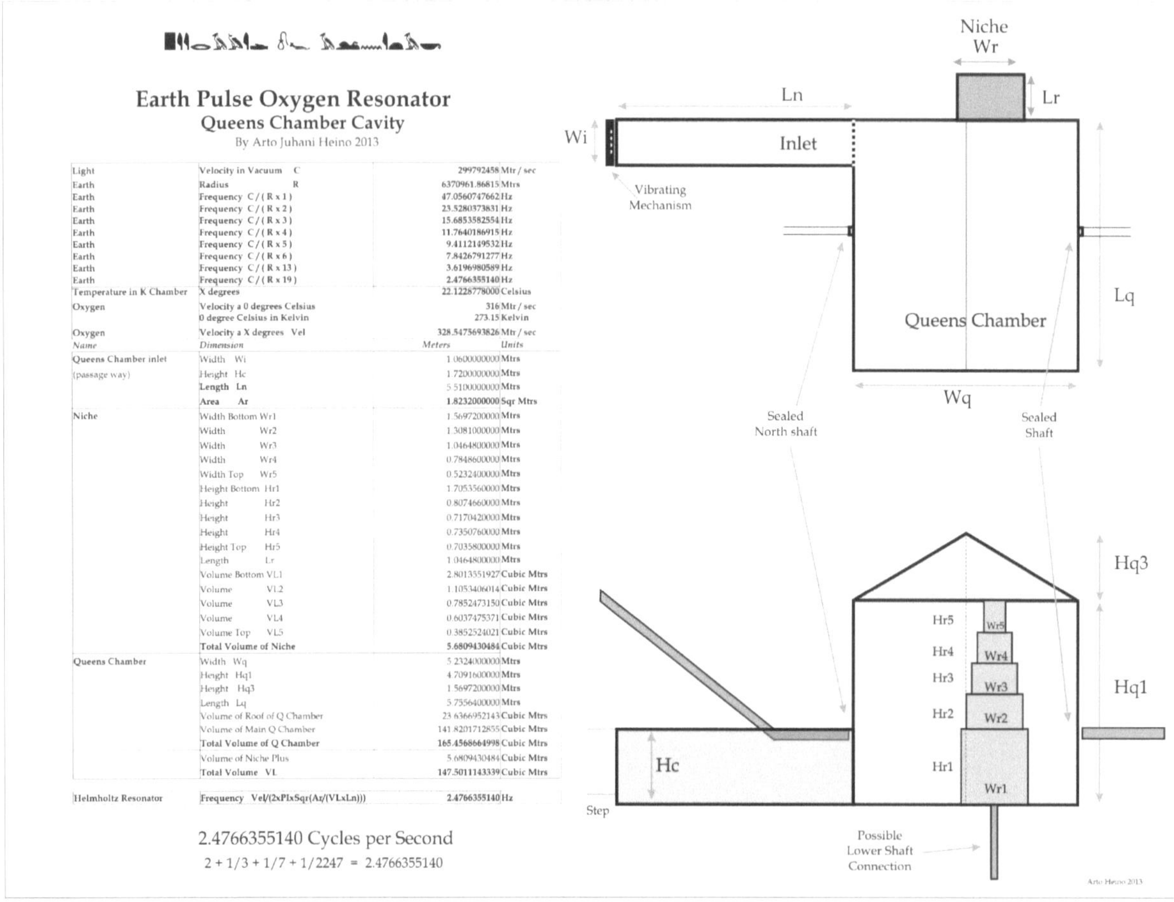

1.2c - Longitudinal Sonics

If you really want to understand how longitudinal waves can be transmitted, received and created through a medium you cannot be happier than discovering the work of George Constantinesco. He was a Romanian inventor who gave us a lasting legacy "The Theory of Sonics", a new (in 1913) science of power transmission though Solids, Liquids and Gases. The same understanding of the Tesla longitudinal electric wave can be engineered using the simple mathematics of Constantinesco and his understanding. I have drawn a diagram of a possible impulse system of transmission, based on one of his patents, this could be built and replace a gearbox, drive shaft and differential from a vehicle. This is a design model for experimental purposes, all dimensions should be reasonably close to actual build sizes.

Synchronous Alternating Liquid Current Motor

Gogu Constantinesco

from Patent in 1916

Theory of Sonics

v = velocity of medium at any point in mtrs per sec
n = number of revolutions of the crank per second
l = wavelength in mtrs
t = time in seconds
w = area of pipe section in square mtrs
r = the equivalent length of the driving crank in mtrs
a = angular velocity of the crank in radians per sec.
p = angle of phase
M = piston section in sqr mtrs
I = maximum alternating flow in mtrs cubed per second or the Amplitude of the flow
= r * a * M
I = flow of medium in sqr cubic mtrs per second
= v * w
= I * sin(a * t + p)
l = v / n

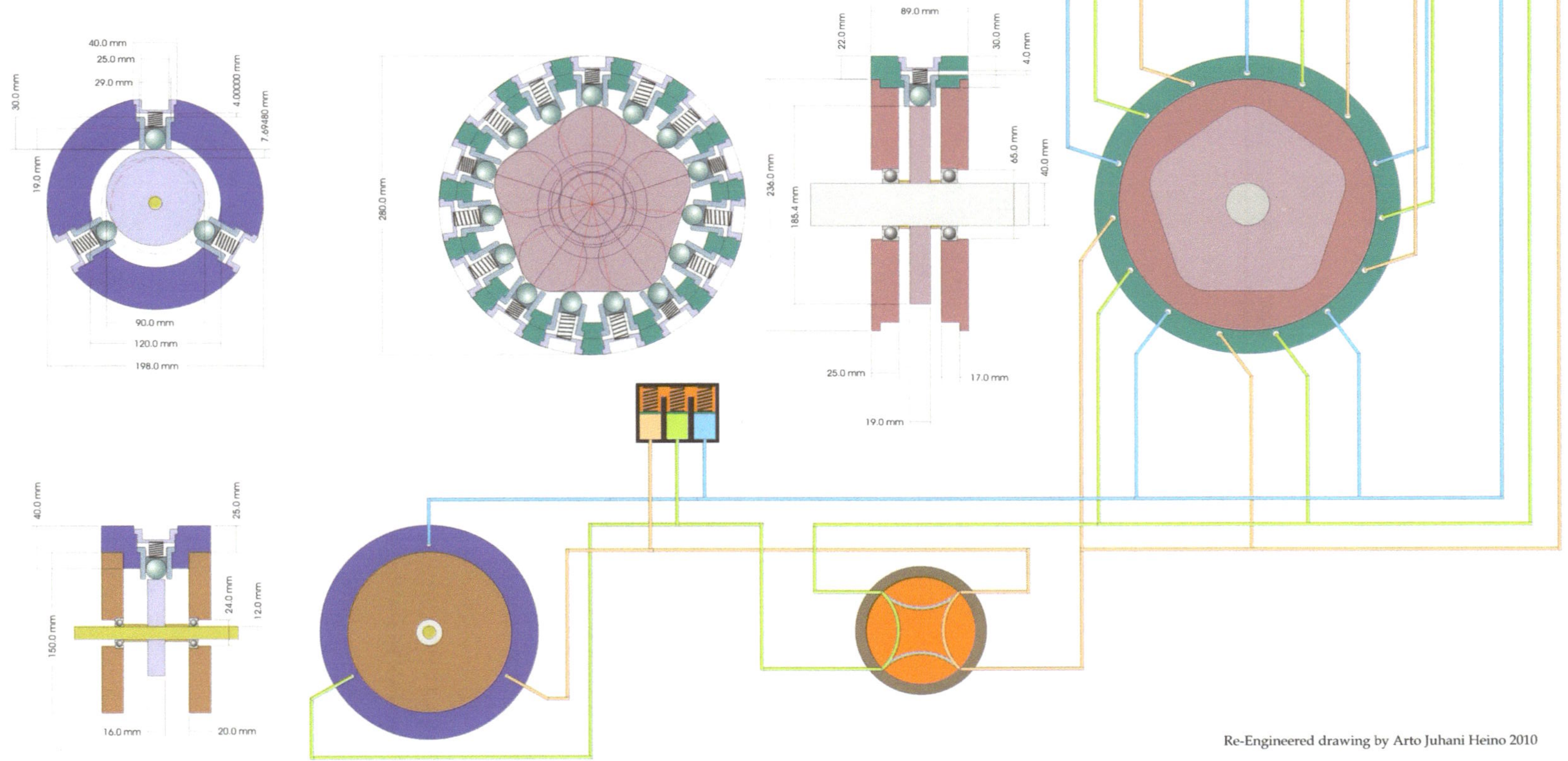

Re-Engineered drawing by Arto Juhani Heino 2010

1.2d – Home Made Keely

In 1999 I was asked by a colleague, Weston Fisher, to help figure out an invention by Worrell Keely, his hydro motor. We understood the basic principals of the water-hammer backlash that would help drive the flywheel, as I was only in possession of the unit for a very short period the completion of the device was not fully realized.

Here is what I revealed to the public on my website in 1999:

1 Energy by water hammer effect, a case of Water Hammer=Back EMF

Parametric impulse system.

2 19-6-99 - The 2 pistons still need to have a better vacuum operation, Weston started this generator with a 1 1/2 HP motor until it reached 1400 rpm then the generator ran without assistance and generated 10 amps, it ran for about 3 minutes until the leaky system starts to slow down the generator. It incorporates a Bourke engine crank system.

3 10-4-00 - Improved piston seals

Specifications

Stroke Length	4 inches (adjustable).
Piston Diameter	1 3/8 inches
Phase Angles between Piston operations	0 degrees(adjustable)
Vacuum obtained	not completed

The system Weston designed was partly understood as he intuitively figured the rotating valve as the lynch pin of the system. Here are some design parameters based around water hammer phenomenon.

Gravity Acceleration	*g*	9.80665500	m s2
Velocity of Sound in Fluid (water)	*c*	1431.78210633	m s
Density of Fluid (Water)	*p*	1000.00000000	Kg / m3
Bulk Modulus Fluid (water)	***Bem***	2050000000.00000000	pascal
Bulk Modulus Pipe (copper)	*E*	117000000000.00000000	pascal
Bulk Modulus Pipe (SS 302)	*E*	180000000000.00000000	pascal
Bulk Modulus Pipe (Iron)	*E*	210000000000.00000000	pascal
Bulk Modulus Pipe (PVC)	*E*	2833745246.19000000	pascal
Conversion	***Pa-Psi***	6894.75729000	
Conversion	***Ltr-Gal***	3.78541180	

The Keely Weston Hydro-Vaco Motor

Artwork and conceptualization aided by Arto Heino 1999-2013

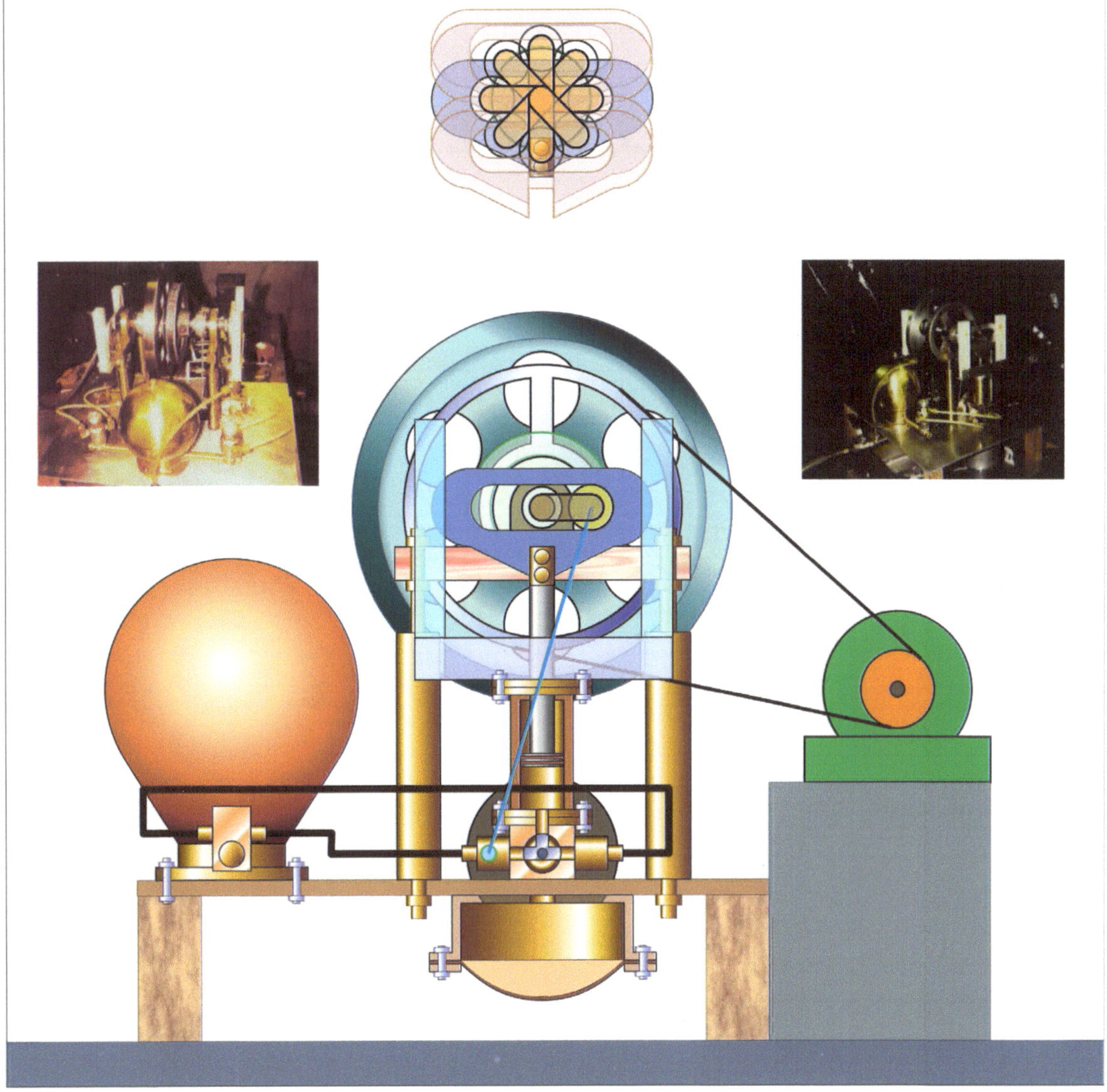

The basic equations are:

Velocity of Pressure wave in pipes = sqrt(1 / (p x { (1/Bem)+ (Dp/(A x E) }))
in mtrs per second

Over Pressure = $\boldsymbol{vp}$ x $\boldsymbol{c}$ x $\boldsymbol{vc}$ in Pascals

Where *vc* is the velocity change in the fluid, this might be a sudden valve shutting off or a diversion to to different ratio pipe or chamber. Here are some examples, when *vc* = 1 m per second.

Mean diameter of pipe	*Dp*	0.02000000	mtr
Cross section Area of Pipe Wall per unit Length	*A*	0.00125664	sqr mtr per mtr
Wall Thickness Pipe	*pt*	0.01000000	mtr
Flow Velocity Change	*vc*	1.00000000	m s

Where $A = (pi \times ((Dp / 2) + pt) \wedge 2 - (pi \times ((Dp / 2) - pt) \wedge 2$

Velocity of pressure wave in pipe	*vpPVC*	404.74813175	m s
Velocity of pressure wave in pipe	*vpCu*	1266.09187168	m s
Velocity of pressure wave in pipe	*vpSS*	1317.35923089	m s
Velocity of pressure wave in pipe	*vpFe*	1332.03987431	m s

Over Pressure	***PovPVC***	404748.13175260	pascal
Over Pressure in Psi	*PimPVC*	58.70375341	lbs per sqr inch
Over Pressure	***PovCu***	1266091.87168293	pascal
Over Pressure in Psi	*PimCu*	183.63110091	lbs per sqr inch
Over Pressure	***PovSS***	1317359.23088621	pascal
Over Pressure in Psi	*PimSS*	191.06680271	lbs per sqr inch
Over Pressure	***PovFe***	1332039.87430662	pascal
Over Pressure in Psi	*PimFe*	193.19605002	lbs per sqr inch

As you can see by these examples that Iron has the stiffest wall parameter, this promotes higher velocity of the longitudinal pressure wave. The wall thickness also aides in this reaching near sound velocity when the walls are very thick, this is just what is required to promote water hammer. The schematic design of the Home Made Keely is a tentative layout based on the hydraulic multi-vibrator.

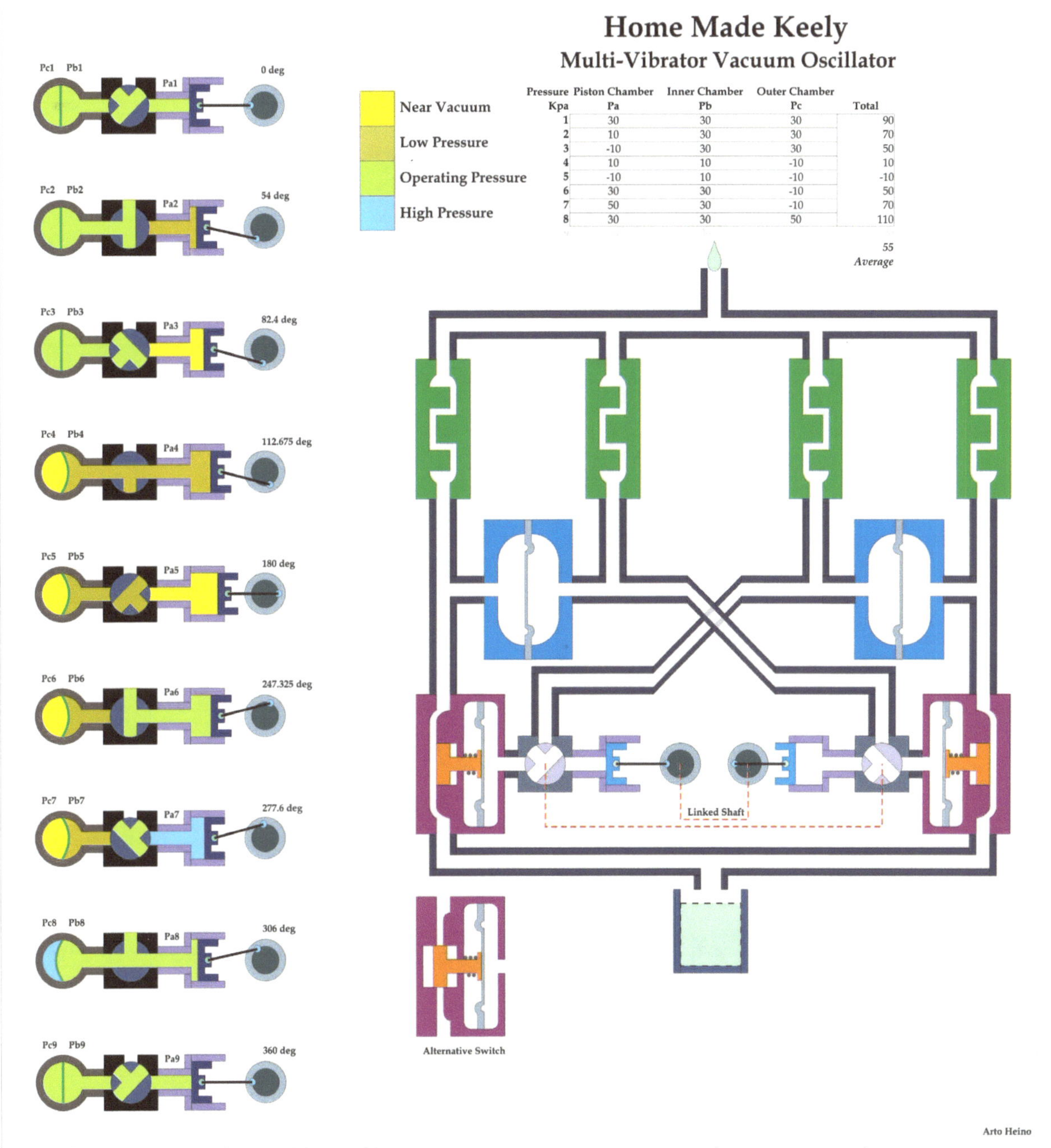

Pressure Kpa	Piston Chamber Pa	Inner Chamber Pb	Outer Chamber Pc	Total
1	30	30	30	90
2	10	30	30	70
3	-10	30	30	50
4	10	10	-10	10
5	-10	10	-10	-10
6	30	30	-10	50
7	50	30	-10	70
8	30	30	50	110
				55 *Average*

1.2e - Earth as a Half Wave Resonator

I have developed and followed a simple methodology, one of observation, study, imagination, testing and analogy. While I was developing the short distance Tesla Transceiver in 2011, limiting it to 1 Kilowatt and 8.16 Kilometer transmission for use in farm land and difficult terrain where power-lines would be an expense beyond the return of the investment. I had decided to build a prototype that would transmit 100 watts of electrical energy at a distance of 680 mtrs, this achievement seemed a reasonable first step and a plausible engineering feat. As my financial situation was only meager, the Tesla oscillator was put together from salvaged microwave parts(MOT/Caps) and 25 ltr/300mm bucket and 100 mm PVC, the primary was welded aluminum strip. The unit was rough and cheap, this was a slow and painful first step as I would rather have designed from ground up caps and all. Here are some pics of the build.

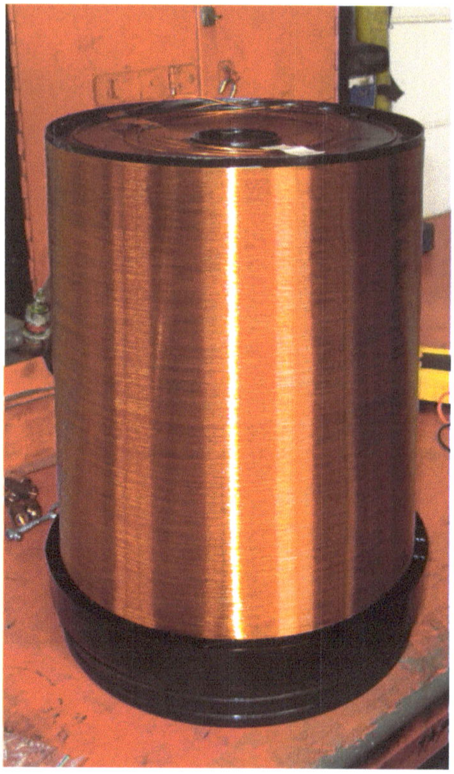

Tesla Secondary Hand Wound

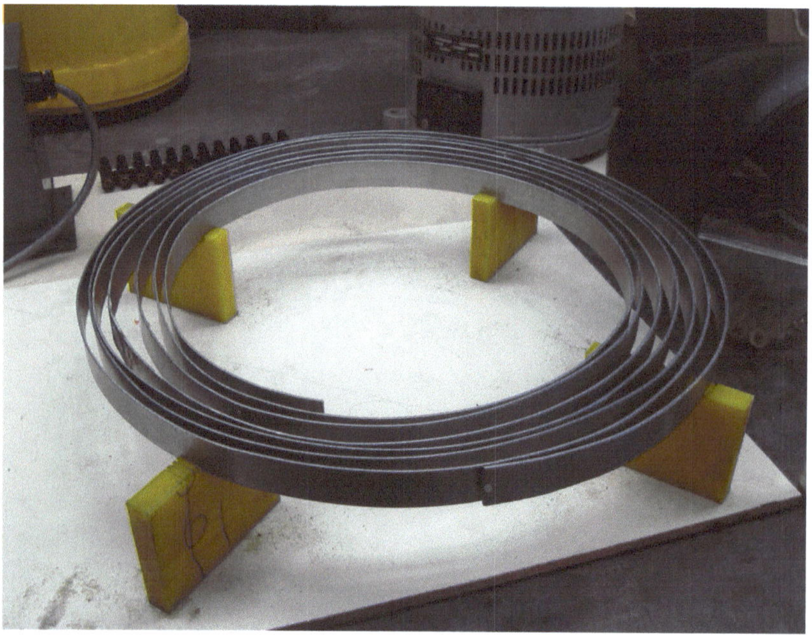

Here are some important parameters I used when I designed this system.

Vel	**299792458.0000**	**mtrs sec**
Tesla Vel	**470912891.8272**	**mtrs sec**
Tesla Freq	**345575.1919**	**hz**
Freq A	**220000.0000**	**hz**
Tesla wv length	**1362.6930**	**mtrs**
¼ wv Wire Leng A	**340.6732**	**mtrs**

Pri wire len	5.4947	mtrs	5494.73	mm
Pri diam tube 0a	0.3300	mtrs	330.0000	mm
Pri wire diam 0a	0.0060	mtrs	6.0000	mm
Pri circum	1.0367	mtrs	1036.73	mm
Pri turns	5.3001			
Pri Turns Spacing	0.0060	mtrs	6.0000	mm
Pri coil len	0.0636	mtrs	63.6010	mm
Volume of Cu Prim	0.000155359825	cubic mtrs	155359.8250	cubic mm
Inductance of Prim	0.000014376239	Henries	14.3762	uH
Cap at Freq A	0.000000036404	Farads	36.4040	nF
Cap at Tesla Freq	0.000000014754	Farads	14.7540	nF

Pri wire len max	17.0217	mtrs	17021.6705	mm
Pri wire len min	0.9135	mtrs	913.5186	mm
Pri diam spiral outer 0b	0.5458	mtrs	545.7820	mm
Pri diam spiral inner 0b	0.2878	mtrs	287.7820	mm
Pri wire diam 0b	0.0030	mtrs	3.0000	mm
Pri circum outer	1.7146	mtrs	1714.6247	mm
Pri circum inner	0.9041	mtrs	904.0938	mm
Pri turns max	13.0000			
Pri turns min	1.0000			
Pri Turns Spacing	0.0075	mtrs	7.5000	mm
Pri coil height	0.0200	mtrs	20.0000	mm
Pri Average radius (max)	0.2084	mtrs	208.3910	mm
Pri Average radius (min)	0.1814	mtrs	181.3910	mm
Pri width of spiral (max)	0.1290	mtrs	129.0000	mm
Pri width of spiral (min)	0.0030	mtrs	3.0000	mm
Inductance of Prim max	0.000093626108	Henries	93.6261	uH
Inductance of Prim min	0.000000872823	Henries	0.8728	uH
Inductance required at Freq A	0.000035471949	Henries	35.4719	uH
Inductance required at Tesla Freq	0.000014376239	Henries	14.3762	uH

OD	Radius Ave mm	Width mm	Turn	Inductance	Wire Leng mm
293.78	181.3910	3.0000	1	0.000000872823	913.5186
314.78	183.6410	13.5000	2	0.000003283121	1893.0106
335.78	185.8910	24.0000	3	0.000006992100	2938.4761
356.78	188.1410	34.5000	4	0.000011831165	4049.9151
377.78	190.3910	45.0000	5	0.000017678704	5227.3274
398.78	192.6410	55.5000	6	0.000024445507	6470.7133
419.78	194.8910	66.0000	7	0.000032065307	7780.0725
440.78	197.1410	76.5000	8	0.000040488444	9155.4052
461.78	199.3910	87.0000	9	0.000049677525	10596.7114
482.78	201.6410	97.5000	10	0.000059604371	12103.9910
503.78	203.8910	108.0000	11	0.000070247839	13677.2441
524.78	206.1410	118.5000	12	0.000081592229	15316.4706
545.78	208.3910	129.0000	13	0.000093626108	17021.6705

Sec wire len	340.6732	mtrs	340673.2477	mm
Sec diam tube 1a	0.2718	mtrs	271.7800	mm
Sec wire diam 1a	0.0008	mtrs	0.7570	mm
Sec circum	0.8538	mtrs	853.8221	mm
Sec turns	398.9979			
Sec coil height	0.3110	mtrs	311.0000	mm
Volume of Cu Sec	0.000153327365	cubic mtrs	153327.3646	cubic mm
Inductance of Sec	0.027050895644	Henries	27.0509	mH
Cap at Freq A	0.000000000019	Farads	19.3470	pF
Cap at Tesla Freq	0.000000000008	Farads	7.8410	pF

Ter wire len	340.6732	mtrs	340673.2477	mm
Ter diam tube 2a	0.1080	mtrs	108.0000	mm
Ter wire diam 2a	0.0008	mtrs	0.7570	mm
Ter circum	0.3393	mtrs	339.2920	mm
Ter turns	1004.0710			
Ter coil len	0.7601	mtrs	760.0817	mm
Volume of Cu Ter	0.000153327365	cubic mtrs	153327.3646	cubic mm
Inductance of Ter	0.014494182309	Henries	14.4942	mH
Cap at Freq	0.000000000036	Farads	36.1078	pF
Cap at Tesla Freq	0.000000000015	Farads	14.6339	pF

As I have no longer any more funds to continue with the Tesla Transceiver Project, I have released all the information that might help those who could take it further.

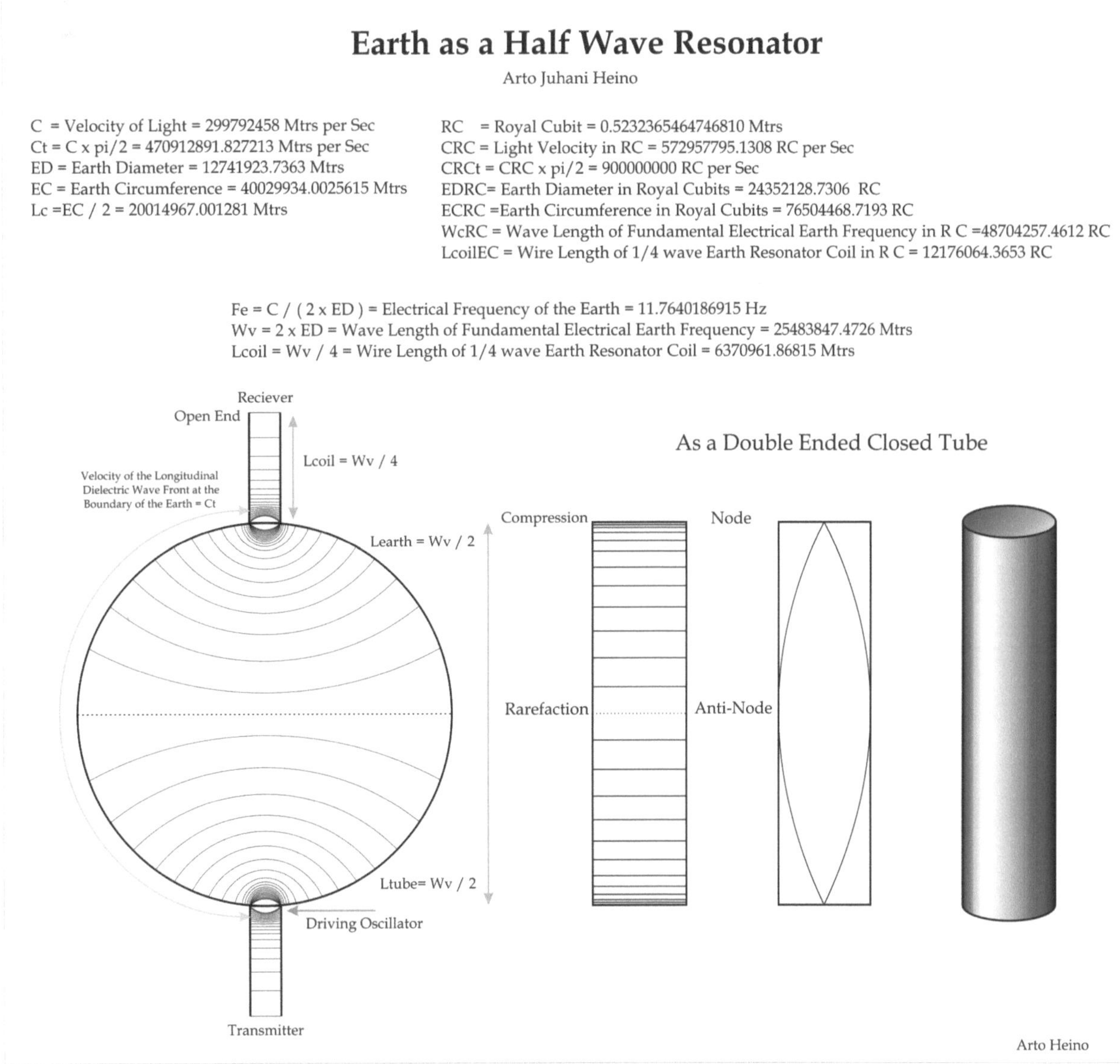

Tesla Coil as Resonant Tube

And here is the Longitudinal Dielectric wave description that I had to conceive as a way to start engineering the Tesla Transceiver Project.

Tesla Coils as Cylindrical Acoustic Resonators

Arto Juhani Heino

A = Electrical Input AC Driving Current
B = Transducer - Electric to Mechanical Transformer = Step up Transformer
C = Half Wave Acoustic Conduction Chamber = Primary Coil
D = Primary Vibration transfer plate = Primary Induction
E = Transmission Pipe = Earth
F = Secondary 1/4 wave Resonator
G = Secondary 1 /2 wave Resonator
H = Amplifying Chamber - Helmholtz Resonator
K = Floating Tertiary 1/4 wave Resonator
M = Tertiary 1/4 wave resonator

T0 = 4th Order Dual Reflex Radiator
T1 = 4th Order Single Reflex Radiator
T2 = 5th Order Single Reflex Resonator
T3 = 6th Order Series Tuned Resonator
T4 = 6th Order Dual Reflex Resonator

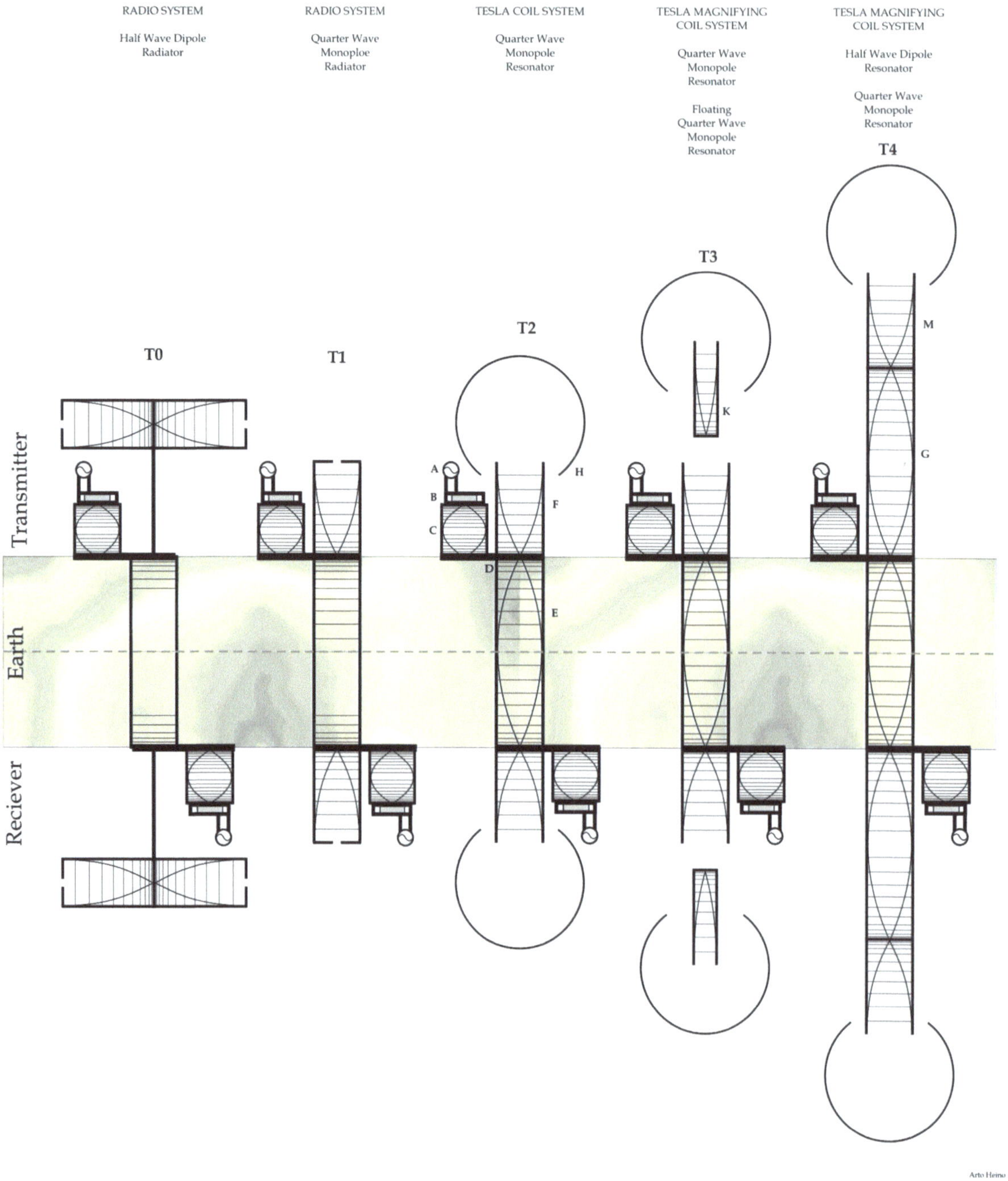

Tesla Transceiver Wave Description

Tesla Transceiver Standing Wave Structure

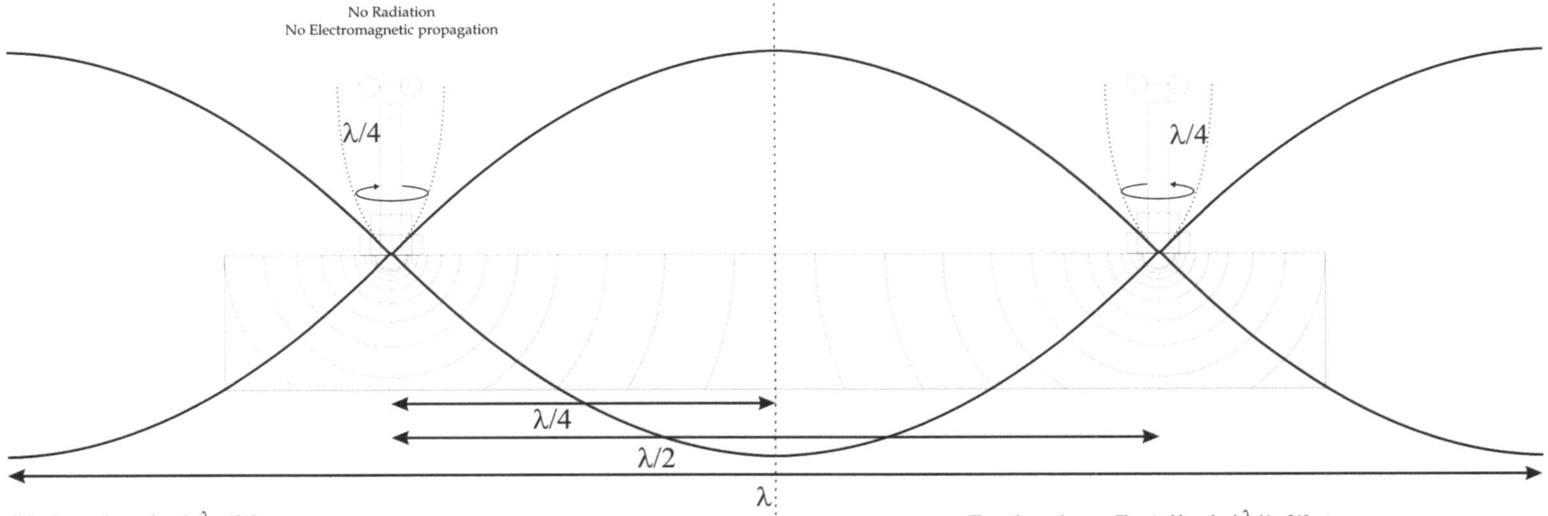

If fundamental wave length λ = 1360 mtrs
then a half wave $\lambda/2$ = 680 mtrs
then the distance between Tesla Transceivers must also be 680 mtrs
and any multiple of this length.
As you have $\lambda/4$ transmitter and $\lambda/4$ receiver plus the $\lambda/2$ equal 1 complete wave
Magneto Dielectric Longitudinal waves propagating through the etheric medium
not physical Electromagnetic waves though the Earth
The interface of the Earth and the Crust act like a wave guide, allowing the velocity to be C x pi/2.

The coil must have an Electrical length of $\lambda/4$ = 340 mtrs
This assures that a standing wave structure is maintained
The Topload structure must be able to hold the electrostatic tension of the impulse wave.

Each Transceiver must be at a physical and electrical Node
All coil windings should also be of opposite helix direction
The opposing helix assures the correct end polarity in standing waves.

Arto Juhani Heino

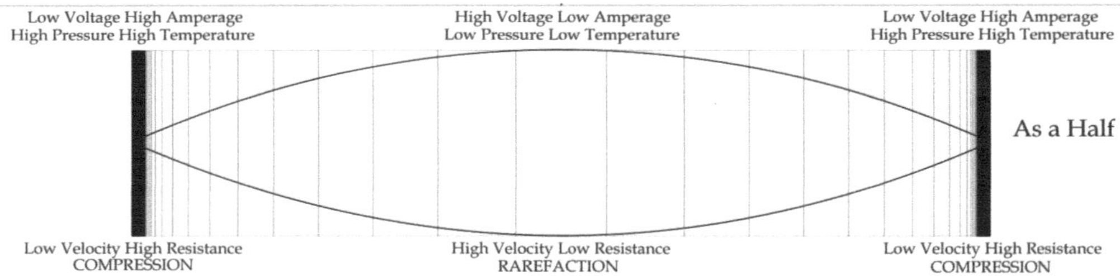

As a Half Wave Resonant Tube

So much has been written about the propagation of the wave that Tesla described in his lectures, patents and books that has been misunderstood by the radio engineering fraternity. The few enlightened individuals like Eric Dollard explain it in their own terminology as there are no electrical terms for this type of wave, this was the reason I choice the acoustic analogy. Using Eric Dollards terminology it would be called the Magneto-Dielectric wave, with the wave moving in counter-space, knowing its name gives you some idea of what has been ignored in scientific literature for the last 100 years.

The task ahead is to relearn all the radio antenna radiation techniques so as to NOT transmit radio waves. Radio , is radiating waves just like a leaky pipe whistling in the air, all loss, all radiation very little to recover, if it were not for amplifiers you would never pickup any signal. The Tesla system is a natural conduction of Magneto-Dielectric Longitudinal polarizations of charge that has small transverse component which we call electromagnetic radiation. As not to confuse matters do not make the mistake that the Schumann Cavity resonant mode is part of this system.

The Tesla Wardenclyffe system has a small radiation component that could be used for the earth ionospheric cavity, but this is the nature of imperfect materials and only a minor aspect of a complete world wireless system of communication and power transmission that has as its main form of transmission the Magneto-Dielectric Longitudinal conduction through the Earth.

The most important factors in designing any prototypical forms would be how close is the analogy and to what degree can we elucidate the Tesla longitudinal velocity wave. A working system of acoustic resonators would help in the understanding of the electrical version. When carefully checking the acoustic analogies and equivalences to electrical circuits the ***Voltage*** analogy seems to be correct with potential difference between the standard atmospheric ***Air Pressure***(Earth) and the ***Over Pressure***(potential) being the equivalent, this would make the ***Amperage*** as equivalent to volumetric flow, which is better described by being call ***Volumetric Velocity***. This last variable requires you to know the particle velocity, this seems straight forward, as the pressure wave exerts a slight density change in the medium , this condition is called ***Condensation***, another term is compression with the less dense part of the wave called the rarefaction. There are also temperature differences in the two opposing conditions as well, this differential engineering will result in the well known acoustic sterling engine design. The movement of particles is ***PI/2*** slower than the ***Sound Velocity*** of the medium, if taken as the electrical current these particles would have the equivalent velocity as the speed of light, thus by analogy arriving at the fundamental longitudinal velocity of light which would be equal to ***Light Velocity x PI/2***.

To gain a better understanding of the wave functions in acoustic pipes here is a diagram of the different harmonics in a tube.

Longitudinal Sound Waves in Tubes

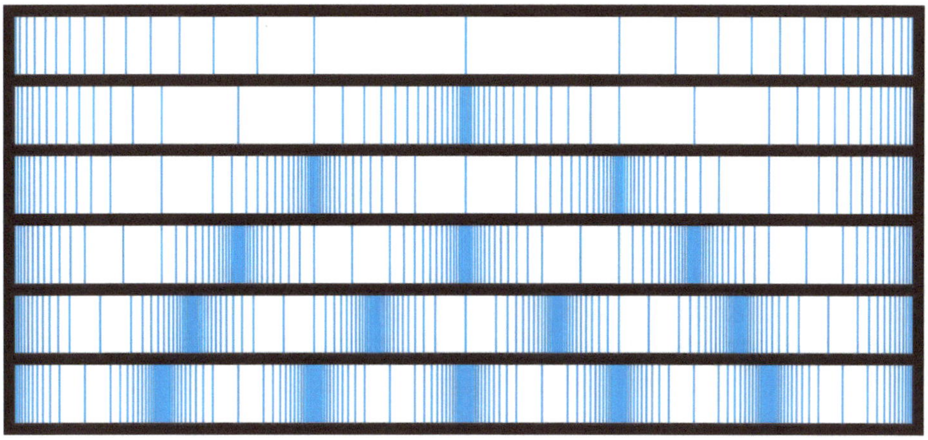

Closed Tube Frequency = Velocity / (2 x Tube Length / Harmonic)

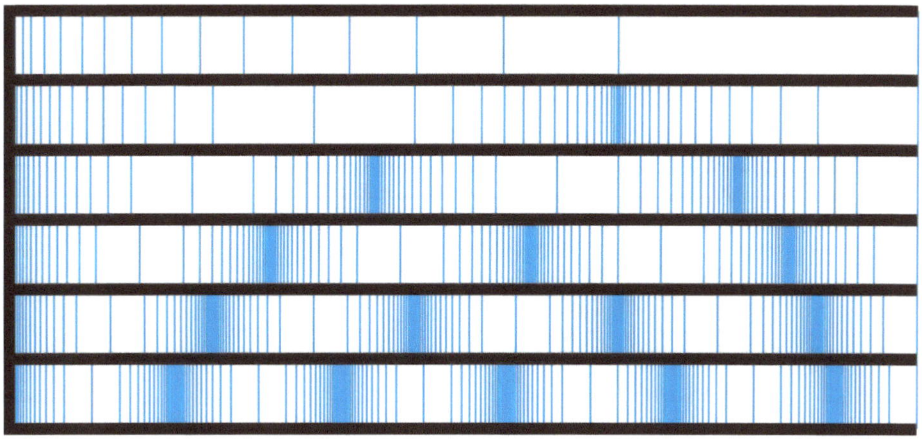

Closed Tube One End Frequency = Velocity / (4 x Tube Length / Odd Harmonic)

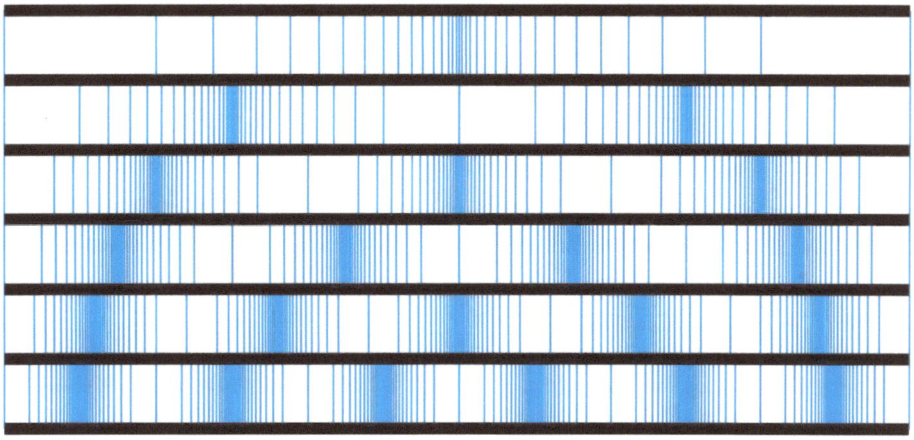

Open Tube Frequency = Velocity / (2 x Tube Length / Harmonic)

Arto Heino

Acoustic circuits require the use of the particle velocity when creating the nuts and bolts of the system, when parameters are then correct and the design completed, the value of the propagation of sound velocity is the end product, meaning the system performs with the trues velocity of sound, or if it is electrical then with the longitudinal velocity of light. A new term should be coined for the true velocity of light, where light is generally shown as a ***C*** the his new term should represent ***PI/2*** faster which is the whole circle not the ***C*** half, putting two of these together would create a number of forms, the simplest would be a ***O***, other possibilities I have contemplated is the Greek letter Theta, this symbol looks promising as one that represents its function. The value of **theta** is 9 and the entomology has many meanings such as meaning of soul in Egyptian and also the sun Ra. The value 9 also correlates perfectly with what I have discovered with the royal cubit measures and light x Pi/2 , with the value 900000000 royal cubits per second as the fundamental longitudinal wave velocity of light.

Θ

Here are some Acoustic and Electric circuit analogs.

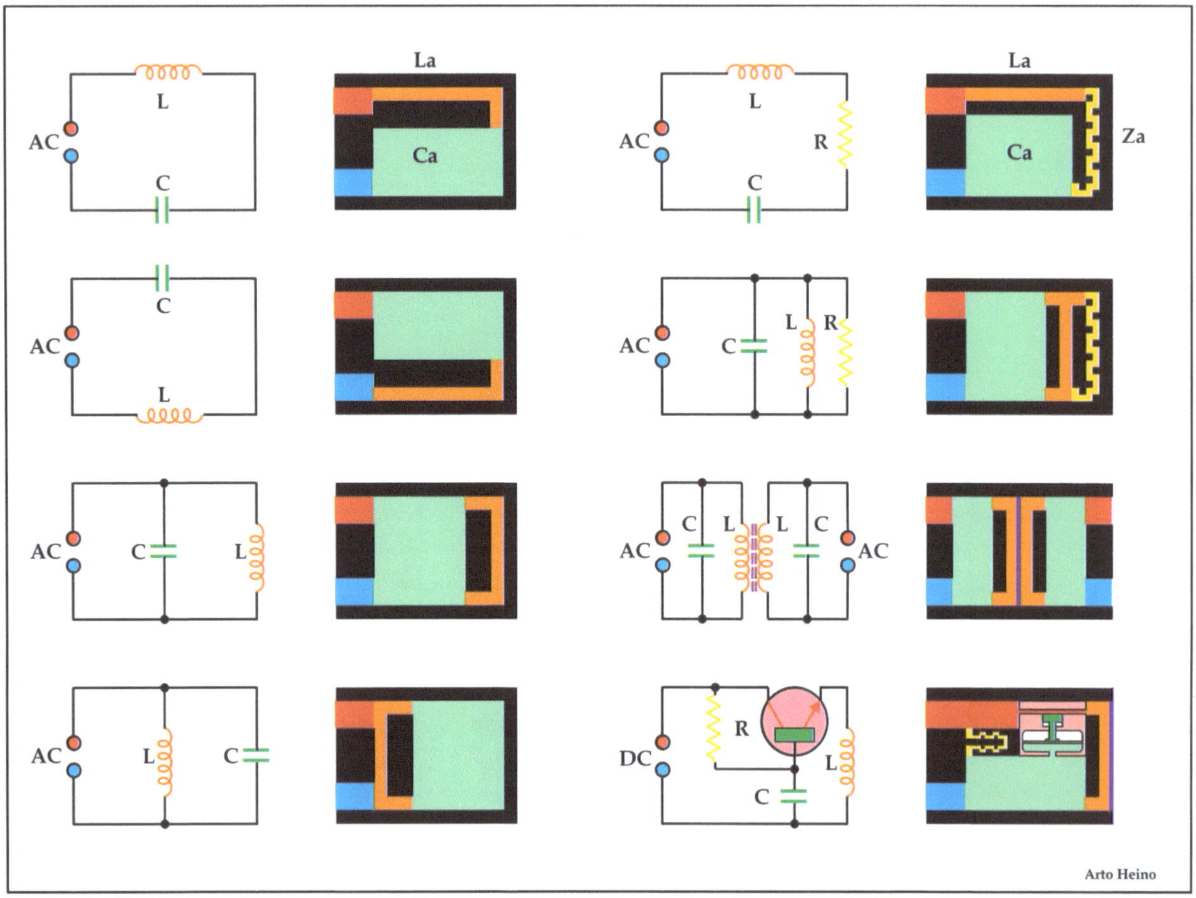

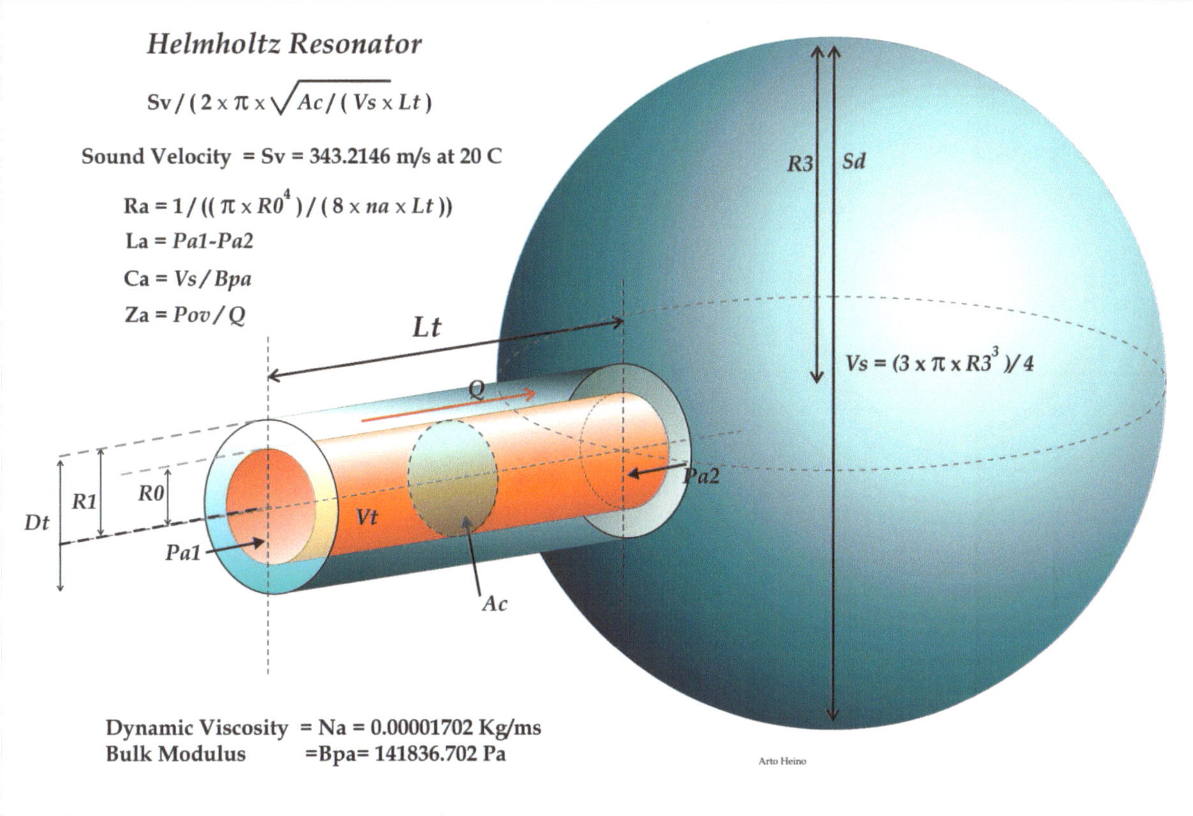

This is a list of the equations that should help to clarify the engineering required to create acoustic circuit analogs.

Over Pressure

Pov Density x Sound Velocity^2 x Condensation

Pov 2 x pi x Frequency x Acoustic Inductance x Volume Velocity

Pov Acoustic Impedance x Volume Velocity

Volume Velocity

Q Particle Velocity x Area of pipe

Q Over Pressure / Acoustic impedance

Acoustic Impedance

Za Over Pressure / Volume Velocity

Acoustic Impedance of Air

Za0 Over Pressure / Sound Velocity

Acoustic Resistance

Ra 1/ ((pi x Pipe Radius^4) / (8 x Dynamic Air viscosity x Length Tube))

Condensation

cn Density at Temperature Tce degrees / Density at 0 degrees

Density of Air at 0 C

da Atmospheric pressure at Height above sea level / (specific gas constant x Temperature 0 degrees Kelvin)

Density of Air at Tce C

da Atmospheric pressure at Height above sea level / (specific gas constant x Temperature Tkv degrees Kelvin)

Temperature in Kelvin

Tkv 273.15 + Tce

Temperatue Celsius

Tce 20 degrees

Gas Constant

R 8.31447

Temp Lapse rate

TLR 0.0065

Gravity acceleration

g 9.80665

Sea Level Temperature

STR 288.15

Atmospheric pressure at sea level

Aps 101325

Atmospheric pressure

Apa Atmospheric pressure at sea level x (1 - ((Temperature Lapse rate x Height of unit base above sea level) / sea level Temperature)) ^ ((Gravity acceleration x Molar mass of air) / (Gas constant x Temp Lapse rate))

Sound Velocity at 0 C

Va0 331.3

Sound Velocity at Tce C

sv Sound Velocity at 0 c x square root (1 + (Temperature C / 273.15))

Particle Velocity

vp 2 x pi x Frequency x Displacement

vp Sound velocity / (pi / 2)

Displacement

dsp (Volume Velocity / pipe area) / (2 x pi x frequency)

Bulk Modulus of air

Bpa Air density x Sound Velocity^2

Acoustic Inductance

La (Air density at Tce x Length of tube) / Area of tube

La Inlet Pressure - End Pressure

Acoustic capacitance

Ca Volume of Chamber / Bulk Modulus

Resonant frequency

fr Sound Velocity / (2 x pi() x square root (pipe area / (Volume of chamber x Pipe Length))))

Frequency of 1/4 wave

f4 Sound Velocity / (Length tube x 4)

Frequency of 1/2 wave

f2 Sound Velocity / (Length tube x 2)

Chapter Two

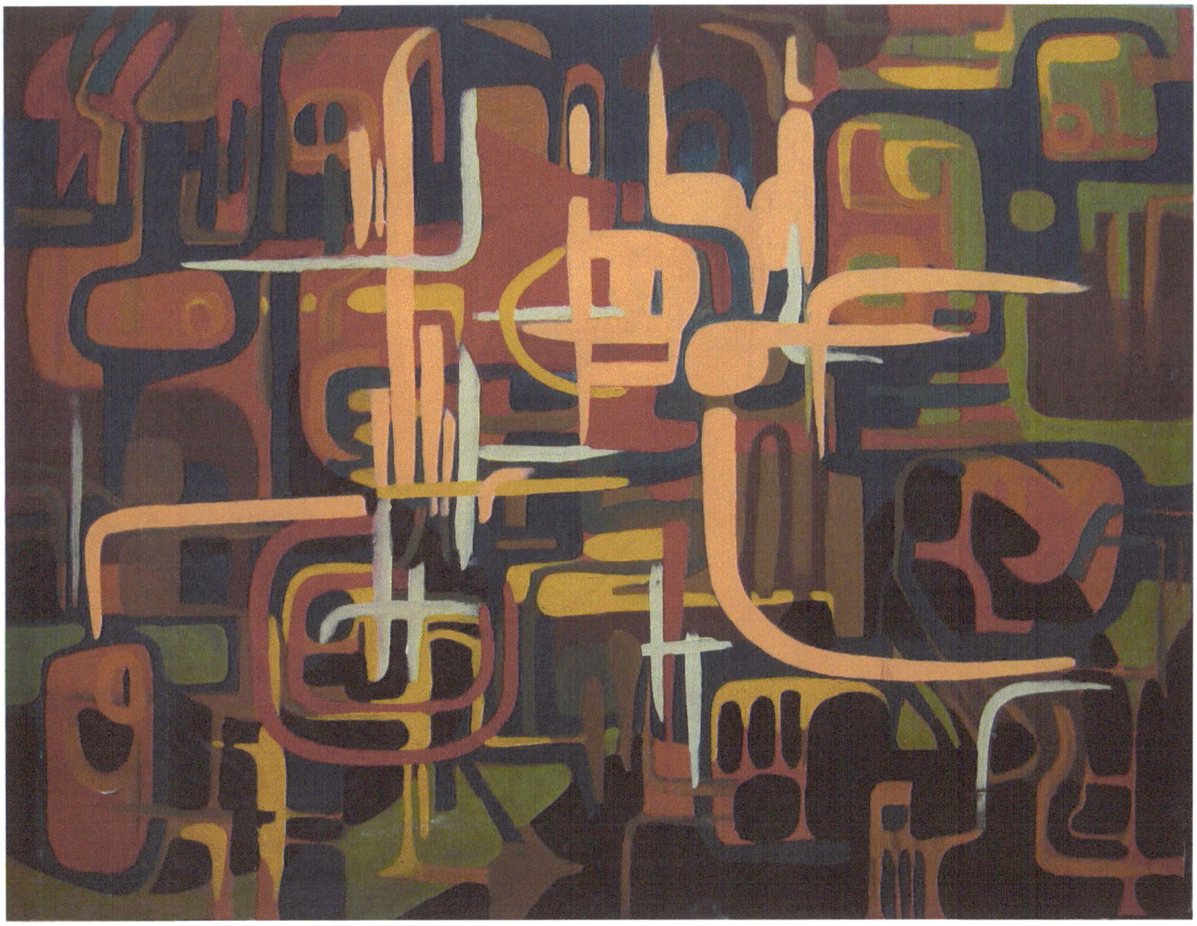

Machines

2.0 - 3000BC - 500 AD

As far as archaeologists can guess the machine age never started until the 16 Century, this is absurd and points to the narrow view point of mankind's development, this shows how uninformed the researchers are about how the ancient civilizations built their amazing monuments. We can find plenty of evidence for a highly developed and mechanized building system, only with a keen engineers mind and an artists eyes you can find evidence at each ancient site.

At the Giza site there are many partly cut granite blocks and drill holes that are beyond the simplistic methods described by archaeologists. The site at Puma- Punka is a classic example of this amazing technique that current stone cutting and engineering can not duplicate easily. Not to mention the jigsaw like stone wall construction that is earthquake proof, a unique geometry that involves prime numbers and Quantum Arithmetic.

2.1- Types of machines

2.1a - Motors

The earliest types of motors that have been designed in the machine age(16th century) were based on clock mechanisms using weights and pendulums , this type of motor would be called a Gravity motor. The dates these devices were invented are only conjecture due to the lack of evidence of the investigator, some mechanistic artifacts have been unearthed by the archaeologists that date back to the Roman times. The understanding of these types of gear trains never happens in isolation, there was some history hidden that has been kept from view, I would suggest the writings in the Alexandrian Library were the sources of the information about this technology.

We can try to find traces of technology in the artworks from earlier civilizations, only when they depict this type of activity. What would the Sumerian, Khemitian or the Toltec cultures have need of motors? The simple answer would be for machining stone as a building material and if they built the cutters in metal I am sure they would have rusted eons ago, resulting in the loss of any artifacts. The clue would be look for the trace rust or the fine edge left by machine tools. I am sure the schematic ideas would have been passed to the artists to preserve for the future generations, time to look for unusual symbols that might represent theses ideas of actions on theses devices.

Nicola Tesla was the first inventor to grasp a new way of seeing the abstractions of the electric and magnetic field and devise motors and generators according to these visual clues based on the newly discovered world of electromagnetism. It was Micheal Faraday's pioneering experiments that started to put the clues together, it took Tesla and other inventors to be creative with the knowledge and discover more possibilities. Still it was the clear imagination of Tesla who understood the whole motor/coil unit to create a new form of rotating machinery, like an artist understands color, this was part of his pallet.

The early exploits of Nikola Tesla were his designs for regulating the DC Dynamo by adding a third brush, this seminal patent showed how he approached this simple problem and found elegant solutions. Refer to the diagram to see how he solved this problem, the patents do not make this understanding very clear. Please refer to the original patents when reviewing my diagram.

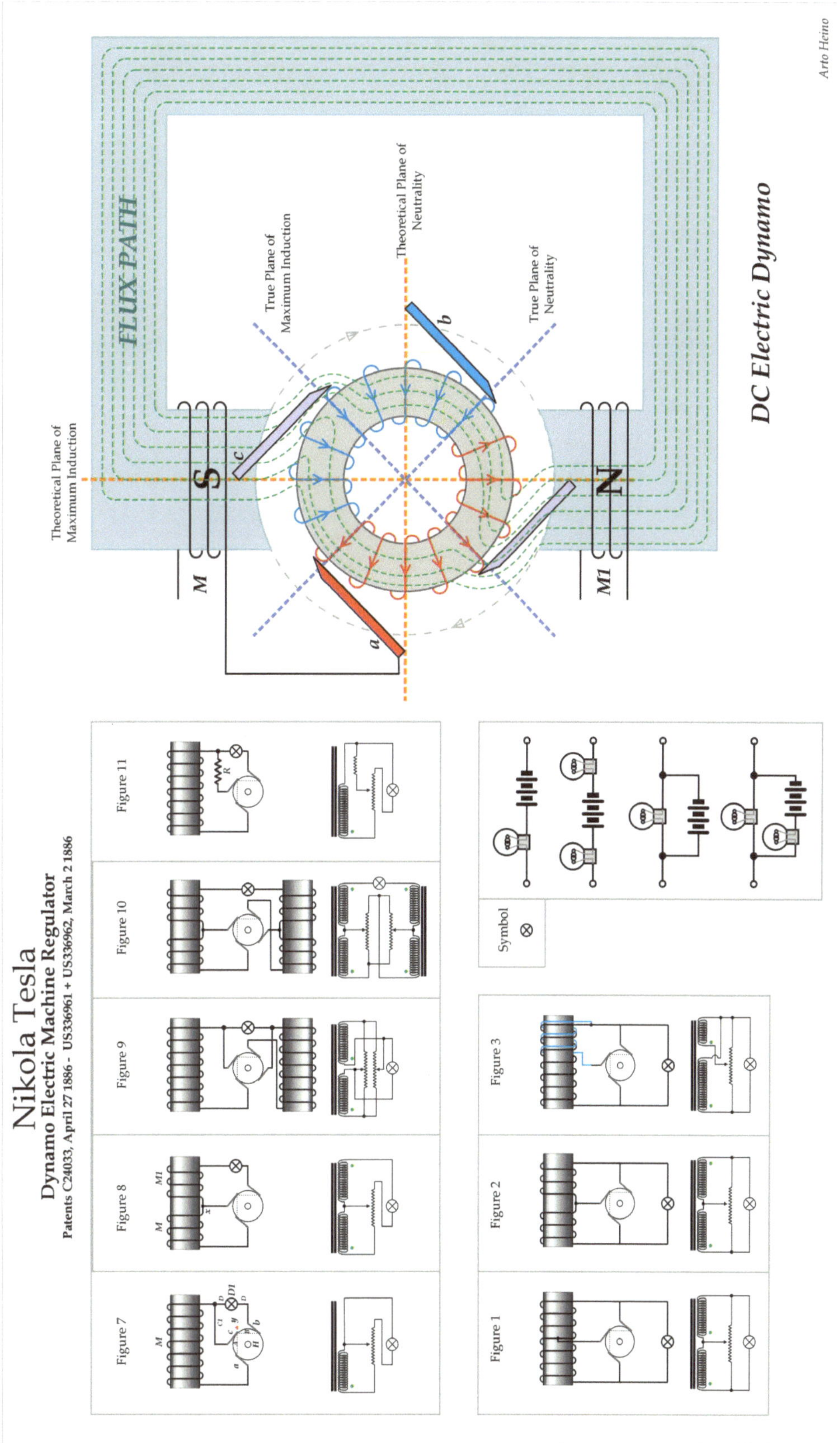
Nikola Tesla
Dynamo Electric Machine Regulator
Patents C24033, April 27 1886 - US336961 + US336962, March 2 1886
Figure 7
Figure 8
Figure 9
Figure 10
Figure 11
Figure 1
Figure 2
Figure 3
Symbol
FLUX PATH
Theoretical Plane of Maximum Induction
True Plane of Maximum Induction
Theoretical Plane of Neutrality
True Plane of Neutrality
S
N
M
M1
a
b
c
DC Electric Dynamo
Arto Heino

Refer to next Diagram , there are six types of AC motor arrangements that Tesla writes about in one of his dissertations about lag effects in his motors.

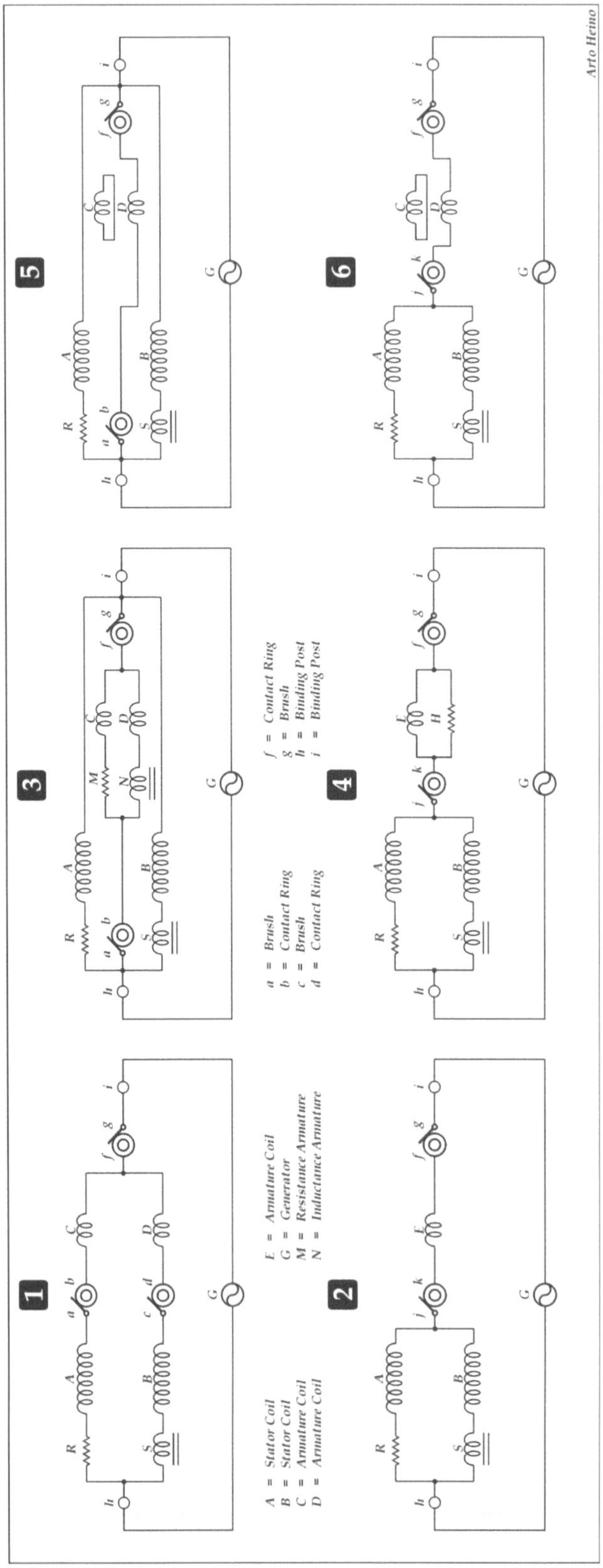

2.2b - Generators

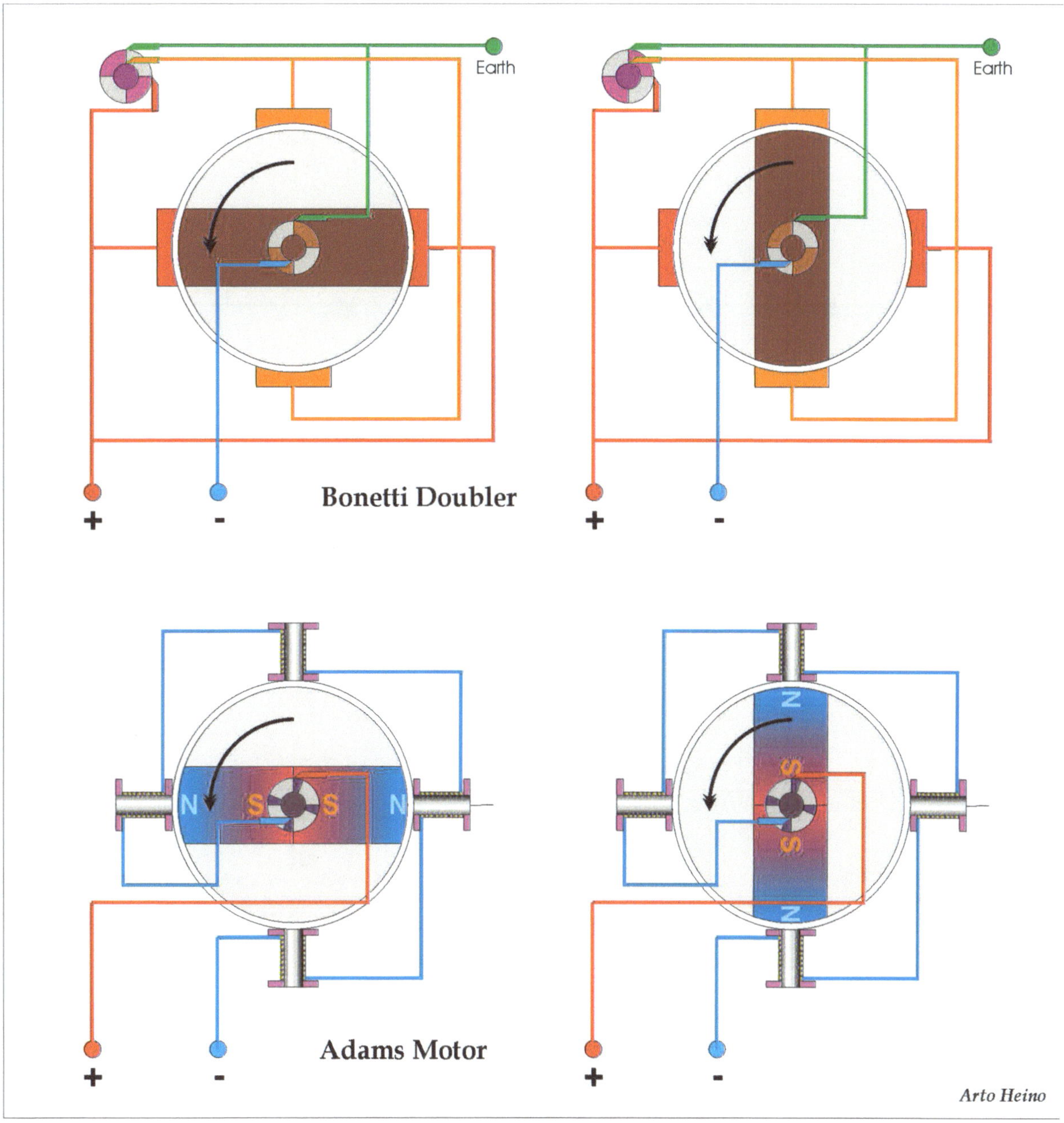

The Bonetti Doubler is a electrostatic generator and the Adams Pulse Motor is an Electromagnetic motor, it is these differences that can make an interesting combination as one act as a charge pump and the other drives this pump action. If they run on the same shaft at the same rotational velocity and they were coupled electrically, such as the doubler charging a capacitor that discharges in to a transformer to charge the battery that runs the motor, this would be considered a closed loop system, sorry no free energy, just an interesting energy transformation system.

2.3c - Levers and Arms

THE COMPOSITION AND RESOLUTION OF FORCES OF THE LEVER

By Arto Heino

F = W x La/Lb	F =(W x L - Lb)/Lb	W = (F x L -La)/La	W = F x Lb/La
F/W = La/Lb	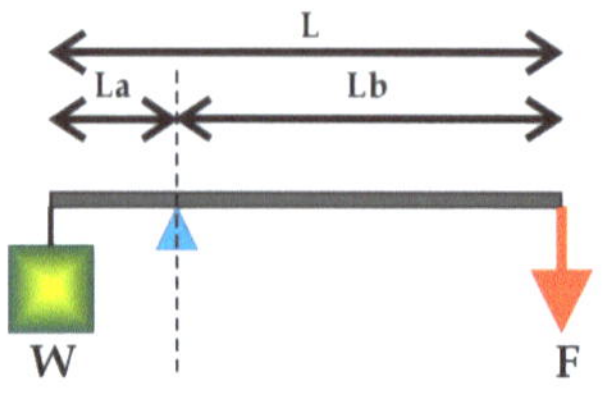		W/Lb = F/ La
L = La+Lb			W x La = F x Lb
La = F x L/W+F	La = F x Lb/W	Lb = W x L/W+F	Lb = W x La/F

F = W x La/L	F =(W x L - Lb)/L	W = (F x La+Lb)/La	W = F x L/La
F/W = La/L	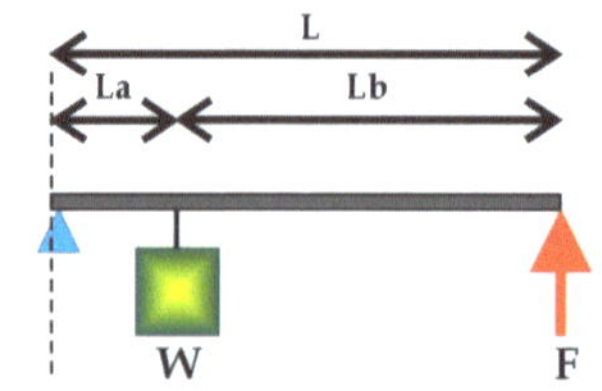		W/Lb = F/ La
L = La+Lb			W x La = F x L
L = W x Lb/W-F	L = W x La/F	La = F x Lb/W-F	La = F x L/W

F = W x L/La	F =(W x La+Lb)/La	W = (F x L -Lb)/L	W = F x La/L
F/W = L/La	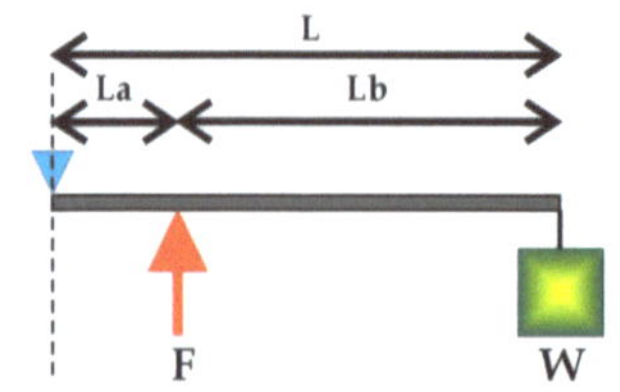		W/La = F/ L
L = La+Lb			W x L = F x La
La = W x Lb/F-W	La = W x L/F	L = F x Lb/F-W	L = F x La/W

THE COMPOSITION AND RESOLUTION OF FORCES OF THE LEVER

By Arto Heino

<table>
<tr><td>F = (W1xLb + W2xLc)/La</td><td>W1 =(FxLa - W2xLc)/Lb</td><td>W2 =(FxLa - W1xLb)/Lc</td><td>La =(W2xLc + W1xLb)/F</td></tr>
<tr><td>Lb = (FxLa - W2xLc)/W1</td><td colspan="2" rowspan="2">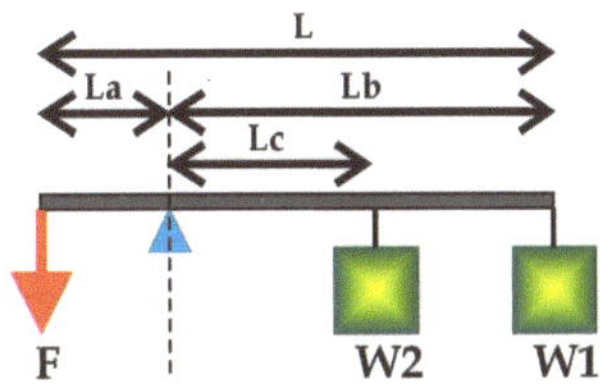
</td><td>FxLa = W1xLb +W2xLc</td></tr>
<tr><td>Lc = (FxLa - W1xLb)/W2</td><td>W1xLb = FxLa - W2xLc</td></tr>
<tr><td>W2xLc = FxLa - W1xLb</td><td>L= La+Lb</td><td>La = L-Lb</td><td>Lb = L-La</td></tr>
</table>

2.4 - Interesting Machines

2.4a – Hydro-Automat

This interesting type of Ram Pump design was described in an early 20th century encyclopedia , The Book of Knowledge, Volume Nine, Page 6601, Indian Edition 1926. I have extrapolated a design form based on a 4 inch pipe, which is 101.6mm. The inventor is Thomas Gaskell Allen, US Patent 1597664, 1926.

The geometric values of the water tanks

	Width	Height	Length	Surface Area	Max Volume	Max Weight Kg	Max Pressure Pa
Wt1	1.117600	0.254000	0.304800	0.340644	**0.086524**	848.507622	2490.889100
Wt2	0.508000	0.203200	0.304800	0.154838	**0.031463**	308.548226	1992.711280
Wt3	0.558800	0.254000	0.304800	0.170322	**0.043262**	424.253811	2490.889100
Wt4	0.558800	0.254000	0.304800	0.170322	**0.043262**	424.253811	2490.889100
Wt5	0.558800	0.254000	0.304800	0.170322	**0.043262**	424.253811	2490.889100
Wt6	0.838200	0.254000	0.304800	0.255483	**0.064893**	636.380717	2490.889100

The geometric values of the water pipes

	Height 1	Height 2	Pipe Diam	Pipe Vol 1	Pipe Vol 2	U bend OD	U bend ID	U bend Volume	Volume Totals
Wp1	1.028700		0.101600	0.008340					**0.008340**
Wp2	0.762000	1.828800	0.101600	0.004633	0.013282	0.381000	0.177800	0.002264	**0.020180**
Wp3	0.838200	1.879600	0.101600	0.005457	0.013900	0.330200	0.127000	0.001617	**0.020974**
Wp4	0.825500	1.739900	0.101600	0.005354	0.012767	0.330200	0.127000	0.001617	**0.019739**
Wp5	0.838200	1.701800	0.101600	0.005457	0.012459	0.330200	0.127000	0.001617	**0.019533**
Wp6	0.812800	1.651000	0.101600	0.005251	0.012047	0.330200	0.127000	0.001617	**0.018915**

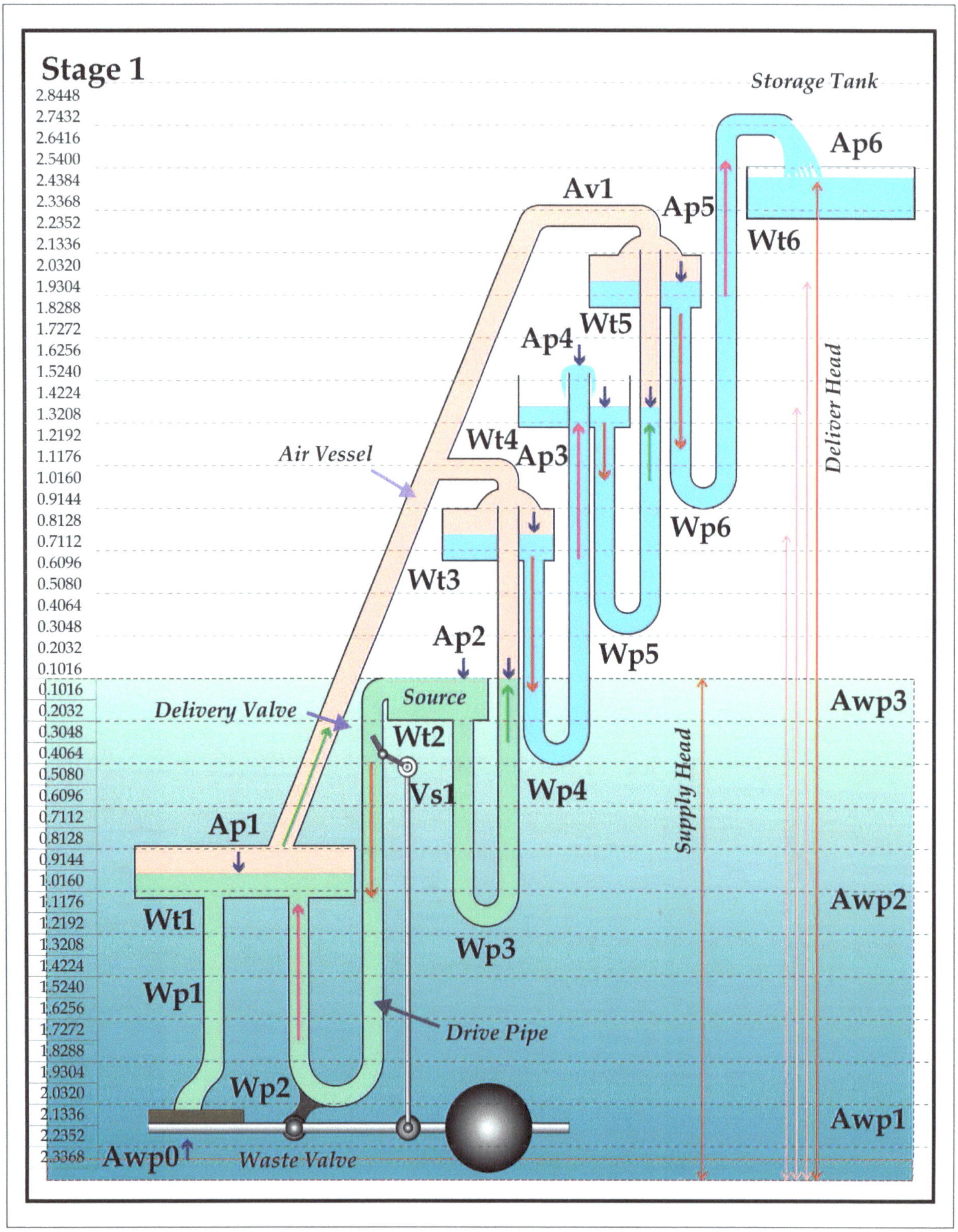

The geometric values of the joining cavity pipe

	Length 1	Length 1	Length 1	Pipe Diam	Pipe Vol	OD	ID	Volume	Totals
Cv1	3.302000	0.584200	0.381000	0.101600	0.034596	0.160655	0.094234	0.004259	**0.043113**

Total Volumes

Water Tanks		Water Pipes	
	0.086524		0.008340
	0.031463		0.020180
	0.043262		0.020974
	0.043262		0.019739
	0.043262		0.019533
	0.064893		0.018915
	0.312665		**0.107681**
Pipe Cavity Volume	**0.043113**	**TOTALS**	**0.463459**

To give this clear information usable, I have also included a few equations. First a few variables that can help in establishing atmospheric pressure to altitude, just so we have an accurate values.

Variable		At Sea level
APa	Atmospheric Pa	101325
TLR	Temp Lapse Rate	0.0065
STR	Sea Level Temp	288.15
g	Gravity acceleration	9.80665
Mma	Molar Mass of air	0.0289644
R	Gas Constant	8.31447

Using the equation:

Pa = APa x (TLR x h)/ SLT)) ^ ((g x Mma)/(R x TLR))

where h = Height above sea level in mtrs

Tank Number	Bottom of Tank Height From Sea Level	Top of Tank Height From Sea Level	Atm Air Pa Per Max Height	Atm Air Pa Per Min Height	Total Max Pressure Pa	Total Min Pressure Pa
Wt1	-1.066800	-0.812800	101337.816041	101334.764483	103828.705141	103825.653583
Wt2	-0.203200	0.000000	101327.441049	101325.000000	103320.152329	103317.711280
Wt3	0.571500	0.825500	101318.134804	101315.083726	103809.023904	103805.972826
Wt4	1.219200	1.473200	101310.354704	101307.303816	103801.243804	103798.192916
Wt5	1.803400	2.057400	101303.337773	101300.287056	103794.226873	103791.176156
Wt6	2.235200	2.489200	101298.151599	101295.101009	103789.040699	103785.990109

	Surface Area	Max Volume	Max Weight	Max Pressure
Wp1	0.008107	0.008340	81.787458	10088.100855
Wp2	0.008107	0.020180	197.896518	24409.610784
Wp3	0.008107	0.020974	205.688353	25370.697264
Wp4	0.008107	0.019739	193.571692	23876.163804
Wp5	0.008107	0.019533	191.552249	23627.074894
Wp6	0.008107	0.018915	185.493919	22879.808164

The **R** description is the Right hand side of the U-pipe and **L** represents the left hand side of the same pipe.

Pipe Number	Bottom of Pipe L Height From Sea Level	Top of Pipe L Height From Sea Level	Atm Air Pa Per Max Height	Atm Air Pa Per Min Height	Total Max Pressure Pa	Total Min Pressure Pa
Wp1	-2.095500	-1.066800	101350.175609	101337.816041	111438.276464	111425.916896
Wp2	-2.082800	-1.066800	101350.023014	101337.816041	125759.633798	125747.426825
Wp3	-1.206500	0.000000	101339.494429	101325.000000	126710.191693	126695.697264
Wp4	-0.419100	0.571500	101330.034717	101318.134804	125206.198521	125194.298608
Wp5	0.215900	1.219200	101322.406437	101310.354704	124949.481332	124937.429598
Wp6	0.825500	1.803400	101315.083726	101303.337773	124194.891891	124183.145937

Pipe Number	Bottom of Pipe R Height From Sea Level	Top of Pipe R Height From Sea Level	Atm Air Pa Per Max Height	Atm Air Pa Per Min Height	Total Max Pressure Pa	Total Min Pressure Pa
Wp1			101325.000000	101325.000000	111413.100855	111413.100855
Wp2	-2.082800	0.000000	101350.023014	101325.000000	125759.633798	125734.610784
Wp3	-1.206500	0.825500	101339.494429	101315.083726	126710.191693	126685.780991
Wp4	-0.419100	1.485900	101330.034717	101307.151273	125206.198521	125183.315078
Wp5	0.215900	2.082800	101322.406437	101299.981989	124949.481332	124927.056883
Wp6	0.825500	2.641600	101315.083726	101293.270691	124194.891891	124173.078855

All theses values should be enough to engineer this design. This unit is unique in that there is 1 main delivery head and 3 minor delivery heads and 1 main drive pipe with 3 auxiliary drive pipes. The tank **Wt1** is main the surge tank with each smaller delivery tanks acting as a smaller surge tank due to the interconnected air vessel **Cv1**. The complex equation needs to find the flow rate of the Ram Pump would entail all 4 tanks acting in sequential timing , which its action will be a slower rate due to the multistage pipes and tanks. Here is an tentative equation, which still needs more work to satisfy the accuracy required to build such a device.

Prate = (Eff x SF) / (VE / VF) Ltrs per Minute

Where

Eff = 0.6 Efficiency, generally about 60%

SF	=	50	Source Flow, tentative value 50 Ltrs per minute
VL	=	2.2352	Vertical Lift from pump to destination, **Wt6**
VF	=	1.8288	Vertical Fall from Source to the Drive Pipe, **Wp2**

Prate = (0.6 x 50) / (2.2352 / 1.8288)

= **24.545454** Ltrs per Minute This would be only for the lower stage

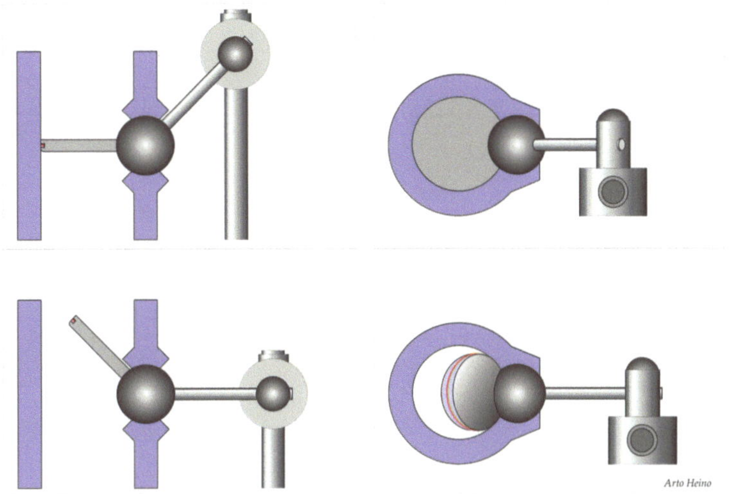

The valve mechanism is crucial to this design as any other type of system used must have the correct leverage and geometric parts equal to the task required. To not comply with this would welcome a failure into the replication.

Volume Stages	**Full**	**Stage 1**	**Stage 2**	**Stage 3**	*Negative vacuum*	**Stage 4**	*Negative vacuum*
Wt1	**0.086524**	0.043262	0.086524	0.043262		0.000000	
Wp1	**0.008340**	0.008340	0.008340	0.008340		0.000000	
Wt2	**0.031463**	0.031463	0.031463	0.031463		0.031463	
Wp2	**0.020180**	0.020180	0.020180	0.013453	**0.006727**	0.013453	**0.006727**
Wt3	**0.043262**	0.021631	0.000000	0.021631		0.043262	
Wp3	**0.020974**	0.013983	0.013983	0.020974		0.013983	
Wt4	**0.043262**	0.021631	0.043262	0.021631		0.000000	
Wp4	**0.019739**	0.019739	0.013159	0.013159		0.013159	
Wt5	**0.043262**	0.021631	0.000000	0.021631		0.043262	
Wp5	**0.019533**	0.013022	0.013022	0.019533		0.013022	
Wt6	**0.064893**	0.064893	0.000000	0.000000		0.000000	
Wp6	**0.018915**	0.018915	0.012610	0.012610		0.012610	
Cv1	**0.043113**	0.143139	0.143139	0.129636		0.151479	

1. These pictures show how a river is made to raise itself by the apparatus called a hyrdrautomat. It consists of an operating chamber at the bottom, and a series of tanks alternately opened and closed and connected with pipes. An air-pipe also links them all. Water is here shown running into the operating chamber from the river at the higher level, compressing the air and forcing it through the air-pipe to the closed overhead tanks.

2. The closed overhead tanks are full of water at the beginning of the operation, but the pressure of the air from the operating chamber forces the water out the closed tanks, and in each case it passes through a bent outlet pipe into the tank immediately above. At the close of the operation, therefore, as shown here, the operating chamber and the open tanks are full of water and the closed tanks are full of air.

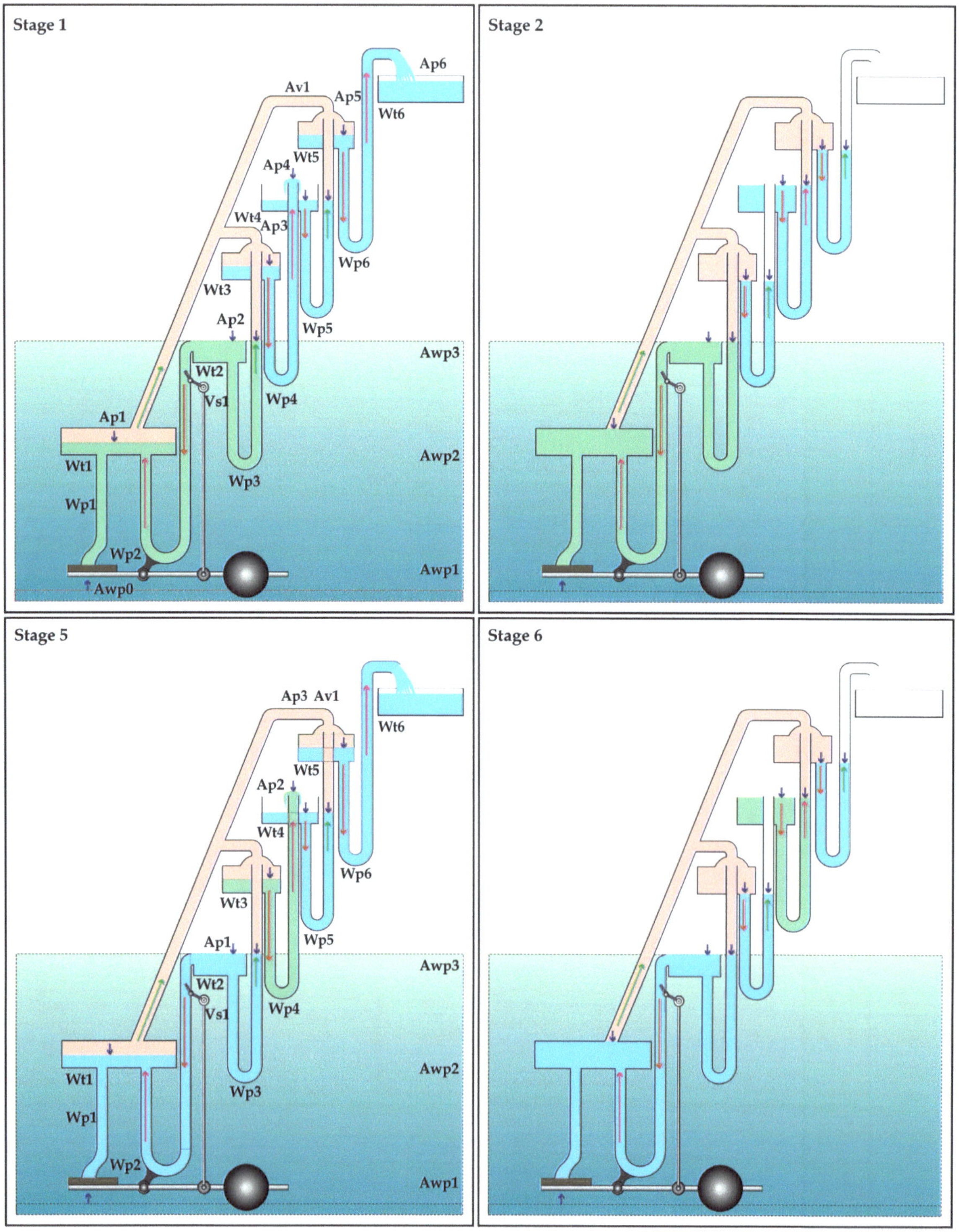

3. The operating chamber now discharges its water into the river at the lower level, and in doing so automatically closes, by a simple lever mechanis, the outlet into the chamber from river. When the water has all run out at the lower level, practically a vacuum is left, which extends by the air-pipe to the various closed tanks above.

4. The vacuum in the closed tanks causes the water to be sucked up one flight from the open tanks below, and at the end of the movement the operating chamber and the open tanks are empty of water, while the closed tanks are full. The inlet from the upper level of the river then automatically opens and the operation is repeated.

Stage 3

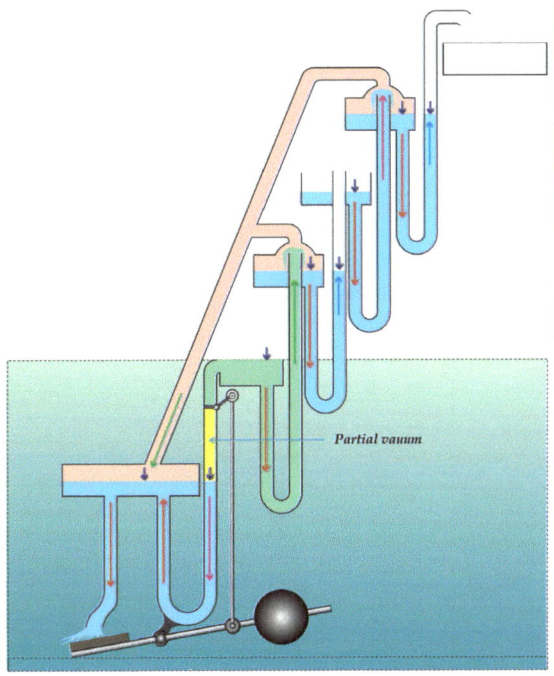

Stage 4

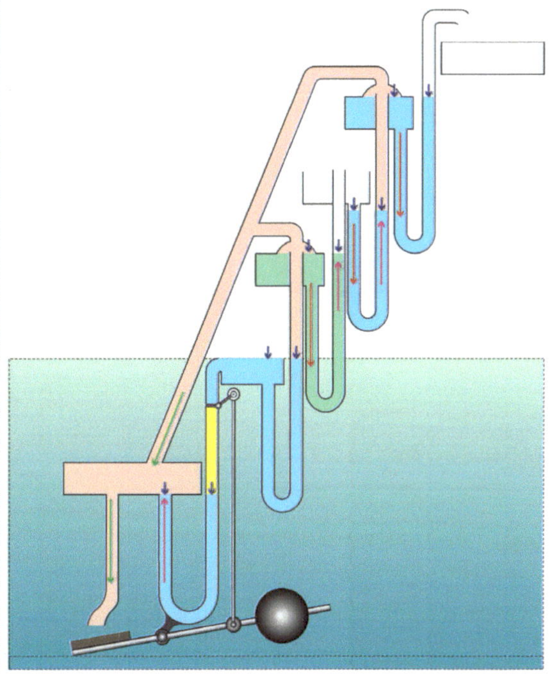

Stage 7

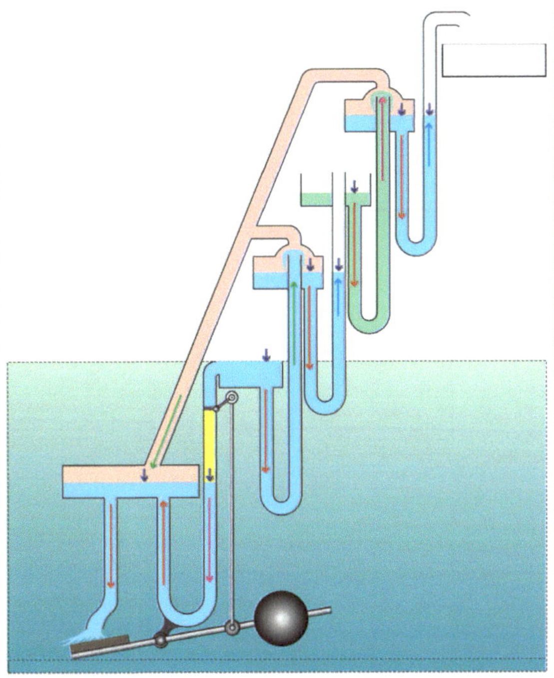

Stage 8

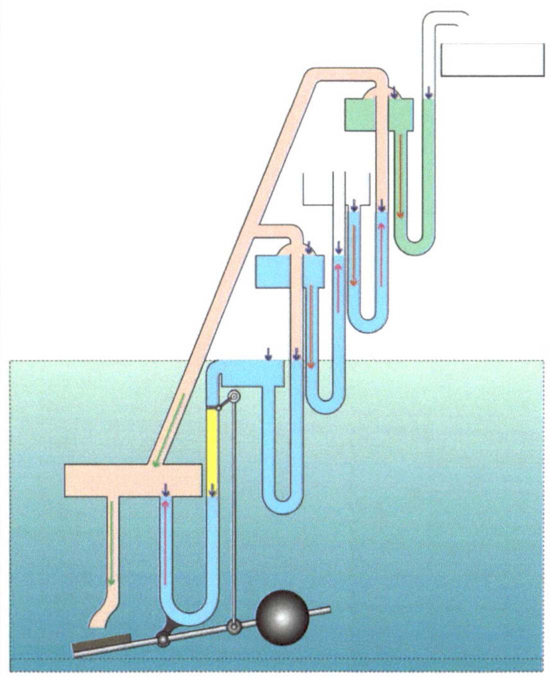

2.4b - Tesla Turbine

I had parts of this unit made in 2000 and finally run out of funds for any further development, here is as far as I got with this project.

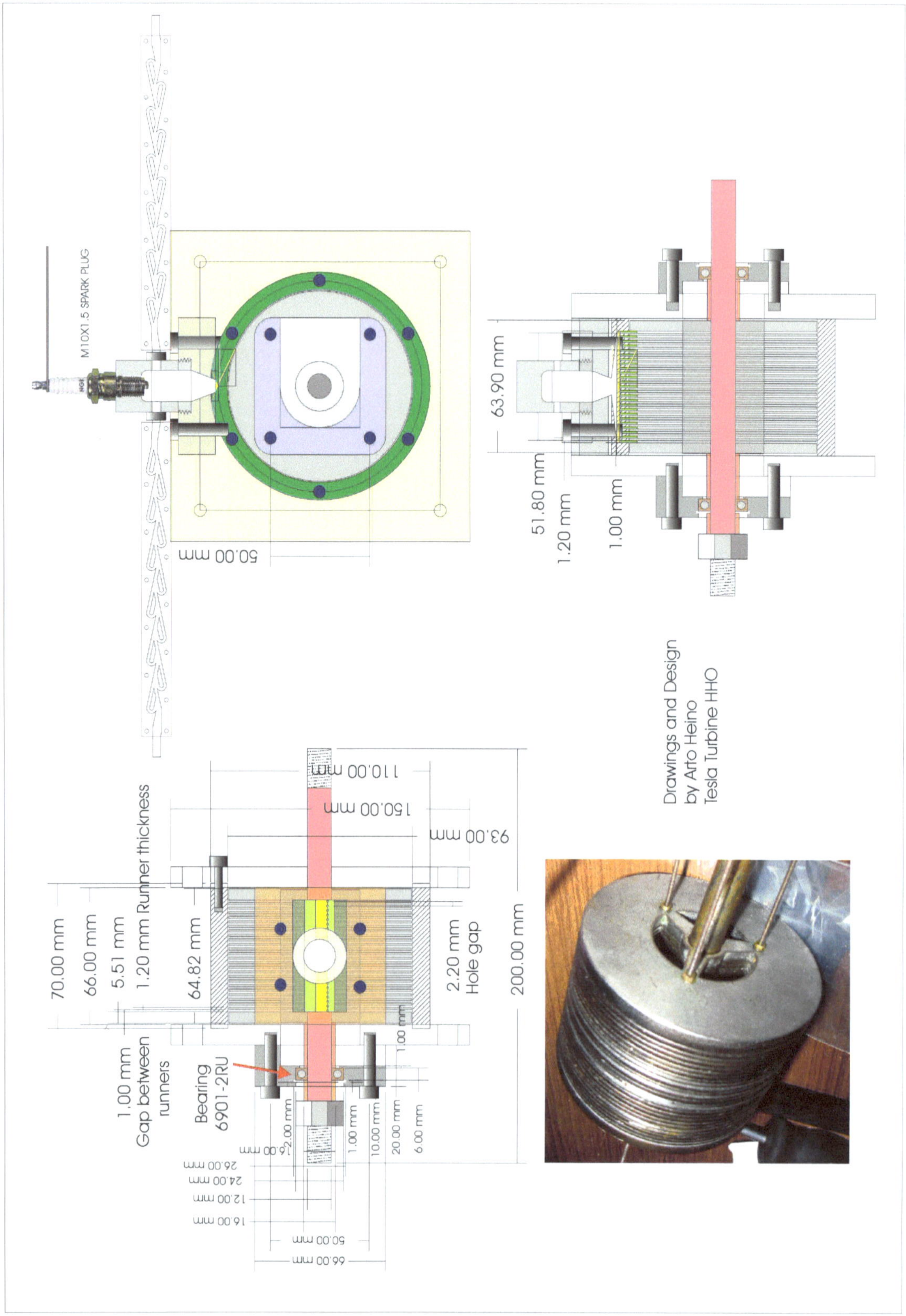

2.4c - Mechanical Oscillator

To really understand how this device operates you must find all the parameters that make it a functioning unit, guess work is too costly and the mathematics is not very complex. This means success of a functioning unit is at easy reach.

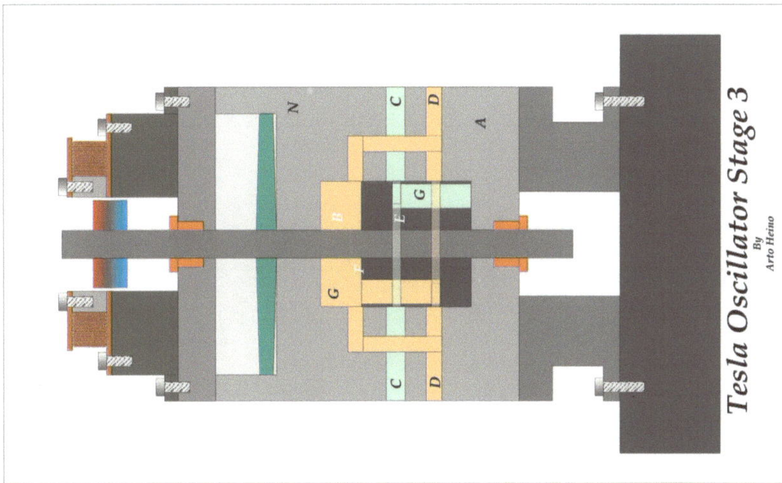

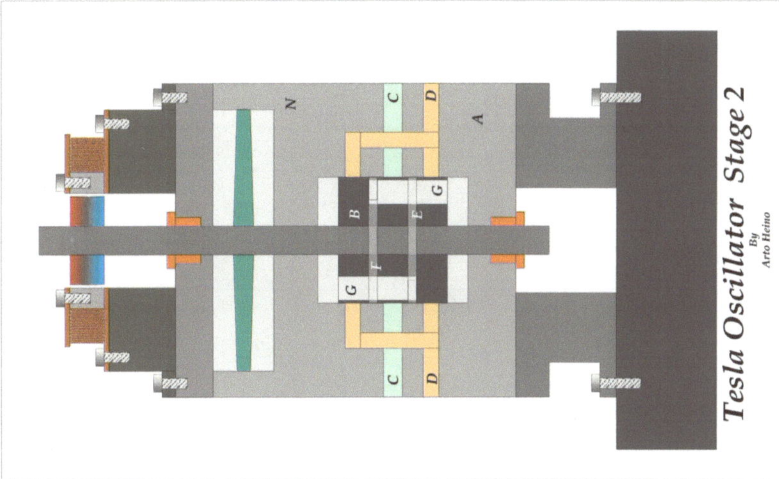

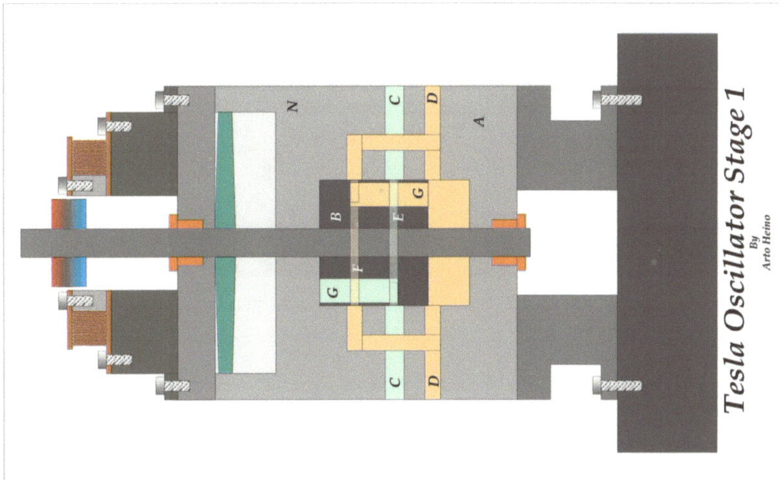

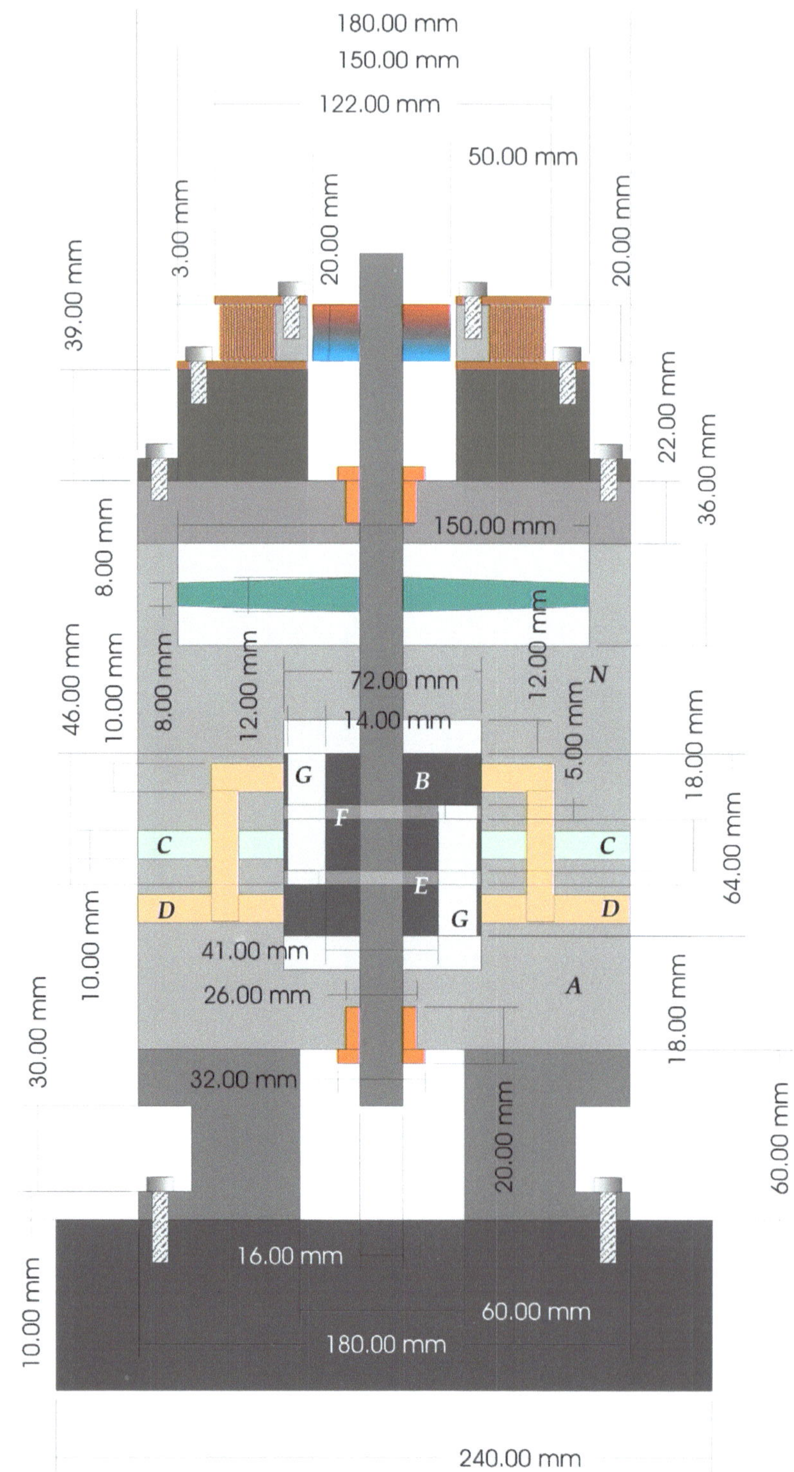

Tesla Oscillator

By
Arto Heino

2.4d - Torque Converters

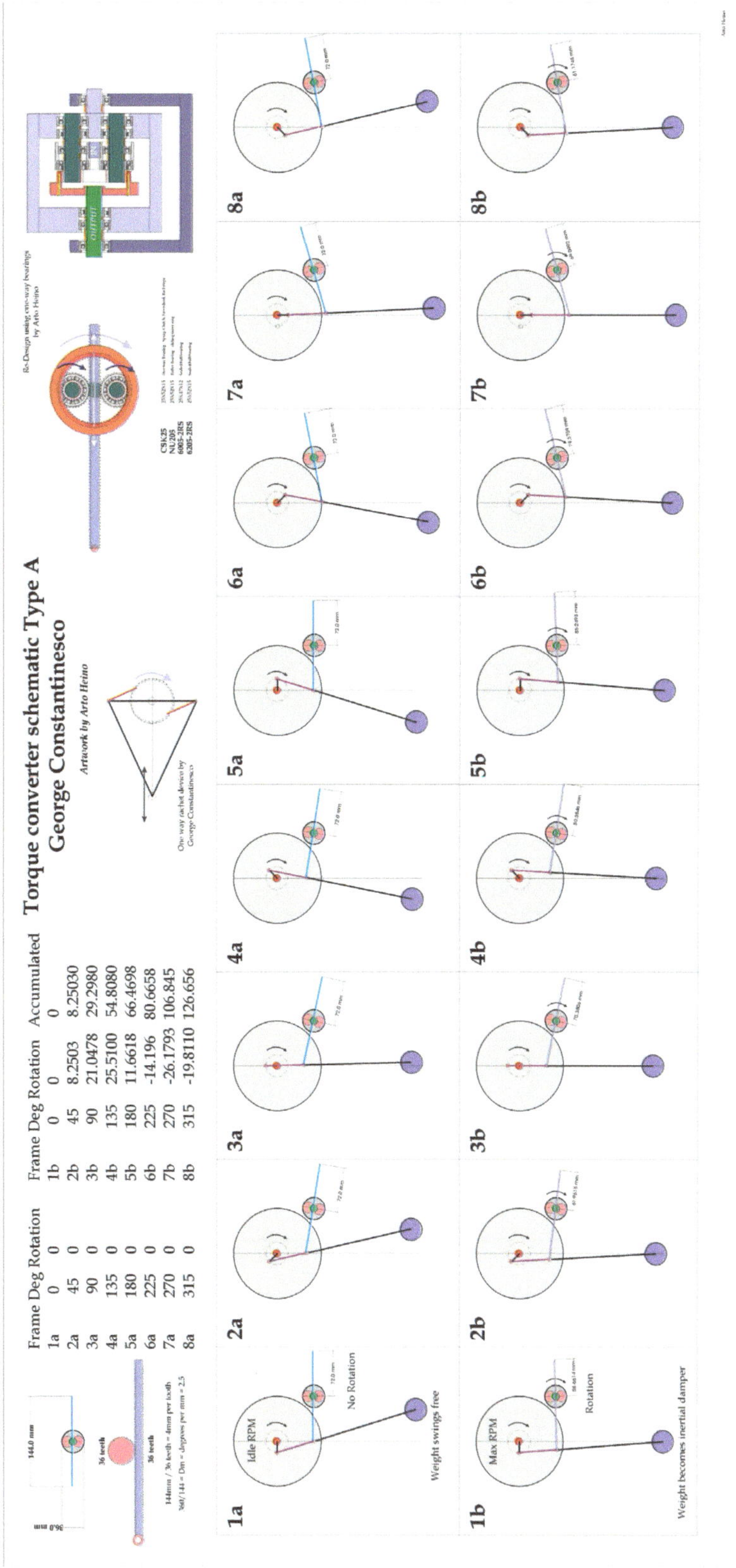

One of the most important devices created to utilize the modern torque converter is the one way rotating mechanical diode. the current terminology is a Sprag clutch or One-way bearing. Here are some close ups of the mechanisms.

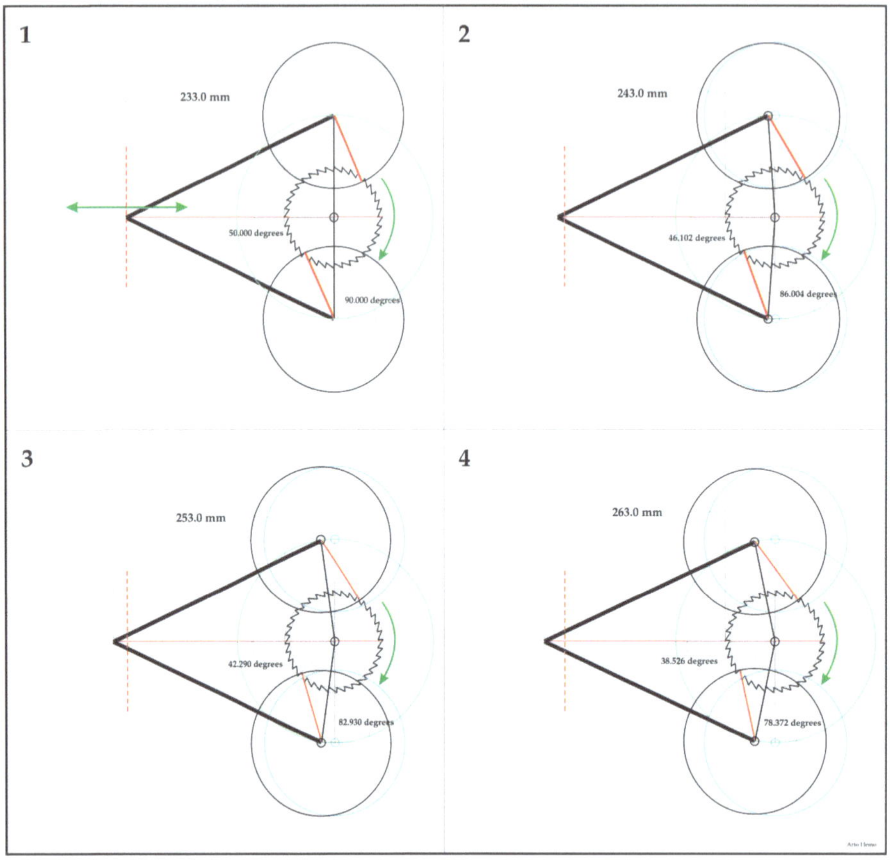

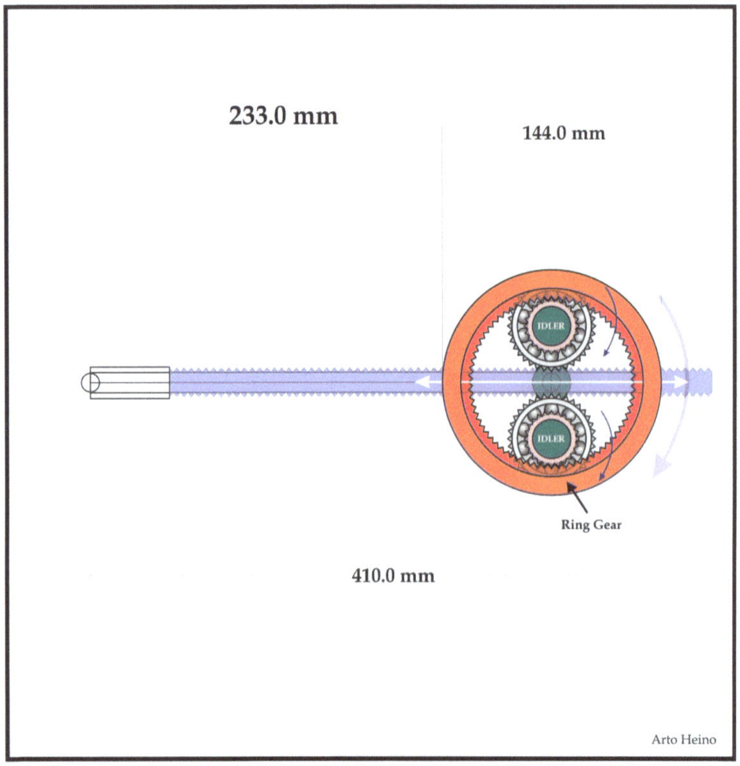

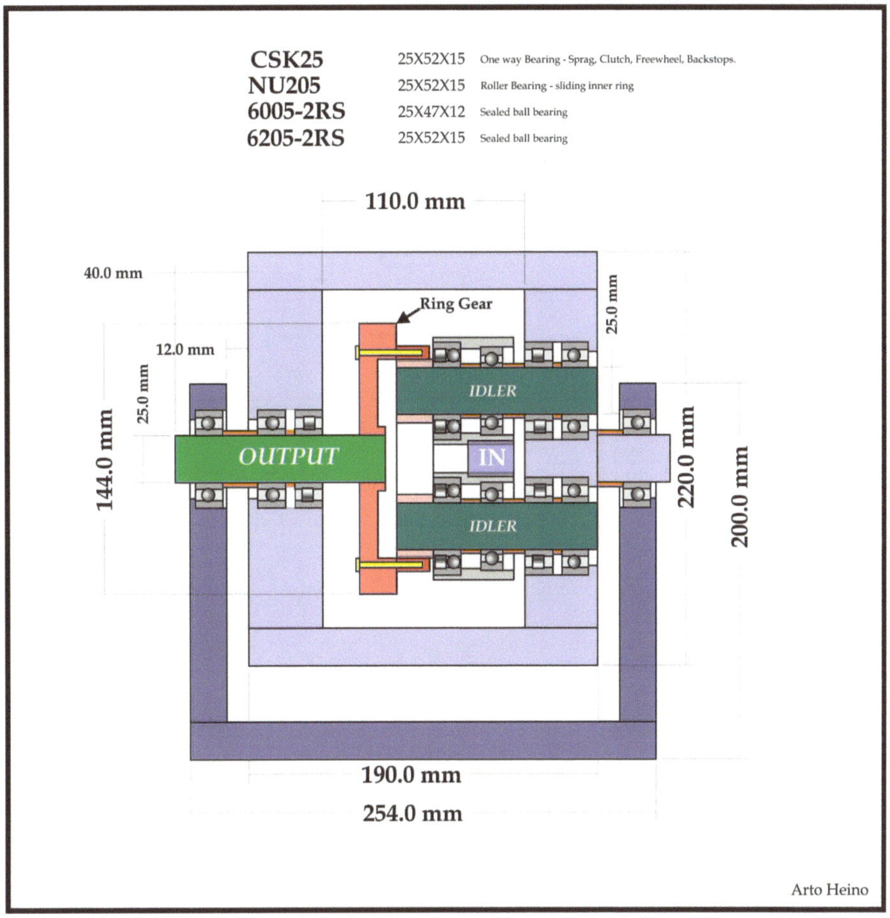

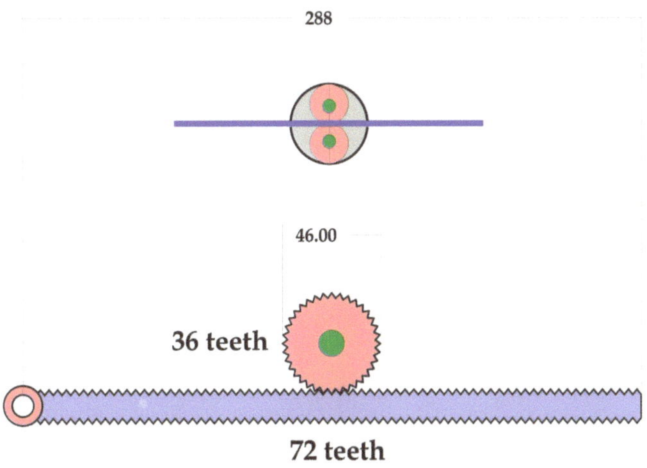

144mm / 36 teeth = 4mm per tooth

360/144 = Dm = degrees per mm = 2.5

Arto Heino

1a

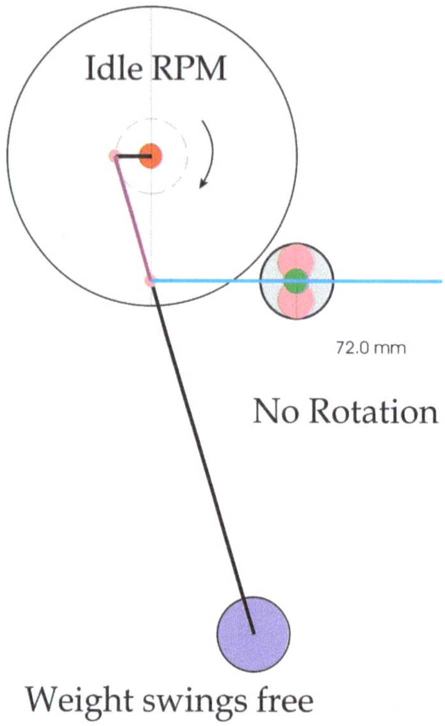
Idle RPM
72.0 mm
No Rotation
Weight swings free

1b

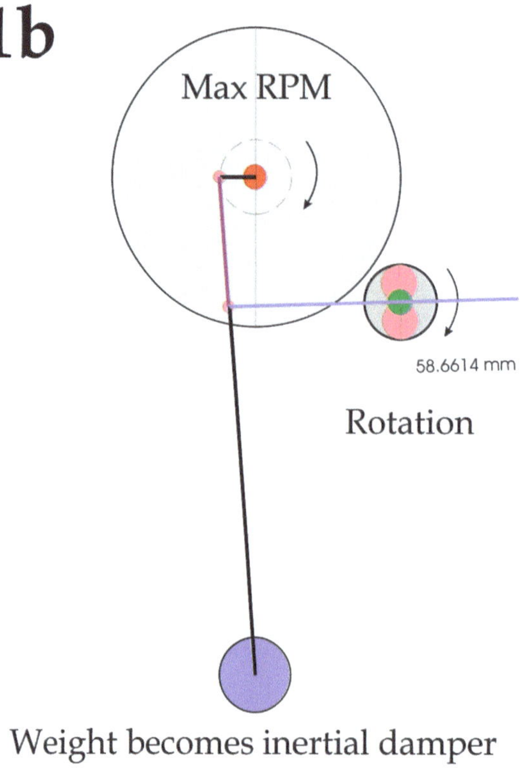
Max RPM
58.6614 mm
Rotation
Weight becomes inertial damper

2.4e - Contra Rotational Magnetic Repulsion Vacuum Oscillator

This unit is a design concept that utilizes a multitude of ideas in a simple system of opposing forces.

1 Magnetic repulsion
2 Electromagnetic coil attraction
3 Spring compression of gases
4 Adiabatic cycling
5 Acoustic standing waves
6 Twin counter rotating flywheels.

The basic operation is similar to that of a relaxation oscillator, the flywheel inertia that starts the system working, forces the the 2 opposing pistons into a condition of compression like that of a spring, with each opposing part to the complete a highly evolved compression force, magnetic repulsion, gas compression which includes heat expansion and the opposing inertia from two flywheels. The magnetic repulsion can be controlled by the coils on the body of both halves of the cylinder and the gas compression can be adjusted so the pump action is at acoustic resonance to the cavity. This can be finely tuned so when the system operates, the pump frequency is at twice the frequency of the signal frequency thus creating a parametric oscillator. The retarding forces are small due to the action of the flywheel. This system of different actions should result in an acceleration of all events, only the impedance of each event retard this and create a balanced set of actions by a varying load from each part, the final result is a single operational frequency of resonance.

This is a schematic of the actions of the opposing pivots that hold the main piston pair on the flywheel.(red and blue)

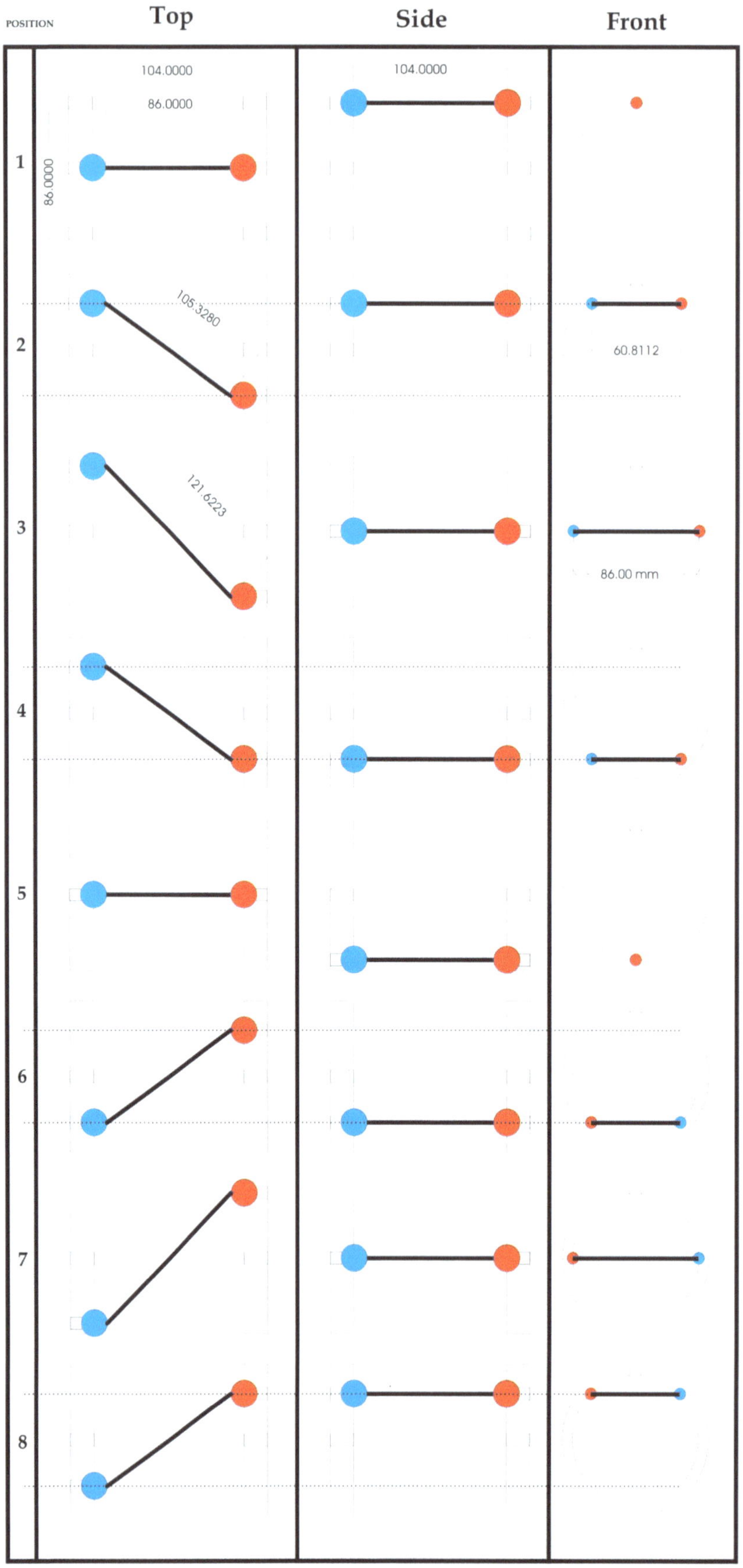
POSITION
Top
Side
Front
104.0000
86.0000
104.0000
1
86.0000
2
105.3280
60.8112
3
121.6223
86.00 mm
4
5
6
7
8

CRMRVO

Contra Rotational Magnetic Repulsion Vacuum Oscillator

By Arto Juhani Heino (c) 2007-2013

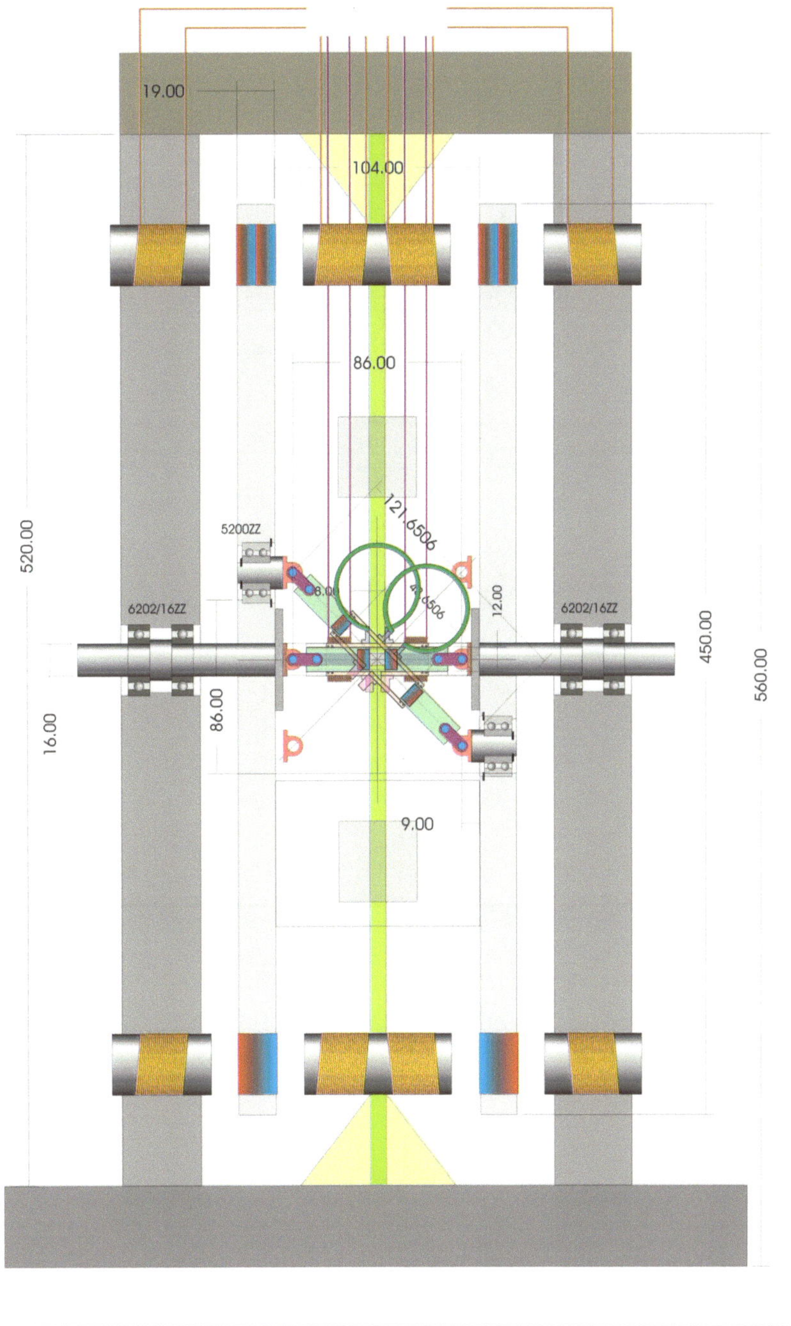

CRMRVO
Contra Rotational Magnetic Repulsion Vacuum Oscillator

By Arto Juhani Heino (c) 2007-2013

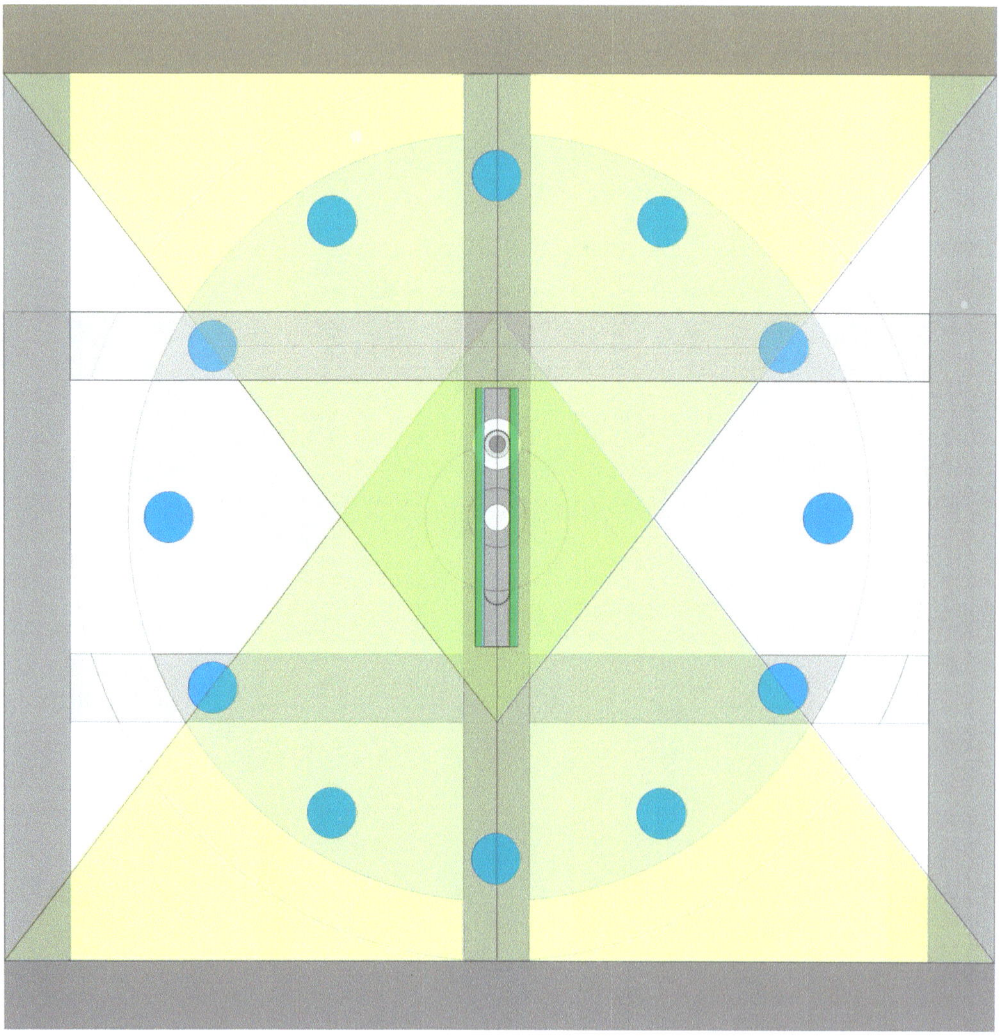

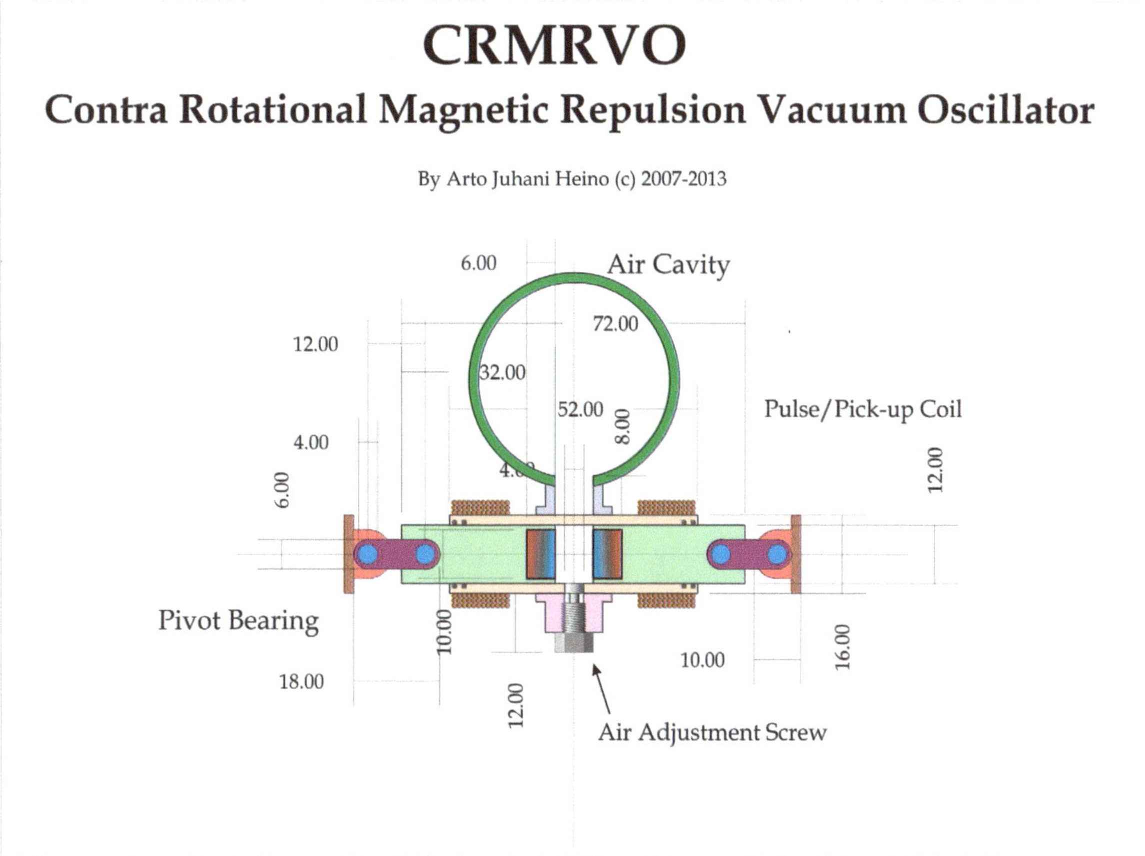
CRMRVO
Contra Rotational Magnetic Repulsion Vacuum Oscillator
By Arto Juhani Heino (c) 2007-2013
6.00
Air Cavity
72.00
12.00
32.00
52.00
8.00
Pulse/Pick-up Coil
4.00
12.00
6.00
Pivot Bearing
10.00
10.00
16.00
18.00
12.00
Air Adjustment Screw

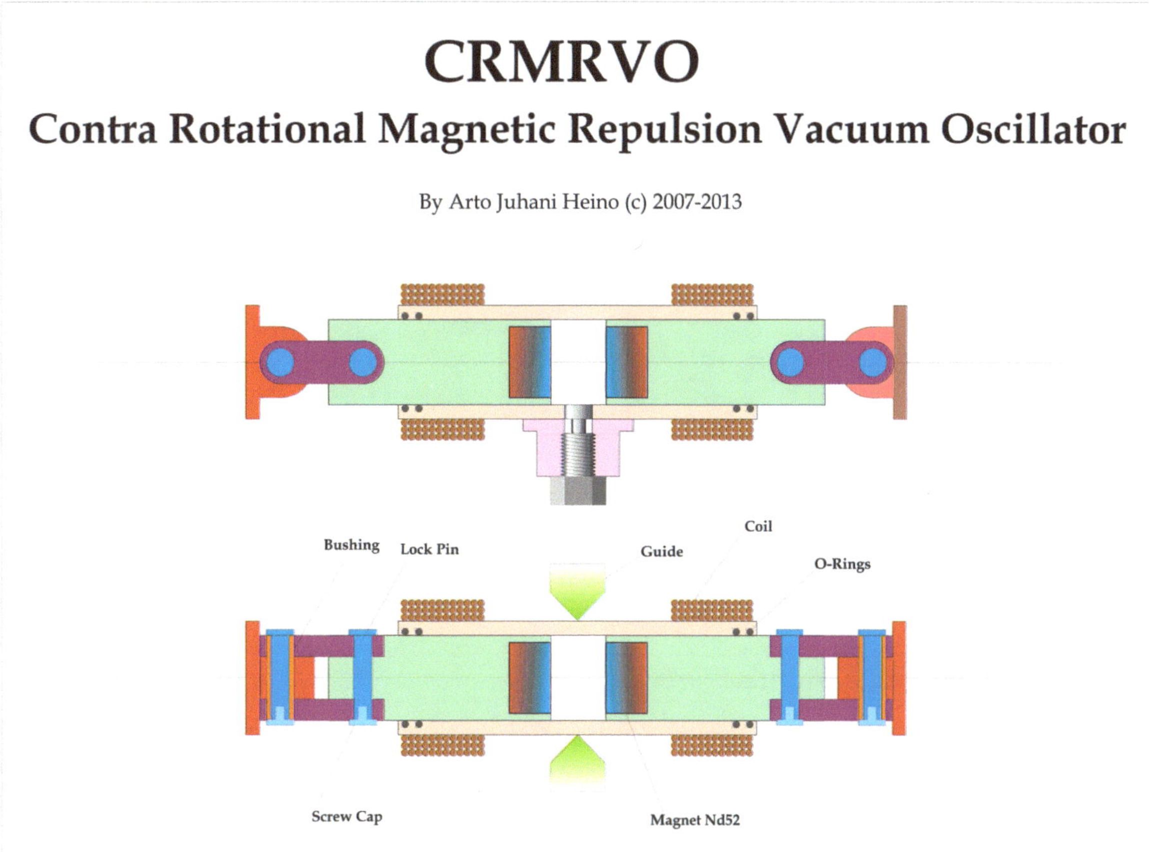
CRMRVO
Contra Rotational Magnetic Repulsion Vacuum Oscillator
By Arto Juhani Heino (c) 2007-2013
Coil
Bushing
Lock Pin
Guide
O-Rings
Screw Cap
Magnet Nd52

CRMRVO

Contra Rotational Magnetic Repulsion Vacuum Oscillator

By Arto Juhani Heino (c) 2007-2013

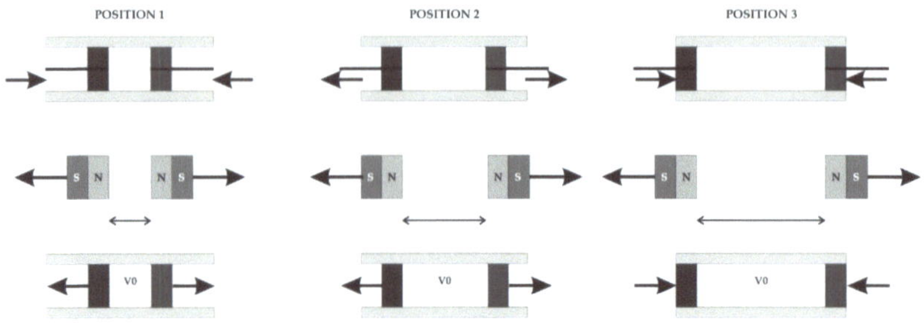

FD = Forward Direction Disk
BD = Backward Direction Disk
SD = Stationary Direction Disk

FD1+FD2+000
FD1+FD2+180
FD1+BD2
BD1+BD2
BD1+FD2

OPERATIONAL MODES

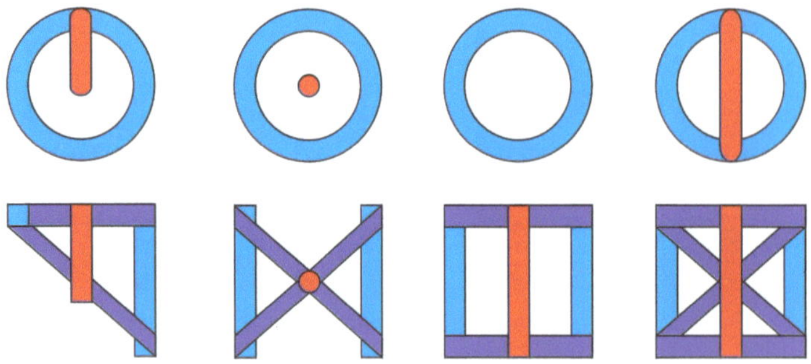

SD1+FD2+P---+R-:X FD1+FD2+P180+R1:1 FD1+FD2+P000+R1:1 FD1+BD2+P180+R1:1

2.4f - Hyperbolic Impulse Drive

This unit uses the constant change of the fulcrum to gain advantage when rotating a flywheel. The action begins at the rotation of the input shaft, as the flywheel turns the the joining arm moves in an arc while sliding inside the opposite wheel which begins to rotate as it acts like a sliding gear peg. This action becomes interesting when the pivot arm base is at the lower part of the cycle, this is when the action of the passing arm locked into the sliding hole creates a momentary impulse as the velocity of the outer rim exceeds the rotation of the opposite arm holder. To prevent the retarding force of deceleration when passing this point , a one-way bearing is forced into a free wheel action by its internal negative forces.

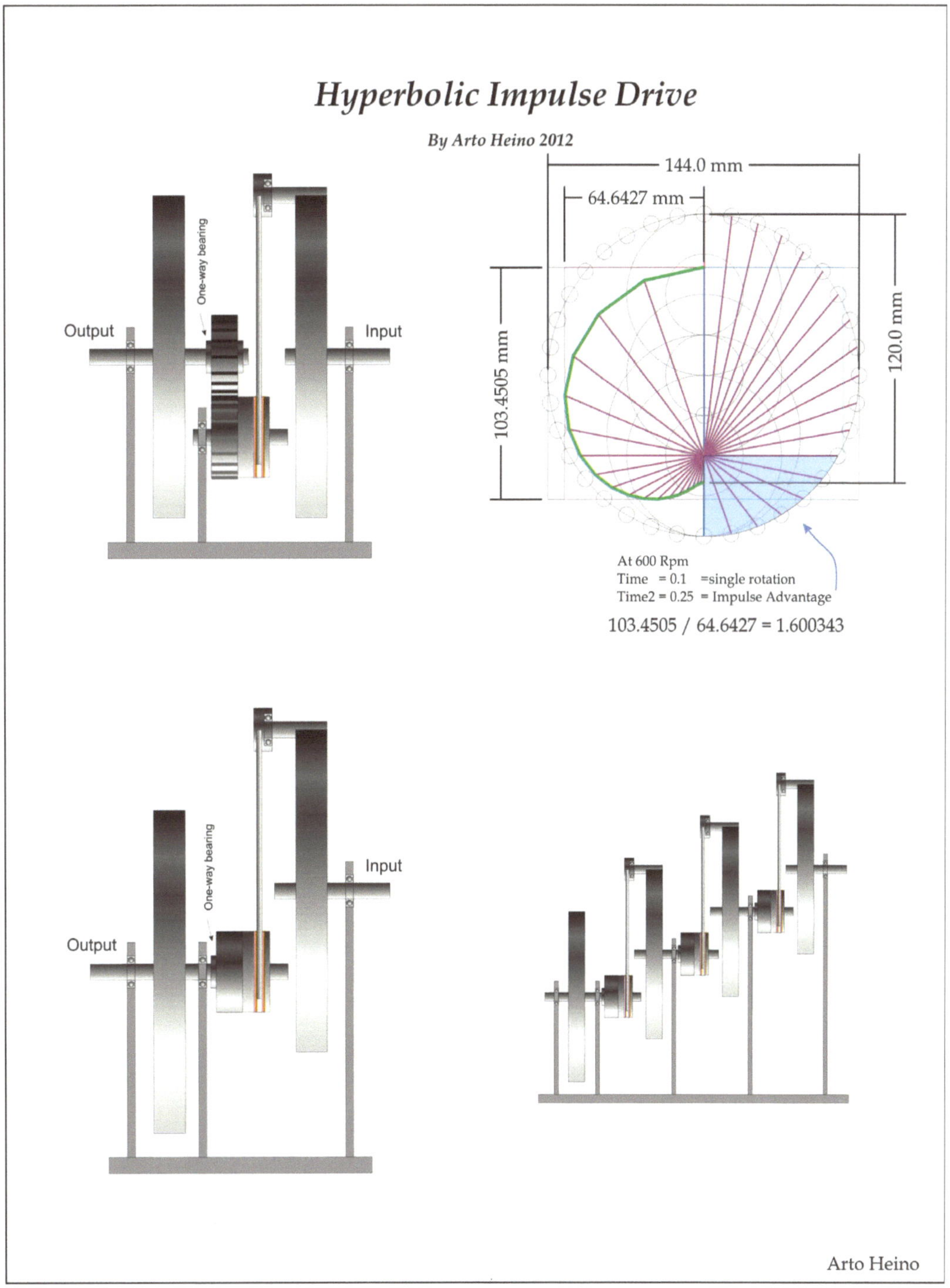

2.4g – Vortex Tube

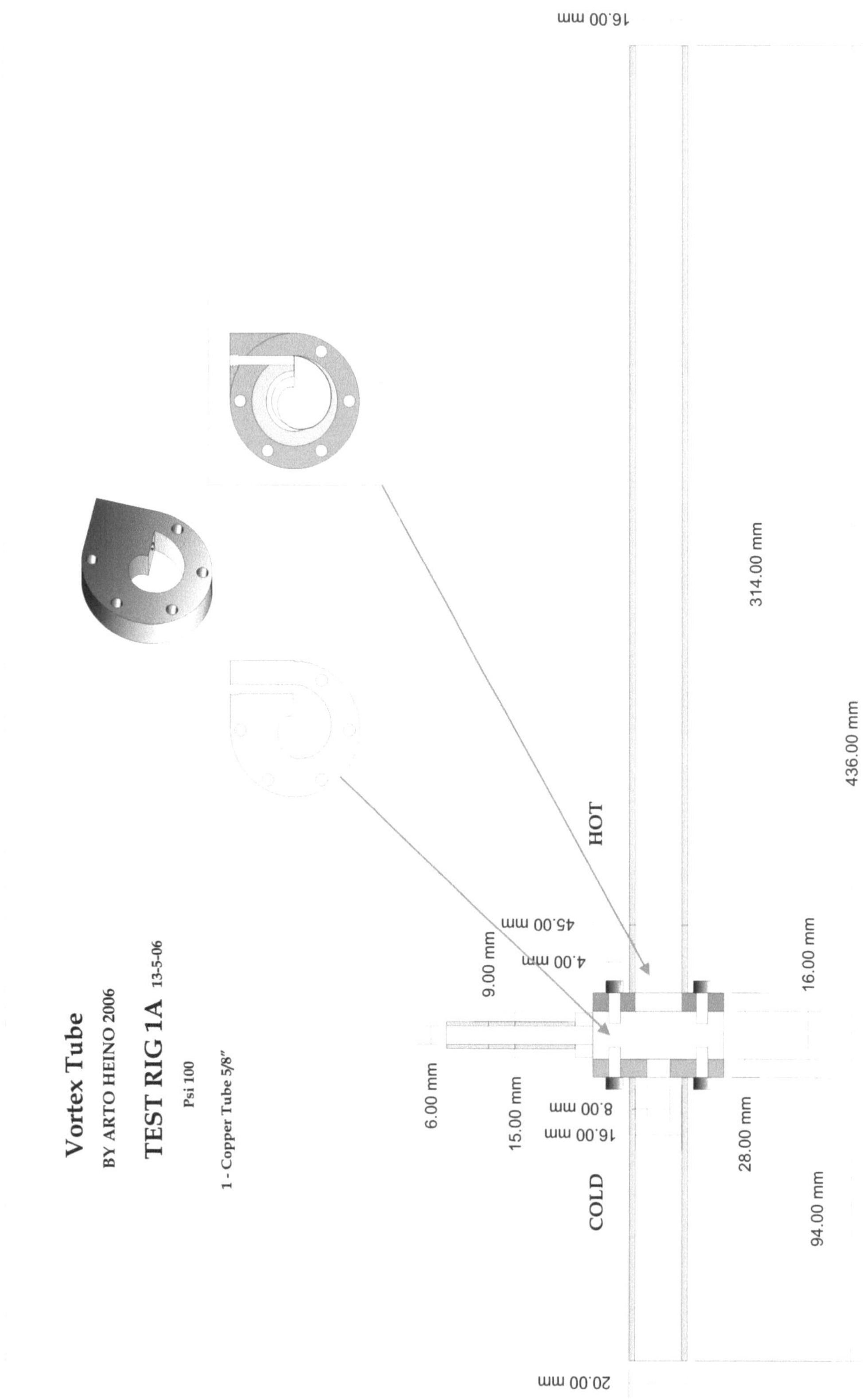

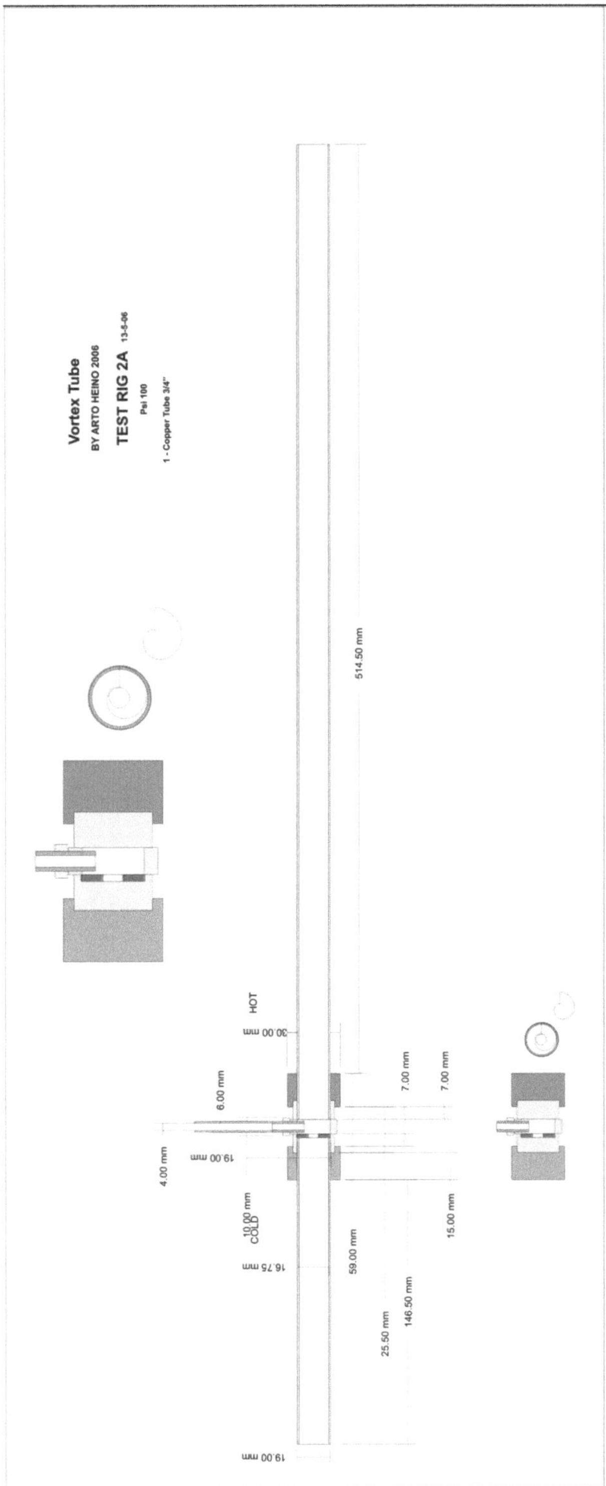
Vortex Tube
BY ARTO HEINO 2006
TEST RIG 2A 13-5-06
Psi 100
1 - Copper Tube 3/4"
514.50 mm
HOT
COLD
30.00 mm
6.00 mm
4.00 mm
19.00 mm
10.00 mm
16.75 mm
7.00 mm
7.00 mm
15.00 mm
59.00 mm
25.50 mm
146.50 mm
19.00 mm

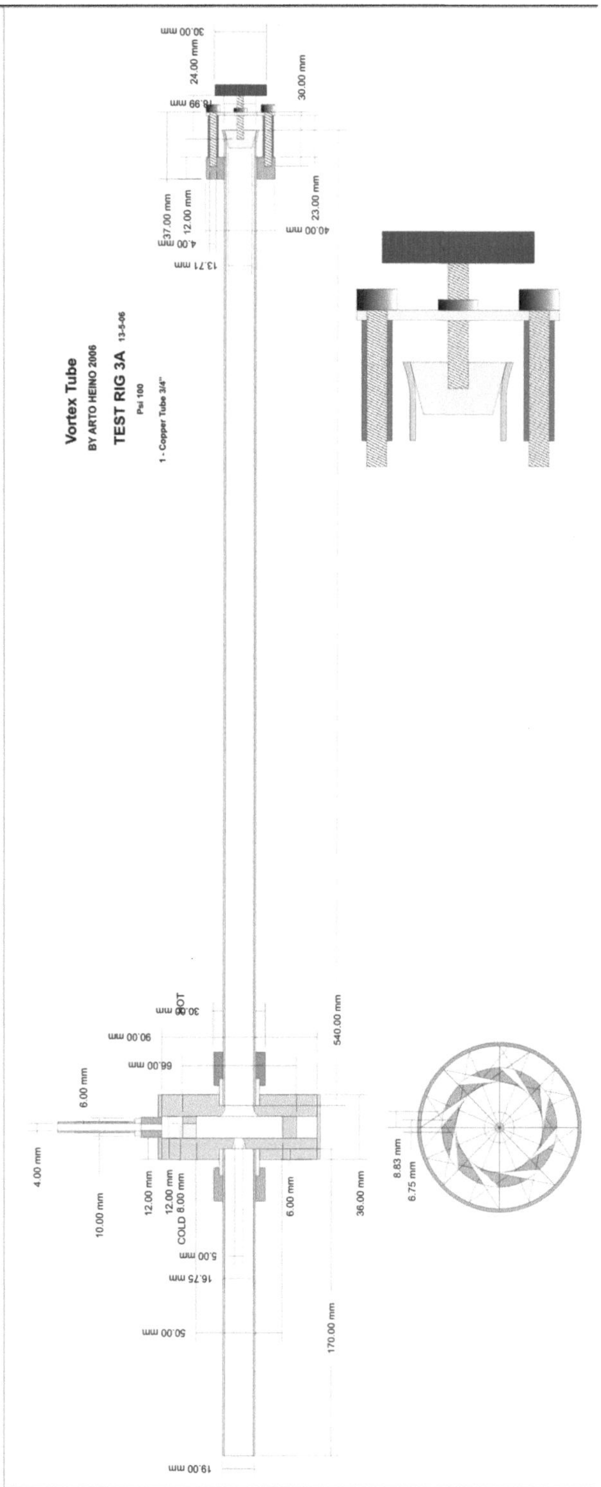
Vortex Tube
BY ARTO HEINO 2006
TEST RIG 3A 13-5-06
Psi 100
1 - Copper Tube 3/4"
30.00 mm
24.00 mm
30.00 mm
18.99 mm
37.00 mm
12.00 mm
4.00 mm
23.00 mm
40.00 mm
13.71 mm
HOT
COLD
540.00 mm
90.00 mm
68.00 mm
6.00 mm
4.00 mm
10.00 mm
12.00 mm
12.00 mm
6.00 mm
6.00 mm
36.00 mm
8.83 mm
6.75 mm
5.00 mm
16.75 mm
50.00 mm
170.00 mm
19.00 mm

Chapter Three

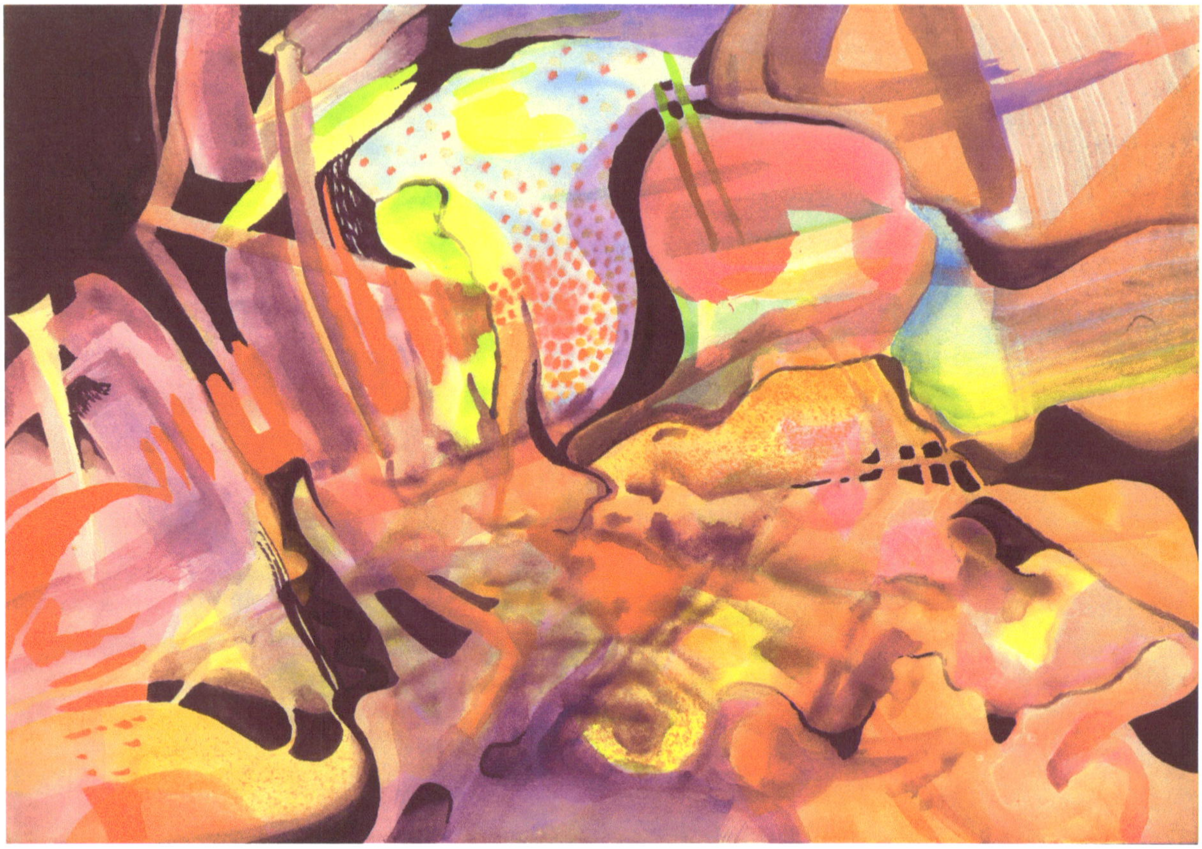

Electronics

3.1 - Electricity

3.1a – What is Electricity?

The perpetuated error coming from the corporate electrical industry for the recent generations of electrical engineers is a carefully crafted mindset that has its basic assumptions that man can generate electricity, and create the charge from batteries and turbines etc. This is a fallacy that has been perpetrated by the corporate electrical industry so mankind we will be forever tied to the systems they control.

Electricity is never created but its properties revealed and gated by stress applied with machinery such as a rotating generator or a chemical cell, electricity is always present as a part of the fabric of the universe, just like a hydraulic system where the fluid is ever present and its properties are used to transfer motion to a different points in space and time. Electricity appears when ever the conditions are correct, such as a motion or separation between two different polarities where stressing the etheric medium has occurred. The Ether was viewed as a true identity in science during the latter part of the 19th century and was considered *REAL* but was rejected because of the small minded few who had to define its ownership in their narrowly defined terms of a profit motive based on the utilitarian usage so it could be accessed by a system of control, such as a corporation.

Going back to the understanding of electricity that the electrical engineers had in the late 19th and early 20th century, we find the pursuits of that generation was to reveal all the possible avenues of discovery and define electricity in the mechanistic equivalents that would make engineering electrical devices as simple as designing a purely mechanical or hydraulic system such as a gear train or a hydraulic lift. As we know from history this sort of scientific and engineering endeavor was co-opted by the theoretical physicists. The minders of the false paradigm have been misdirecting the completion of the task started by Faraday, Heaviside and Tesla by creating overly mathematical and highly complex models that cannot be understood or even comprehended by any person thus making it into a metaphysical belief system not an art of scientific research and discovery. This methodology of indoctrination has made successive generations of followers into a bureaucratic system that has been funded by the corporate and fascist elite to perpetuate the model so a hierarchy of control can infiltrate all aspects of the populace, it has been built so each level of this tower of babel does not have knowledge of the whole system.

As of the year 2013 we still use the equations developed by Micheal Faraday 160 years ago to make a simple generator. The foundation laid by these early scientific and electrical discoverers have been waiting for the sowers of the next generation, the last great flurry of activity was when Tesla revealed his Wardenclyffe tower, then the panic by the corporate heads acted to shut down all the possible ways to rediscover the natural electrical understanding that Tesla had revealed. By 1917 they had banned all radio and terminated any new thoughts by concerted effort to make Tesla a non-person and took down Marconi and others that used the same ideas that could reveal its true nature as a free and natural resource. Today this movement has been co-opted by fringe groups that exclaim it as Free Energy, which is a twist on words as the cost is never born by those exclaim its freedom, as energy has never been imprisoned or lost only converted and directed, this cost is in the machinery that act in its usage.

As is now known it was Steinmetz that was recruited by General Electric to reverse engineer the two phase system of Nikola Tesla and re-engineer the system to reduce the 4 wire copper wire version to a 3 wire reduced copper 3-phase, just to save money and generate higher profits thus forever sealing our fate to an unnatural electrical distribution system. If you only have two polarities thus making the system an even number of conductors, then what energy battle must exist in the surrounding space around the cabling when you have an odd number of conductors.

The simple four wire system of Tesla is easy to follow and simple to engineer, here is the a diagram that shows his original intention for power distribution:

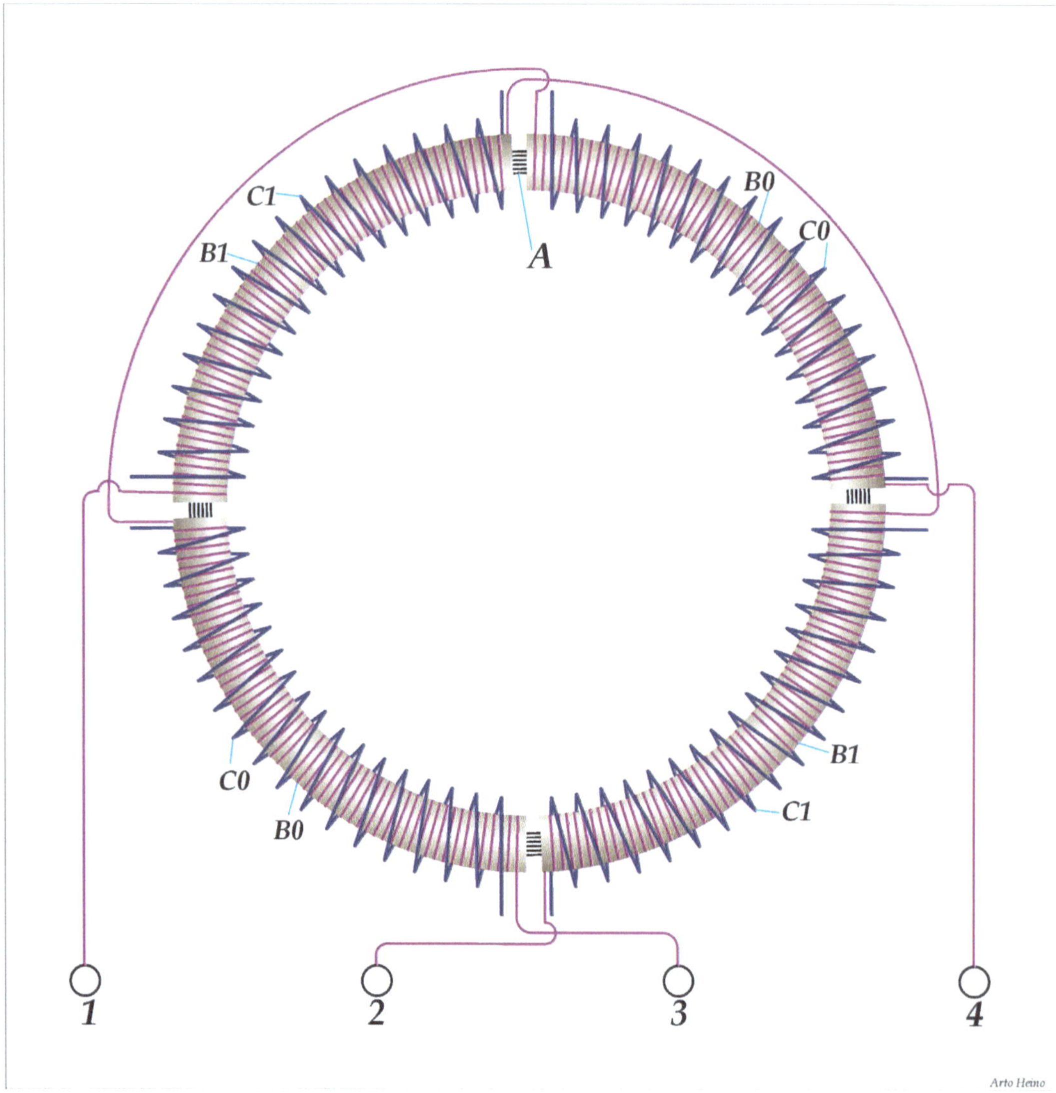

To simplify the diagram I will straighten out the circular torus so the wire system can be easily seen;

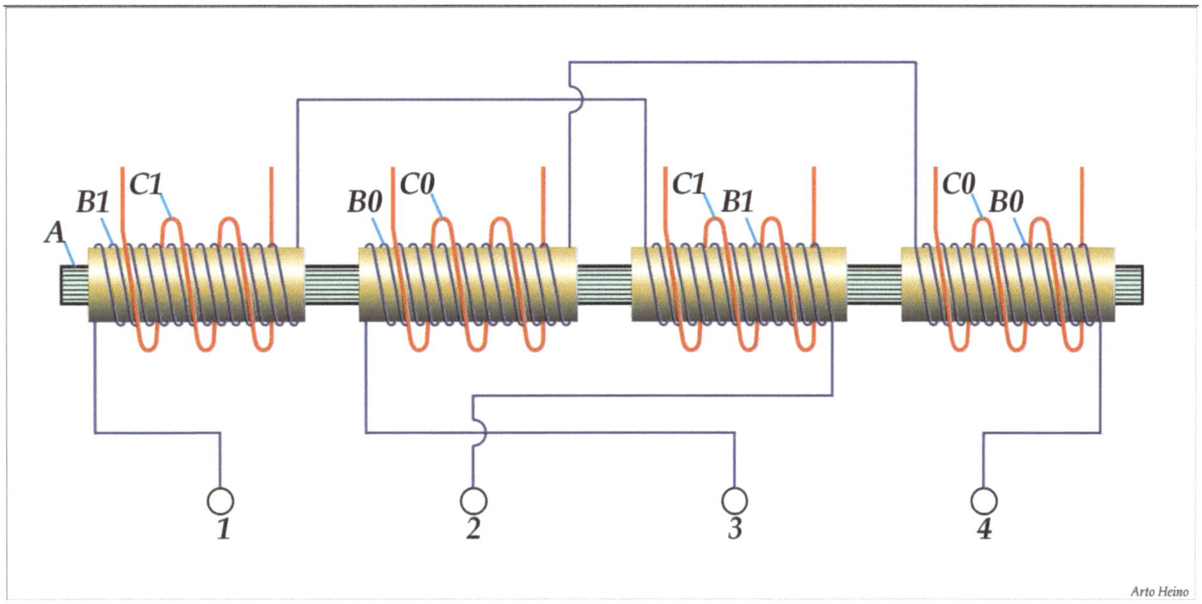

The primaries can be connected as per his patent(US381970), The complex nature of the three phase system can be seen when you try to figure the current bifurcation.

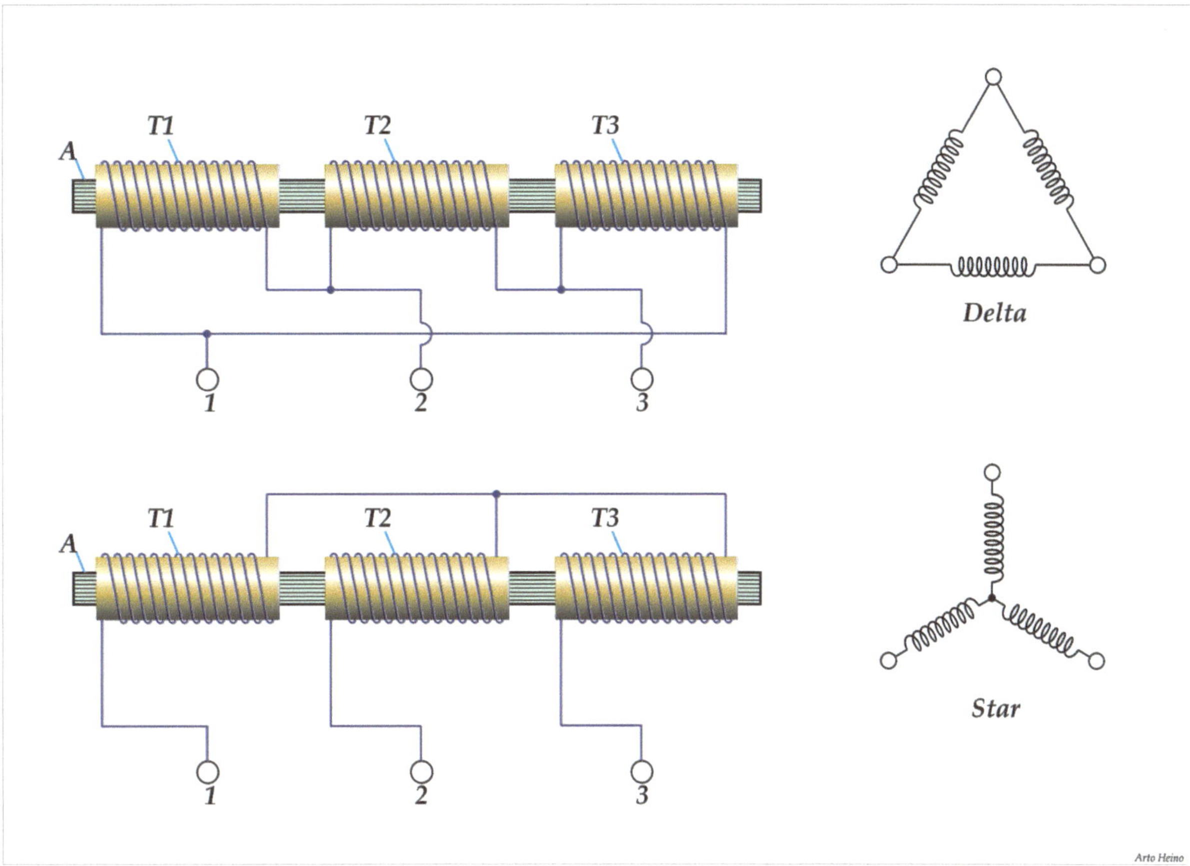

This is the sort of extra burden of complexity that made AC a difficult task to engineer. To make matters narrow in it conception the elimination of Tesla type oscillating and pulsed systems that incorporate the logarithmic wave forms has been codified as a degenerate sine wave and relegated to part of the lossy radiative system confined to the velocity of light and the transverse wave functions. Luckily Steinmetz included a simpler version of transient waves within his writing allowing the future generations to decipher his work and clarify the work of Tesla. As an example of revealed knowledge missing or hidden, many inventors of the last 100 years have found anomalies that cannot be understood by utilizing the standard equations of the corporate codifications, only by great suffering that any one person can dare to go against the system and try to set the records straight again, I refer to people like Moray, Royal Raymond Rife, Schauberger, Reich and the current man that has nefarious elements trying to derail his efforts the highly esteemed Eric Dollard.

The speakers and the mouthpieces of a co-opted public debate system along with the controllers of all the types of public media play a role in keeping the system of ignorance running it course, this has been happening for the last hundred years, included in this public brainwashing are the publishers, educators and their hoards of student minions. This activity is promoted by keeping the popular literature clear of any revealing truths that might emerge from the populace, if they did not engage in this constant stage show we might have a well educated public that is capable of critical thinking and be informed of some real history and science.

This is the position we are in at present, as a creative individual I cannot proceed forward with tools that are inadequate to the task of trying to make clear revelations about phenomenon that I observe in the experiments I perform. The touted and glorified icon of the dogmatic 20th century science the Electron seems to fail in all aspects of understanding when you try to explain its existence and the obtuse complexity when explained by modern physics. The scientists must create even smaller particles to explain the larger one, then after they explain their existence they go another level smaller again, surely the explanation of these smaller particles are really a property of the geometry of space itself as expressed so eloquently by the works of Walter Russell . The modern physicist has poor understanding and skills in engineering, he only has skills in clever metaphysical mathematics and theoretical ideas that very rarely lead to any true simplification and understanding of new complex events, they create papers that only others indoctrinated believers can read.

3.1b - Is Ether, Electricity?

To introduce a new way of understanding the electrical condition of conductors is to see electricity as stress of ether that exists around metal reflectors. To begin to visualize you must assume that each atomic element has layers of ether at its outer boundary, the highest stress is at the actual boundary and at some distance the ether is at equilibrium with its Etheric environment. The better the conductor the greater the reflection of ether stress and the equilibrium boundary is at a greater distance, where insulators have elements that the outer boundaries are closer to the geometric elemental boundary thus a smaller radius of action, this will allow the ether to move within the gross geometric boundary of the material.

To illustrate the general concept here are a few diagrams;

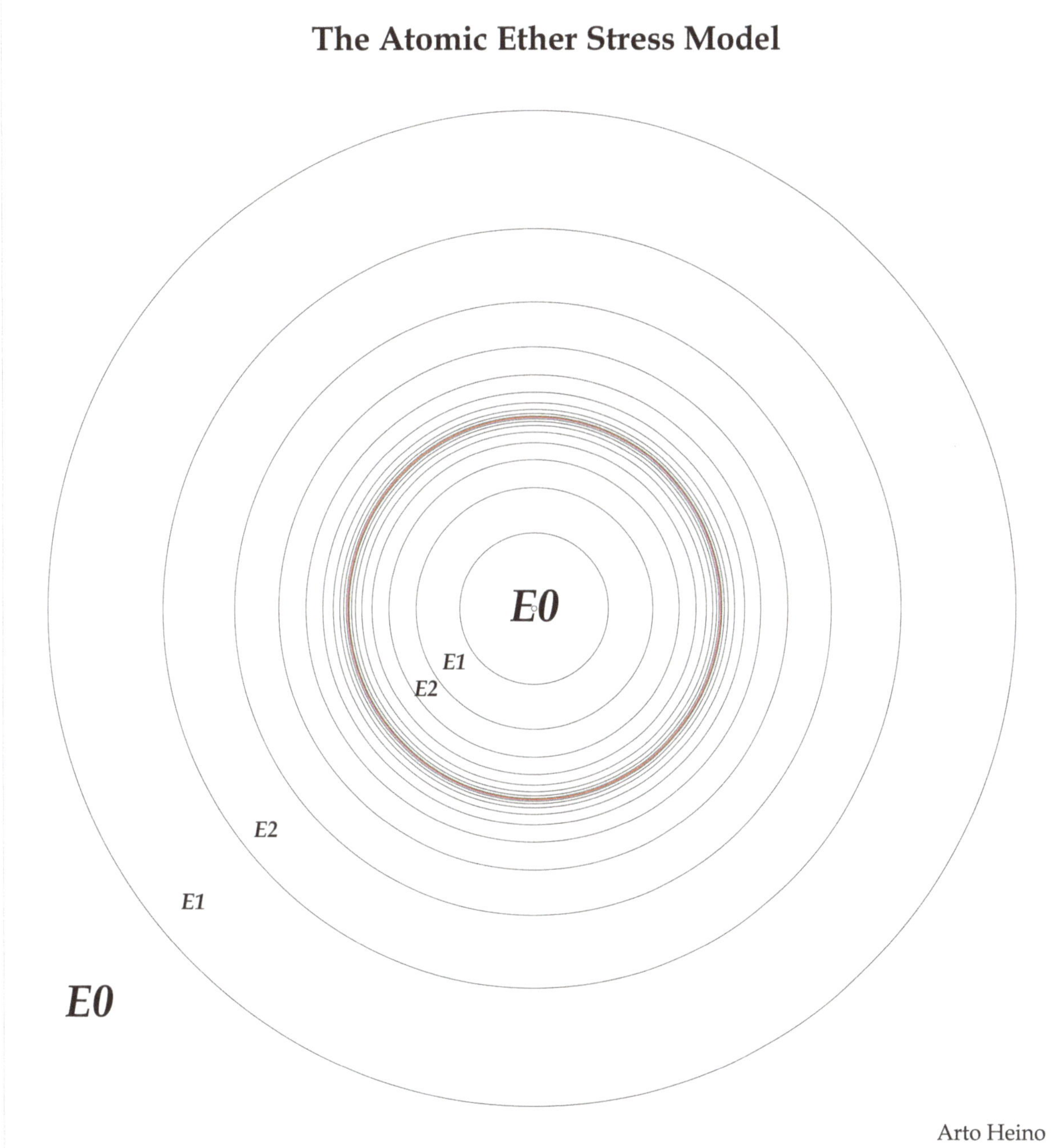

Semi-conductors allow partial Ether stress to be contained in and around the geometric gross boundary. The area on the conductor that is used to describe the resistivity of a material is the displacement area outside of the conductor which is equal to the area inside the conductor, this is why the formula for resistance works accurately. An example of this is to look at three materials that have different resistance values and then correlate this to the expanded or reduced field geometry.

The basic gross material volumes will be the same in the first case, such as

Wire Radius	0.000500000000 mtrs
Wire length	1.000000000000 mtrs
Wire area	0.000000785398 sqr mtrs
Volume	0.000000785398 cubic mtrs

Material	Resistivity	Ohms
Copper	0.000000017200	0.0219
Germanium	0.460000000000	585690
Rubber	10215000000000	1.300E x 19

The second case the density of the material will adjust the volumes so all the masses are equal.

Material	Density	Mass per Vol1	Volume per Mass Copper
Copper	8950	0.0070293	0.000000785398 cubic mtrs
Germanium	5320	0.0041783	0.000001321299 cubic mtrs
Rubber	1200	0.0009424	0.000005857761 cubic mtrs

Without further analysis, I can see that these are roughly the correct masses when designing electronic circuits, where the insulators are the greatest volume.

Referring to the atomic Ether stress model the value E0 is the general neutral Ether pressure of the universe, as you see it also resides in the center where the theoretical particles called neutrons would exist, as this model only concerns itself with the Ether outside of the interface, we will not labor the simplicity presented with overly complex notions, even though they will have to be addressed at some future date.

The actual sequences that are present in this diagram are modeled using the Fibonacci series, thus allowing the natural order of growth to be revealed, as stress is reduced then the volume is increased. As this is not a formal deposition I am using arbitrary diameters to illustrate some possibilities.

Outer Boundary Diameter		Differences Between Diameters
E0 - E1	375	89
E1 - E2	286	55
E2 - E3	231	34
E3 - E4	197	21
E4 - E5	176	13
E5 - E6	163	8
E6 - E7	155	5
E7 - E8	150	3
E8 - E9	147	2
E9 -E10	145	1
E10-E11	144	0
Inner Boundary Diameter		
E9 - *E8*	143	1
E8 - E7	141	2
E7 – E6	138	3
E6 – E5	133	5
E5 – E4	125	8
E4 – E3	112	13
E3 – E2	91	21
E2 – E1	57	34
E1 – E0	2	55

If we insert a dielectric between two copper plates, the instruments will indicate an increase in capacitance over that of the ether. This increase in capacitance is in the exact proportion to the square (second power) of the decrease in the velocity of light through the same identical piece of dielectric. It is then given, the change in the velocity of light through a dielectric medium is the square root of the inverse of the change in capacitance effected by this dielectric medium. Hence capacitance exists in a direct relationship with the velocity of light in a given medium. Zero capacitance, infinite velocity.

The following diagrams will give you an idea of the Ether effect of differing atomic arrangements.

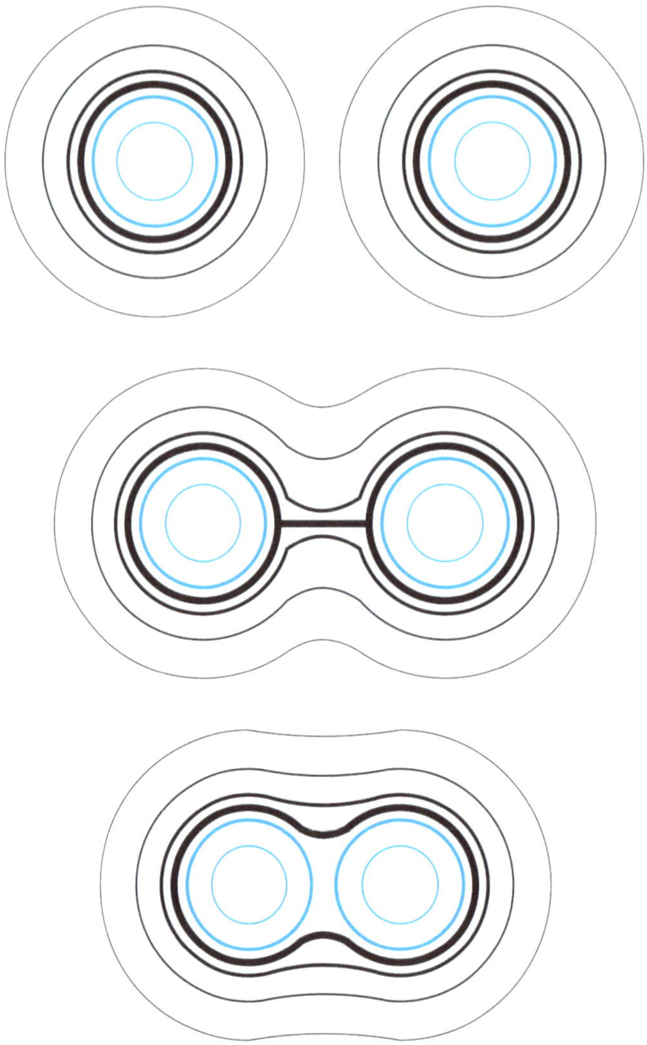

Arto Heino

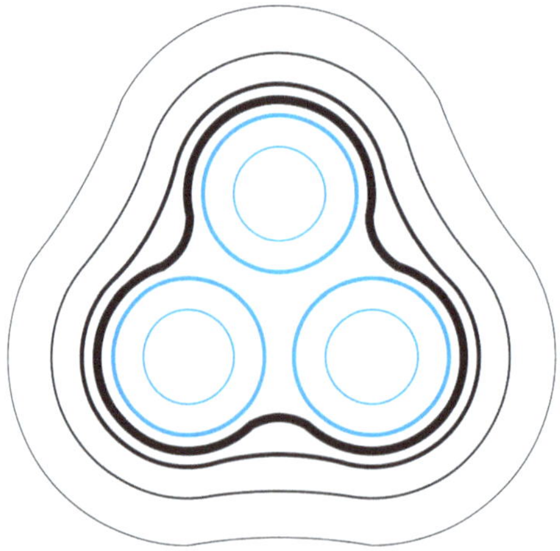

Arto Heino

This sort of conjecture will bring you further refinements of the idea. The semi conductor has presented a very interesting array of models to define how diodes, transistors and such operate. The engineering requirements are quite rigid and precise, allowing the creation of micro circuits using the abstractions that have been abound in technical literature in the last 50 years.

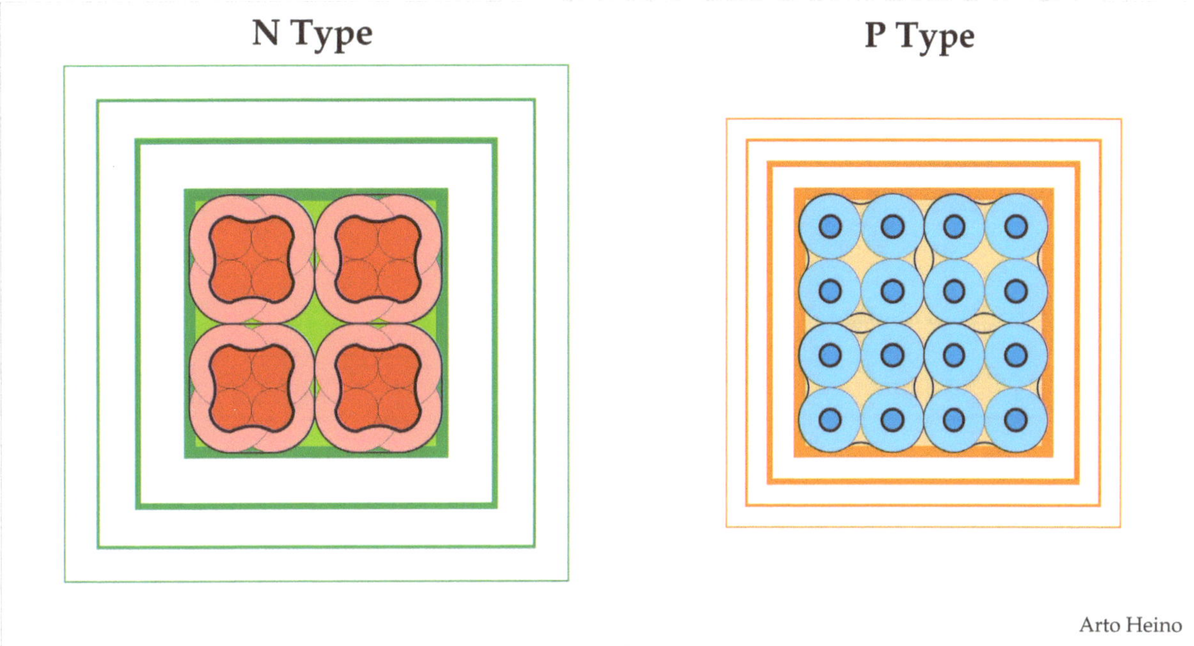

Here are some models to further illustrate this models possibilities.

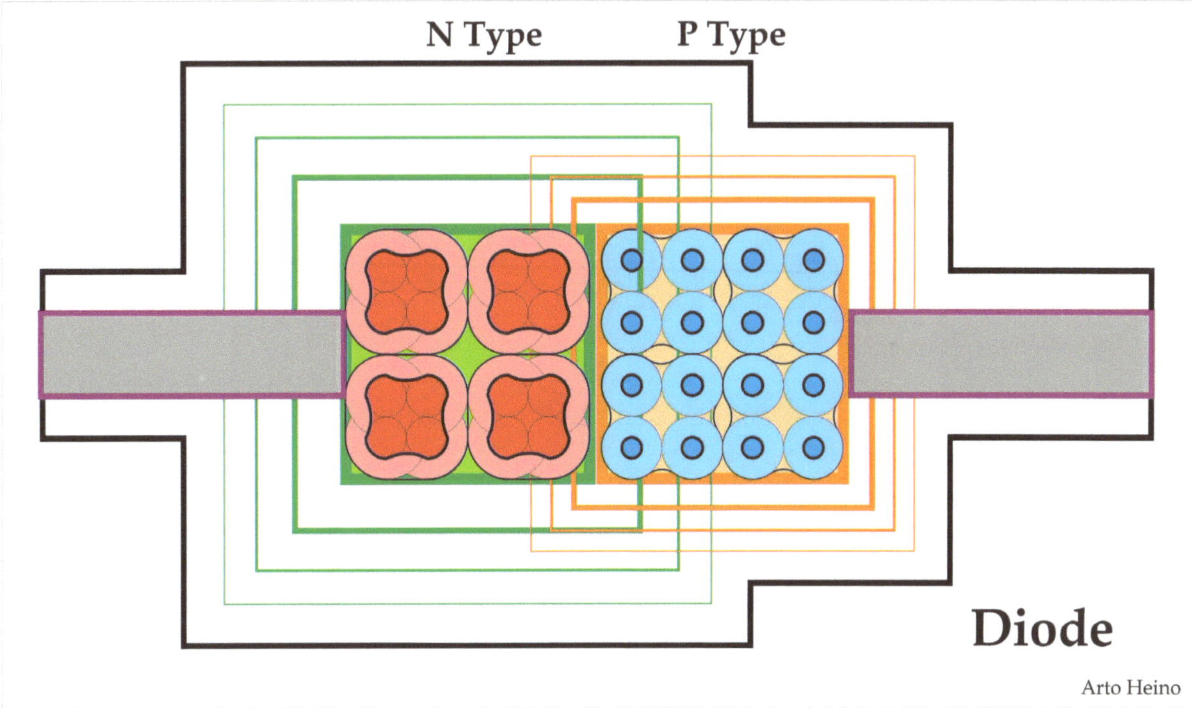

As you can see the boundary layer follows the commonly used graphical descriptions of how diodes and transistors function.

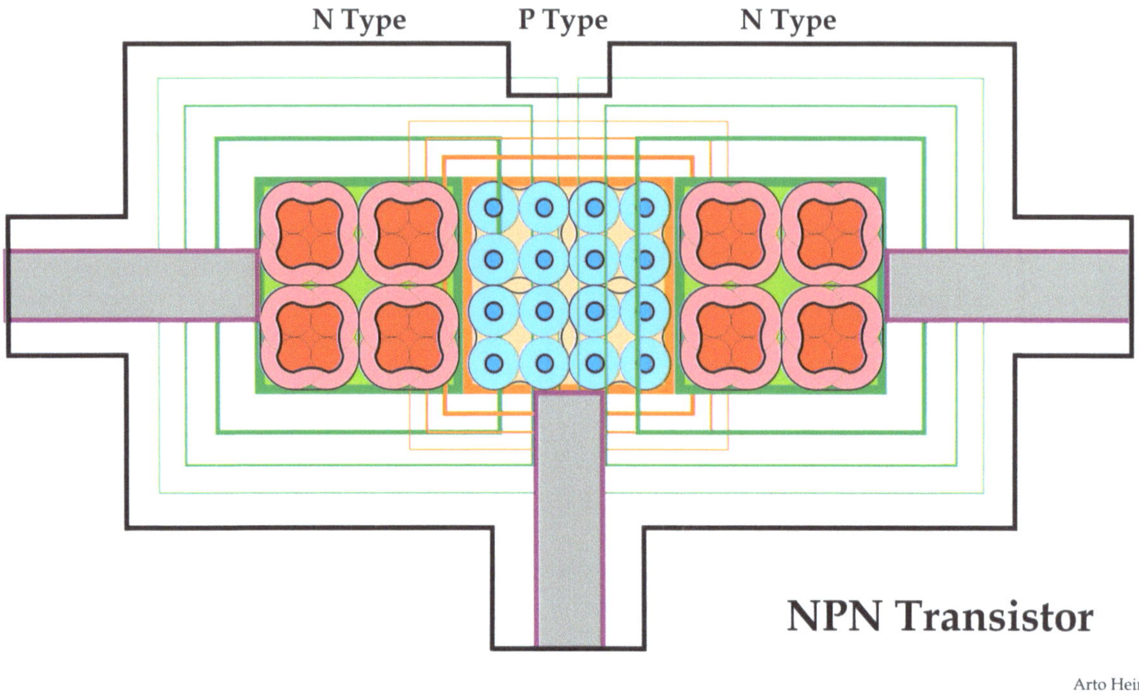

A final diagram just to guide those who are mathematically inclined into possible re-writing of our electrical books.

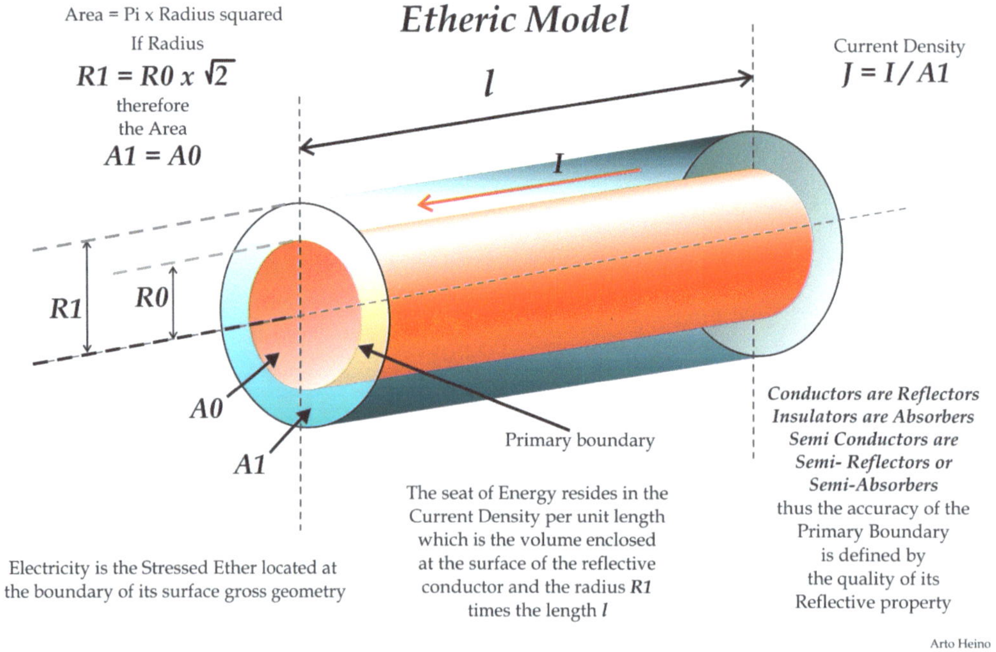

3.2 - Schematics

3.2a - History of Symbols

The symbolic representation of mechanical or abstract ideas goes back to the dawn of writing, it was mans way of passing down knowledge in a simplified and cohesive way. The sandal maker kept samples of his craft at hand to teach his apprentice or family member, this was the common method of instruction. The knowledge would soon be forgotten if the sandal maker and his wares were destroyed by some natural disaster. The retention of a knowledge base was best served by visual and verbal instruction and recorded on a durable tablet/wall or leather/paper scroll. This method was used for thousands of years by all ancient cultures and much was kept to improve the lives of those who retained a library of sorts.

The Library at Alexandria was one of those repositories of knowledge that housed millions of scrolls that covered lost and forgotten technologies from the past. As I will explain in another chapter it was the loss of mankind's greater self that was becoming diminutive due to the cyclic nature of our golden ages. The scrolls had true secrets that only now we could appreciate, when in the past it was treated as sorcery and against the established doctrine of the time and subsequently destroyed. What scrapes we have left, leaves us agape with the sheer wisdom and clarity the ancients had shown.

This leads me to the symbols that have been used in the past to explain some types of technology, such as the Egyptian hieroglyphs ;

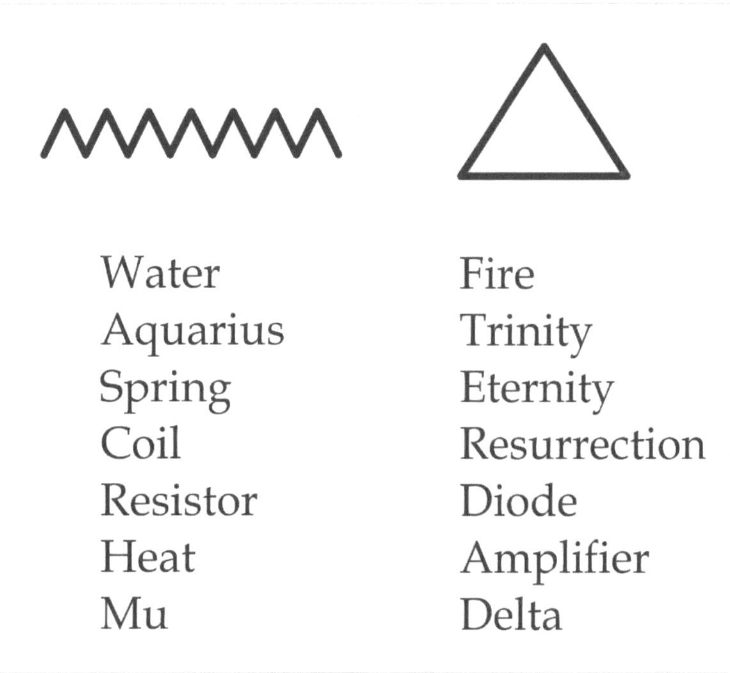

The use of symbols in the retention of methods and materials was the job of the scribe and he would choose and design new symbols that would convey these ideas by a form of simplification and a strict form guidance taken from the library of symbols in use for similar methods and designs.

3.2d - New Schematic Ideas

When designing circuits and looking for better ways to layout your ideas, you start to develop a feel for the right way to start and the right placement of parts both in the schematics and the physical layouts. This sort of knowledge is hard won, meaning at least 20-30 years of working through circuits of different types to arrive at the possible ways to achiever better forms. The modern system incorporates computer algorithms and excludes the designer completely, the main reason is the extreme complexity of the circuits. While the older electrical engineers required good memory and maths skill sets, while having a sense of clean linear forms and some artistry as well.

The need to refine a system based on a natural form has always been in the back of my mind, as we are dealing with electricity and some of the varied ways it can be manipulated by materials and geometry, a need arises to let the function match the form., This is how I arrived at a idea for a new basic layout system, place the complete idea on a square matrix with connection limited to a geometric boundary based on a few simple premises.

Here are some of my worked samples to see if this could be accomplished. Make a note that only 3 to 4 point components reside in the central region, all others are 2 point components.

Square Circuits I

By Arto Heino

Schematics of Electronic Circuits utilizing ordered design

The order of Square 3 Principle layout and connections

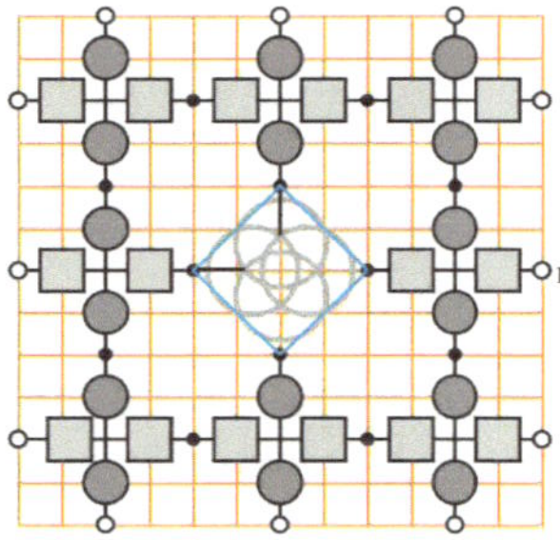

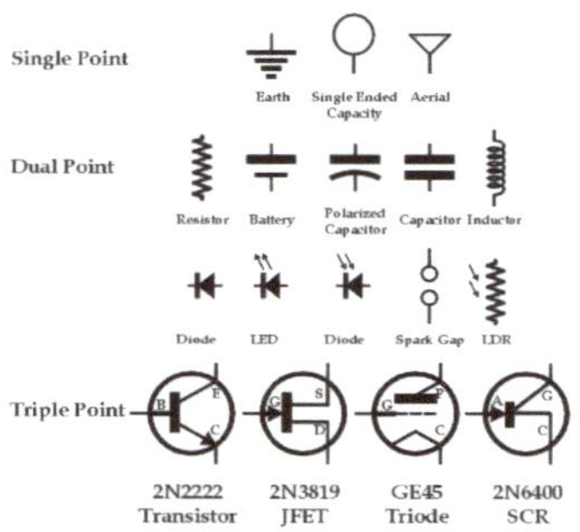

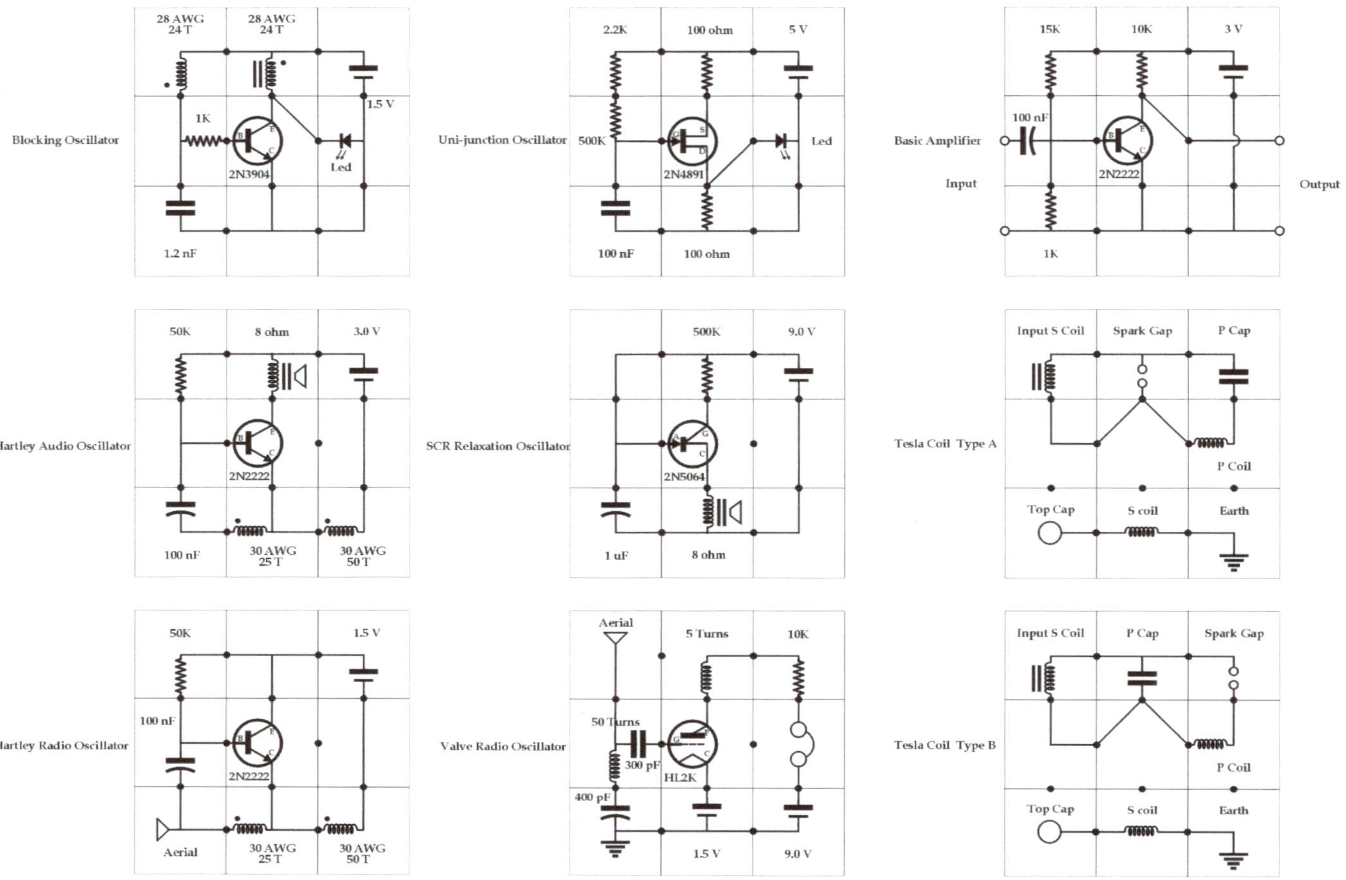

Arto Heino

This idea intrigued me as a great way to simplify complex ideas, I thought I might include some of the work of Nikola Tesla and his early researches into this type of layout. As Tesla did not have any 3-point elements to deal with, I thought the central area in the square would be only used for cross connections.

Square Form Circuit Arrangements based on the designs of Nikola Tesla Pages 89 to 92 in his Colorado Springs notebook

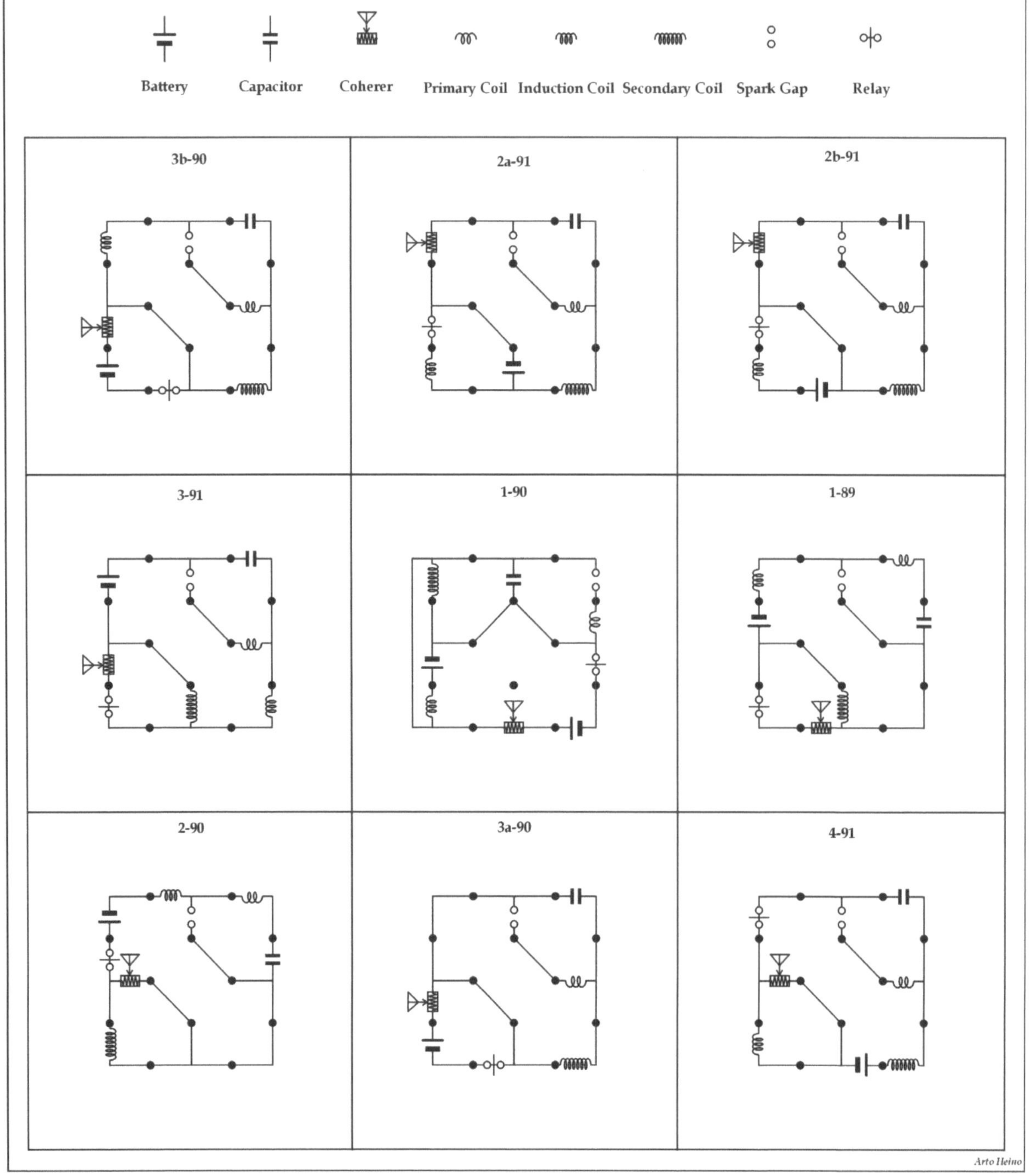

3.3 - Circuits

3.3a - Fibonacci Circuits

There are many different ways to connect the basic elements of a circuit, let us begin at the true beginning, the resistor. The first two you are all very familiar with, the third is some thing nature gave us but is ignored by engineers.

Series Connection	1+1 +1	= 3 ohms	3
Parallel Connection	1/(1/1 + 1/1 + 1/1)	= 1/3 ohms	0.3333
Fibonacci Connection	1/(1/1 + 1/1+1)	= 1/1.5 ohms	0.75

This seems a little simplistic, just series /parallel, or is it something unique, let us secure the idea with more visual data.

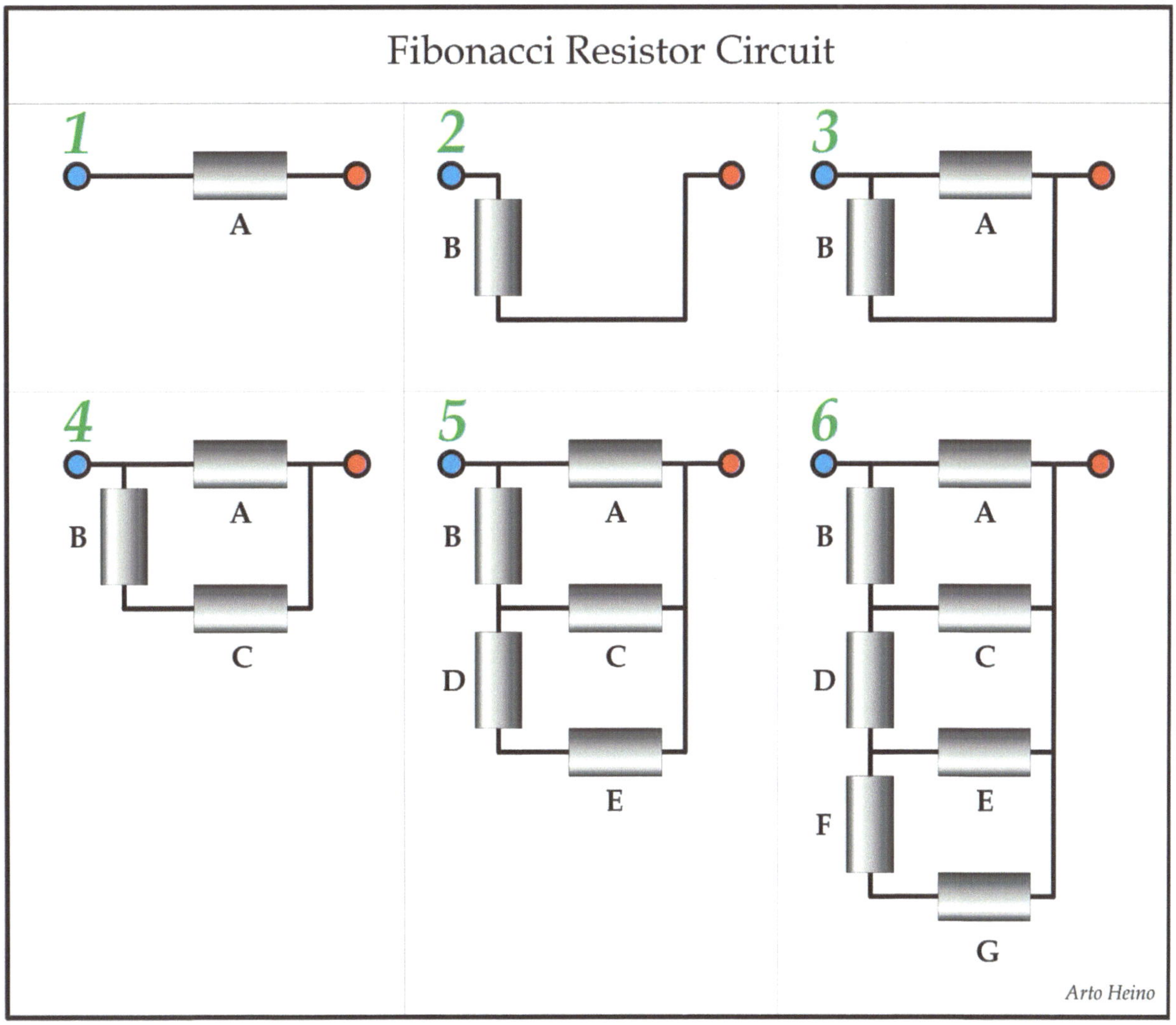

Here are the equations for the PHI related Fibonacci circuit:

1) R = 1/(1/A)

2) R = 1/(1/B)

3) R = 1/(1/A + 1/B)

4) R = 1/(1/A + 1/(B+C))

5) R = 1/(1/A + 1/(B + 1/(C + 1/(D+E))))

6) R = 1/(1/A + 1/(B + 1/(C + 1/(D + 1/(E + 1/(F+G))))))

All resistors must be the same value, if

	A= 1	A=2	A=3	A=4
1)	R =1	R=2	R=3	R=4
2)	R =1	R=2	R=3	R=4
3)	R =0.5	R=1	R=1.5	R=2
4)	R =0.666666	R=1.333333	R=2	R=2.666666
5)	R =0.625	R=1.1	R=1.575	R=2.058823
6)	R =0.619047	R=1.094017	R=1.572082	R=2.057330

Each consecutive values are simply the multiple of the initial resistance eg (3) A=1 and R =0.5 then in (4) A=4 then 0.5 x 4 = 2. So we only need to consider the initial A=1 value and multiply any value by the new amount, saying that let us extend beyond line (6) and you will see they all correspond to the Fibonacci sequences exactly.

0,1,1,2,3,5,8,13,21,34,55,89,144,233,377,610,987

			1/R =	
1)	R =1	1/1		1
2)	R =1	1/1		1
3)	R =0.5	1/2		2
4)	R =0.666666	2/3		1.5
5)	R =0.625	5/8		1.6
6)	R =0.619047	13/21		1.615384
7)	R = 0.618181	34/55		1.617647
8)	R = 0.618055	89/144		1.617977
9)	R = 0.618037	233/377		1.618025
10)	R = 0.618034	610/987		1.618032

If you are using capacitors in this arrangement just use the 1/R values the get your base values.

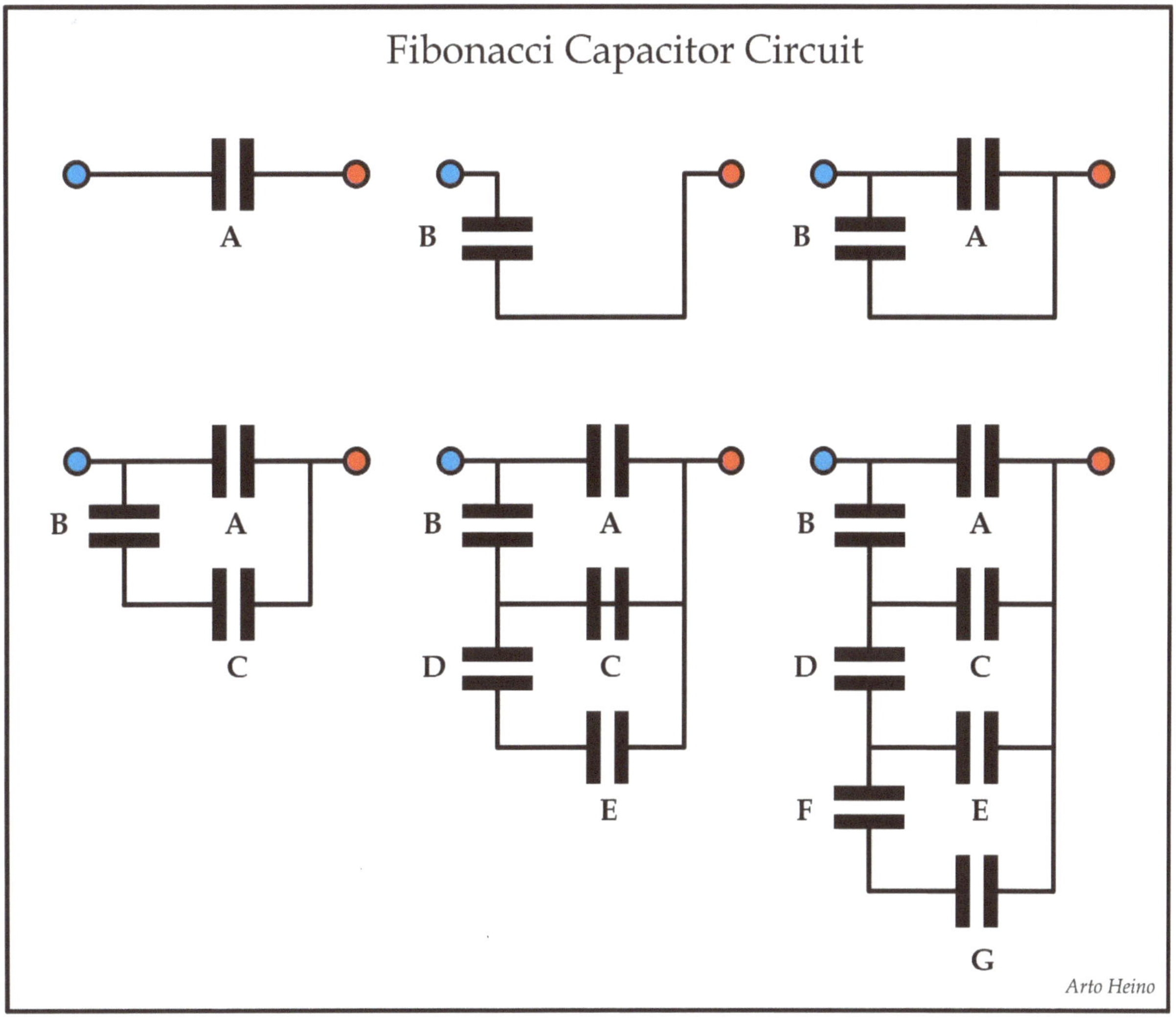

You can also extend these ideas to resonant circuits, either as filters of as resonators. The use of resonant circuits and the Z impedance product will give you another avenue for exploration. The filter created by the parallel resonant design will also give you another aspect of how to use these ideas.

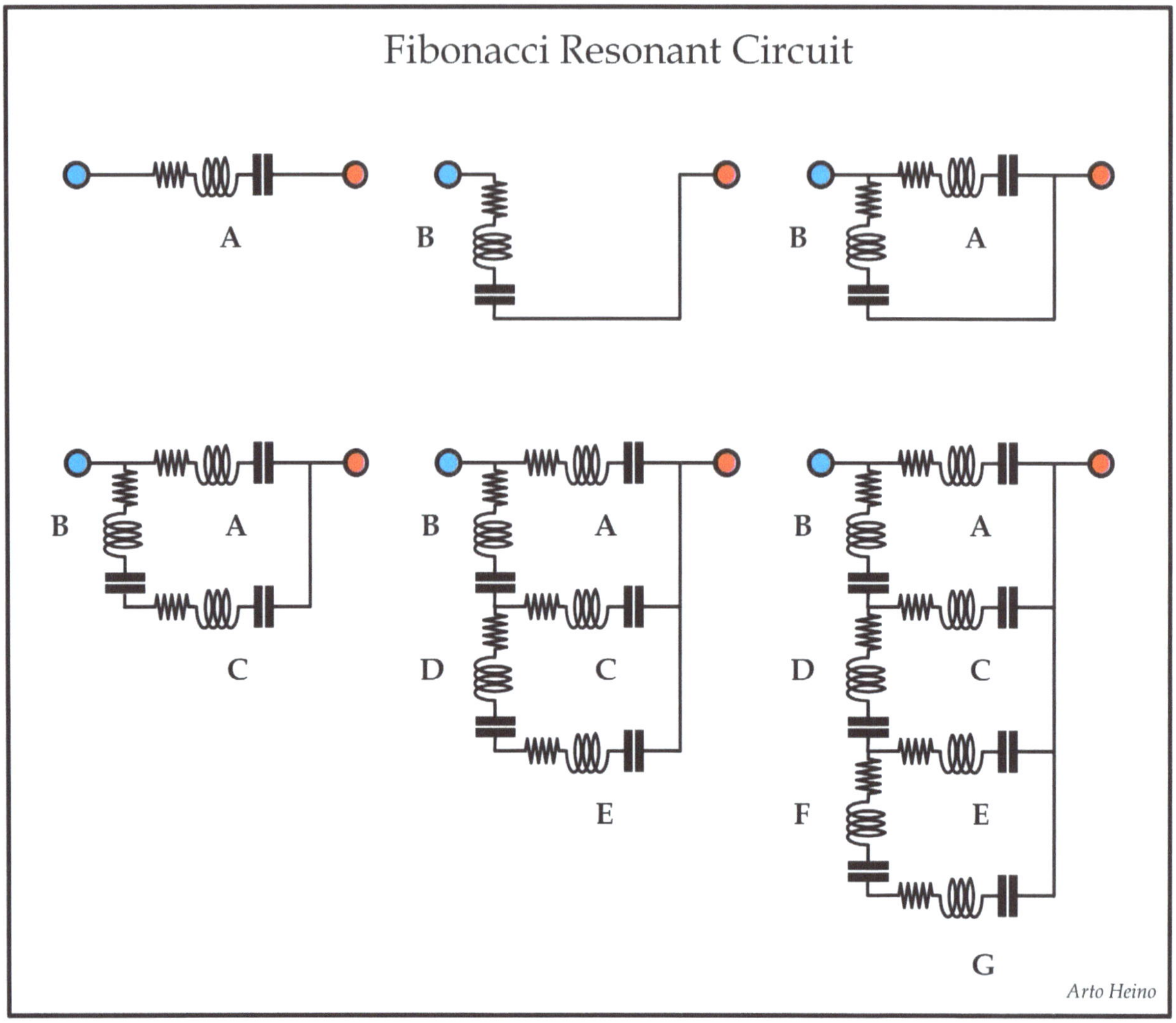

3.4 - Interesting Circuits

3.4a – Differential Galvanometer

One of the earliest instruments for measuring current activity in a circuit, a semi-passive simple device that can be placed in a circuit without too much interference.

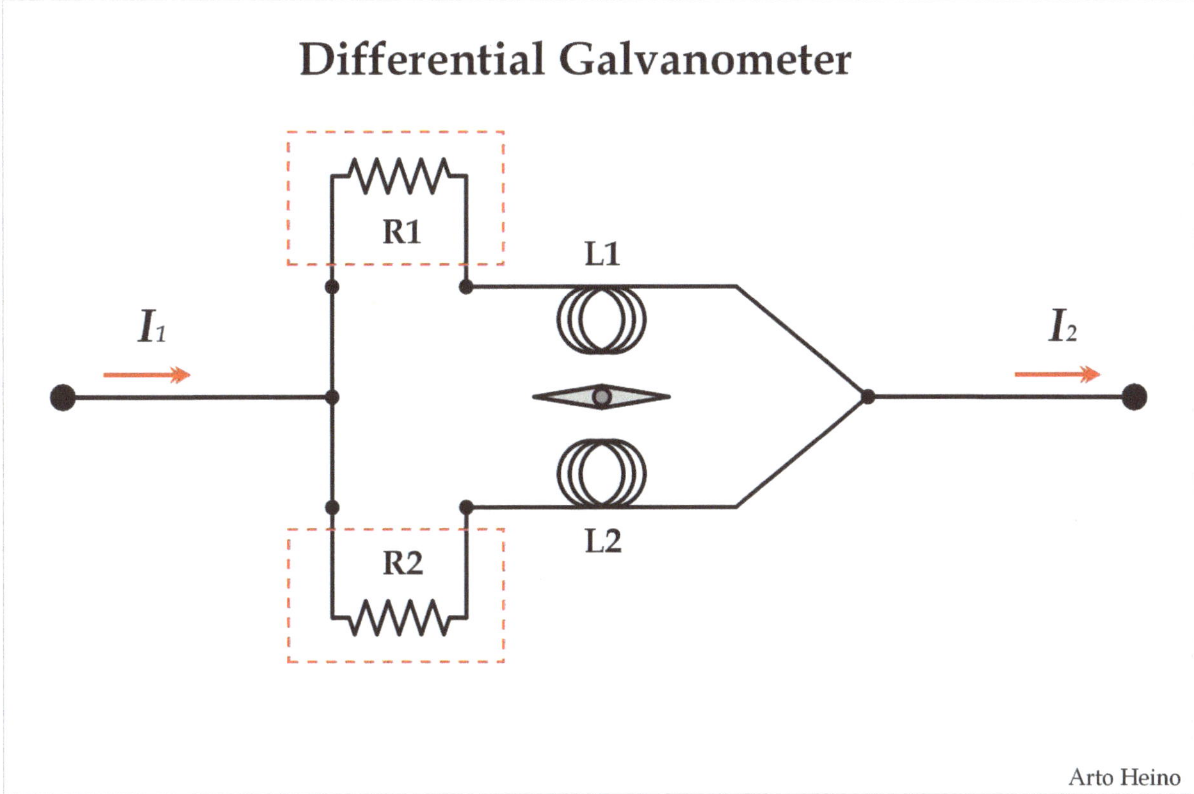

As you can see I have placed the variables in a simple geometric arrangement, showing the parallel set of series elements, splitting a current carrying wire into two resistance branches to reveal its magnetic nature.

R1	1	1	2
R2	1	2	1
L1	0.001		
L2	0.001		
Deflection	0	to L1	to L2

The reason I have included this particular circuit design, is its visual understanding based on symmetry and symbolic placement.

3.4b - Vibrator Power Supply

Early in 1920's the radio circuits were in need of some size reduction, these units used alternators to create the AC wave forms required for those early spark gap systems, when the Diode and Triode valves were in use, the possibility of a portable radio was to be born. The vibrating switch was the equivalent to our switch mode systems we have now.

The variety of input out combinations are based on need, Here are some basic design types.

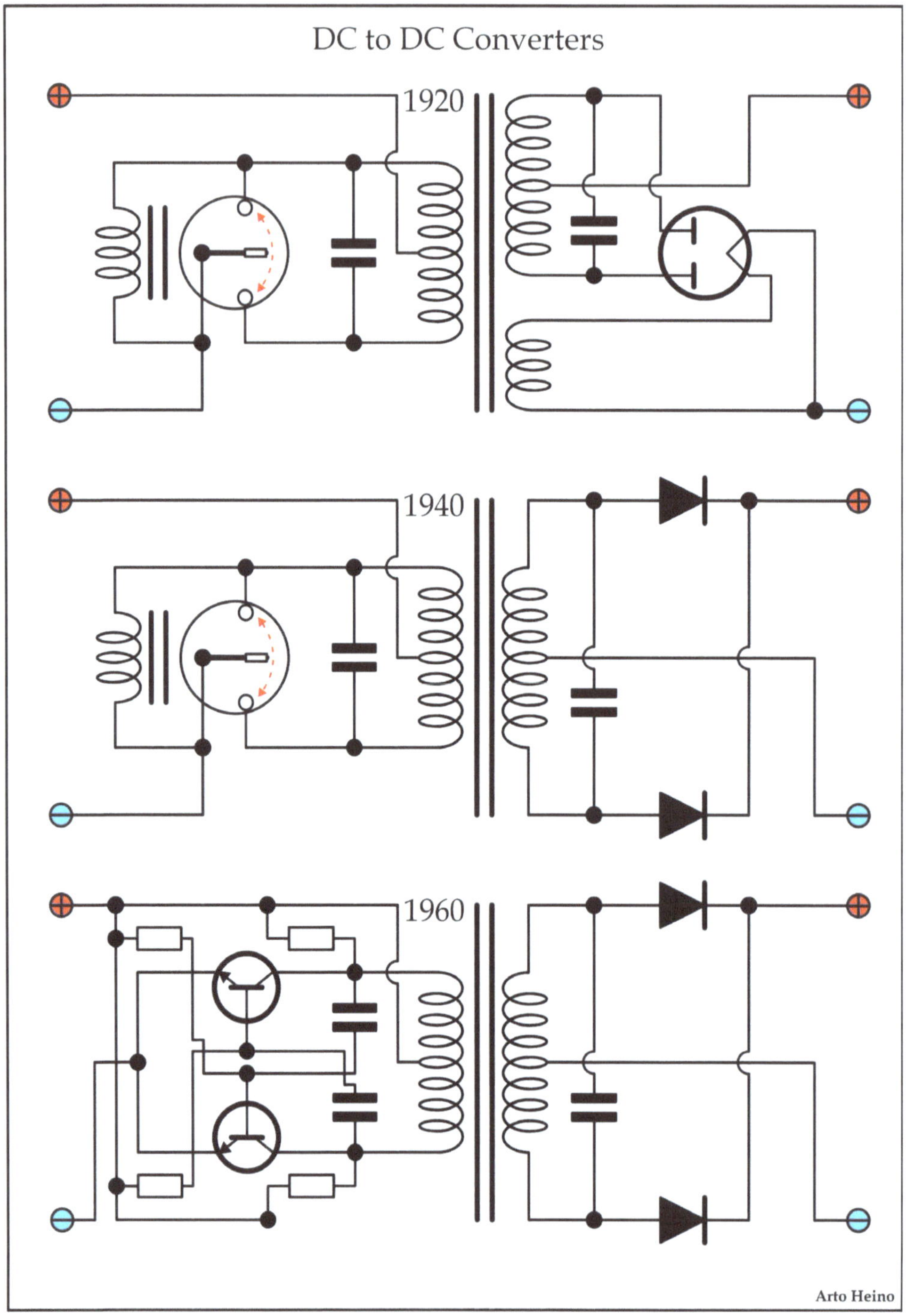

This breakdown was my introduction into this type of unit.

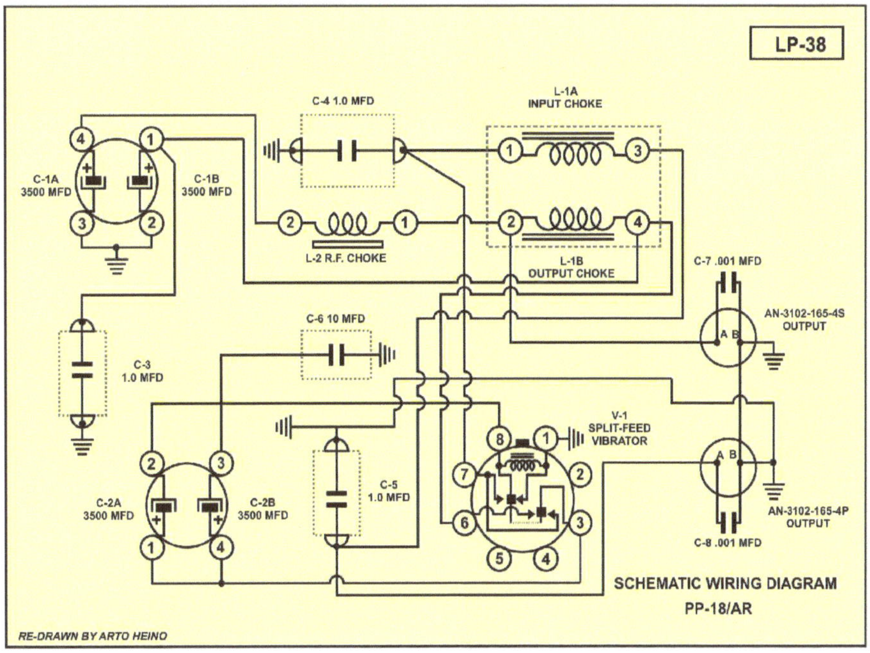

C1A = 3500 uF 500V	C3 = 1 uF 500V	L1A = Input Choke 7 H
C1B = 3500 uF 500V	C4 = 1 uF 500V	L1B = Output Choke 1mH
C2A = 3500 uF 500V	C5 = 1 uF 500V	L2 = RF Choke 1uH
C2B = 3500 uF 500V	C6 = 10 uF 500V	L3 = Vibrating Coil 1mH
P1 = + cycle	C7 = 0.001 uF 1200V	AC = Output AC 240
P2 = Bemf cycle	C8 = 0.001 uF 1200V	DC = Input DC 12

P3 = - cycle
60 Hz C1A +C1B
85 Hz C1A
60 Hz C2A+C2B
85 Hz C2B

SID100 19mmx38mm FE Dust core
1.00 mH 1mm wire

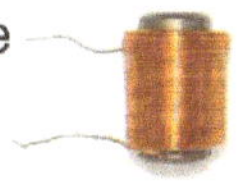

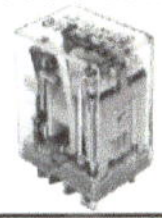

HF13F/012 160 ohm
1.00 mH coil 10 amp

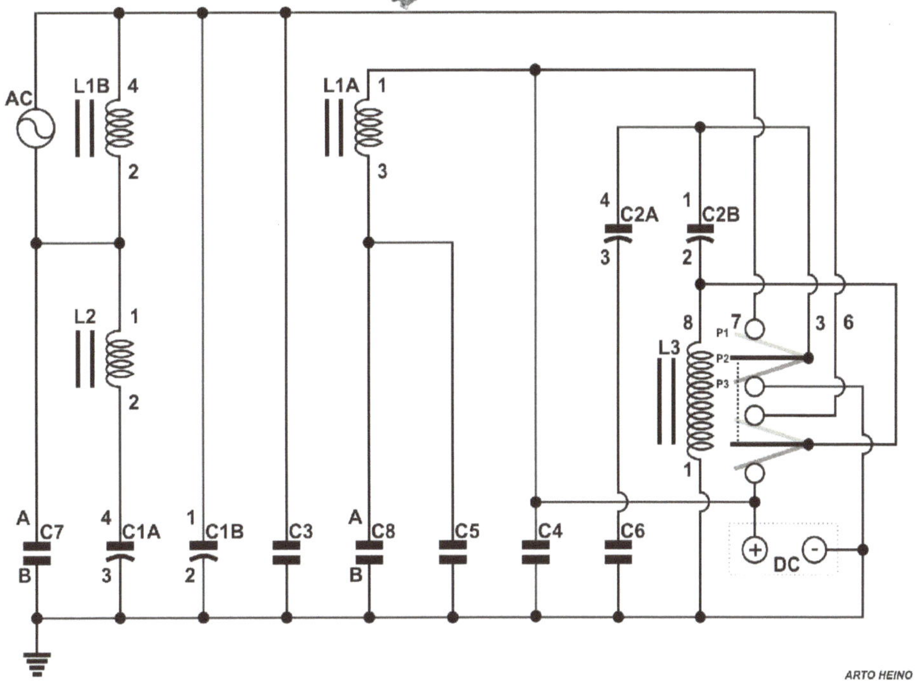

3.4c – Moray and the Nuclear Battery

This brings us to the work of Henry T Moray and his radiant energy device, included in this category would be the Hubbard Coil device and the Hendershot device. As each one of these devices ran without moving parts and incorporated a hidden part not revealed easily without detailed explanation. Henry Moray ran a laboratory that used radioactive substances, Ron Hubbard is on record saying he acquired radioactive radium from a supplier and Lester Hendershot could have acquired radium from the aircraft company he worked for.

Henry Moray wrote detailed dissertation on the phenomena of the types of rays he used, the only problem was he wrote this with an understanding that few could grasp. He states categorically that his mentor was Tesla on matters that are related to cosmic rays, where Tesla explains his work in terms that was not accepted by the corporate scientists, thus making Morays work quite difficult to unravel.

The first light of understanding must come from the discovery of Radium, Uranium and other radioactive elements, where anomalous events would reveal that these substances are in fact a radiator of very high frequency energy. Tesla understood this as a converter of cosmic rays of the universe to gamma rays, a lowering of the frequency. This is also how Moray explained his work, the idea that these were substances that had energy in the matter was simply ridiculous, Tesla knew that the universal environment was the only source of energy, these substances were mere geometric converters of frequencies from the terahertz region and converters of cosmic velocities of 50 or more times the speed of light.

As we have been taught that the radioactive decay of a substance are due to the alpha, beta particles and gamma ray photons, these are mere conventions of a science rooted in non-etheric understanding. The manipulation of results of experiments in nuclear experiments are geared to confirm an abstract theory based on metaphysics not on observational understanding that is hard won by the natural scientists personal gestalt art-forms.

The Moray device was simply a radio device that had radium as the energy amplifying material, his valves were masterfully designed to radiate the current wire known as the heater element, this in turn increased the charge going to the plate and amplifying the energy. Moray knew these elements radiated in the gamma ray region , this is why he designed valves that could be carefully made to oscillate in sub-millimeter regions.

3.4d – Lester Hendershot

The inclusion of the Hendershot generator was a last minute inclusion as this technology has never been proven or understood in normal electrical engineering terms. The amazing feat of stroking wires until a charge appeared in the circuit thus lighting a lamp was truly amazing. How did Lester achieve his results? This is the question researchers have been asking for at least 80 years.

The speculation surrounds the device are numerous, still we can possibly make a few guesses based on the way he tuned the device. One suggestion was the use of wax that had radium salts impregnated in it, which was packed around the basket weave coils . Another is he tuned in to the local 60 Hz power lines, as the L8/C3/C13 and L6/L3/C13 combination are laid out like 60 Hz resonators. The one that Lester mentions is the magnetic earth lines of force traveling in the East to West direction as the Earth spins, here he has included a set of resonators that will give you a frequency based on the Earth spin.

The L7 coil (capacitor wrap coil)is the hand made capacitor it also constitutes an inductance, with 5.8 turns with a diameter of 127mm, a height of 69.85mm, giving you approx 8.353 uH. I consider this vital as the combination of a parallel inductance L7 and L4 coil with the C1 capacitor will give you a low frequency resonator of about 5.45 Hz. This might seem just another number until you realize parametrics require some type of harmonic relationship between the signal frequency and the pump frequency.

Here are some interesting figures:

Fa	Frequency	L6,L3,C3	approx 60 Hz
Fb		L8,C3,C13	approx 60 Hz
Fc		L4,L7,C1	approx 5.45 Hz
Fd		L6,L3,C2	approx 43 Hz
Fe		L8b,C3,C13	approx 120 Hz
Fg		L2,C2	approx 8246 Hz
Fh		L4,C1	approx 236000 Hz

60/5.45	= 11 (whole number)
43/5.45	= 7.89 (approx Earth freq, refer chart)
120/60	= 2 (double freq required as pump for parametrics)
1570/5.45	= 288
5.45^3	= 161.8 (phi)
60/43	= 1.395 (1.333 Browns constant, Brownian motion)
7.89/6	= 1.315

As Hendershot said very clearly that the magnetic field he utilizes is the one traveling to the West, As we know when we rotate a magnet on its axis aligned with the pole the magnetic field does not rotate with the magnet, this is the principle of the Faraday disk generator. Now this means on the surface of the Earth(magnet) the magnetic lines are traveling in the opposite direction as the spin of the Earth, this is what Hendershot spoke of, here is the equation.

Circumference of the earth = 40029934 mtrs (based on Ancient Metric)

Velocity of Earth Spin = 40029934/24/60/60 = 463.30942 mtrs per second

5.6548 mtrs is 12 turns of L1 also L2 coils

463.30942/5.6548 = 82 Hz , very close to 43 x 2

30.15928 mtrs is 64 turns of L3 coil

463.20942/30.15928 = 15.362 Hz , very close to 60/4

These ideas begin to open understanding of what could actually going on. As I prefer to work from a frequency that is uniform to the whole earth, we will have to see other possibilities.

if freq = V / L

V = 463.3 m/s at equator

if freq = 82 hz at equator

as the frequency domain is uniform due to the magnetic field being held stationary relative to the sun(refer Faraday generator)

then if freq at 40 S lat = 82 Hz

and V = f x L and L = V / f

there for L and V are variable per curvature of earth towards the poles

so if V = 355 m/s

then L = 4.32 mtrs

If L1 length L = 5.65 mtrs

and V = 355 m/s

then freq = 62.76 Hz

Then for all wire lengths must be in accord to the velocity factor at any particular latitude. As Hendershot built these in that Lat 40 S this could be why he had spurious results when he moved locations.

My Location is at Lat 39.903 N

My Velocity factor is then 384.537 m/s

This will change my frequency due to my wire length.

If L =5.65 mtrs

and V = 384.537 m/s

then Freq = 68 Hz

So we either build to a set frequency and vary the wire lengths or we have variable frequencies and have the same wire lengths. These are only approximations and the magnetic field intensity varies, the circuits operations would have variables along with 1000 different parameters that goes into the build and the local environment as well. The equations above only represent the passing of one field line per meter, this will not be the correct value to compute voltages based on Faraday flux cutting rule. Thus these frequencies only represent a base value before precision can be attained.

Here is a chart showing the velocity variances per the Latitude change.

Degrees	Cosine of Degrees	Velocity M/S
0	1.00000000	463.30942133
5	0.99619470	461.54638910
10	0.98480775	456.27071017
15	0.96592583	447.52253562
20	0.93969262	435.36844436
25	0.90630779	419.90093636
30	0.86602540	401.23772868
35	0.81915204	379.52085962
40	0.76604444	354.91560765
45	0.70710678	327.60923361
50	0.64278761	297.80955548
55	0.57357644	265.74336681
60	0.50000000	231.65471066
65	0.42261826	195.80302229
70	0.34202014	158.46115469
75	0.25881905	119.91330201
80	0.17364818	80.45283671
85	0.08715574	40.38007674
90	0.00000000	0.00000000

I consider the magnetic vibrator(L8) as the signal frequency and Fc as a sub harmonic pump frequency. The inclusion of the operator into the circuit would also be part of a complete chaotic system that would also have other external field interactions that cannot be quantified as simply as we would like. The new paradigm of open system engineering has not yet come to terms with what creates a living circuit that has clear definitions that includes the parameters needed for consistent replication.

As an artist I am making great efforts in making a bridge for a new understanding that encompass both the science of the measurable with the art of intuitive concepts of the ether. Remember true knowledge is born from leaps of intuitive understanding not from dry equations or from circuits deplete of a new type of logic. This is what I see in some of Hendershots work. What I read in many forums are many inquiring minds working towards a great new understanding of the energy that surrounds us, we will eventually understand how these subtle energies affect our living systems and how we can live in harmony with it.

To understanding the ideas of wire forms from straight to coiled it is more like a transmission line series LC acceptor as apposed to a closed parallel LC rejector, these are some of the ways we should be developing to come to terms with open system engineering. As all straight wires have both L , C and R its will also have many other properties, the coil is just one geometry as you know, it is just a new line of inquiry that will provide useful concrete results. The basket weave coil has some of these unique properties that Lester was looking for, as I have also realized the combinatorial reactive components are oneself with a geometry of intended action that activates energy from the environment, from particular understanding of it nature. This last statement implies we are also part of the circuit, only by being at one with the gating of the energy conduit can we hope to operate the device and direct the environmental energy chaos in a synchronized system of oscillators and such to create an imbalance in a circuit enough to polarize the capacitors and direct the charge through the load.

To understand the implications much was discovered when working with the Robert Adams motor way back in the early 90's. This type of motor was another way to gate energy from our environment, by plotting the voltage, current and temperature curves, you will find it followed the time interval of the earth and sun with peaks and troughs matching exactly, also the predictions of weather changes and other earth activity was also in the data. The motor is like the Hendershot device, acting as a gating device, an instrument of the environment, moving energy in the circuit at differentiated inertial moments.

Another plausible scenario is the vibrating piece that is hinged between the poles is carefully balanced and begins to vibrate when the two coils have a differential voltage due to the earths magnetic field being slightly different from moment to moment due to the earths rotation. This vibration acts like a voltage vibrator that I have mentioned in an another section, this along with the multi-vibrator like layouts of his circuits could be that he is implementing a form of parametric oscillator.

The two small transformers need to have some deign parameters attributed to them so some of the calculations can have some validity.

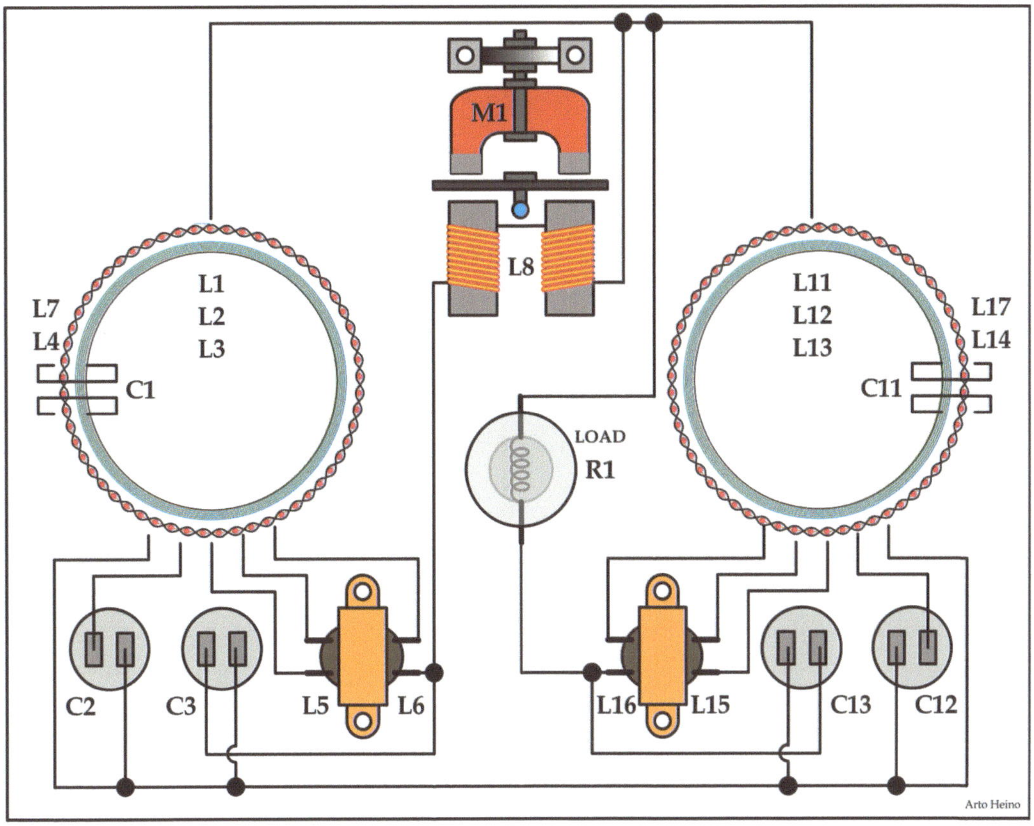

Frequency 60.000000000000

Description	Name	Inches	Values	Units
L5	**Transformer Primary Coil**		**0.068218705620**	Henry
Turns			240.000000000000	
Diam		0.590551	0.015000000000	Mtr
Length		2.362205	0.060000000000	Mtr
Wire Diameter		0.020118	0.000511000000	Mtr
W Length		445.2651	11.309733552923	Mtr
W Area			0.000000205084	Sqr Mtr
Resistance			**0.926467040184**	Ohm
copper volume			0.000002319445	Cub Mtr
ur			320.000000000000	
XL5			25.717846129708	Ohm
R1			1.396027700504	Ohm
X1 reactance			25.679928275031	Ohm
Z1 impedance			25.717846129708	Ohm
V1 voltage			90.000000000000	Volt
I1 amperage			3.499515454991	Amp
P1 Power			314.956390949214	VA
L5 Self Inductance			0.068118125387	Henry
Flux per Amp			0.000283825522	Webers
Max Flux per Amp			0.000401389903	Webers
Max Flux			**0.001404670170**	Webers

Description	Name	Inches	Values	Units
L6	**Transformer Secondary Coil**		**1.705467640508**	Henry
Turns			1200.000000000000	
Diam		0.590551	0.015000000000	Mtr
Length		2.362205	0.060000000000	Mtr
Wire Diameter		0.012638	0.000321000000	Mtr
W Length		2226.326	56.548667764616	Mtr
W Area			0.000000080928	Sqr Mtr
Resistance			**11.739016507992**	Ohm
copper volume			0.000004576383	Cub Mtr
ur			320.000000000000	
XL6			642.946153242697	Ohm
R2			34.900692512603	Ohm
X2 reactance			641.998206875784	Ohm
Z2 impedance			642.946153242697	Ohm
V2 voltage			450.000000000000	Volt
I2 amperage			0.699903090998	Amp
P2 Power			314.956390949214	VA
L6 self inductance			1.702953134674	Henry
Flux per Amp			0.001419127612	Webers
Max Flux per Amp			0.002006949516	Webers
Max Flux			**0.001404670170**	Webers

To make sure all these values are realistic I implemented a double check on the whole Magnetic flux of the transformer, as this would be only one value for the complete unit, either primary or secondary. Here are the equations:

XL = 2 x Pi x *f* x L5

f = Frequency = 60 Hz

X1 = sqrt(Z1^2 - R1^2)

R1 = Rp + Rs x (Np/Ns)

Rp = resistance Primary Rs = resistance Secondary

Np = Number of turns Primary Ns = Number of turns Secondary

Z1 = V / I

Then

L5 = (ur * 4 * Pi *0.0000001* Np^2 * Pi * Radius^2)/ Length

L5si = X1/(2 * Pi * *f*)

Then L5 = L5si

and the Maximum Flux of L5 = Maximum Flux of L6

Hendershot Device Design #1 Schematic Breakdown

By Arto Heino

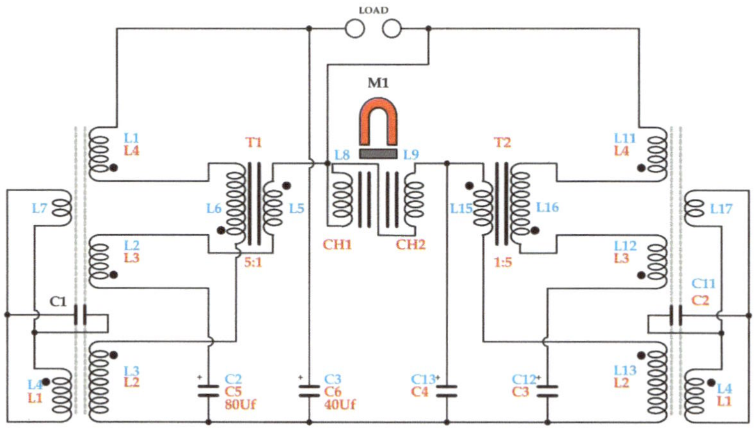

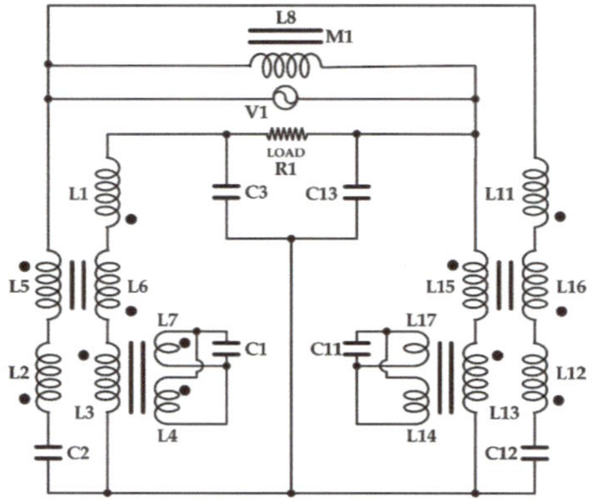

ED SKILLING SCHEMATIC

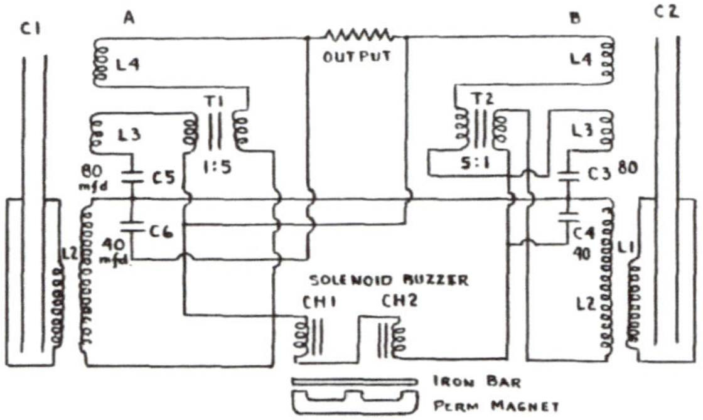

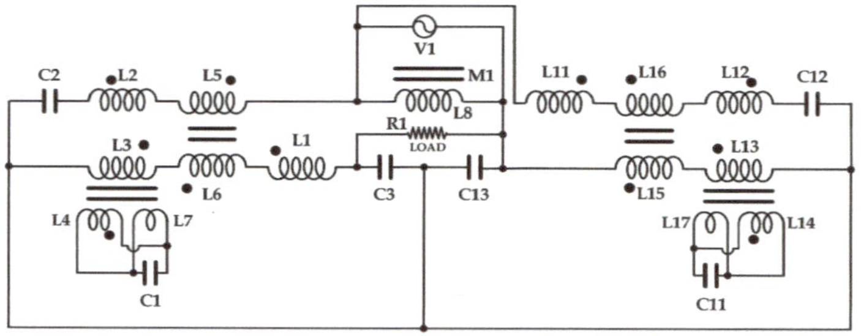

Arto Heino

Hendershot Device Design #2 Schematic Breakdown

By Arto Heino

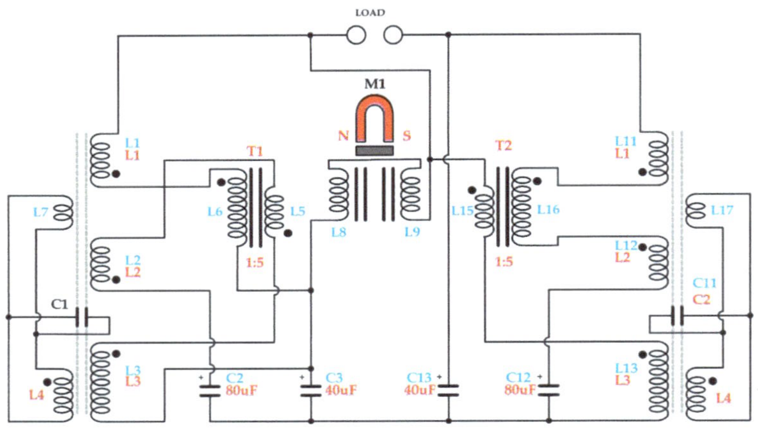

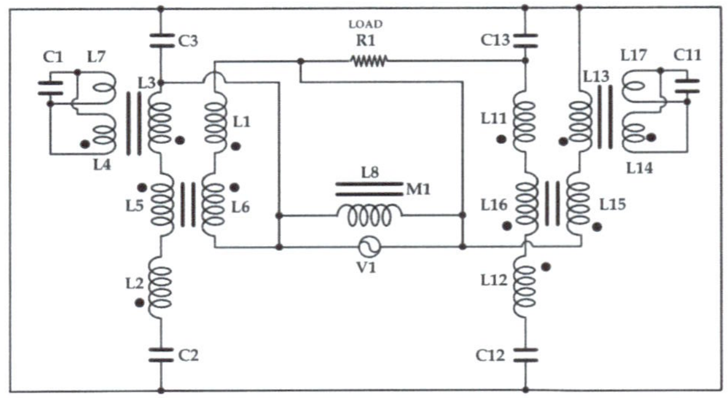

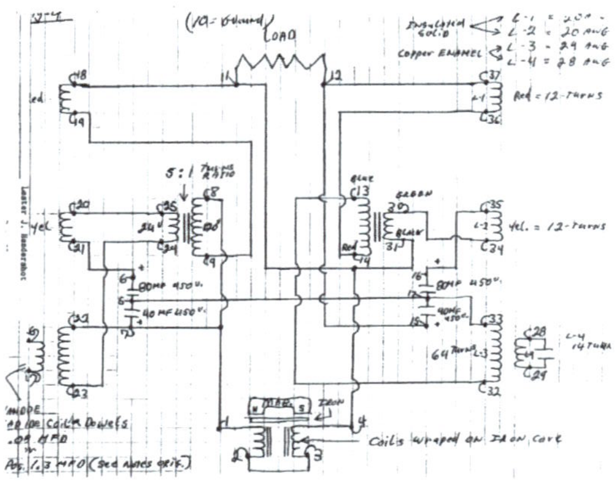

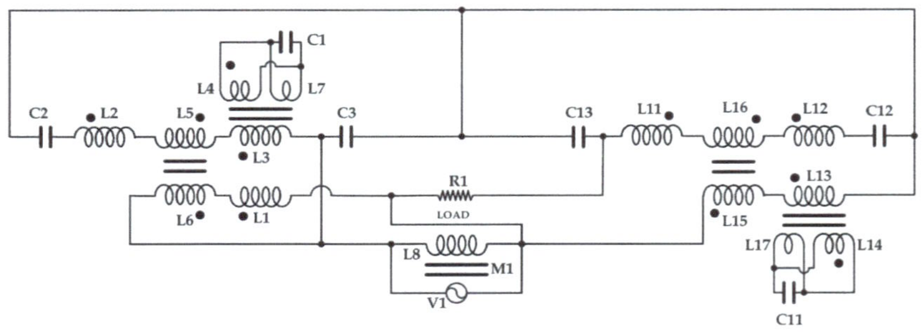

Arto Heino

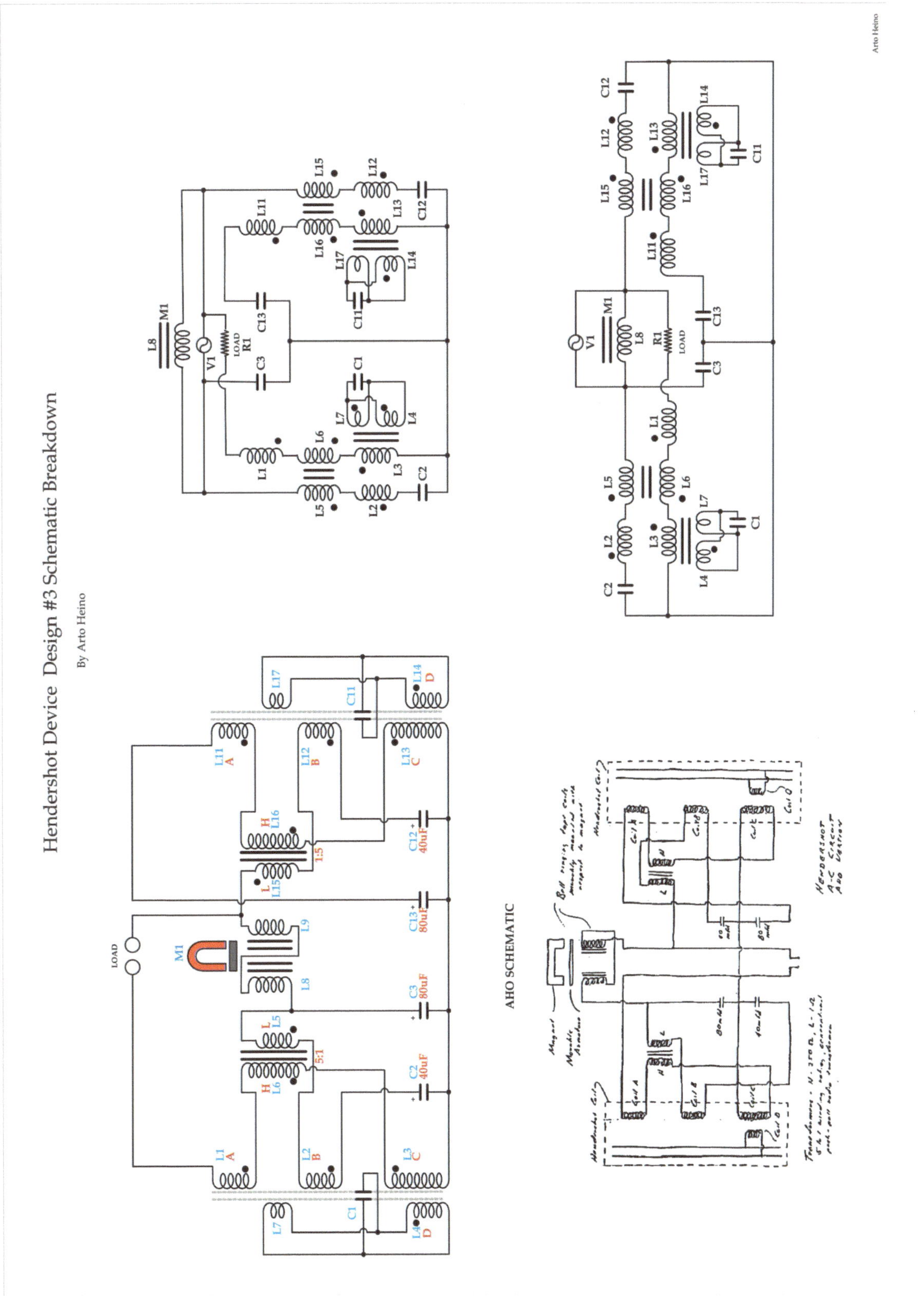
Hendershot Device Design #3 Schematic Breakdown
By Arto Heino
LOAD
M1
L8
L9
L1 A
L2 B
L3 C
L4 D
L5 L
L6 H
L7
5:1
C1
C2 40uF
C3 80uF
C13 80uF
C12 40uF
L11 A
L12 B
L13 C
L14 D
L15 L
L16 H
L17
C11
1:5
AHO SCHEMATIC
V1
R1
Arto Heino

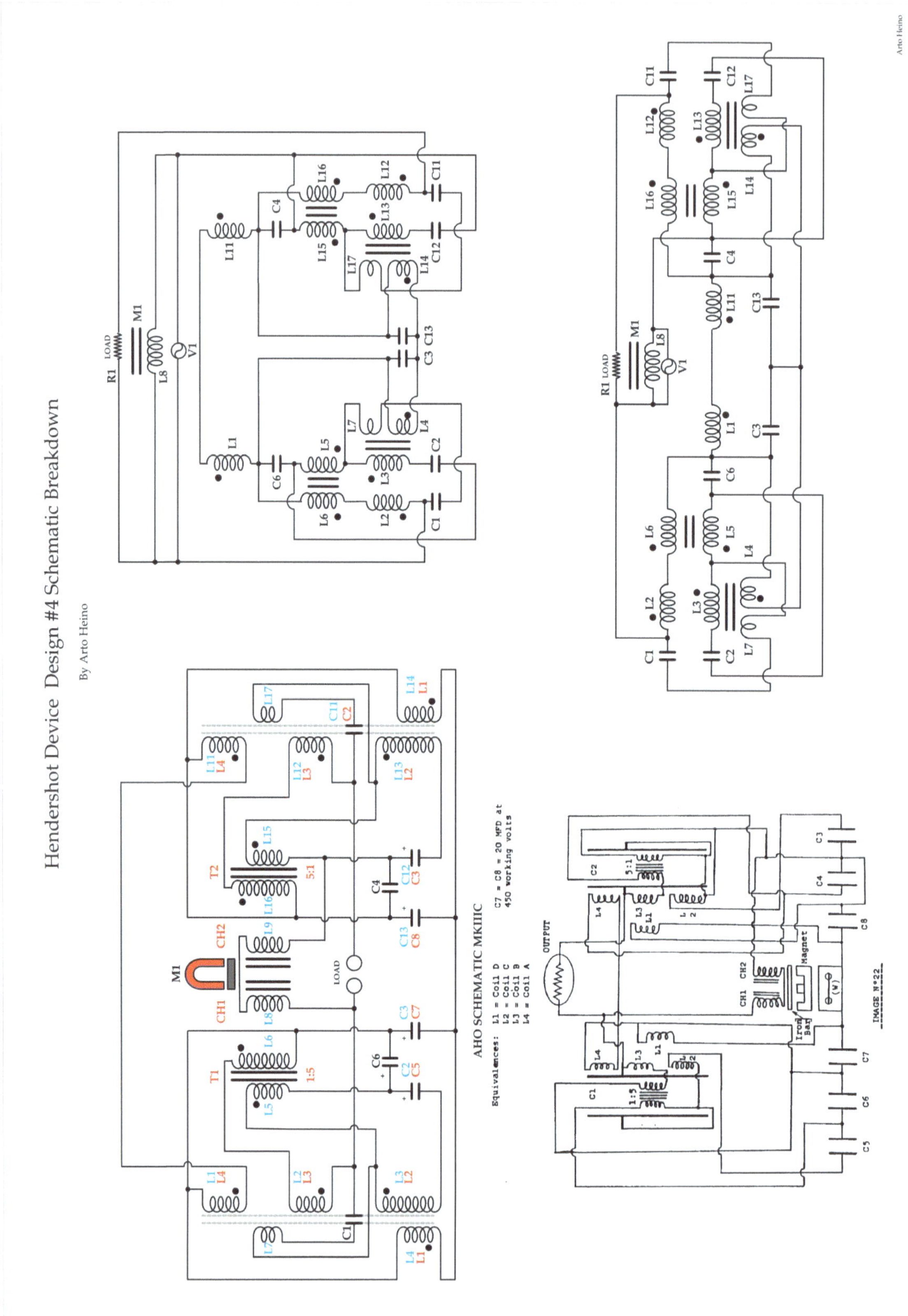
Hendershot Device Design #4 Schematic Breakdown
By Arto Heino
AHO SCHEMATIC MKIIIC
Equivalences: L1 = Coil D
L2 = Coil C
L3 = Coil B
L4 = Coil A
C7 = C8 = 20 MFD at
450 working volts
IMAGE N°22
Arto Heino

Hendershot Device Design #5 Schematic Breakdown

By Arto Heino

SIMPLE SYSTEM

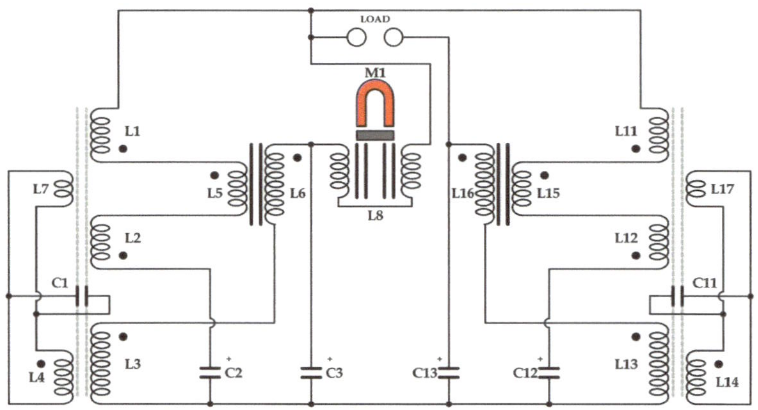

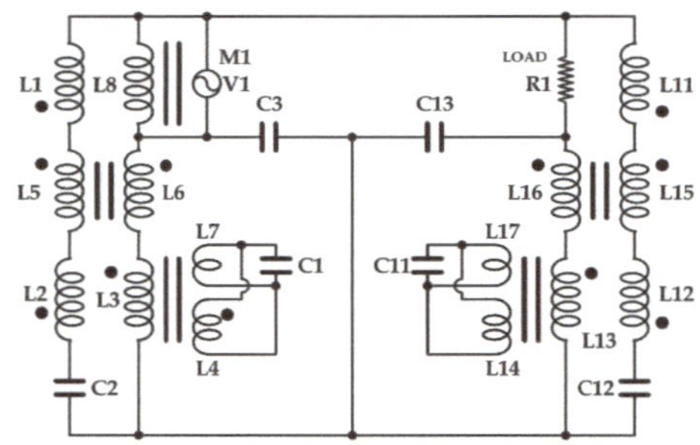

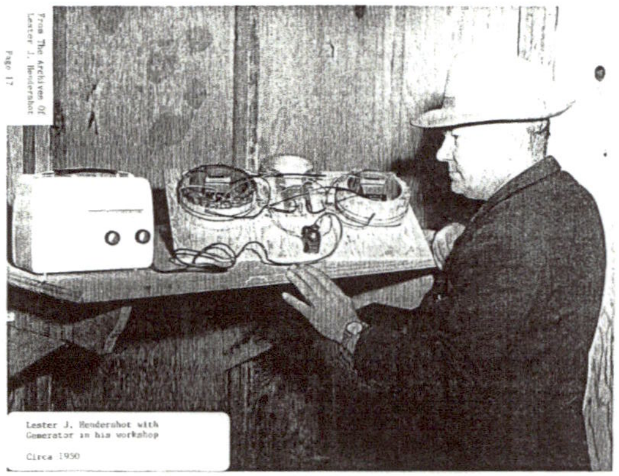

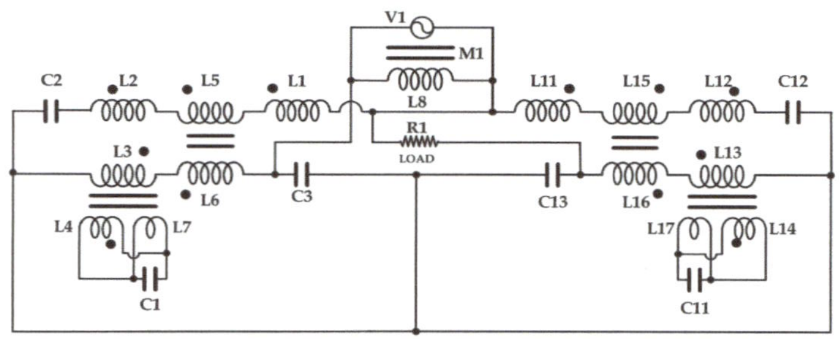

Arto Heino

Here are some of my early details of the variable parameters used to define the frequencies in my versions of the Hendershot device. The capacitors are non polarized electrolytic types packaged in a aluminum can.

Name	Description	Inches	Value	Unit	m	u	n	p
Handmade Capacitor	C1		**0.000000007120**	*Farad*	7.120E-006	0.0071201	7.1201443	7120.1443
	Roll Length	91.25000000	2.317750000000	*Mtr*				
	Roll Width	2.75000000	0.069850000000	*Mtr*				
	Area		0.027868911350	*Sqr Mtr*				
	Inner Diameter	5.00000000	0.127000000000	*Mtr*				
	Layers		5.809155422854					
	Gap	0.01968504	0.000500000000	*Mtr*				
	er		3.000000000000	*Permittivity*				
Oscillating Capacitor	C2		**0.000080000000**	*Farad*	0.08	80	80000	80000000
Driving Capacitor	C3		**0.000040000000**	*Farad*	0.04	40	40000	40000000

The coils are numerous and require a degree of precision to build.

Name	Description	Inches	Value	Unit	m	u	n	p
Top Coil	L1		**0.000046571805**	*Henry*	0.0465718	46.571805	46571.805	46571805
	Turns		12.000000000000					
	Coil Diam	5.90551181	0.150000000000	*Mtr*				
	Coil Length	0.38362205	0.009744000000	*Mtr*				
	Wire AWG20	0.03196850	0.000812000000	*Mtr*				
	Wire Length	222.63255025	5.654866776462	*Mtr*				
	Wire Section Area		0.000000517848	*Sqr Mtr*				
	Resistance		**0.183455070494**	*Ohm*				
	Copper Volume		0.000002928359	*Cub Mtr*				

These are basket weave designs like the early radios from the 1920's when this design was at its heyday.

Name	Description	Inches	Value	Unit	m	u	n	p
Middle Coil	L2		**0.000046571805**	*Henry*	0.0465718	46.571805	46571.805	46571805
	Turns		12.000000000000					
	Coil Diam	5.90551181	0.150000000000	*Mtr*				
	Coil Length	0.38362205	0.009744000000	*Mtr*				
	Wire AWG20	0.03196850	0.000812000000	*Mtr*				
	Wire Length	222.63255025	5.654866776462	*Mtr*				
	Wire Section Area		0.000000517848	*Sqr Mtr*				
	Resistance		**0.183455070494**	*Ohm*				
	Copper Volume		0.000002928359	*Cub Mtr*				

Name	Description	Inches	Value	Unit	m	u	n	p
Bottom Coil	**L3**		**0.001281731648**	*Henry*	1.28173165	1281.7316	1281731.6	1.28E+009
	Turns		64.000000000000					
	Coil Diam	5.90551181	0.150000000000	*Mtr*				
	Coil Length	1.28755906	0.032704000000	*Mtr*				
	Wire AWG24	0.02011811	0.000511000000	*Mtr*				
	Wire Length	1187.37360136	30.159289474462	*Mtr*				
	Wire Section Area		0.000000205084	*Sqr Mtr*				
	Resistance		**2.470578773825**	*Ohm*				
	Copper Volume		0.000006185186	*Cub Mtr*				

Name	Description	Inches	Value	Unit	m	u	n	p
Outside Coil	**L4**		**0.000063879169**	*Henry*	0.06387917	63.879169	63879.169	63879169
	Turns		14.000000000000					
	Coil Diam	5.90551181	0.150000000000	*Mtr*				
	Coil Length	0.17692913	0.004494000000	*Mtr*				
	Wire AWG28	0.01263780	0.000321000000	*Mtr*				
	Wire Length	259.73797530	6.597344572539	*Mtr*				
	Wire Section Area		0.000000080928	*Sqr Mtr*				
	Resistance		**1.369551925932**	*Ohm*				
	Copper Volume		0.000000533911	*Cub Mtr*				

Name	Description	Inches	Value	Unit	m	u	n	p
Capacitor Wrap Coil	**L7**		**0.000008353041**	*Henry*	0.00835304	8.3530413	8353.0413	8353041.3
	Turns		5.809155422854					
	Diam	5.00000000	0.127000000000	*Mtr*				
	Length	2.75000000	0.069850000000	*Mtr*				
	Alm Thickness	0.01968504	0.000500000000	*Mtr*				
	Alm Length	91.25000000	2.317750000000	*Mtr*				
	Alm Area		0.001158875000	*Sqr Mtr*				
	Alm Resistance		**0.000054000000**	*Ohm*				
	Aluminium Vol.		0.002685982531	*Cub Mtr*				

Name	Description	Inches	Value	Unit	m	u	n	p
Buzzer Coil	**L8**		**3.5826663976**	*Henry*	3582.6664	3582666.4	3.58E+009	3.58E+012
	Turns		2200.0000000000					
	Diam	0.59055118	0.0150000000	*Mtr*				
	Length	2.36220472	0.0600000000	*Mtr*				
	Wire AWG28	0.01263780	0.0003210000	*Mtr*				
	Wire Length	4081.59675466	103.6725575685	*Mtr*				
	Wire Section Area		0.0000000809	*Sqr Mtr*				
	Resistance		**21.5215302647**	*Ohm*				
	copper volume		0.0000083900	*Cub Mtr*				
	ur		200.0000000000	*Permeability*				

3.3e - Blocking Oscillator

The discovery of the blocking oscillator was not a ground breaking discovery, as this type of relaxation oscillator was first used with valves in 1918 and has become an interesting pursuits of scientific investigation since its inception. The properties of this simple device are one that includes the random states and the fractal natures of energy, which makes this a compelling investigation. The efficiency of this design is amazing as very low energies are required to begin the oscillations .

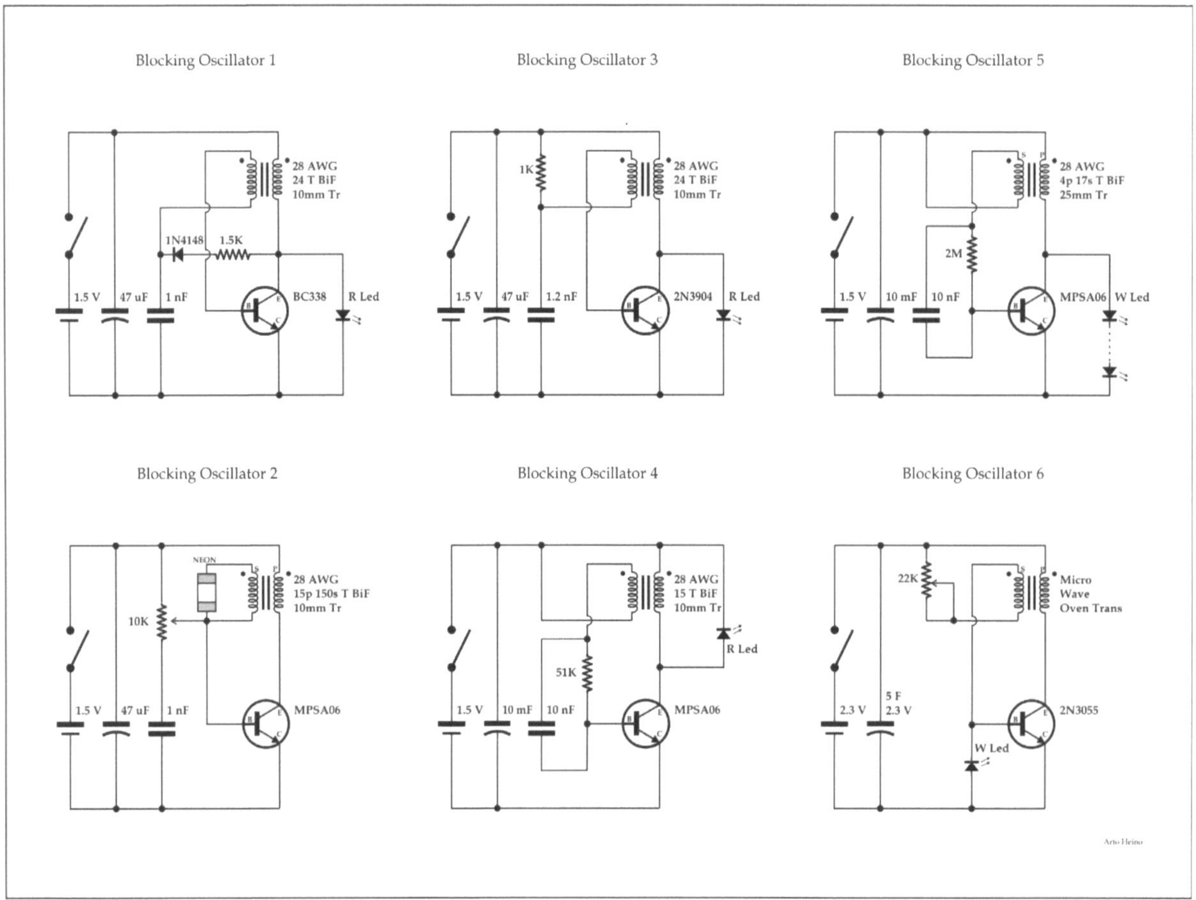

Blocking Oscillator 7

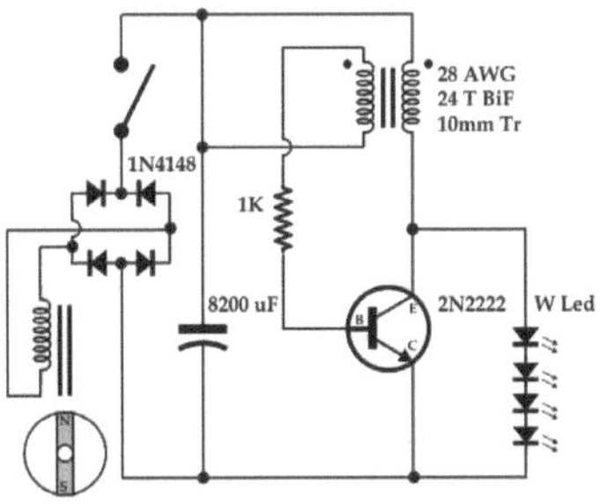

Blocking Oscillator 9

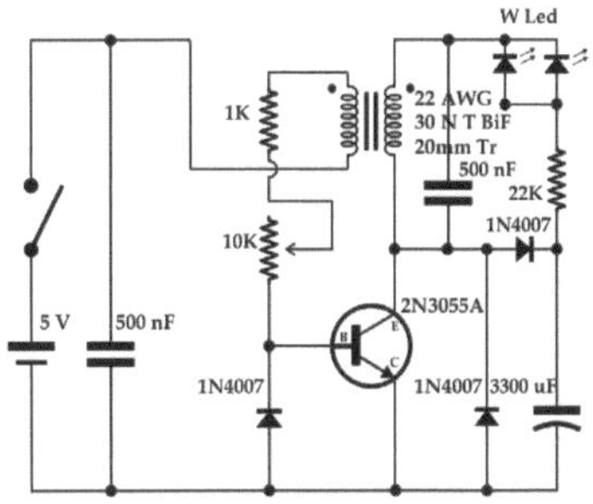

Blocking Oscillator 11

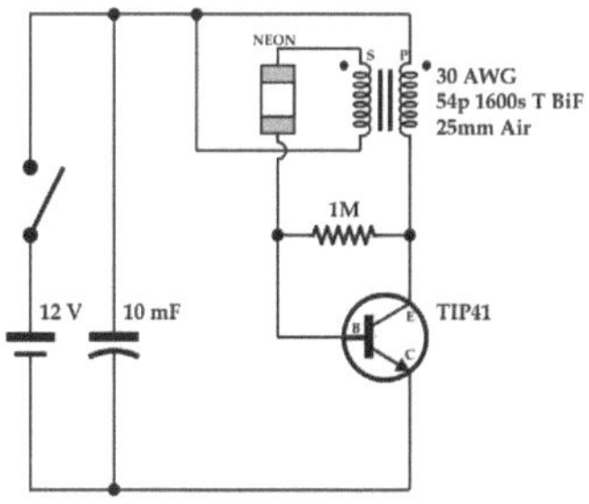

Blocking Oscillator 8

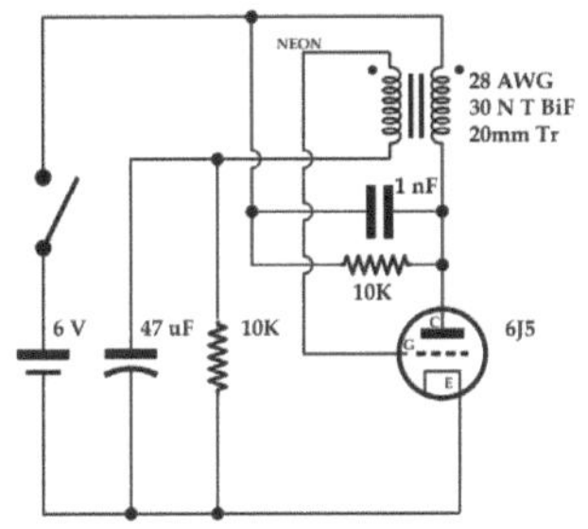

Blocking Oscillator 10

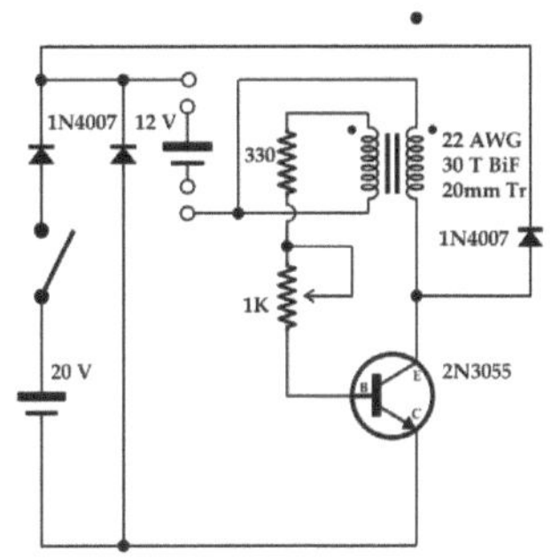

Blocking Oscillator 12

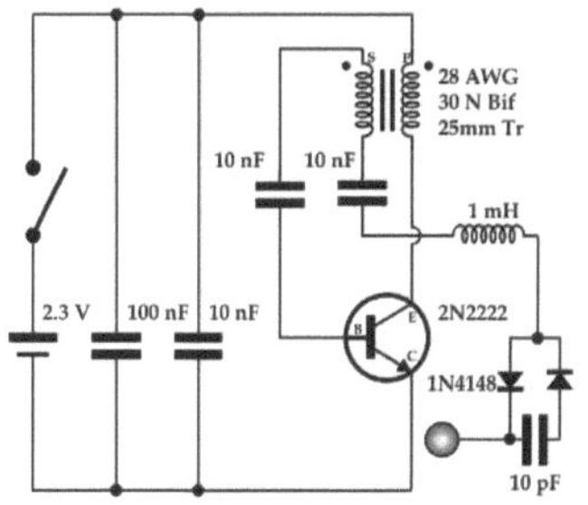

Arto Heino

Blocking Oscillator 13

Transformer specifications

Turns	Pulse uS
20-20	0.1-0.5
125-125	1 - 5
150-150	1 - 5

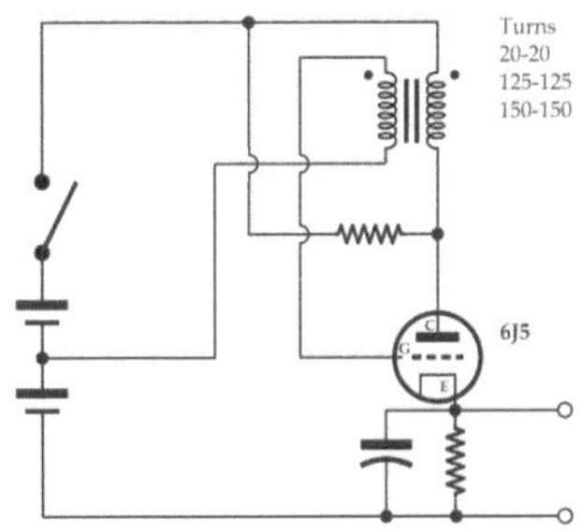

Blocking Oscillator 14

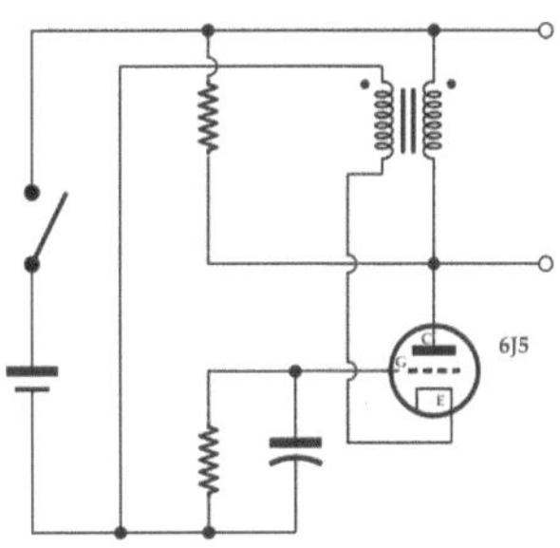

Blocking Oscillator 15

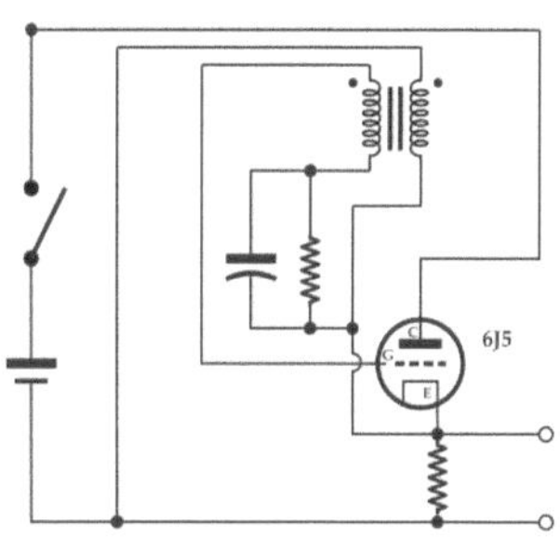

Arto Heino

3.4f – Don Smith

As we move into the modern era a few speakers of the new energy concept have risen to give many interesting but not new ideas about electricity and the energy equation. Don's nuts and bolts demonstrations and his display of his work had too many hidden priority parts, never allowing a true understanding of the science behind this work. He could have easily created misinformation and lead researcher down the proverbial garden path. He propounds the the word "resonance" and "ambient energy " like a priest at the pulpit without revealing its true nature to the audience, he shows contempt for those who dare to ask revealing questions. Many times during his talks he leads the audience astray with his meandering logic, then he says he cannot reveal any real details without fear of death by third parties. So by his own words, his devices have many non standard parts, these are not available to the public, some that are in catalogs are at premium prices, always supporting the corporate electronics industry with many mentions of brands and parts sources. My guess was he misleading only to promote his personal commitments to the corporate suppliers.

The matter that must be cleared is, "does his circuits work?". Looking deeply at all his schematics, he reveals a poor understanding of circuit layout and his circuit diagrams are amateurish to say the least. He displays a poor working knowledge of radio theory and is vague on specifics when it comes to Tesla. He mentions a few circuits by Tesla and d'Arsonval that were his inspiration, some of these these were half wave resonators, which he never mentions. The circuit he shows from Tesla is his DC Motor/Generator controllers, which he explains dismally leading the audience into a false understanding of the work of Tesla and his circuit designs.

It looks like he was trying to create a description of Free Energy that people wanted to hear, unwittingly or intentionally misleading his audience down the experimental path that tries in joining the donkeys mouth to his rear end. As of 2013 not one experimenter has duplicated any circuit by Don that actually works as he says it does. We all have certainly got parts of his circuits to work, this would be an intentional goal so you would keep working on it looking for some mad logic or "Dons Secret", which is another form of indoctrination into his dead end agenda.

I have made my excursion into his madness only once before I realized he was hiding the true nature of his circuits behind special components. So here is my page of work that led me to walk away from any ideas this gentleman has to offer. He had not even explained the frequencies of all his circuits sections clearly enough so those who are skilled enough, allowing you to only duplicate the simpler parts.

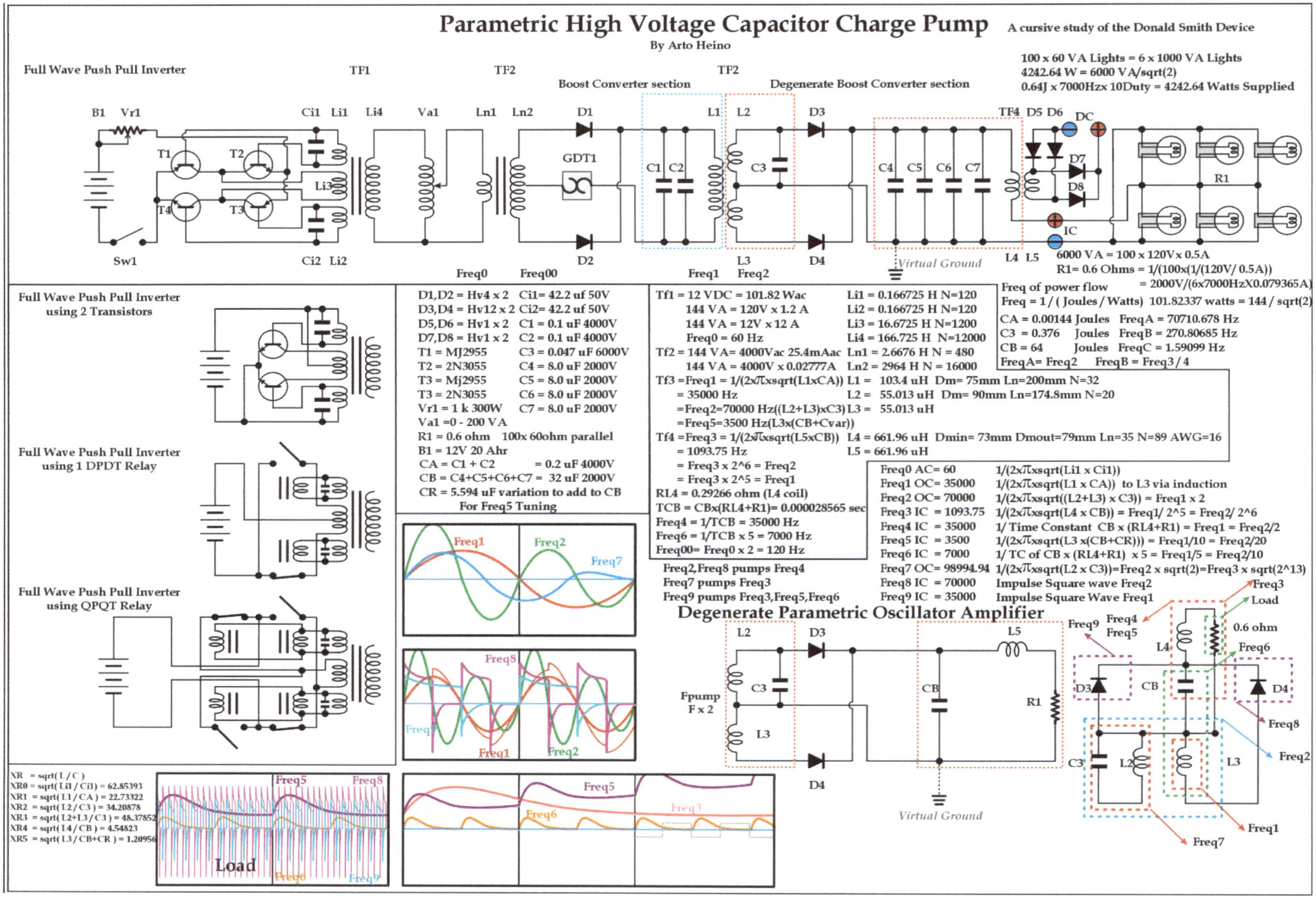
Parametric High Voltage Capacitor Charge Pump
A cursive study of the Donald Smith Device
By Arto Heino
100 x 60 VA Lights = 6 x 1000 VA Lights
4242.64 W = 6000 VA/sqrt(2)
0.64J x 7000Hzx 10Duty = 4242.64 Watts Supplied
Full Wave Push Pull Inverter
TF1
TF2
Boost Converter section
TF2
Degenerate Boost Converter section
B1 Vr1 T1 T2 T4 T3 Sw1 Ci1 Li1 Li3 Ci2 Li2 Li4 Va1 Ln1 Ln2 D1 GDT1 D2 C1 C2 L1 L2 C3 L3 D3 D4 C4 C5 C6 C7 TF4 D5 D6 DC D7 D8 IC R1 L4 L5
Freq0 Freq00 Freq1 Freq2
Virtual Ground
6000 VA = 100 x 120V x 0.5A
R1= 0.6 Ohms = 1/(100x(1/(120V/ 0.5A))
= 2000V/(6x7000HzX0.079365A)
Freq of power flow
Freq = 1 / (Joules / Watts) 101.82337 watts = 144 / sqrt(2)
CA = 0.00144 Joules FreqA = 70710.678 Hz
C3 = 0.376 Joules FreqB = 270.80685 Hz
CB = 64 Joules FreqC = 1.59099 Hz
FreqA= Freq2 FreqB = Freq3 / 4
Full Wave Push Pull Inverter using 2 Transistors
Full Wave Push Pull Inverter using 1 DPDT Relay
Full Wave Push Pull Inverter using QPQT Relay
D1,D2 = Hv4 x 2 Ci1= 42.2 uf 50V
D3,D4 = Hv12 x 2 Ci2= 42.2 uf 50V
D5,D6 = Hv1 x 2 C1 = 0.1 uF 4000V
D7,D8 = Hv1 x 2 C2 = 0.1 uF 4000V
T1 = MJ2955 C3 = 0.047 uF 6000V
T2 = 2N3055 C4 = 8.0 uF 2000V
T3 = Mj2955 C5 = 8.0 uF 2000V
T3 = 2N3055 C6 = 8.0 uF 2000V
Vr1 = 1 k 300W C7 = 8.0 uF 2000V
Va1 =0 - 200 VA
R1 = 0.6 ohm 100x 60ohm parallel
B1 = 12V 20 Ahr
CA = C1 + C2 = 0.2 uF 4000V
CB = C4+C5+C6+C7 = 32 uF 2000V
CR = 5.594 uF variation to add to CB
For Freq5 Tuning
Tf1 = 12 VDC = 101.82 Wac
144 VA = 120V x 1.2 A
144 VA = 12V x 12 A
Freq0 = 60 Hz
Tf2 = 144 VA= 4000Vac 25.4mAac
144 VA = 4000V x 0.02777A
Tf3 =Freq1 = 1/(2xπxsqrt(L1xCA))
= 35000 Hz
=Freq2=70000 Hz((L2+L3)xC3)
=Freq5=3500 Hz(L3x(CB+Cvar))
Tf4 =Freq3 = 1/(2xπxsqrt(L5xCB))
= 1093.75 Hz
= Freq3 x 2^6 = Freq2
= Freq3 x 2^5 = Freq1
RL4 = 0.29266 ohm (L4 coil)
TCB = CBx(RL4+R1)= 0.000028565 sec
Freq4 = 1/TCB = 35000 Hz
Freq6 = 1/TCB x 5 = 7000 Hz
Freq00= Freq0 x 2 = 120 Hz
Li1 = 0.166725 H N=120
Li2 = 0.166725 H N=120
Li3 = 16.6725 H N=1200
Li4 = 166.725 H N=12000
Ln1 = 2.6676 H N = 480
Ln2 = 2964 H N = 16000
L1 = 103.4 uH Dm= 75mm Ln=200mm N=32
L2 = 55.013 uH Dm= 90mm Ln=174.8mm N=20
L3 = 55.013 uH
L4 = 661.96 uH Dmin= 73mm Dmout=79mm Ln=35 N=89 AWG=16
L5 = 661.96 uH
Freq0 AC= 60 1/(2xπxsqrt(Li1 x Ci1))
Freq1 OC= 35000 1/(2xπxsqrt(L1 x CA)) to L3 via induction
Freq2 OC= 70000 1/(2xπxsqrt((L2+L3) x C3)) = Freq1 x 2
Freq3 IC = 1093.75 1/(2xπxsqrt(L4 x CB)) = Freq1/ 2^5 = Freq2/ 2^6
Freq4 IC = 35000 1/ Time Constant CB x (RL4+R1) = Freq1 = Freq2/2
Freq5 IC = 3500 1/(2xπxsqrt(L3 x(CB+CR))) = Freq1/10 = Freq2/20
Freq6 IC = 7000 1/ TC of CB x (RL4+R1) x 5 = Freq1/5 = Freq2/10
Freq7 OC= 98994.94 1/(2xπxsqrt(L2 x C3))=Freq2 x sqrt(2)=Freq3 x sqrt(2^13)
Freq8 IC = 70000 Impulse Square wave Freq2
Freq9 IC = 35000 Impulse Square Wave Freq1
Freq2,Freq8 pumps Freq4
Freq7 pumps Freq3
Freq9 pumps Freq3,Freq5,Freq6
Degenerate Parametric Oscillator Amplifier
L2 D3 L5 C3 CB R1 L3 D4
Fpump F x 2
Virtual Ground
Freq9 Freq4 Freq5 Freq3 Load 0.6 ohm Freq6 L4 D3 CB D4 Freq8 Freq2 C3 L2 L3 Freq1 Freq7
Freq1 Freq2 Freq7
Freq8 Freq1 Freq2
XR = sqrt(L / C)
XR0 = sqrt(Li1 / Ci1) = 62.85393
XR1 = sqrt(L1 / CA) = 22.73322
XR2 = sqrt(L2 / C3) = 34.20878
XR3 = sqrt(L2+L3 / C3) = 48.37852
XR4 = sqrt(L4 / CB) = 4.54823
XR5 = sqrt(L3 / CB+CR) = 1.20956
Freq5 Freq8 Load Freq6 Freq9
Freq5 Freq6 Freq3

As you can see I can easily show all the frequencies and why I chose them and how I arrive at them.

Type	Freq	Section	Frequency		Equation
DC	F	Battery	0	hz	
AC	F0	Inverter	60	hz	1/2xpixsqrt(Li1 x Ci1)
IC	F00	Boost Converter input	120	hz	2/2xpixsqrt(Li1 x Ci1)
OC	F1	Boost Converter output	35000	hz	1/2xpixsqrt(L1 x CA)
OC	F2	Degenerate B/Con Input	70000	hz	1/2xpixsqrt((L2+L3)xC3)
IC	F3		1093.75hz		1/2xpixsqrt(L4xCB)
IC	F4		35000	hz	1/(CBx(RL3+R1))
IC	F5		3500	hz	1/2xpixsqrt(L3x(CB+CR))
IC	F6		7000	hz	1/(CBx(RL4+R1))
OC	F7		98994.94		1/2xpixsqrt(L2xC3)
IC	F8		70000	hz	
IC	F9		35000	hz	

Frequency correlations

F2	=	F1 x 2		
F3	=	F1 / 2^5	=	F2 / 2^6
F4	=	F1	=	F2 / 2
F5	=	F1 / 10	=	F2 / 20
F6	=	F1 / 5	=	F2 / 10
F7	=	F2 x sqrt(2)	=	F3 x sqrt(2^13)

Frequency of Power Flow

Basic Equation for Frequency of Power Flow= 1 / (Joules / Watts)

Watts available to the circuit system based on battery supply from the inverter =

V	=	12		
A	=	12		
VA	=	144		
Watts	=	VA / sqrt(2)	=	101.82337 watts

Joules stored in Capacitor	CA	= C1+C2 = 0.1 + 0.1 = 0.2 uF 4000 volts
	CAj	= ½ C xV = ½ x 888 = 0.00144 joules
	FA	= 1/ (0.00144/101.82337) = 70710.678 Hz
Joules stored in Capacitor	CB	= C4+C5+C6+C7 = 8 + 8 + 8 + 8 = 32 uF 2000 volts
	CBj	= ½ CxV = 0.376 joules
	FB	= 0.376 / = 270.80685 Hz
Joules stored in Capacitor	C3	= 0.047 uF 6000 volts
	C3j	= ½ C x V = 64 joules
	FC	= 1.59099 Hz

FA = F2

FB = F3 / 4

FC = ?

The list of frequencies I have compiled are a foray into the parametrical nature of energy growth and cannot be attributed to anything that Don Smith has revealed by his work or his writing, this is my personal understanding of the nature of frequency coupling.

DC = Direct Current

AC = Alternating Current

OC = Oscillating Current

IC = Impulse Current

The nature of Don's circuits were not revealed by my analysis, only the interesting cross coupling frequencies and harmonics that are present when the circuit is arranged as I have drawn. I can only summarize that the large capacitors either have an extra hidden parameter that can hold the store of energy like a battery or he is not revealing the hidden components that can fulfill the requirements demonstrated to his audience.

So the saga of the Don Smith device will continue until we as observers and participants realize he never intended to reveal his "Tricks" as this would have placed him in the scamming category like many others that are roaming the stages and Internet trying to get you to part with some cash.

3.4 - The Switch

3.4a – Mechanical

The nature of all engineering has in its core a few things that have to be perfected to make things operate, the SWITCH is one of those ideas.

3.4b - Rotating Brush

I hope to make Tesla's Patent revealed as a fundamental high voltage pump circuit drawn 30 years before Walton and Croft, in the Tesla patent 583,953, an apparatus for producing currents of high frequency.

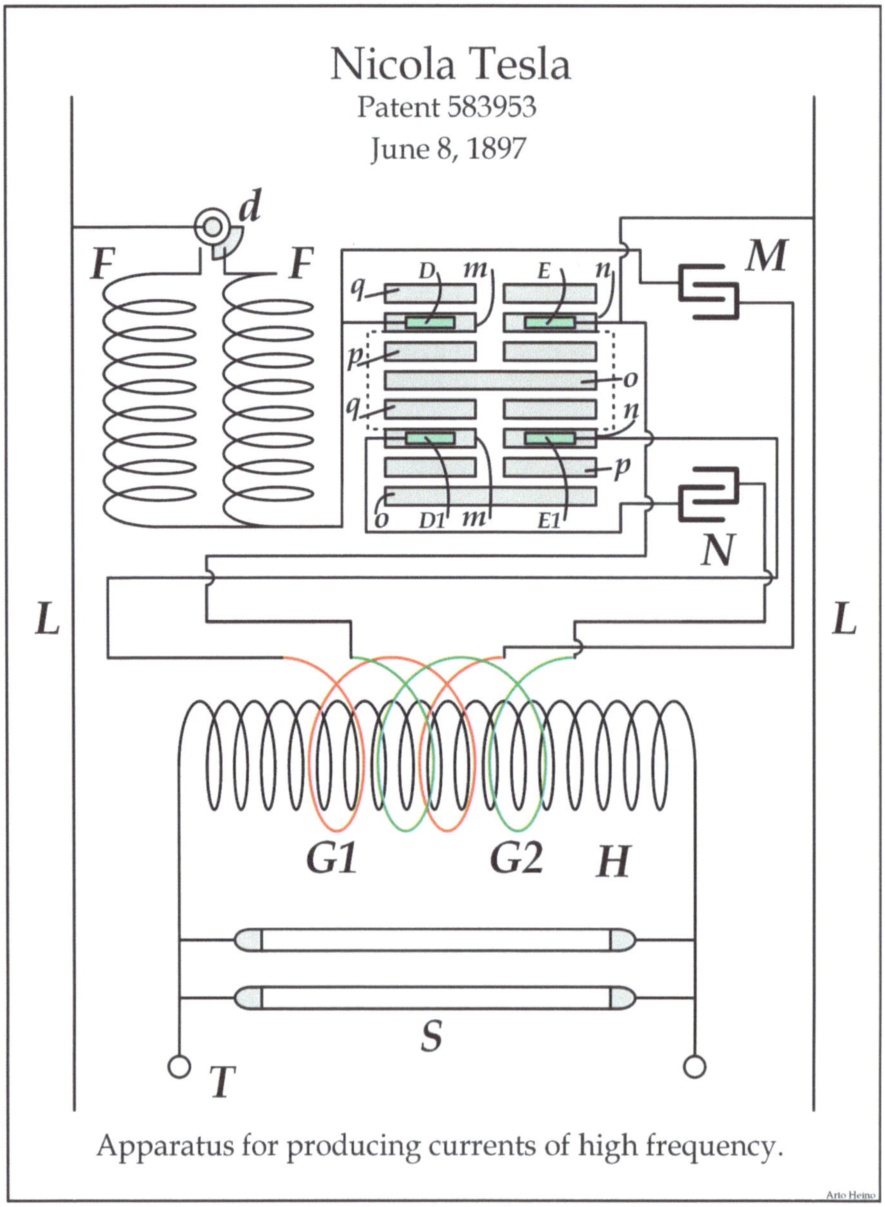

Apparatus for producing currents of high frequency.

Diagram 1 – I redrew the main patent for clarity

As quoted by Tesla in this patent, "In the description of the invention is designed for converting a continuous or direct current into one of high frequency." , this can be interpreted as using 120/240 AC or Battery/Rectified power.

This part is describing how he uses a large induction coil and a rotary commutator for his switching circuits to create his pulsed energy, "I have employed a circuit of high self-induction connected with the mains from a suitable source of current and containing some form of circuit-controller for periodically interrupting it."

The following is very important as Tesla only used a commutator to switch and didn't use the modern equivalents such as transistors and diodes."Around the break or point of interruption I have arranged a condenser, into which the circuit discharges when interrupted, and this condenser is in turn made to discharge through a circuit containing the primary of a transformer, and of such character that the condenser-discharge will be in the form of an extremely rapid succession of impulses."

This part is the essence of his genius, by charging in parallel and discharging in series, Tesla increases the voltage potential available, "I subdivide the condenser necessary for storing the energy required into integral parts or provide independent condensers, and employ means for charging said condensers in multiple and discharging them in series through the primary of the transformer."

The items F F are self-induction coils placed beside the motor. Above these is the transformer, composed, essentially, of primaries G1 and G2 and a secondary H.

L L designate the mains, between which a circuit is formed, including the self-induction coils F F and the commutator. The switch d may be employed to bring either or both of the coils F F into this circuit.

The commutator is built up of insulated plates or segments, upon which the positive and negative brushes bear, and these plates may be considered as belonging to three sets or classes:

1 – the plates m for what may be considered as the positive brushes D D1 in one row, electrically connected together, and the corresponding plates n for what may similarly be considered as the negative brushes E E1 in the other row.

2 - the plates o, which lie in both rows, and hence are conveniently made in single pieces extending across the controller.

3 - the idle or spacing plates p q, which are interposed in each row between the other two sets.

The brush D of one set is connected with one main through the coils F, and each one of the brushes of the same set is connected to one of the terminals of the condensers M N, respectively. Similarly the brush E of the other set of brushes is connected to the opposite main and each of the brushes of said set to the opposite condenser terminals through the primary or strands of a primary G1 or G2.

In the position of the parts shown in which two positive and two negative brushes are shown, the brushes are bearing on plates m m and n n. Consequently the circuit through the coils F F is through the condensers in multiple, and assuming that energy has been stored in said coils, the condensers will thus be charged.

If now by the movement of the commutator plates or brushes the latter are shifted across the idle or spacing plates p q onto the long or cross-connected plates o two results follow:

1 - The mains are short-circuited through the coils F F, which therefore store energy, while the condensers are connected in series through the primary coils G0 or G1.

2 - The condensers being charged in parallel when the brushes are on plates m n and discharged in series when the brushes pass onto plates o.

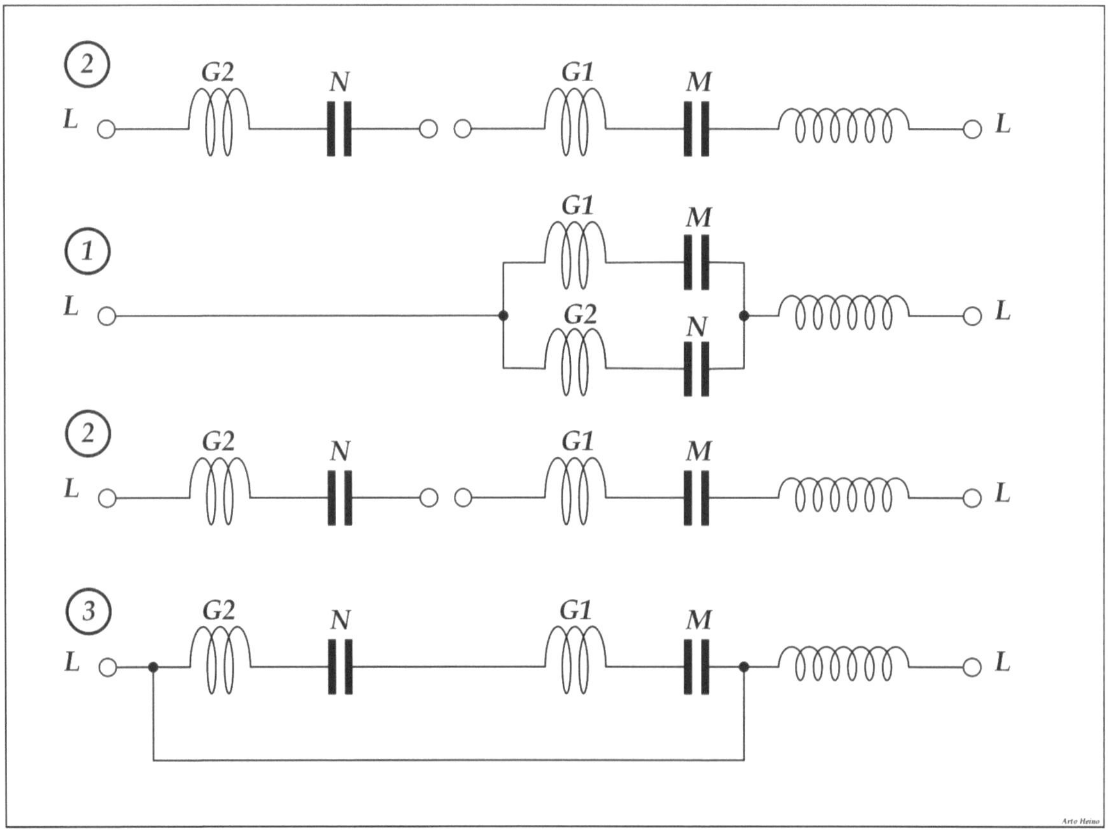

Diagram 2 – These are the basic 3 action as the brushes change the configuration.

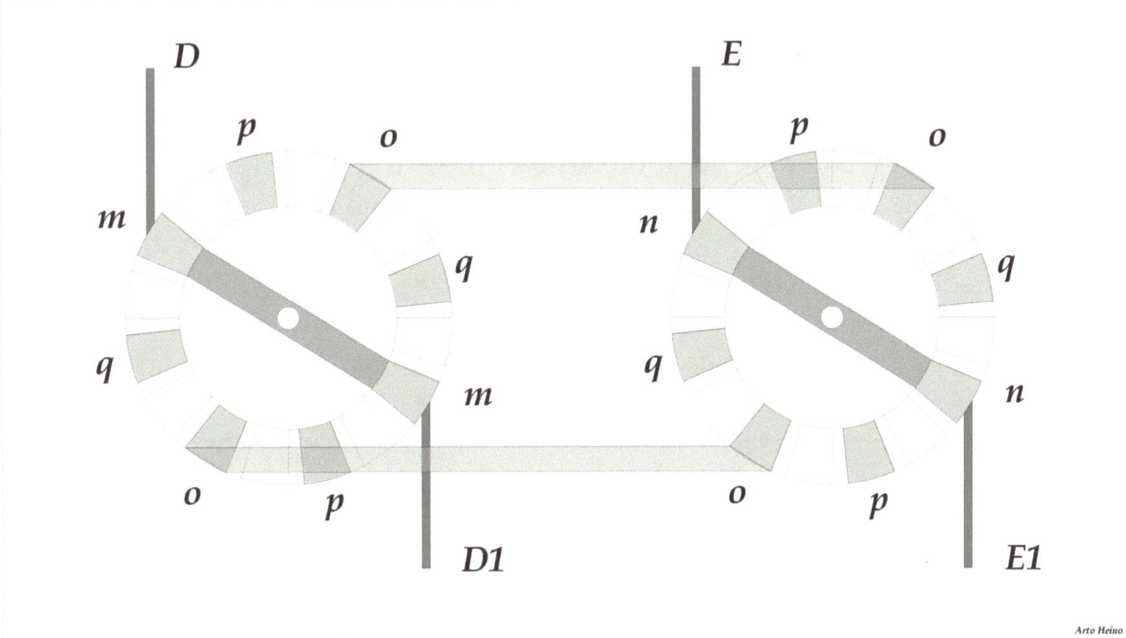

Diagram 3 – The Commutator schematic

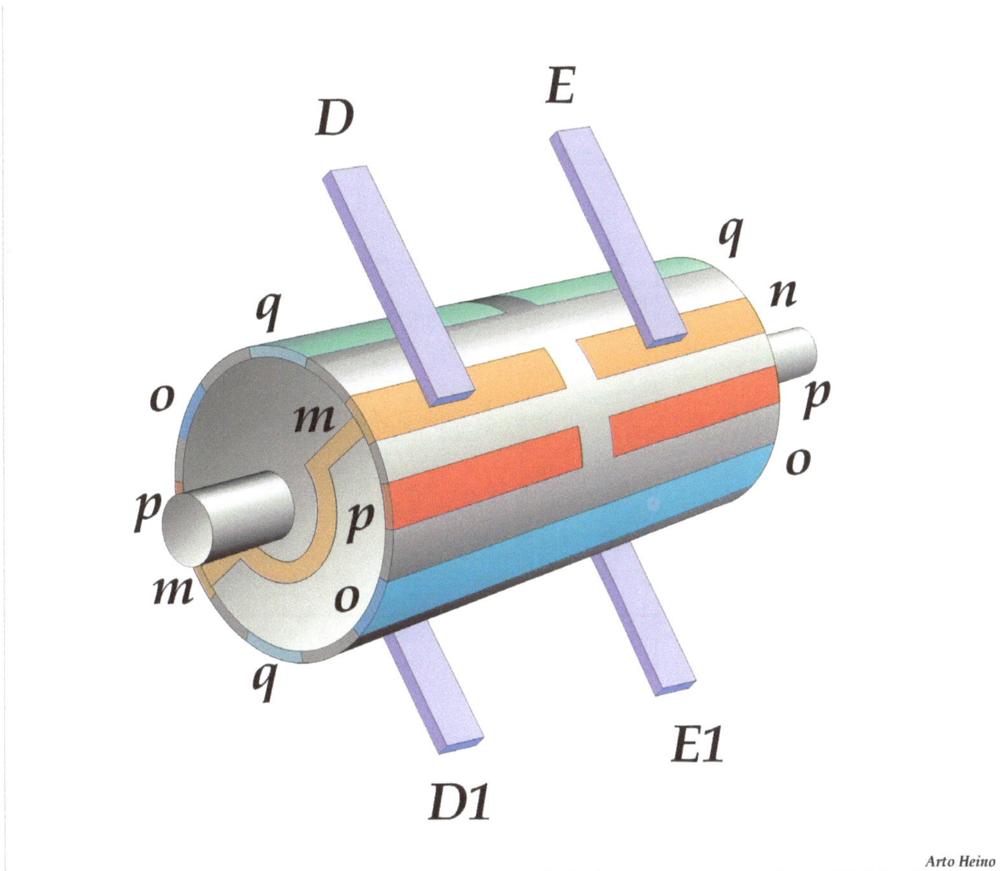

Diagram 4 – How the Commutator is set up in 3D at the end of the motor.

I hope this helps in the understanding of Tesla's methods used to create high frequency and high voltages without transistors, diodes or tubes.

3.5 - Inductors

3.5a - Mutual Inductance

As the layout of any electronics circuit is truly an art form, the mutual inductance's of circuit components and coils become extremely important, I decided to embark on a simple way of discovering the hidden mutual inductance of coil layouts. As I have seen this art is not of importance to most electronic technicians of the general engineering community, only specialists in high frequency, microwave, and microelectronics have explored its consequences. Noting the pioneer Nicola Tesla considered this as a part of the mastery of the electrical art, I am sure most electronic engineers could learn from his examples.

FUNDAMENTAL INDUCTANCE COUPLING FORMS

By Arto Juahni Heino

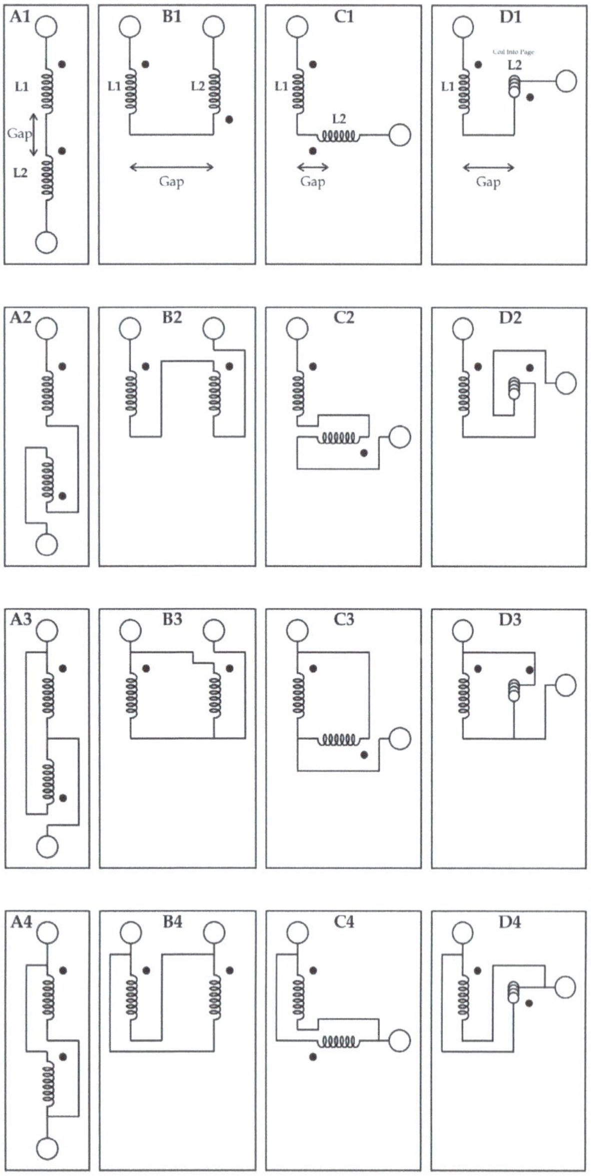

k = Coupling Coefficient

Some Air Core Sample k

	Gap Mm	*k*		Gap mm	*k*
A	6	0.120	**B**	6	0.045
	12	0.080		12	0.030
	24	0.040		24	0.015
C	6	0.010	**D**	6	0.030
	12	0.006		12	0.015
	24	0.002		24	0.0075

Test Coil

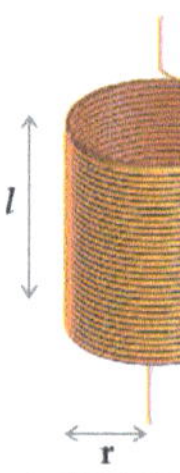

N = 30 turns
l = 0.024 mtr
r = 0.012 mtr
A = π x r²
W= AWG 21

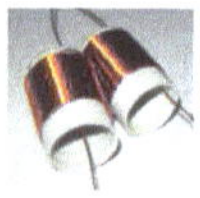

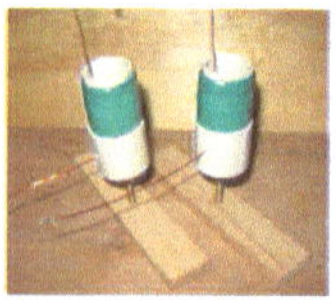

L = (r² x N²) / (228600 x r + 254000 x l)

L = 0.000014662 Henrys
= 14.662 uH

Self Res = 45 mhz .. Approx
Q = 840 (at 45 mhz) .. Approx

Ltest = 15.30 uH   Spacing Apprx = 4-6mm

	Inductance Total A	B	C	D	Coefficient of Coupling A	B	C	D
1	33.65	33.17	32.74	32.37	0.099673	0.083986	0.069934	0.057843
2	29.21	29.80	31.45	31.25	0.045424	0.026143	0.027777	0.021241
3	7.427	7.590	7.920	8.010	0.029150	0.007843	0.035294	0.047058
4	8.546	8.580	7.980	7.870	0.117124	0.121568	0.043137	0.028758

$k = M / \sqrt{L1 \times L2}$ $M0 = \sqrt{L1 \times L2}$

Lm = Measured Total Inductance

To check A1 and A2 Configuration

$k = ((Lm - L1 + L2)/2) / M0$

$M = k \times \sqrt{L1 \times L2}$

$L_{TOTAL}^{A1} = L1 + L2 + 2 \times M$

$L_{TOTAL}^{A2} = L1 + L2 - 2 \times M$

To check A3 and A4 Configuration

$k = 1 - Lm \times (1/L1 + 1/L2)$

$M = k \times \sqrt{L1 \times L2}$

$L_{TOTAL}^{A3} = ((L1xL2) - MxM)/((L1+L2) - 2xM)$

$L_{TOTAL}^{A4} = ((L1xL2) - MxM)/((L1+L2) + 2xM)$

If $L_{TOTAL} = Lm$ then k is correct

ABCD $M = k \times \sqrt{L1 \times L2}$

1 $L_{TOTAL} = L1 + L2 + 2 \times M$

2 $L_{TOTAL} = L1 + L2 - 2 \times M$

3 $L_{TOTAL} = ((L1 \times L2) - M \times M) / ((L1 + L2) - 2 \times M)$

4 $L_{TOTAL} = ((L1 \times L2) - M \times M) / ((L1 + L2) + 2 \times M)$

I am not sure why most physical circuits are laid out so crudely, it could most likely be the educator not emphasizing the need to treat the circuit as a whole unit and explaining the simple fact that the electrical activity has its seat of energy in the space surrounding the conductors, not in the conductors as has been erroneously shown in many inept electrical books.

3.5b - Multi-filar Coils

While making observations of coil designs that have been used over the centuries, I noticed only those that showed some perceived useful results were kept in the arsenal of the scientists, the rest were either ignored or forgotten. There are two basic types and a 3rd being the combination:

1 Circular - Eg Solenoid, flat spiral,spherical, toroidal, mobius

2 Non-circular - Eg Basket weave, geometric, polyhedral, star forms, mobius

The circular types have 3 basic types a resultant 4th and the unique 5th type:

1 Rings adding vertically - Solenoid Coil

2 Rings expanding diametrically - Flat spiral

3 Rings adding in a circular form - Toroidal

4 Rings adding vertically and/or diametrically and/or circular

a) Conical -differing sizes in both direction

b)Multilayer Solenoid -same sizes vertically + differing sizes diametrically

c) Multilayer Flat Spiral - same coils sizes vertically + adding vertically

d) Toroidal and Poloidal -mixed sizes vertically + differing sizes diametrically

e) Spherical rotations -mixed sizes radially + differing sizes diametrically

5 Winding folding into itself, creating a single surface - all possible geometries - Mobuis

The shapes of coils are one aspect the other is the winding order and direction, thus creating multiple poles, cross coupling inductance's, retarding or increasing both the inductance and the capacitance as the design requires, the addition of the mobius windings diversifies the basic options, this is where the art is at in 2012. A good example of combining many geometries and winding types is the:

Magnetic coil (1820 Hans Christian Oersted)

Faraday coil (1831 Micheal Faraday)

Henry coil (1831 Joesph Henry)

Induction coil (1836 Nickolas Callan)

Ruhmkorff coil	(1851 Heinrich Ruhmkorff)
Helmholtz coil	(1869 Herman von Helmholtz)
Cook coil	(1871 Daniel Cook)
Maxwell coil	(1873 James Maxwell)
Flat coil	(1890 Nicola Tesla)
Tesla coil 1/4 wave	(1891 Nicola Tesla)
Tesla Bifilar coil	(1894 Nicola Tesla)
Hubbard coil	(1918 Alfred Hubbard)
O'Leary Coil	(1920 William O'Leary)
Hendershot coil	(1928 Lester Hendershot)
Interceptor coil	(1946 John Wiegand)
Tokamak coil	(1950 Oleg Lavrentiev)
Stellartator coil	(1950 Lyman Spitzer)
Smith coil	(1952 Wilber Smith)
Fusor coil	(1964 Philo Farnsworth)
Hooper coil	(1968 William Hooper)
Rodin coil	(1986 Marco Rodin)
Biaxial Poloid coil	(1990 Bo Atkinson)
Tetra Helix coil	(1991 Bo Atkinson)
Vortex coil	(1992 Ken Gailey)
Double Helix coil	(TM) (2007 AML)

Not to mention my own explorations in coil geometry with:

Mandela coil	(1990 Arto Heino) –
Crown coil	(1994/2011 Arto Heino) –
Infinty coil	(1993/2011 Arto Heino) –

Many toroidal/polyhedral mixtures that have been popping up regularly with names like the Starship coil, Loohan coil, Polish coil, Big Secret coil, some of these are marvels of artistry and human innovation. The properties of these coils have not been explored by any means, by and large these will be no more than a geometric conformation of magnetic and dielectric shapes, only usable if well understood and in the correct placement to other electrical activity. These types of decorative coils were in the creative arsenal of jewelry makers since the first dynasty of Egypt.

Not to mention the many non inductive windings that are used as resistances. This has been looked at for at least 200 years, Nicola Tesla was one engineer/Scientist who's original thinking left a legacy of numerous winding geometries he used to design his motors and coils,one of these is his bifilar design. This design exposed a unique addition to the idea of combining the merits of both coil and capacitor without addition hardware contrivances, the anti-series winding nulls out self-inductance.

Nikola Tesla writes that a standard coil of 1000 turns with a potential of 100 volts across it will have a difference of 0.1 volt between turns.

100/1000 = 0.1

A similar bifilar coil will have a potential of 50 volts between turns.

100/2 = 50

In that the stored energy is a function of the square of the voltages, the ratio of energy in the bifilar to a standard coil will be

E	= ½ C x V^2
sqrt(E/ (1/2C))	= V
if V	=50
50^2/0.1^2 = 2500/0.01	= 250000

Which as stated by Tesla, "the energy stored is 250000 times greater than the standard coil!"

The storage of energy when pulsing a bifilar coil is n amount more than a standard solenoid/flat coil, as Tesla stated and gave you the formula to engineer this type of coil, we should be incorporating this simple idea into many everyday devices. I have drawn a few design methodologies that I hope will spur this well known innovation into more interesting directions. Remember this coil will have one frequency when the inductive reactance and the capacitive reactance will equal each other , thus give you a resonant frequency that will work to magnify the voltage/amperage when designed correctly.

These are basic bifilar solenoid configuration showing the capacitance between each winding. The windings can be rearranged in a different order to change the amount of induction and capacitance you put into the windings of the coil. The winding below shows what can be done to rearrange the capacitive differential to increase the energy storage.

The graphic table below shows how many different arrangements that can be used with a Quadrifilar coil system, each will give a different result for the inductance, capacitance and its resonant frequency spectrum. A great science/engineering project or University paper for someone would be to tabulate and analyze each arrangement, you might just get your doctorate on this.

The flat coil is a dimension less than the solenoid when using round conductors and a lot easier to redefine into mufti-filar arrangements.

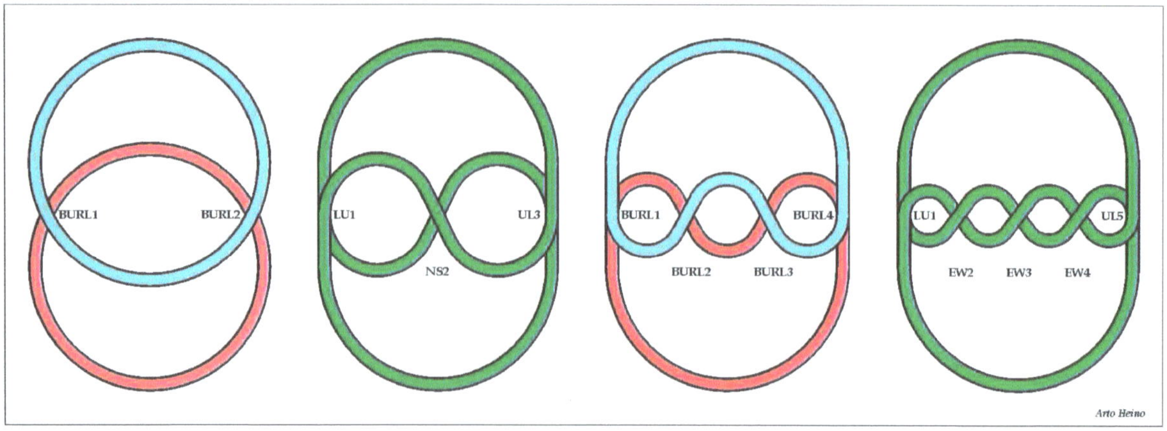

The basic geometries involved in designing coils, transformers and bifilar systems must have a few design rules so the engineering of these items become straight forward and adaptable to the needs of the energy transformations. Here are a few simple systems that can be easily expanded and used creatively.

The cipher I am using here is:

B = Blue *R = Red*

U = Upper *L = Lower*

N = North *S = South* */ is the upper part of the cross*

E = East *W = West* *\ is the upper part of the cross*

The order of the letters is the order from the top to the bottom.

The order of the numbers is from the left to the right.

The first set is based on 2 simple loops, refer Diagram 1. This will give you 8 different forms. The next is the topology of a single loop of wire which loops on itself, like a bifilar or just a plain solenoid:

As you can see by these configurations, the standard engineering practice used in industry only use a small portion of these geometries. Now if you loop 2 coils as used in transformers you will see industry only apply 1 or 2 of these geometries in 98% of all transformers.

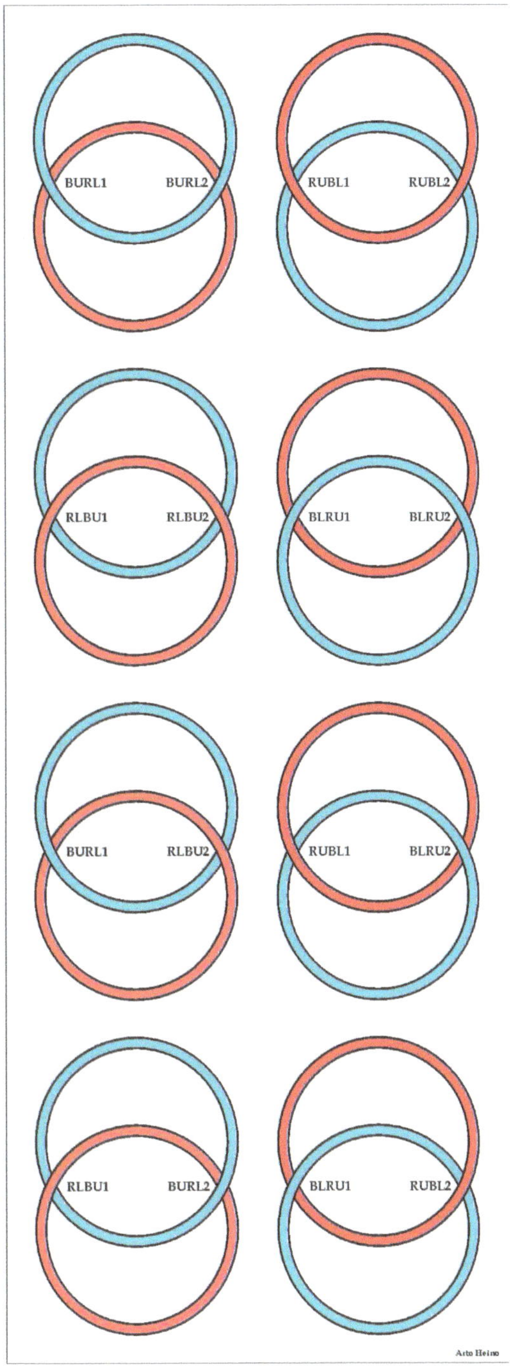

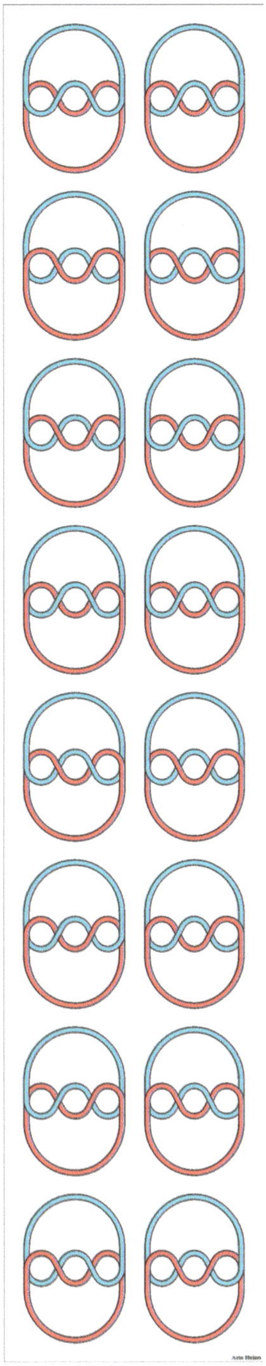

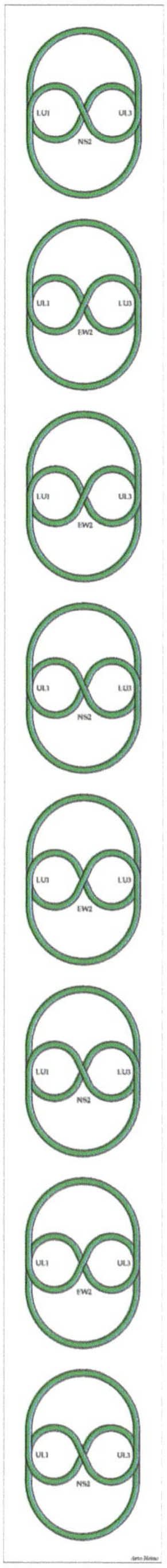

3.5c - The Bifilar Coil Motor Unit

The fusing of Adams motor and the Tesla bifilar coil was one of the first iterations that showed me the correctness of this approach by its exceptional efficiency when all the parameters were engineered in balance.

Below is the original stator coil I wound in 1999, it performed well and required low voltage(12v) and low current(0.03 amp).

LEFT COIL

A1	
A2	B1
A3	B2
A4	B3
A5	B4
A6	B5
B1	A2
B2	A3
B3	A4
B4	A5
B5	A6
B6	C1

RIGHT COIL

B6	C1
D1	C2
D2	C3
D3	C4
D4	C5
D5	C6
C2	D1
C3	D2
C4	D3
C5	D4
C6	D5
	D6

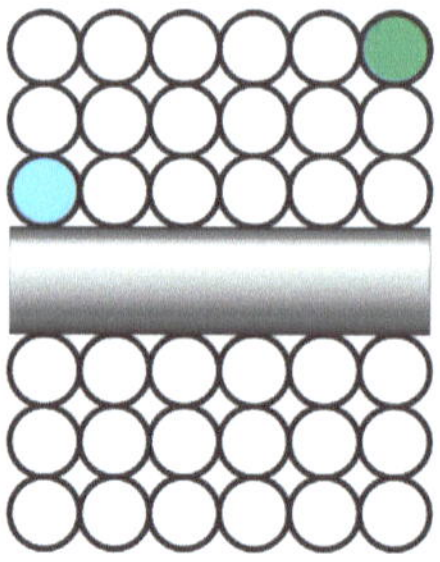

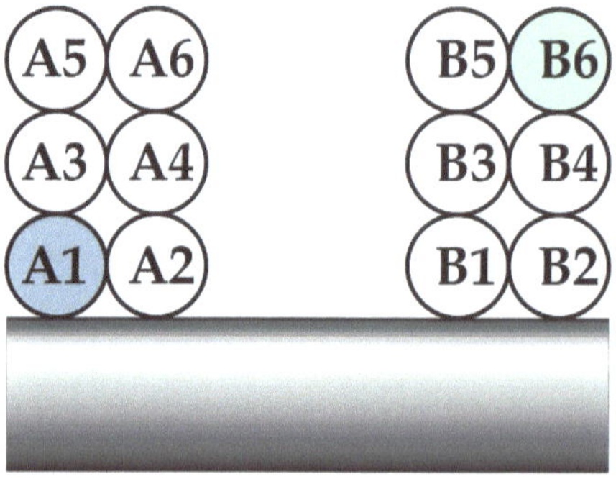

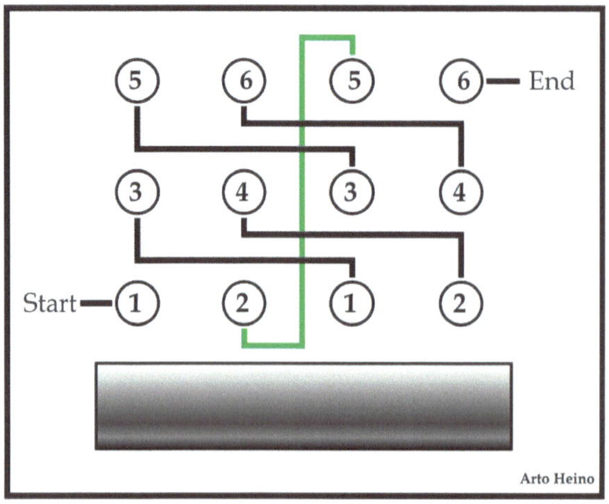

After delving into the unique geometries available , I found the basic Bifilar coil has multiple usage and can substitute for almost any other circuit element. Without pushing this too far left field let me show you the simple logic of my statements.

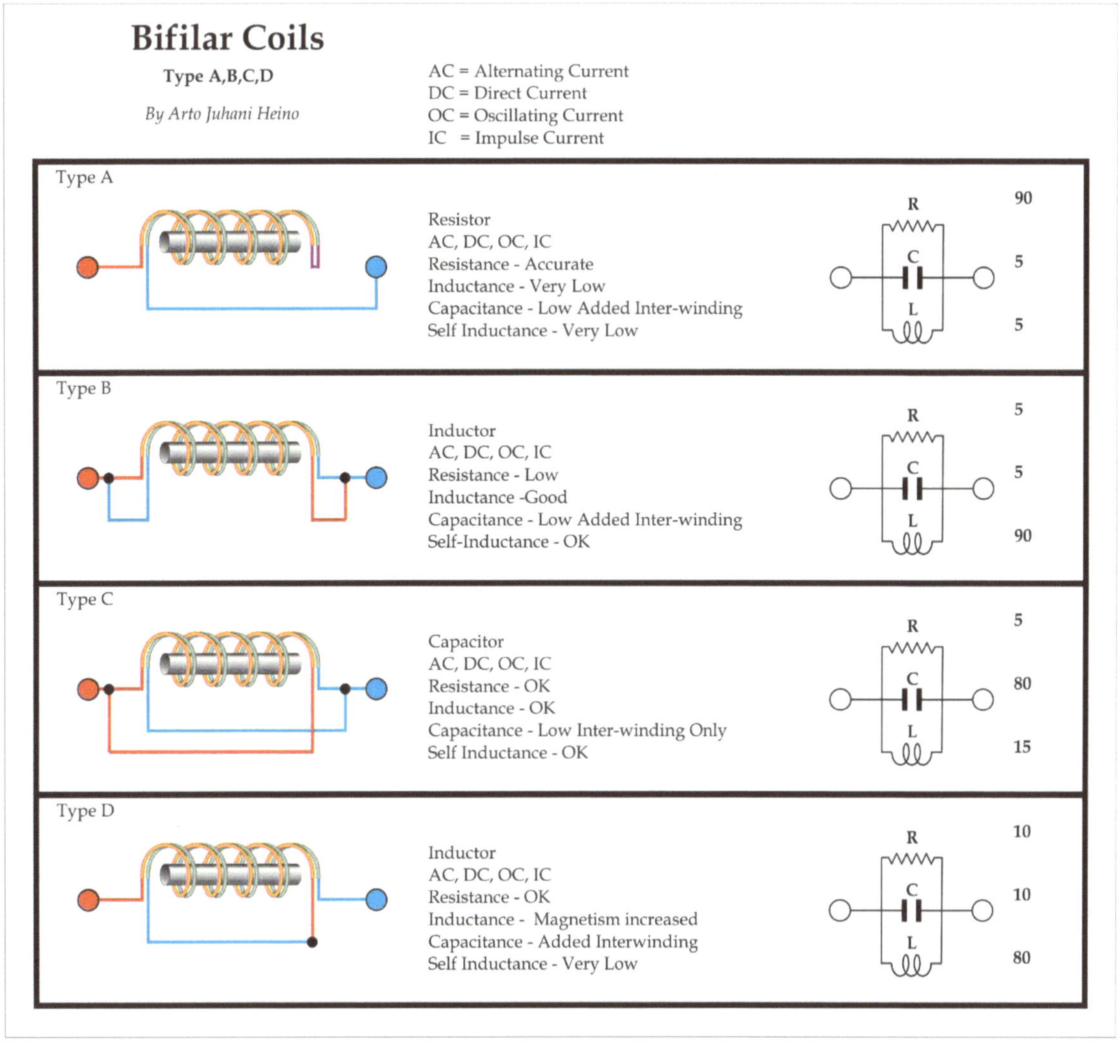

The basic four types can replace the Resistor, Capacitor, Inductor and a unique high magnetic Inductor. There are more types based on the transformer action inherent in this design, the next image is the first of this type.

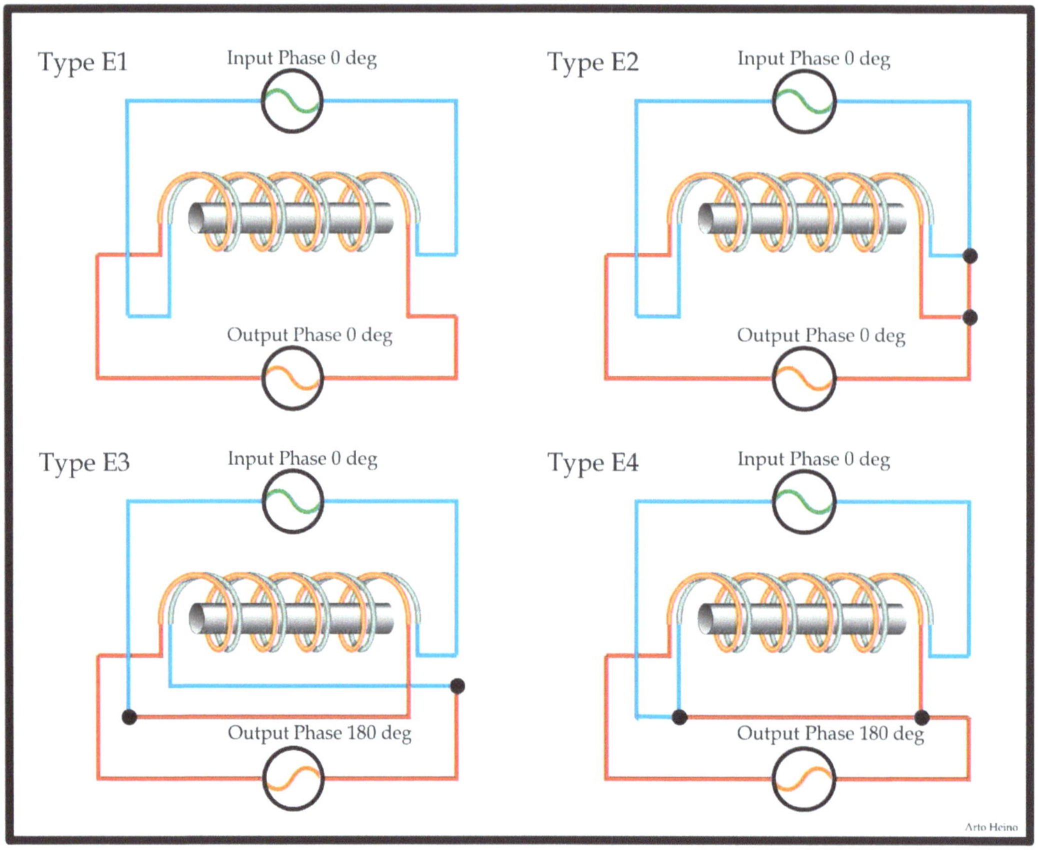

3.5d - The Magic Square Coil Technology

Incorporating both bifilar and loop transformations as shown previously, you can begin to appreciate the transformations I have to create a magic square coil, as originally shown on my Artoworld website in 1999.

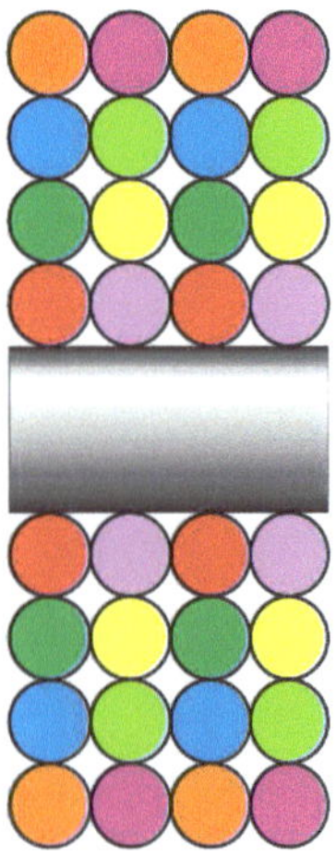

15	9	2	8
6	4	11	13
12	14	5	3
1	7	16	10

1	7	16	10
12	14	5	3
6	4	11	13
15	9	2	8

Most of the original readers of these documents never understood my topology in 1999, even though I built many of these coil constructs as stators for pulsed motors. The current breed of experimenter has less inhibitions to these geometric forms.

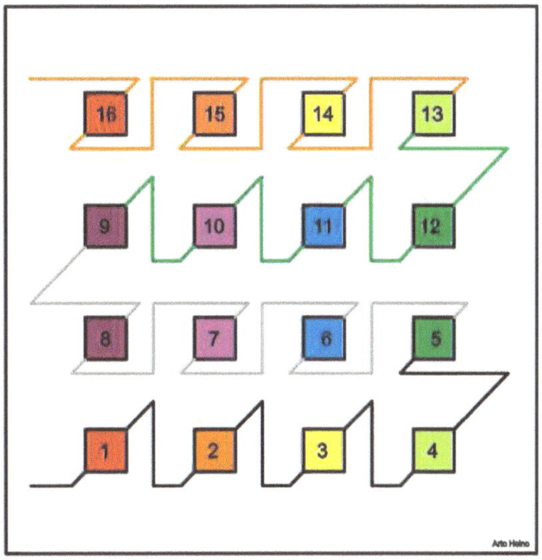

The first arrangement above is the non-magic numeric sequence, just the same as if you wrap a standard solenoid.

The next diagram is the magic square arrangement, which utilizes the the pairing of the "17" as a magic square paring which translates to the bifilar pairing.

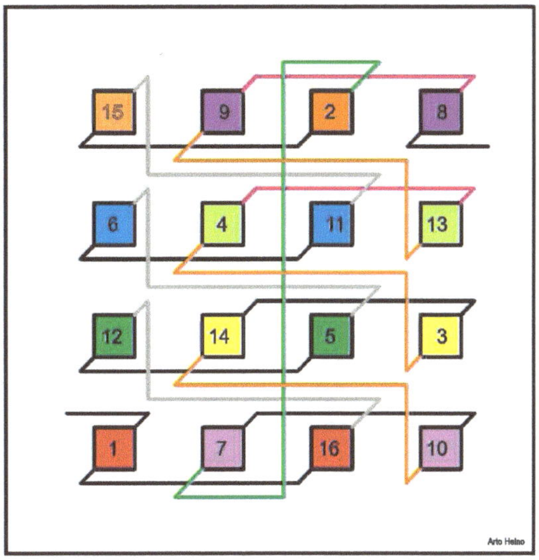

This last arrangement is also non magic sequence which is known as back loop winding, when the winding ends a loop goes back to the start position ready for the next winding loops.

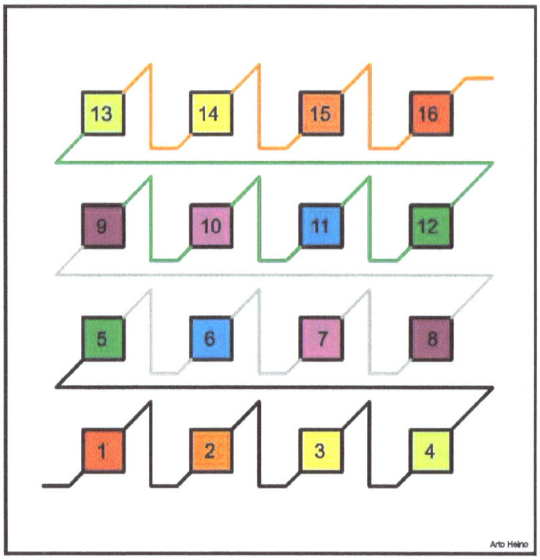

Old photo from 1999

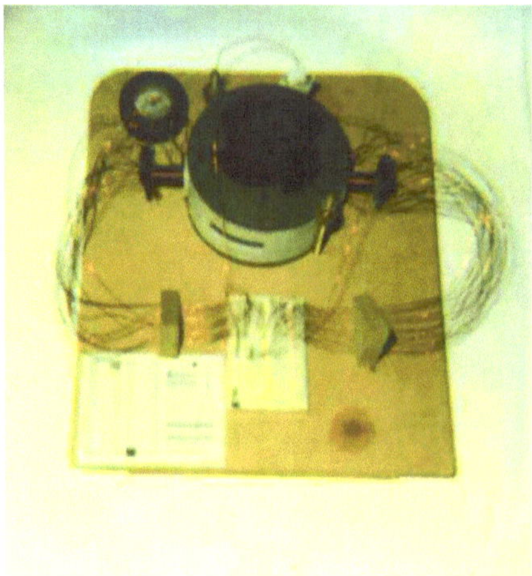

Back in 1996 I was working hard on winding many stator coils for the Robert Adams pulse motor designs, is when I realized the few coils I wound in bifilar geometries seem to function with non linear results. This is when I discovered the varieties of subtle refinements that could be engineered into a basic system, that can give measurable increments of control over the energy storage of the coil while it is pulsed.

Estored = 1/2 * L* I^2 = Energy in Joules stored in the Inductance

Estored = 1/2 * C * V^2 = Energy in Joules stored in the Capacitance

I chose the Magic Square of Four as the prototype for the coils design, refer to the first diagram that reveals my logic in a simplified form. The left stator coil is the left half of the Magic Square, this is how magic squares work, by the mirror reflections that balance the numeric relationships to create harmony, where the wiring is now the pathways that have both Inductance and Capacitance is the same internal geometry as a magic square, the right stator is now the right side of the magic square.

To understand the significance of creating chaotic systems that the outcomes cannot be easily predicted, is to look into the heart of a living organism. The number of square four arrangements available to the constructor becomes 880 basic Magic Square Four patterns. If you expand the system you could have 4 coils each with a complete magic square arrangement giving you millions of 880 combinations. The design parameters can be expanded to include odd squares and very large magic squares. The coils could also be split into the columns and rows giving you say 5 coils each with a in 5 and out 5 winding scheme, this will give you 25 separate wires that give you a square 5 arrangement.

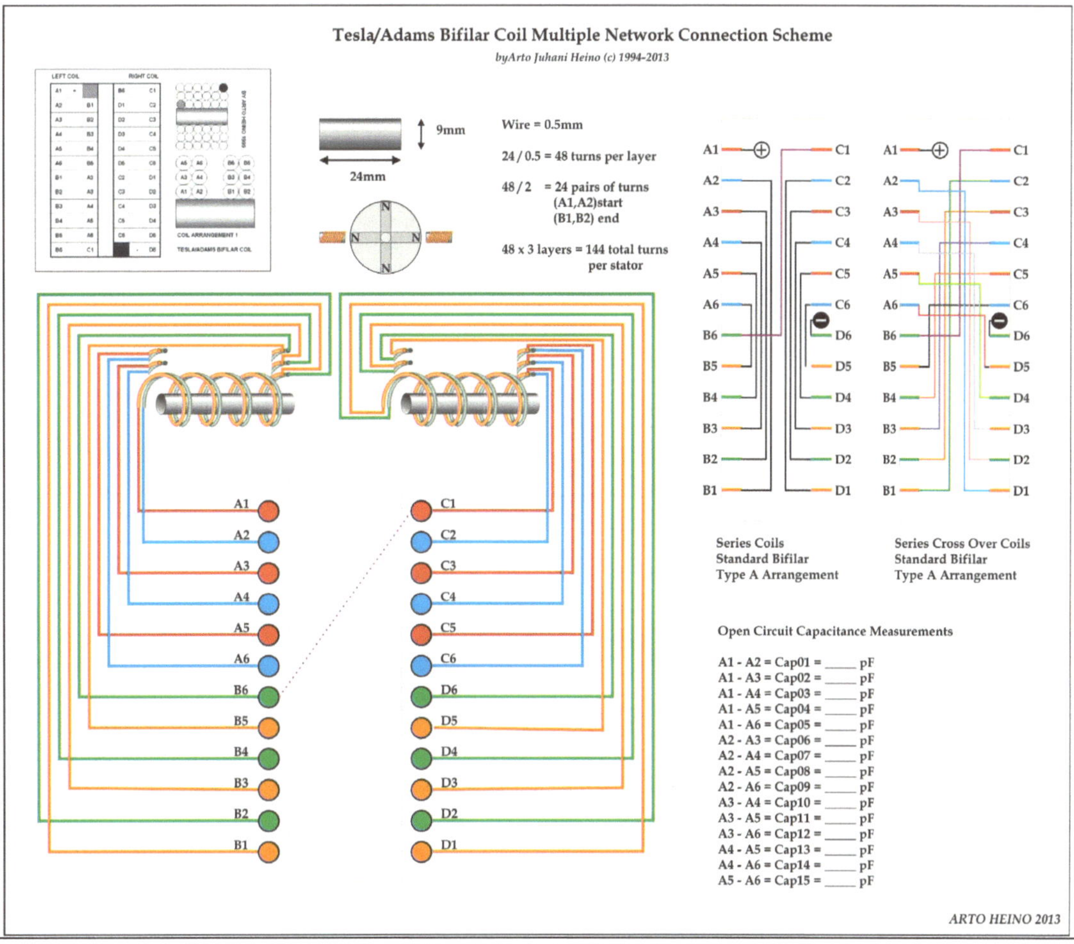

The main criteria for any success is to keep all wires coming from the stators the same length and short as possible, a circuit board design would keep any stray Inductance and Capacitance to a minimum. A small circuit board could be at both ends of the stator with holes matching the wire ends and another attached between the stators, where you could mount a patch board or a special wire coil arrangement.

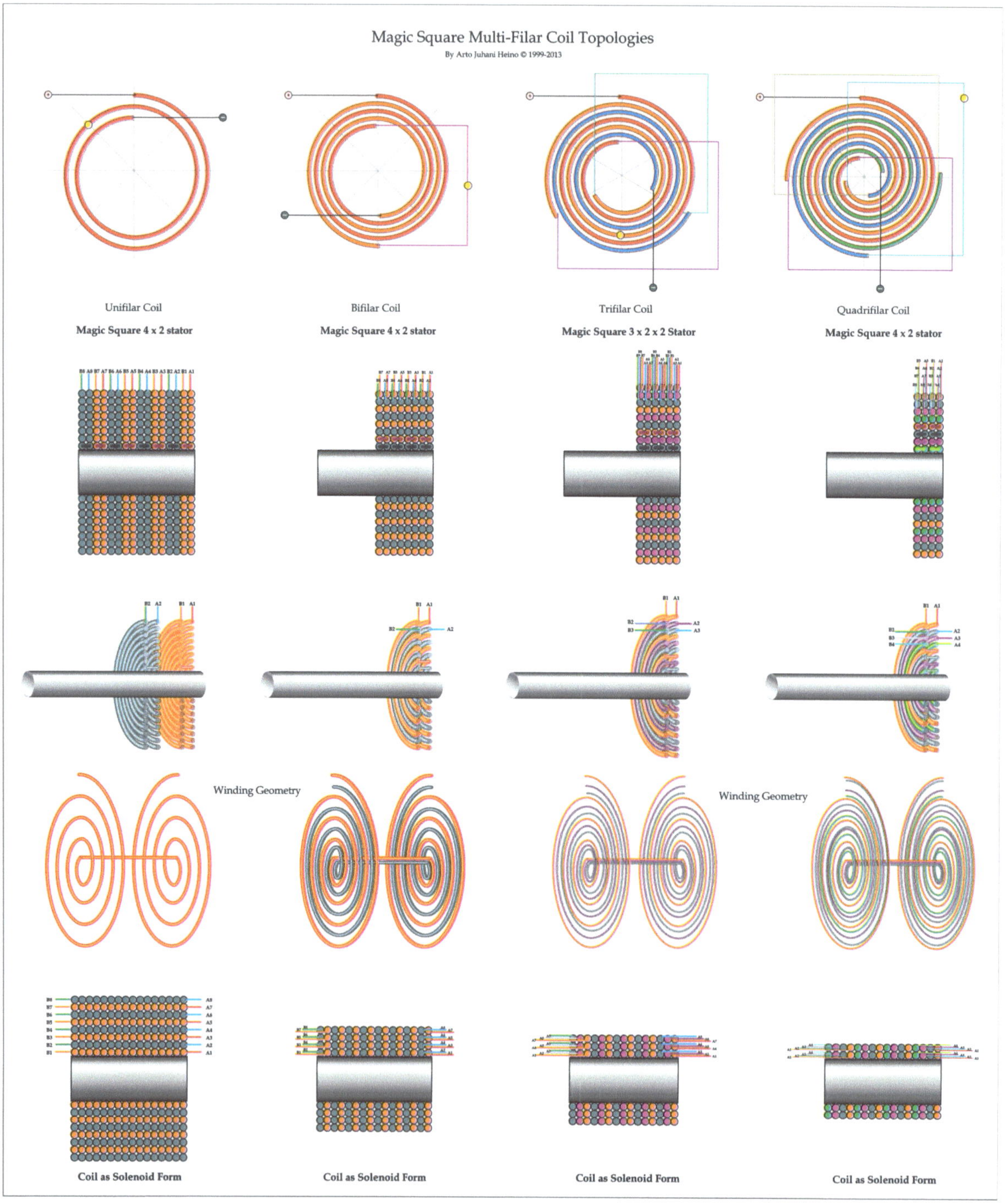

The environmental energy that exists in a chaotic form surrounds every object, a tumult of electrostatic, EM radiation , heat, dielectric charge fields and magnetism, this energy is not available to be absorbed into the circuit unless we can create a couple to its random behavior, and make a gate so the circuit can make use of it.

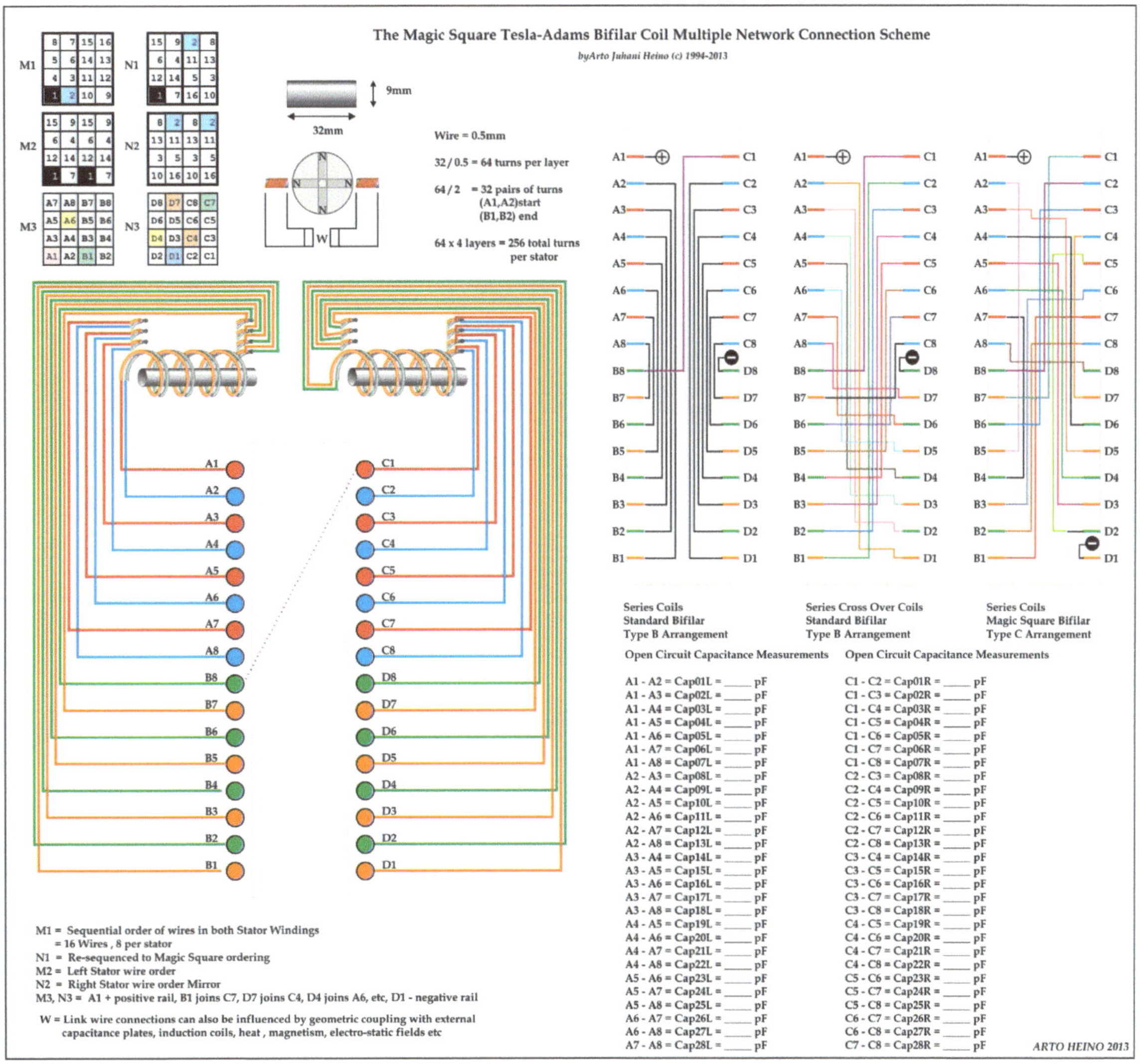

This is the basis of a natural system based on natures gift to humanity Magic Squares, they are but discoveries of a natural system of numbers that can be translated in many different ways. The chaotic nature of magic squares gives you the invisible harmonic coupling that needs to exist in a cyclic system of energy exchange, the gating can now be engineered in the form of magic squares.

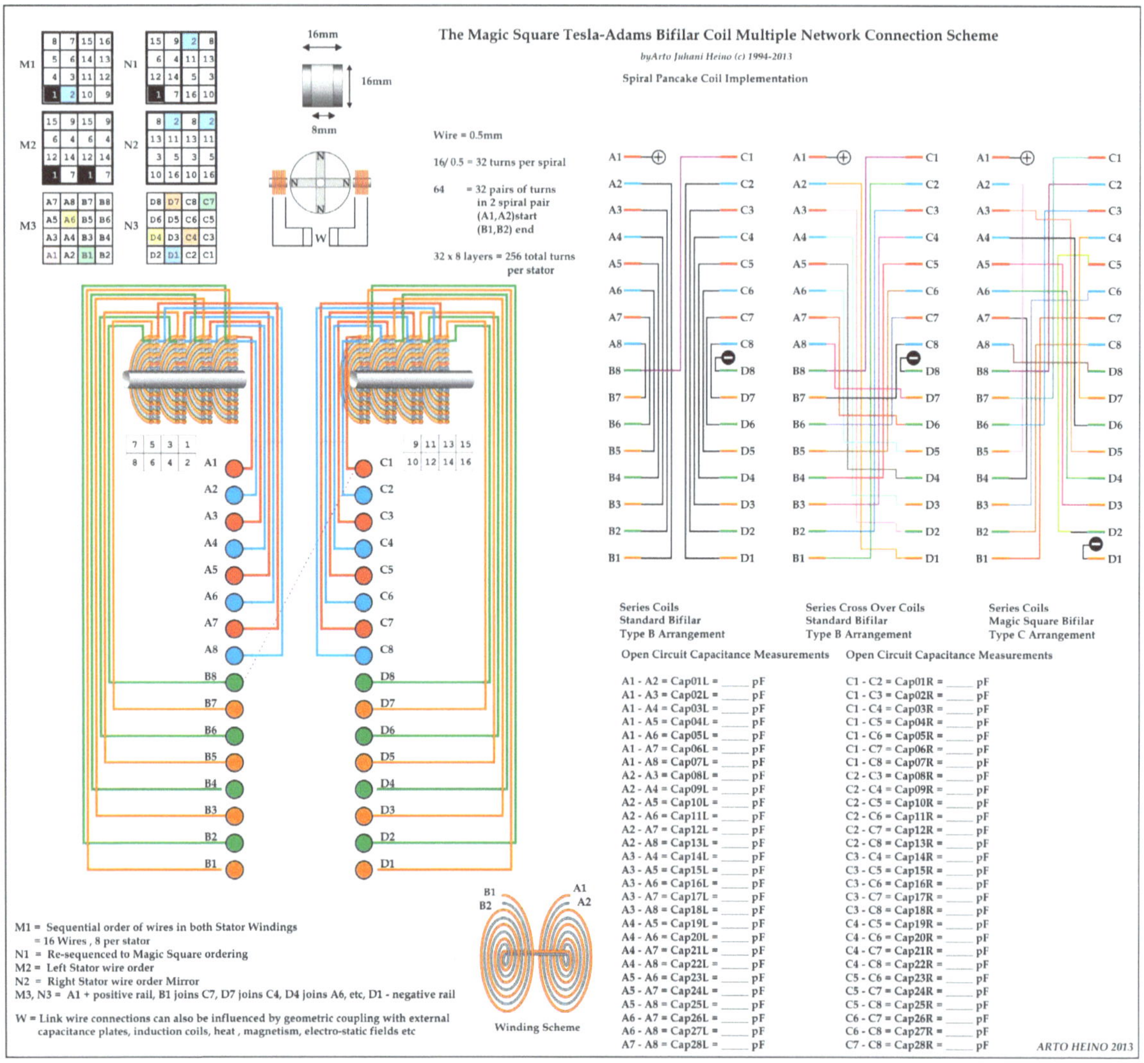

3.5c - The Infinity Coil

When I was working on the many forms of the Robert Adams motor in the 1990's , I was happy to fulfill the basic tenants of the original Robert Adams design. The clear revelations of a simple switched reluctance motor with parametric amplification factors that can go into resonance with itself and the larger environment gave me inspiration in the form of a double coil in the figure eight with two rotors running in opposite directions and two sets of pulse coils.

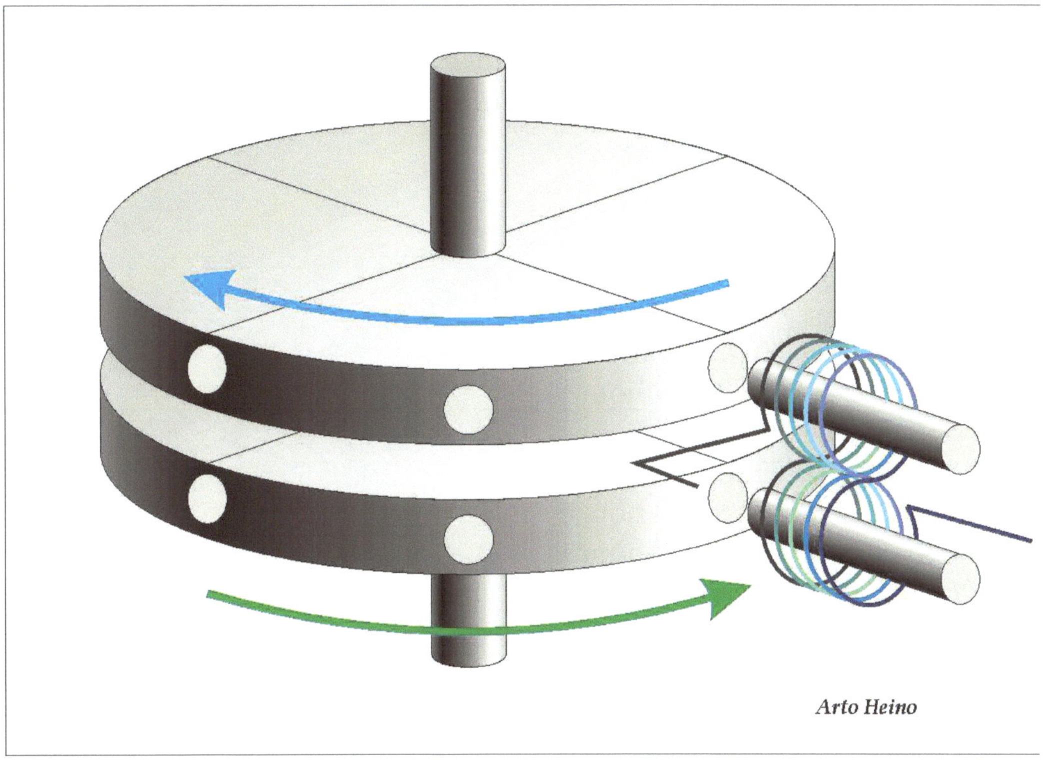

After many years I decided to return back to these designs as they merit greater experimentation. The original Infinity coils were two solenoids in parallel wound in a figure eight fashion, this would also work on two toroid stacked in parallel also adding the bifilar winding scheme, this became my final design as shown below.

As to the functionality of this coil, I put together a simple oscillator circuit, that only uses a 0.5-3volt source and lights a MR16 3W light for about 10 hours starting at 60% illumination, then dropping to 15% at the 5 hour mark, the duty cycle of the blocking oscillator was about 25% as shown below.

The basic equation does not allow for the duty cycle as such;

3 watts x 0.6 light output x 10 hours = 18 watt hours

with the duty cycle added, it should be

3 x 0.6 x 10 x 0.25 Duty cycle = 4.5 watt hours

This is now accounting for the blocking oscillator duty cycle of approx 25%. The figures still looks suspicious to me, as stated the 60% illumination, this was at the start of the run, at the 5 hour mark it was closer to 15%, this means it was running down just like all batteries do. Here is the new equation, a bit closer to the observations at the end of the run.

Let H = 1/sqrt(2) = reduction of illumination per hour

3 x 0.6 x H^0 x 0.25 = 0.450 first hour
3 x 0.6 x H^1 x 0.25 = 0.318 second hour
3 x 0.6 x H^2 x 0.25 = 0.225 second hour
3 x 0.6 x H^3 x 0.25 = 0.159 second hour
3 x 0.6 x H^4 x 0.25 = 0.112 second hour
3 x 0.6 x H^5 x 0.25 = 0.079 second hour
3 x 0.6 x H^6 x 0.25 = 0.056 second hour
3 x 0.6 x H^7 x 0.25 = 0.039 second hour
3 x 0.6 x H^8 x 0.25 = 0.028 second hour
3 x 0.6 x H^9 x 0.25 = 0.020 second hour

This gives a total of 1.488 far from 18 or even 4.5 watts

The COP the becomes 1.488/2.6 = 0.5724 this looks more realistic.

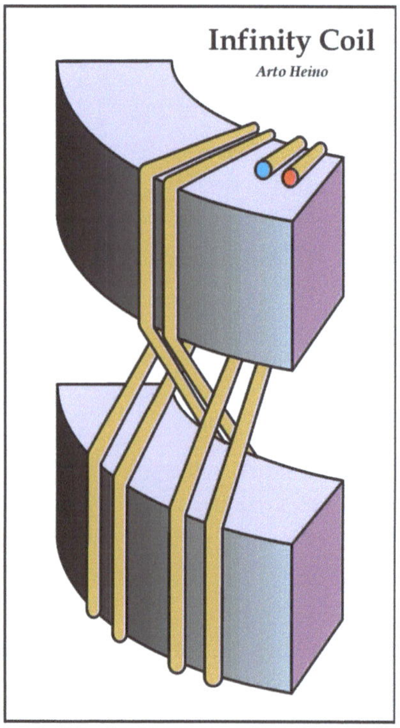

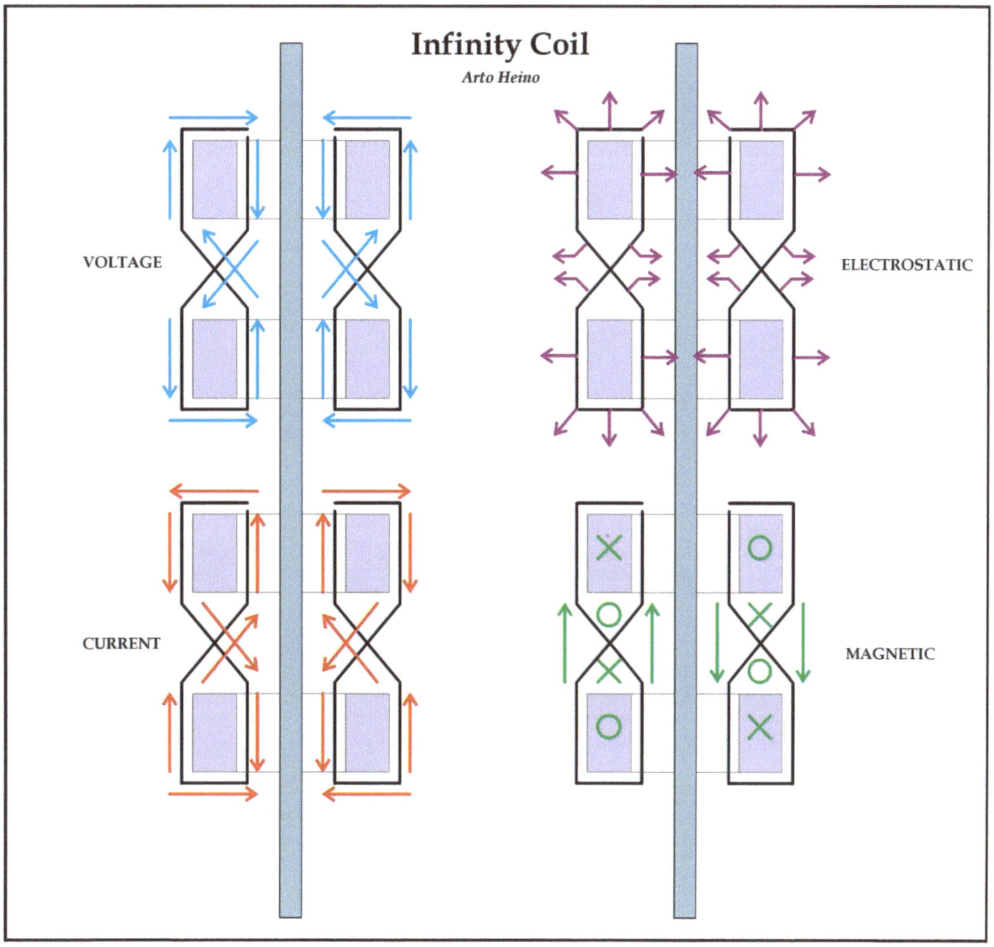
Infinity Coil
Arto Heino
VOLTAGE
ELECTROSTATIC
CURRENT
MAGNETIC

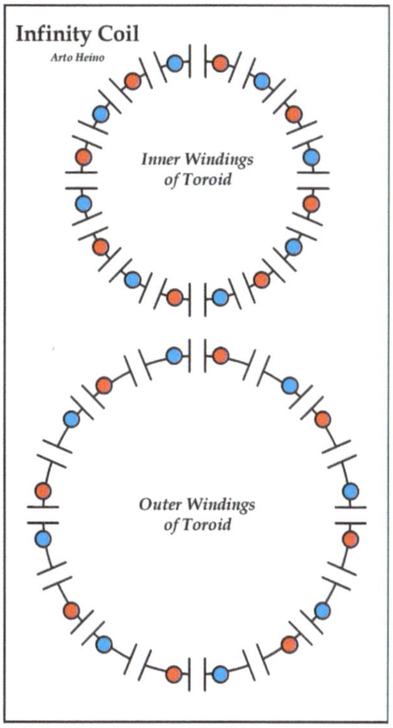
Infinity Coil
Arto Heino
Inner Windings
of Toroid
Outer Windings
of Toroid

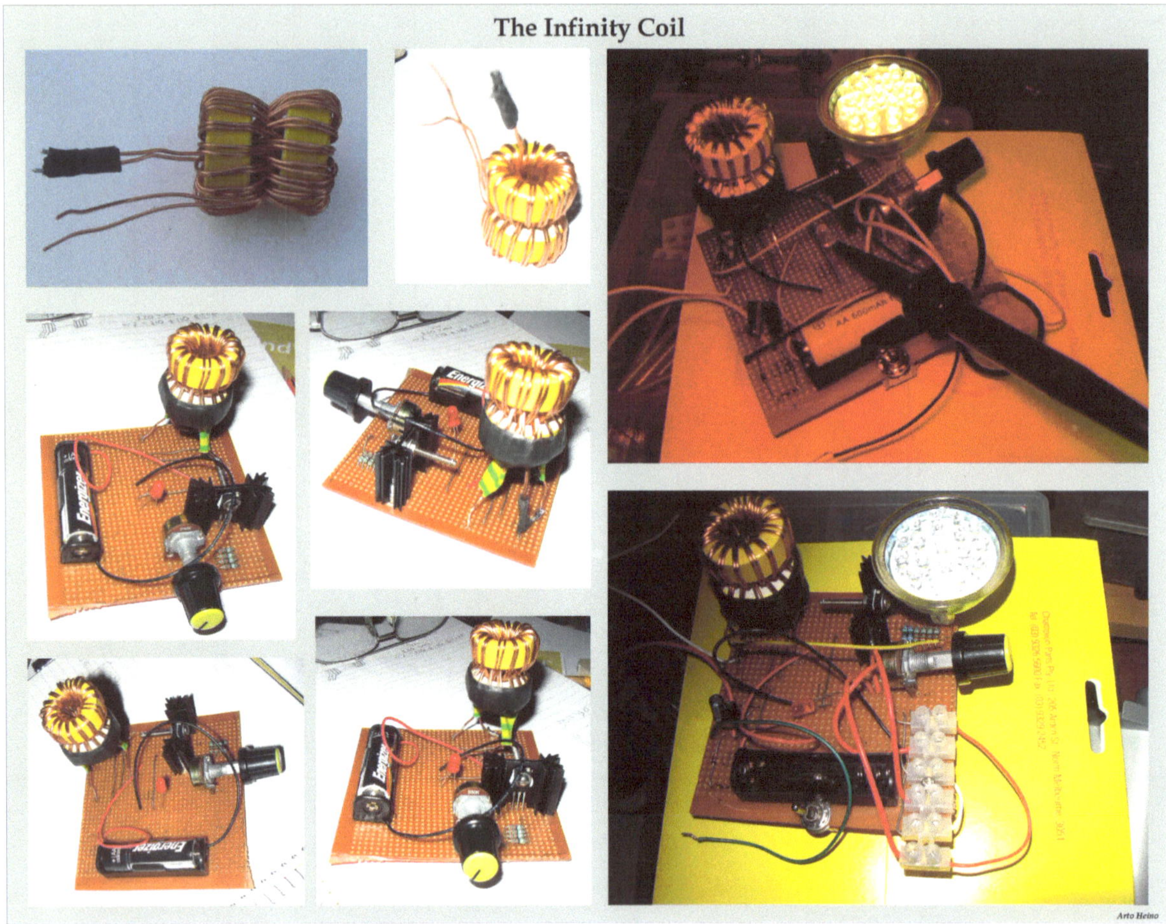

The Infinity Coil

3.6 - Capacitors

3.6a - Water Capacitor

The more unusual design for a capacitor came out of a need to get the most voltage into the smallest package that was build-able from standard industrial supplies. I call it the Cylindrical Capacitor Cell.

I never actually built this unit, its was an exercise to see what could be built with off the shelf parts and still achieve a high density.

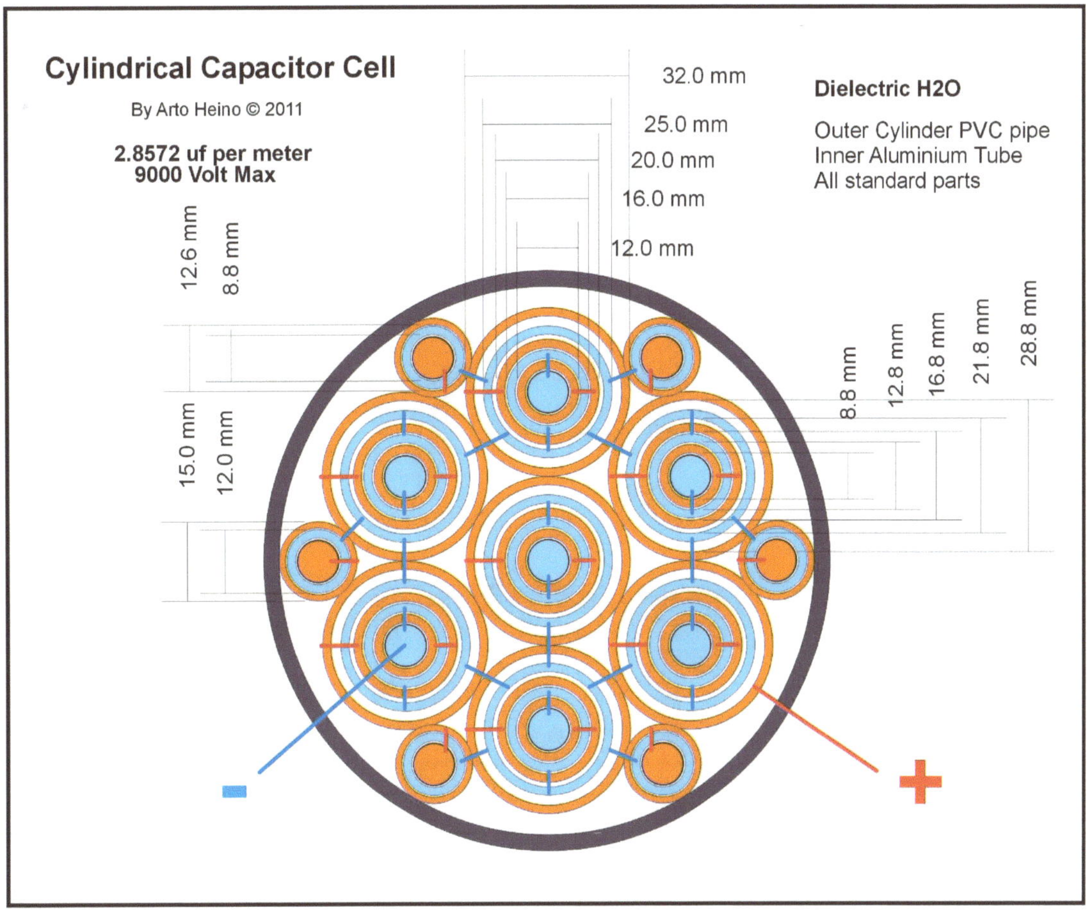

3.6b - Air Capacitor

Another capacitor that was actually built and used for experiments was the massive Air Core Capacitor a 100 plate 6 000 volt unit. This unit varied in capacitance between 5nF to 15nF, graduated with a dial that was calibrated by 3 capacitance meters and mathematically accurate within 1-2% of prediction. Although this was built by hand it was a precision unit.

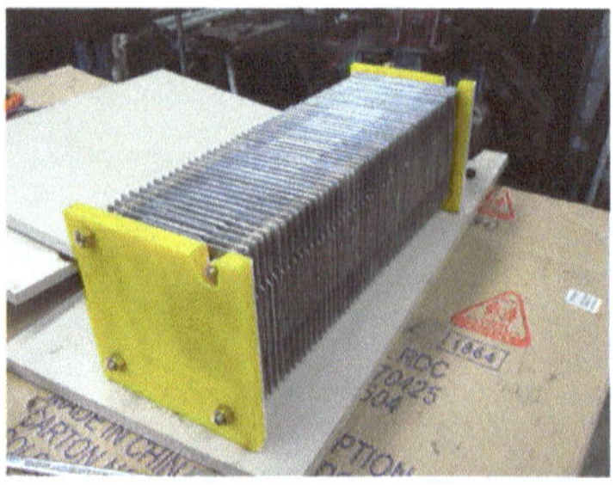
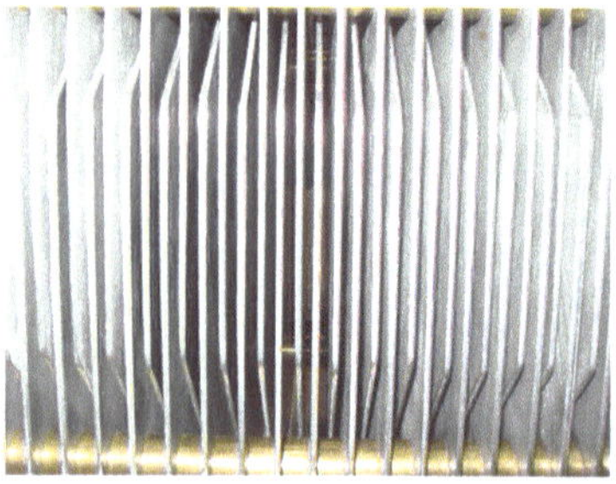
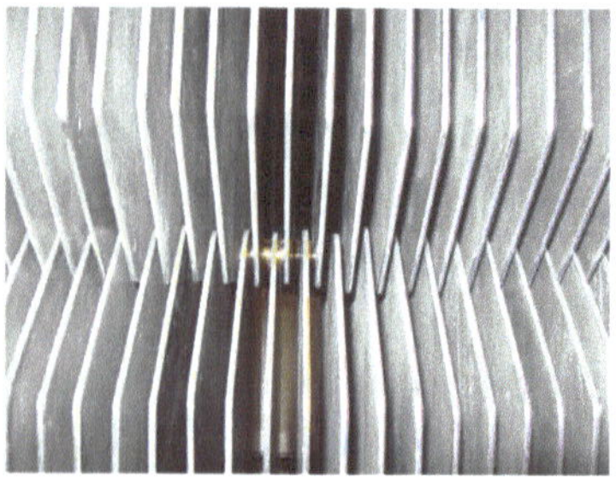
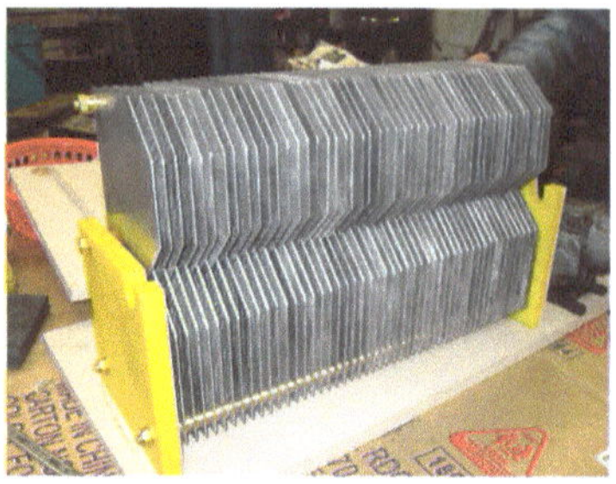
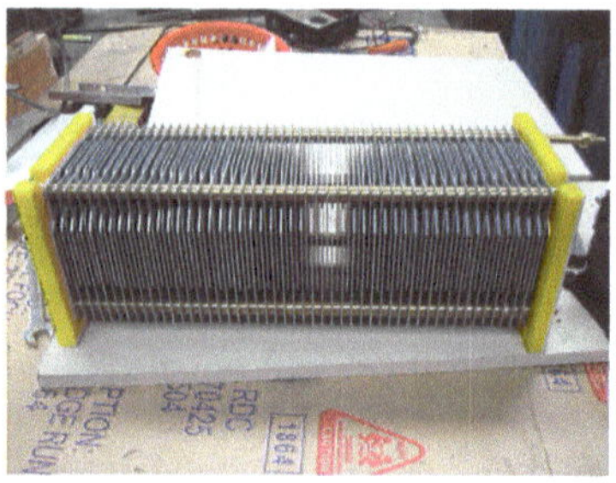
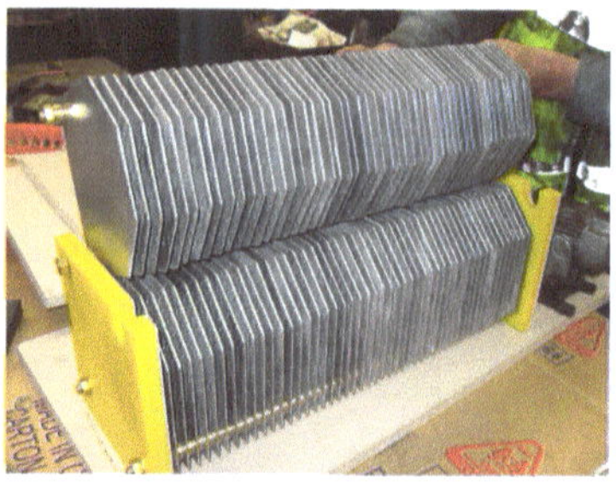
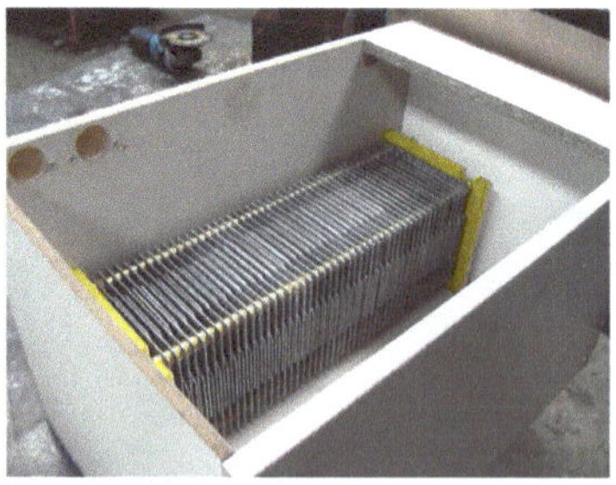
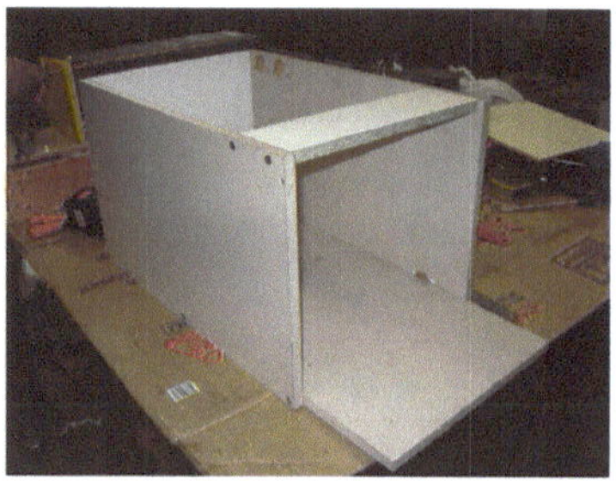
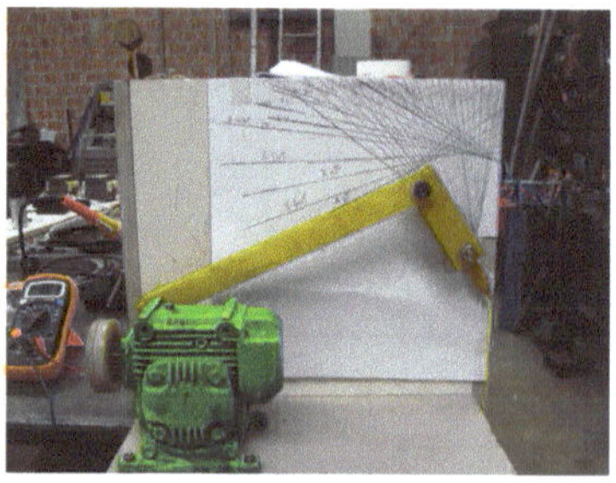
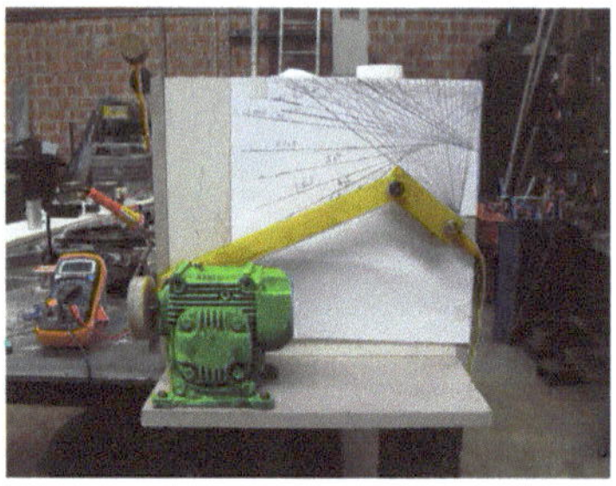
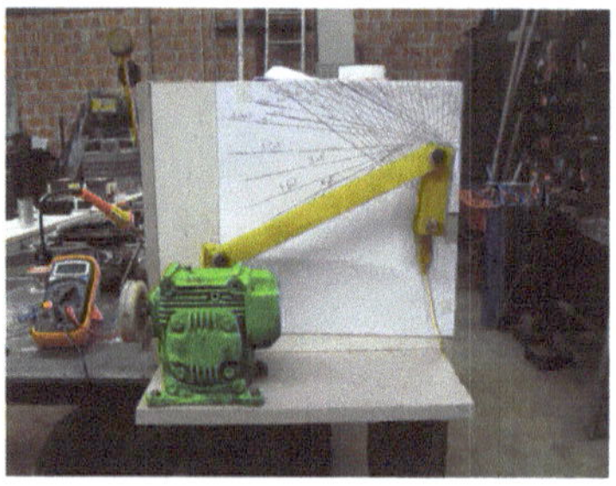
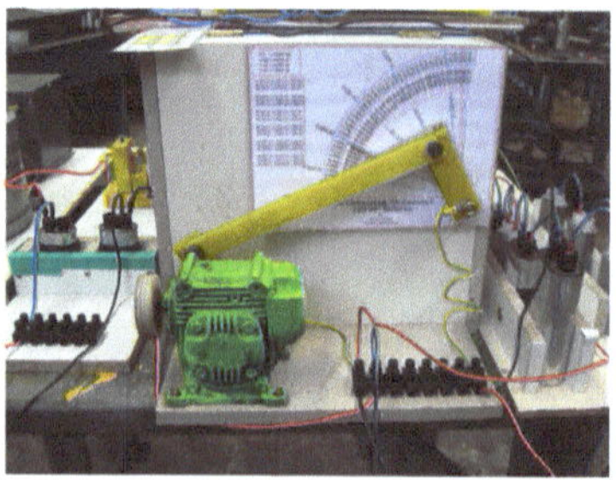
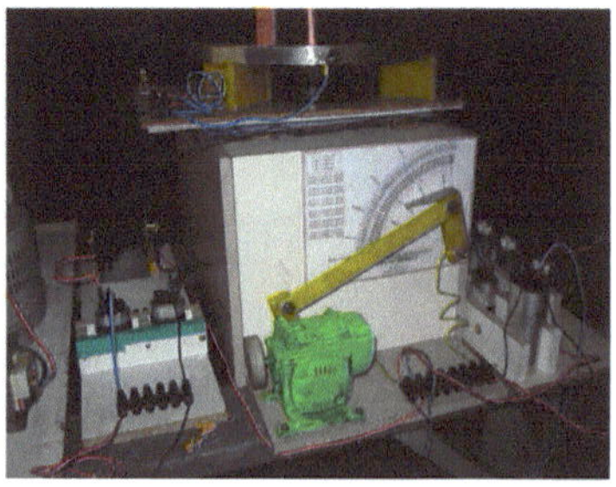
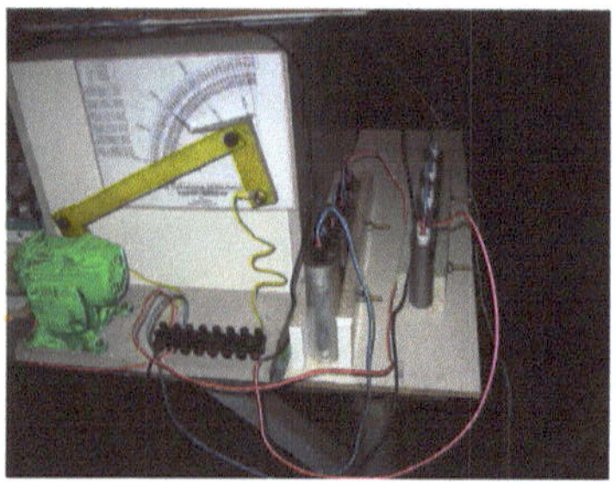
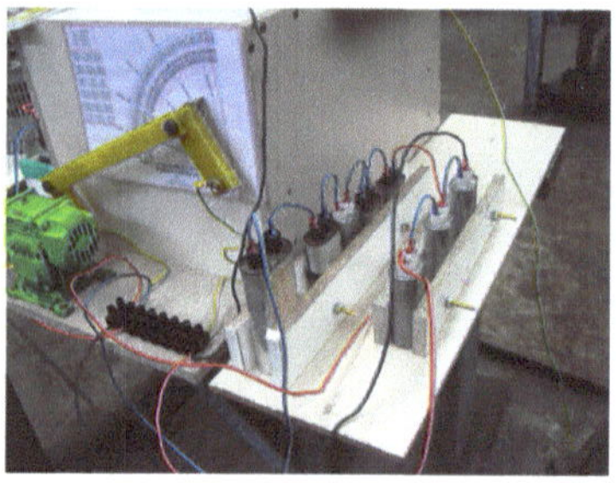

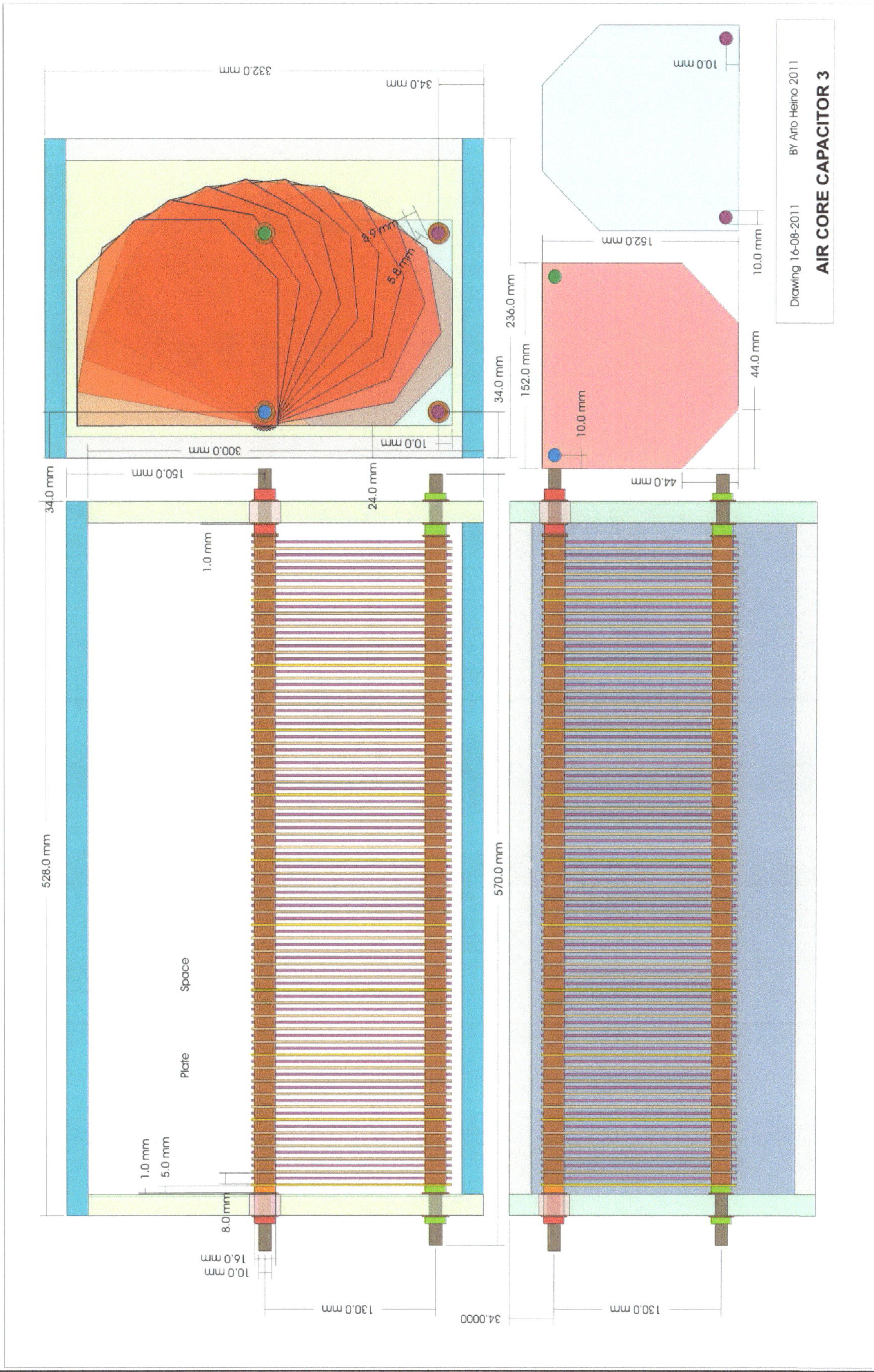
AIR CORE CAPACITOR 3
Drawing 16-08-2011
BY Arto Heino 2011
332.0 mm
34.0 mm
300.0 mm
10.0 mm
150.0 mm
24.0 mm
34.0 mm
236.0 mm
152.0 mm
34.0 mm
152.0 mm
10.0 mm
10.0 mm
44.0 mm
44.0 mm
10.0 mm
8.9 mm
5.8 mm
528.0 mm
570.0 mm
1.0 mm
Space
Plate
1.0 mm
5.0 mm
8.0 mm
16.0 mm
10.0 mm
130.0 mm
34.0000
130.0 mm

3.6c – Multi-Series Capacitor

Yet another large cap I built for a purpose was the 24000 volt pulse capacitor I needed for my experimental Tesla Coil. This was a simple series unit made from 140 2uF 240 volt capacitor I picked up in a bargain bin a few years previously, the pictures do not show the 10 Mega ohm resistors I added, so the time constant and bleed of charge was more manageable and safe with higher frequencies.

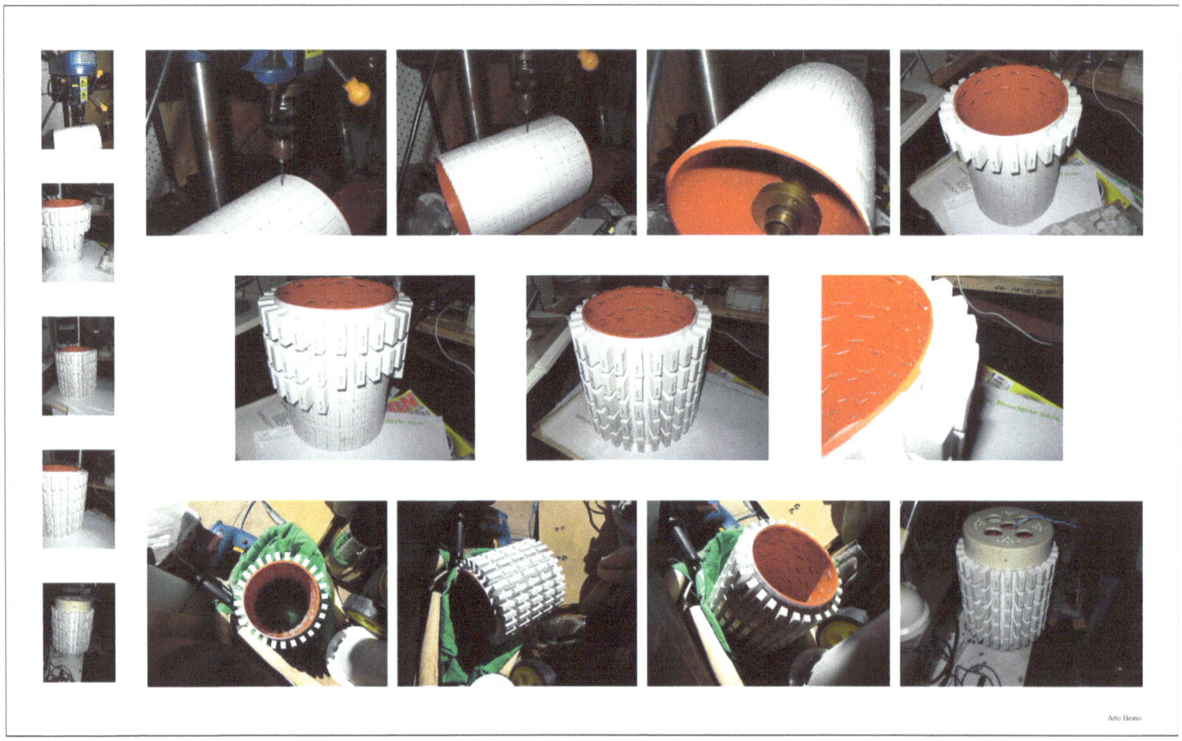

3.7 - High Voltage

3.7a - Voltage Multipliers

The simplest designs used in modern equipment to create high voltage is the basic voltage multiplier that was first invented by Croft and Walton in the 1920's. This simple system involved only capacitors and diodes arranged in a way so the charge on the capacitor array becomes highly polarized.

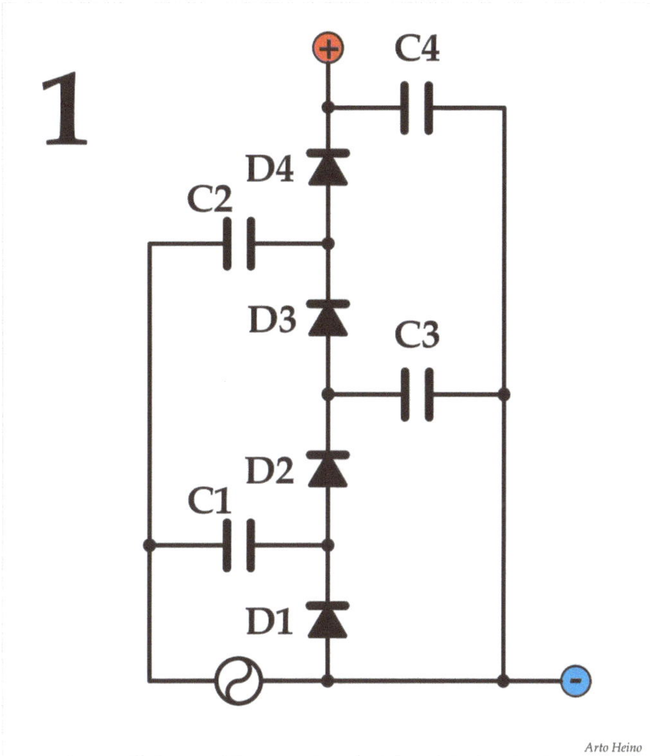

There are a multitude of designs based on this simple concept, here are some more.

2

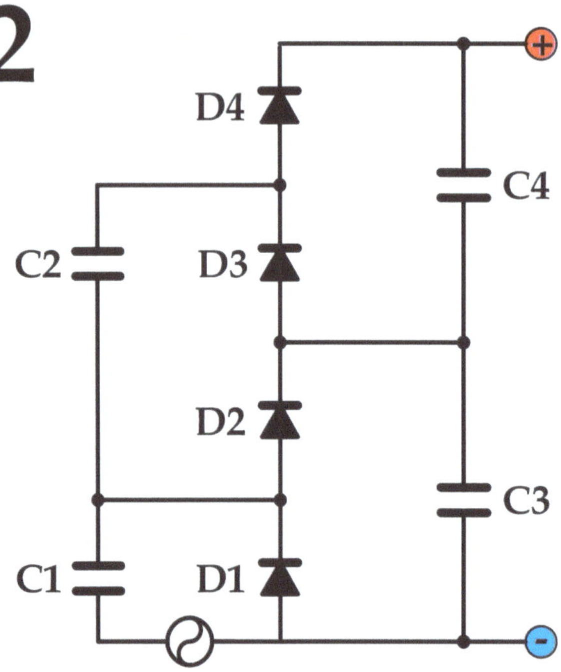

Arto Heino

3

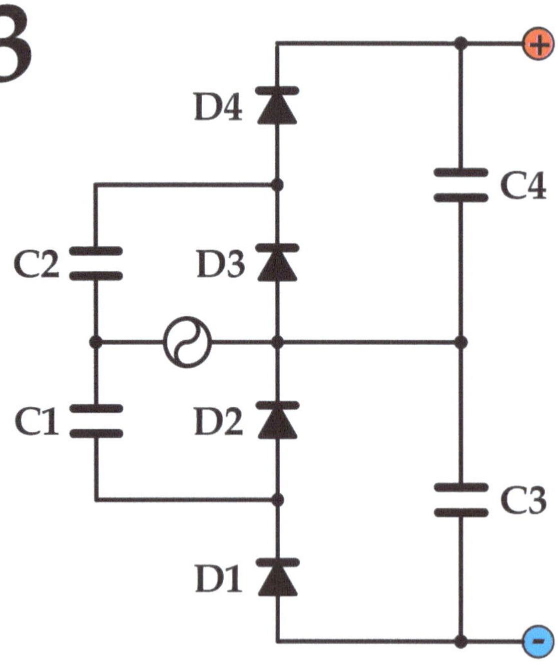

Arto Heino

These types of voltage multipliers rely on the usage of diodes that can withstand high voltages. Diodes are generally made from semiconductor materials in the corporate industry, almost any material has some semiconductors properties, as there is no perfect conductor and there is no perfect insulator.

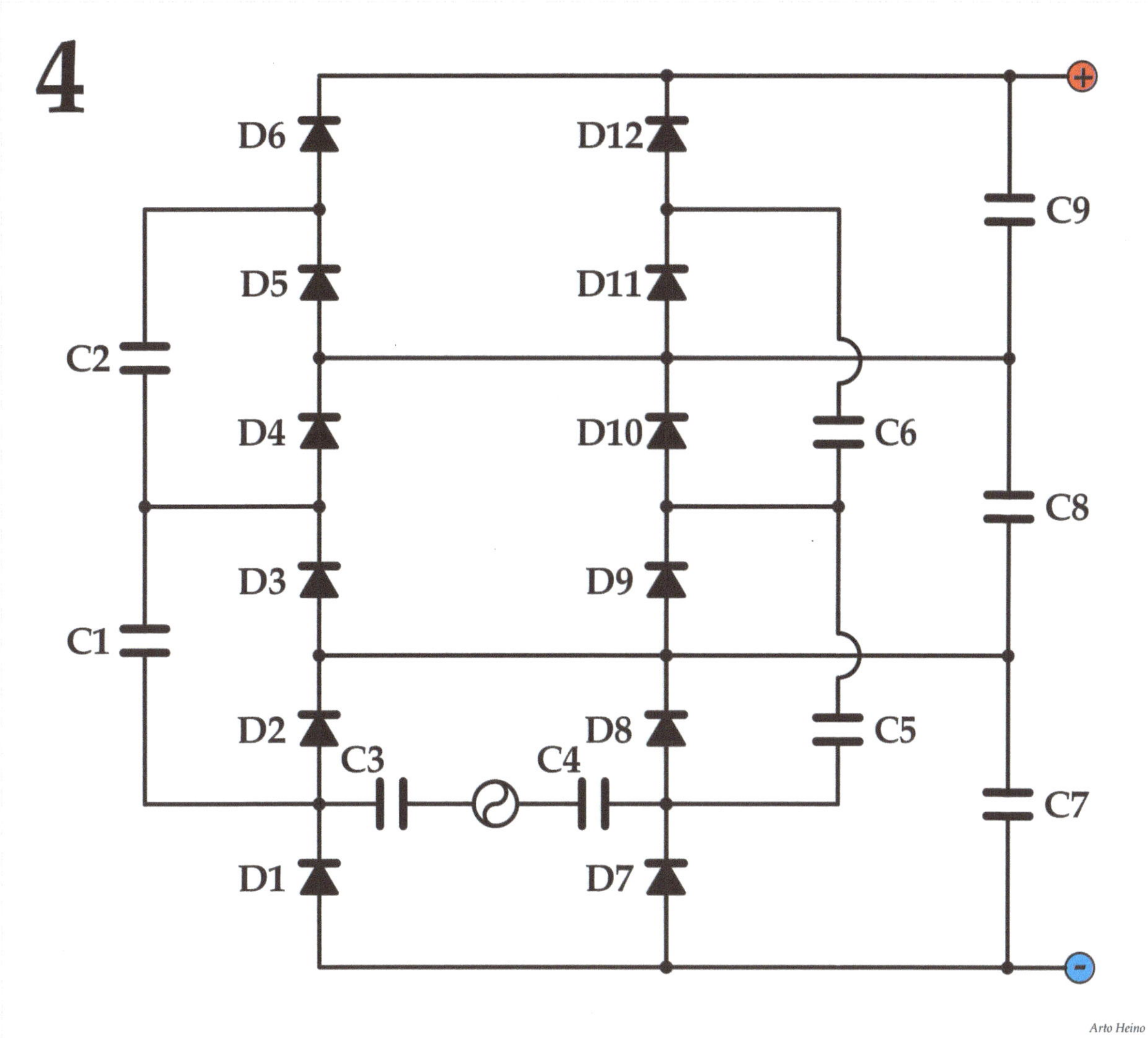

3,7b - Switched Voltage Multiplier

The next type of multiplier is the switchable type that was in use long before Croft and Walton were designed. This type goes back to the days of Nikola Tesla and his investigations into high voltages in about 1895 . Here is the basic schematic and out comes for four capacitors and nine switches.

All Possible Capacitor Arrangements for 4 Capacitors and 9 Switches

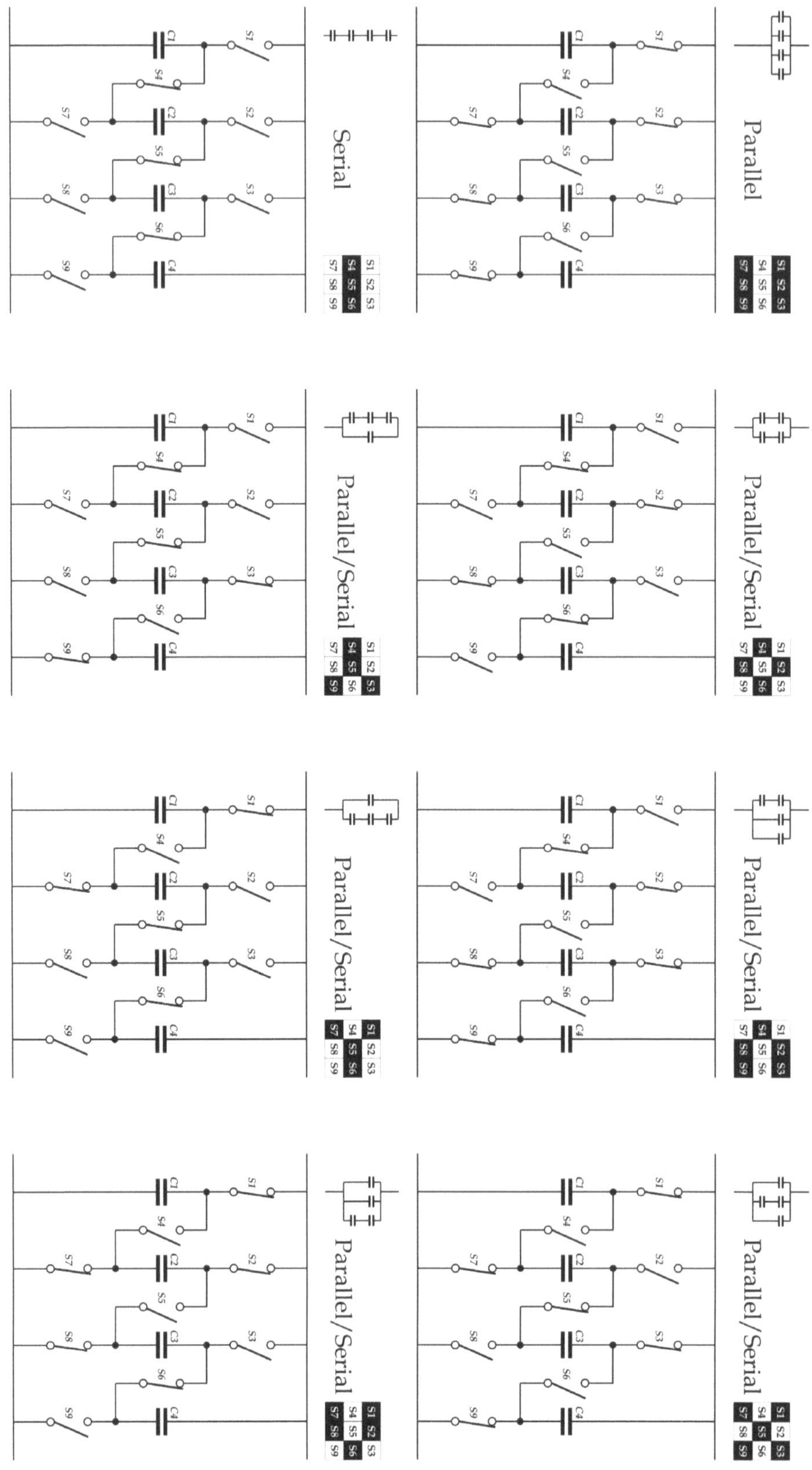

Arto Heino

The last type is a inspiration from the Tesla type commutator switching ideas.

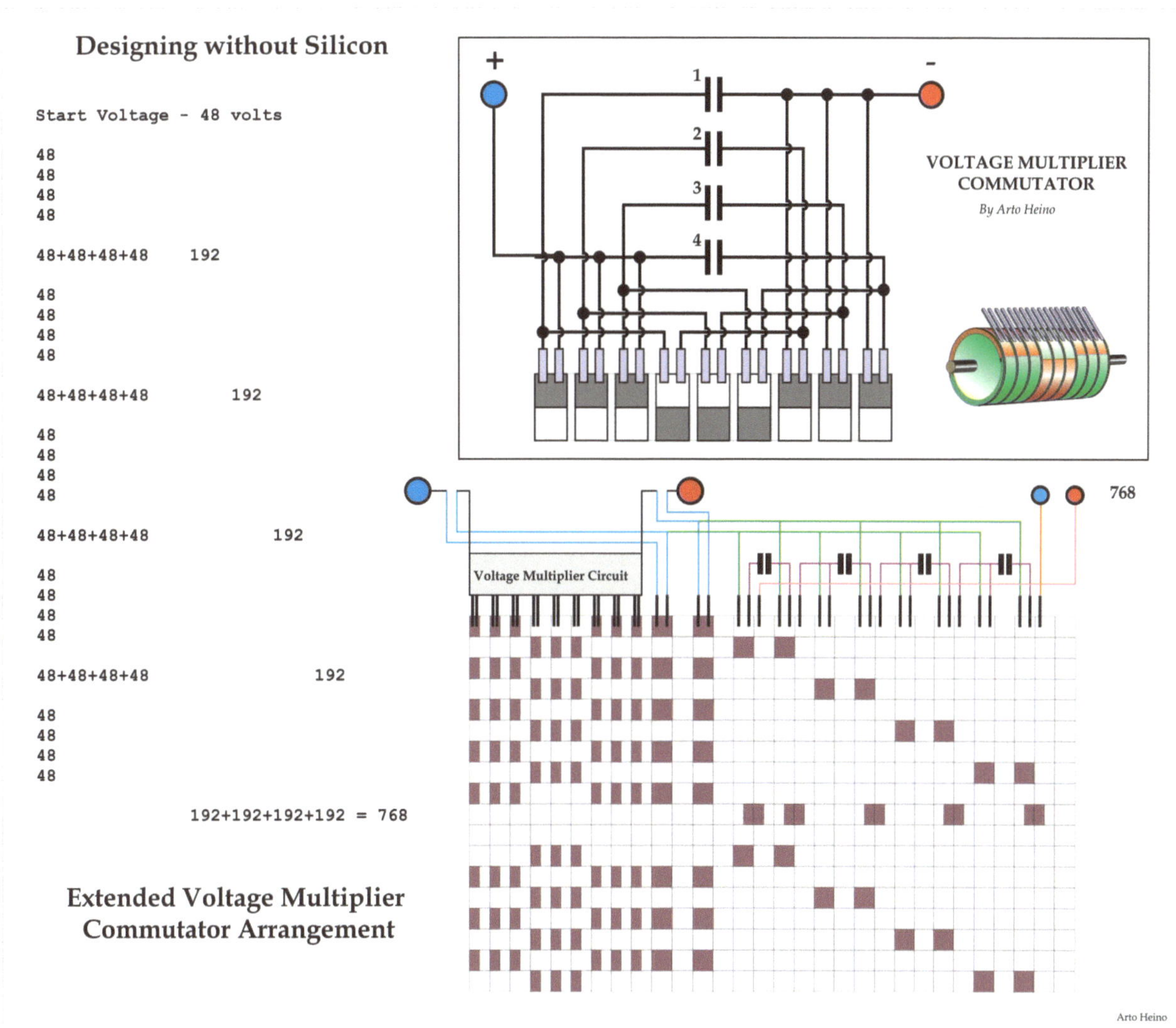

3.7c - Tesla Coil Patent 568176

In the next few pages I have redrawn the Tesla patents in a simplified schematic form that you can easily compare all his designs and reveal the simple ideas that I have illustrated and that now can be seen in terms that our current understanding and interpretations.

Patent
568176
Figure 1

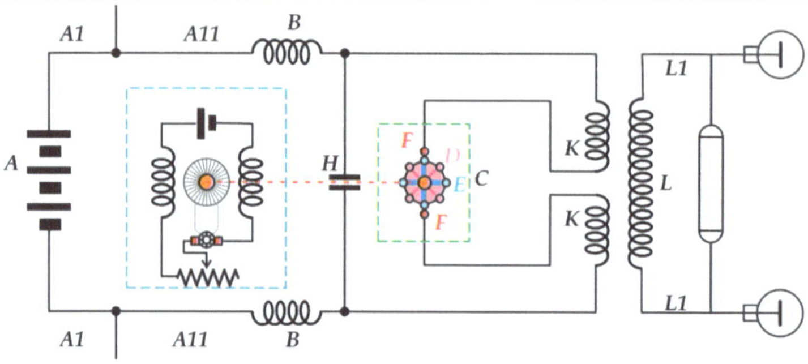

Patent
568176
Figure 2

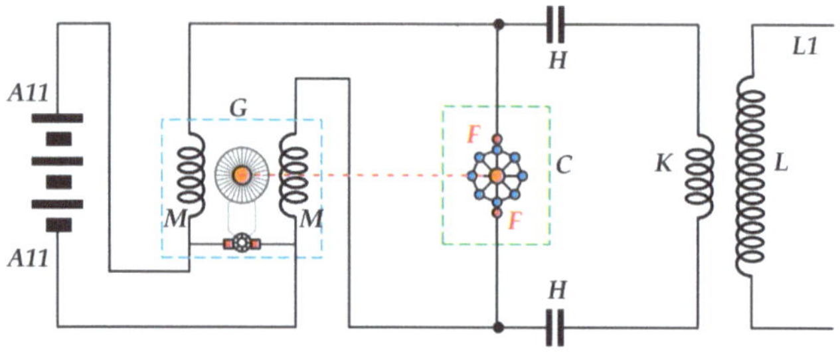

3.7d – Tesla Coil Patent 568177

Patent
568177
Figure 1

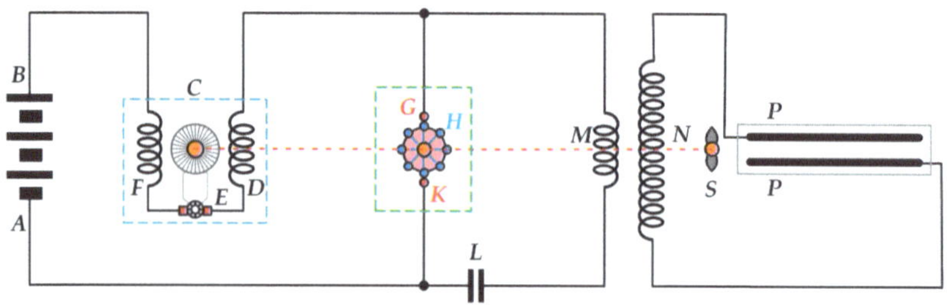

3.7e – Tesla Coil Patent 568178

Patent
569178
Figure 1

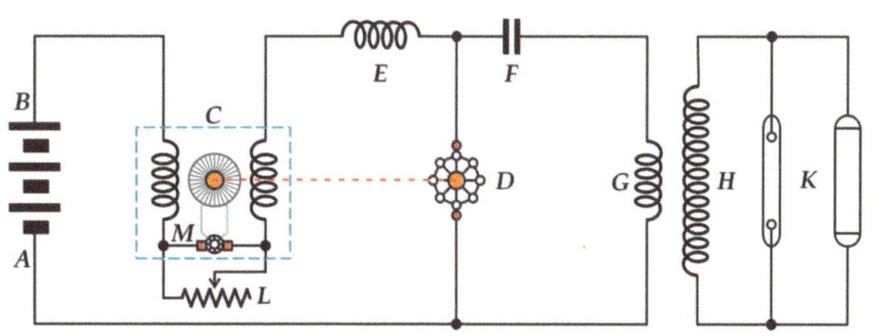

Patent
568178
Figure 2

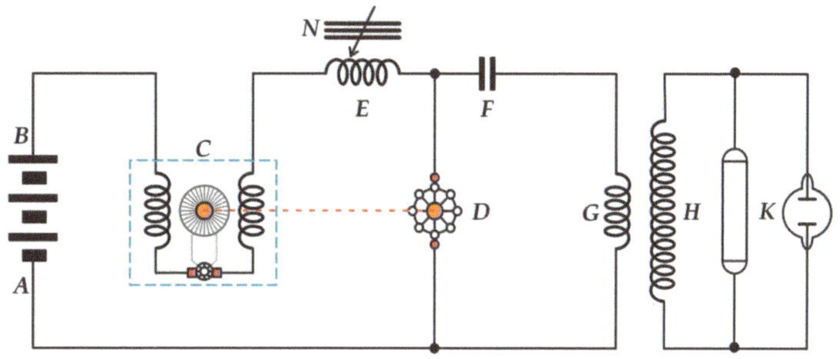

Patent
568178
Figure 3

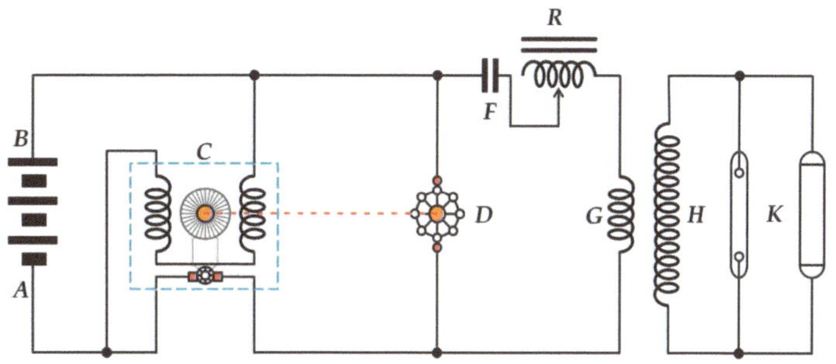

Patent
568178
Figure 4

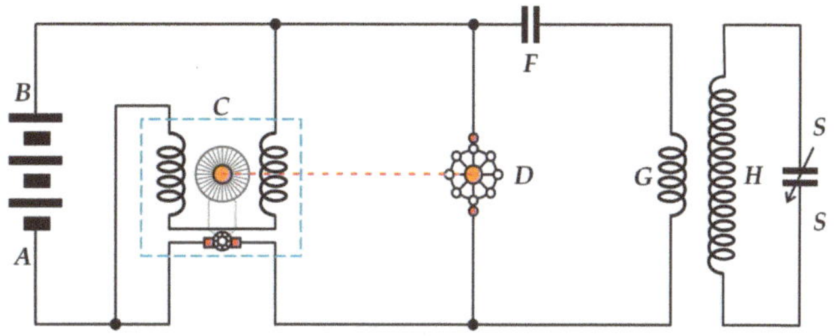

Arto Heino

3.7f - Tesla Coil Patent 568179

Patent
568179
Figure 1

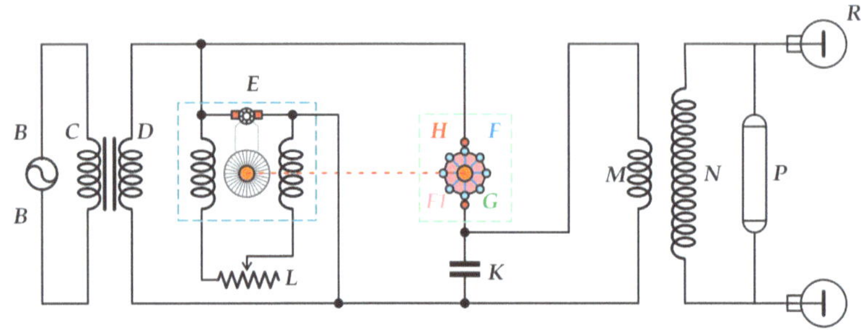

Patent
568179
Figure 2

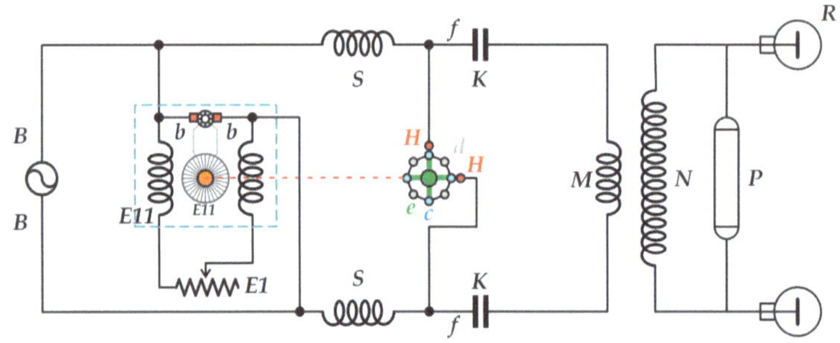

3.7c - Tesla Coil Patent 568180

Patent
568180
Figure 1,2,3

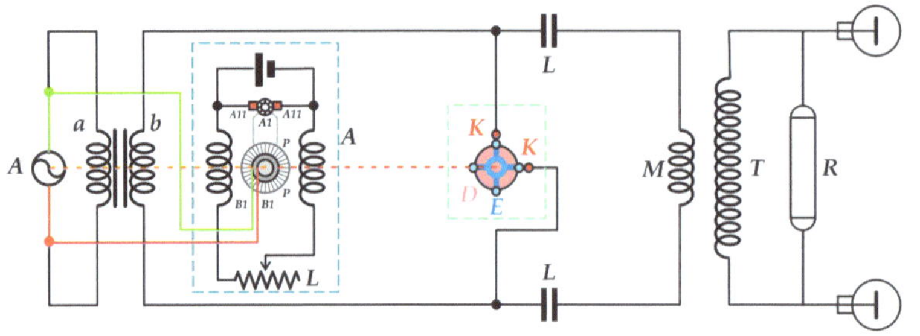

3.8 - Multivibrators

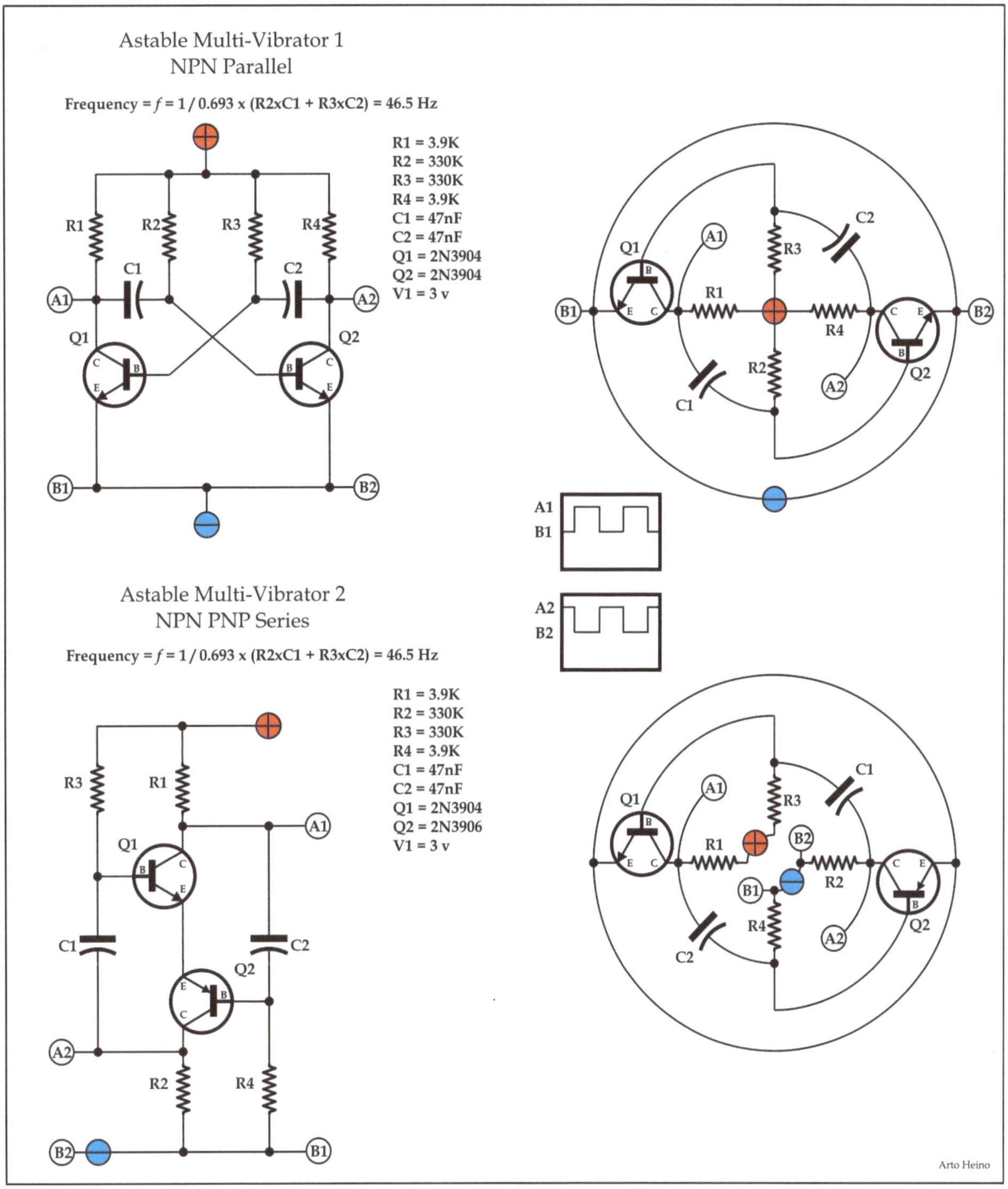

Hydraulic Astable Multivibrator

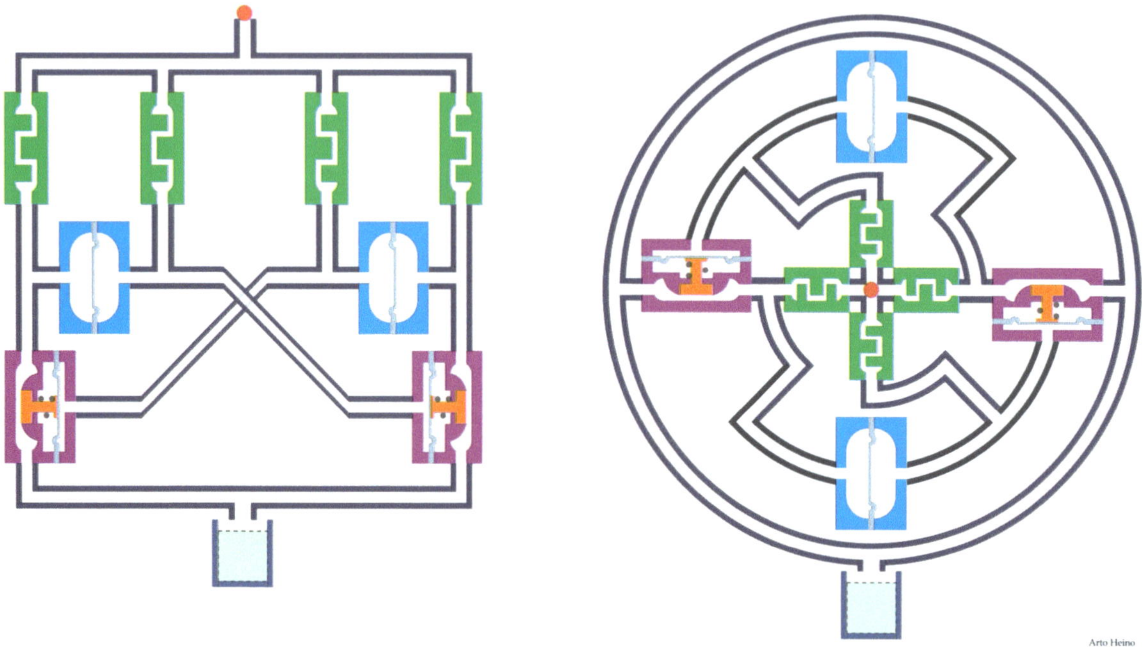

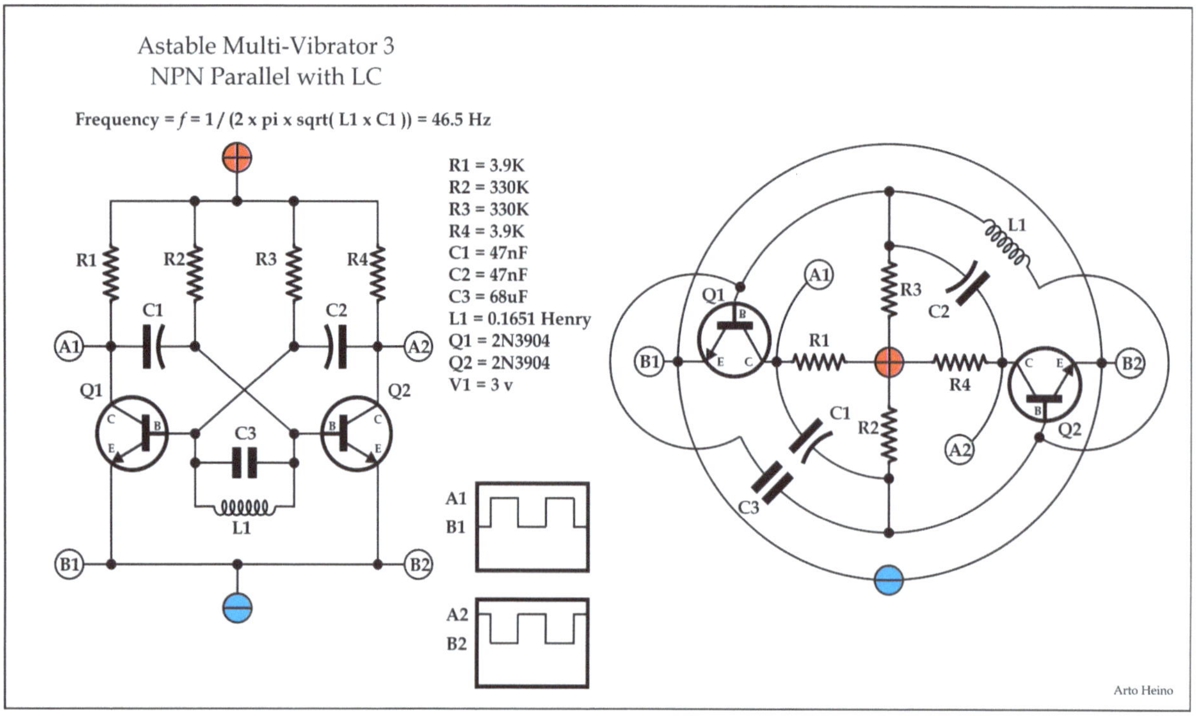

3.9 - Quad Radiation Converter

Here is an interesting project that will give you about 0.5/0.6/1.58 volts of radiation energy conversion. The analogy is like a putting a microphone in a noisy environment and converting the noise into energy, either to recharge a battery or light a led. Some people call this a free energy receiver and they would be wrong in that assumption as the cost was in the original source radiation.

The key to getting the most from this system is to employ 4 separate aerials, each with different lengths and orientations. The other MOST important factor is the earthing that you employ, minimum would be a 1 mtr pipe into the ground, or a more elaborate version would be a radial arm configuration.

Many of the diodes that are usable

1N4148

1N914

1N34A ***** preferred – used in my system

1N60 ***

The smaller antenna caps – frequency dependent

10 pF

100 pf *** preferred

1 nF

10 nF

150 nf – used in my system

200 nf

The main storage Capacitors 12/16/25/35/50/100 V

1 UF

10 Uf **

100 Uf ***- used in my system

1000 Uf **

10000 Uf

Super Caps have been used successfully

The useful aerial wire lengths and coils windings

Wire	Freq 1 wave	1/2 wv	1/4 wv
10 mtrs	30 Mhz	15	7.5
20 mtrs	15 Mhz	7.5	3.75
30 mtrs	10 Mhz	5	2.5
40 mtrs	7.5 Mhz	3.75	1.87
50 mtrs	6 Mhz	3	1.5
60 mtrs	5 Mhz	2.5	1.25
70 mtrs	4.28 Mhz	2.14	1.07
80 mtrs	3.75 Mhz	1.87	937 Khz

As you can see by some of the measurements I have only used 3 types of arrangements

Type 1 = Serial = 1.58 volts DC

A – B – C – D

Type 2 = Parallel = 0.506 volts DC

A

B

C

D

Type 3 = Para / Serial = 0.59 volts DC

```
A   C
| - |
B   D
```

Type 4 = Serial/ Parr

```
A – B
    |
C – D
```

Electromagnetic Radiation Differential Converter with Voltage Doubling Parallel Quad Circuit
4 Antenna
4 Earth

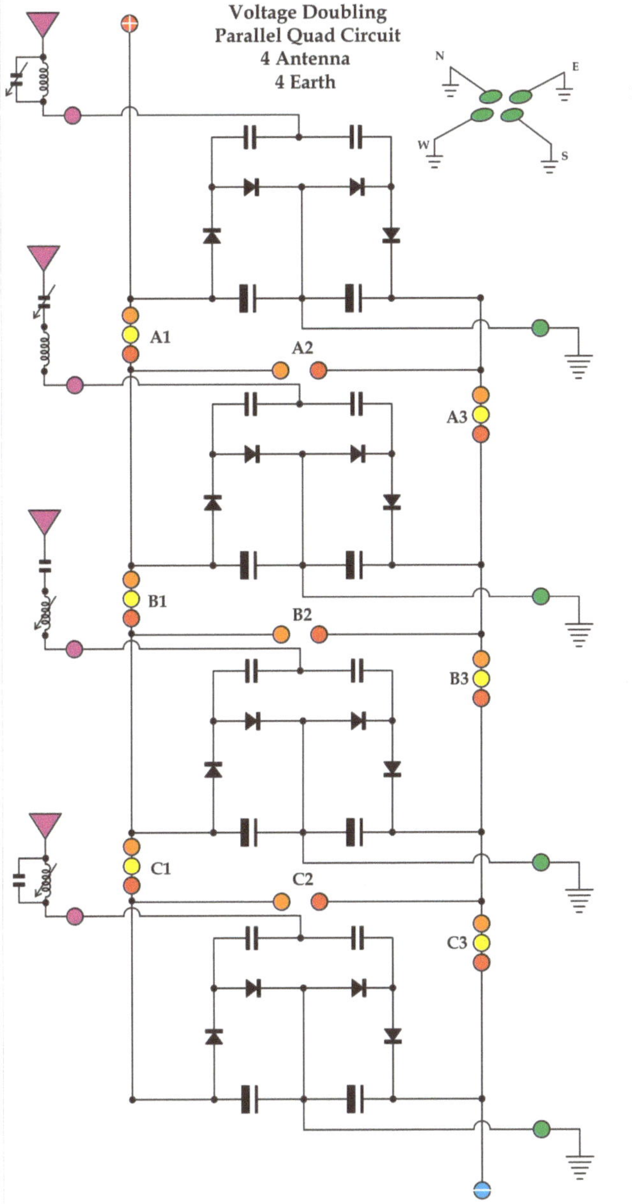

Earth Radial layout

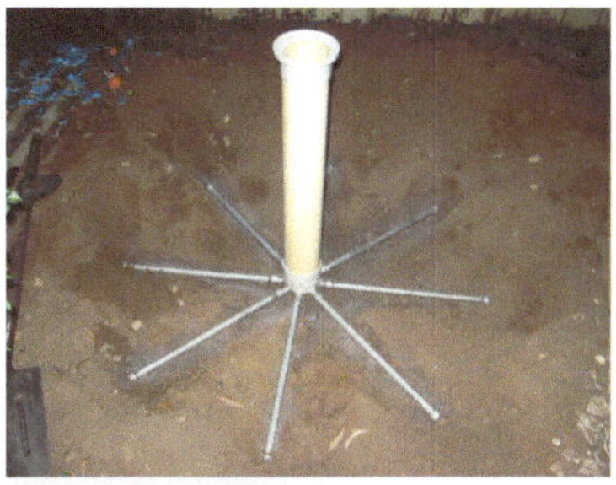

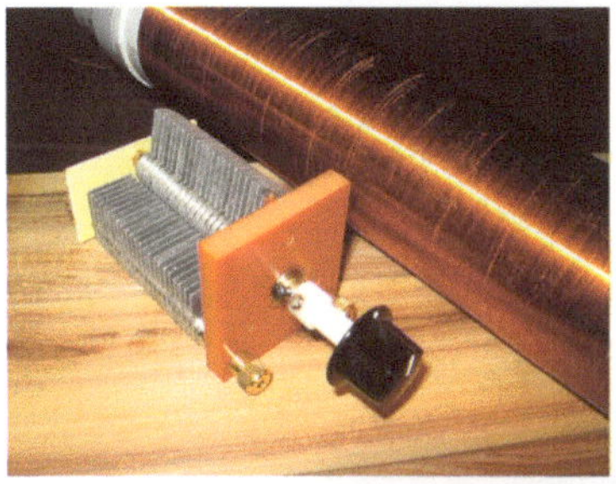

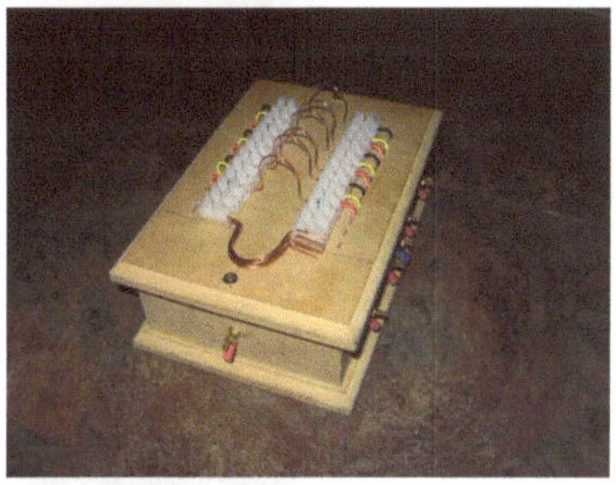

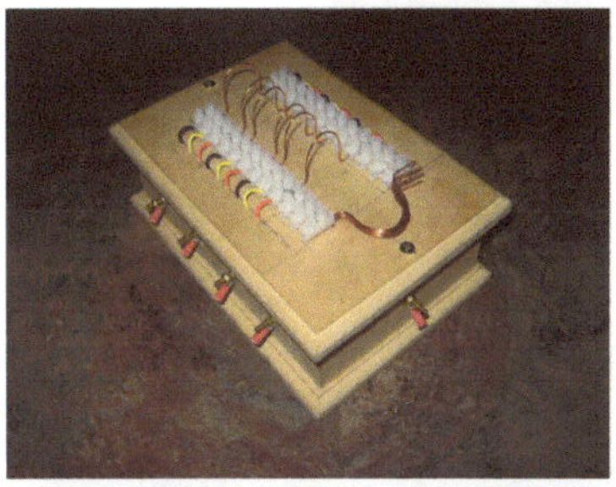

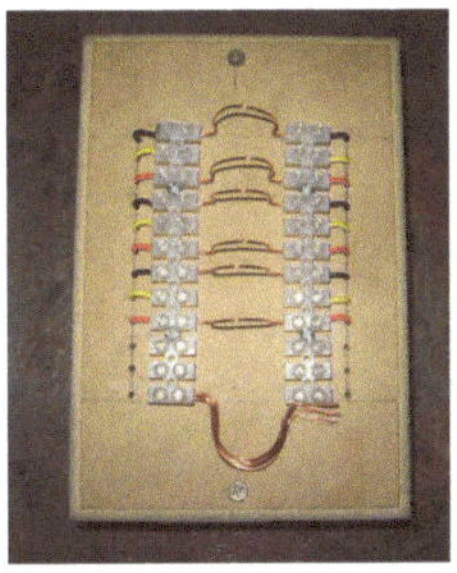

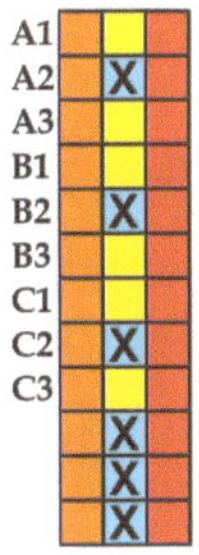

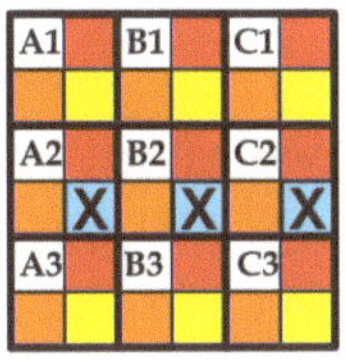

Arto Heino

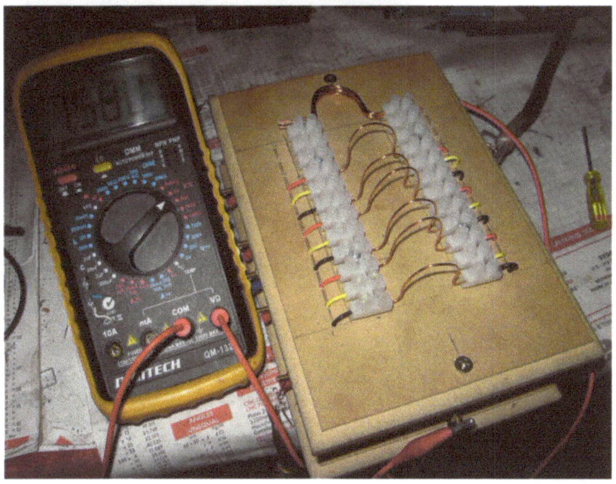

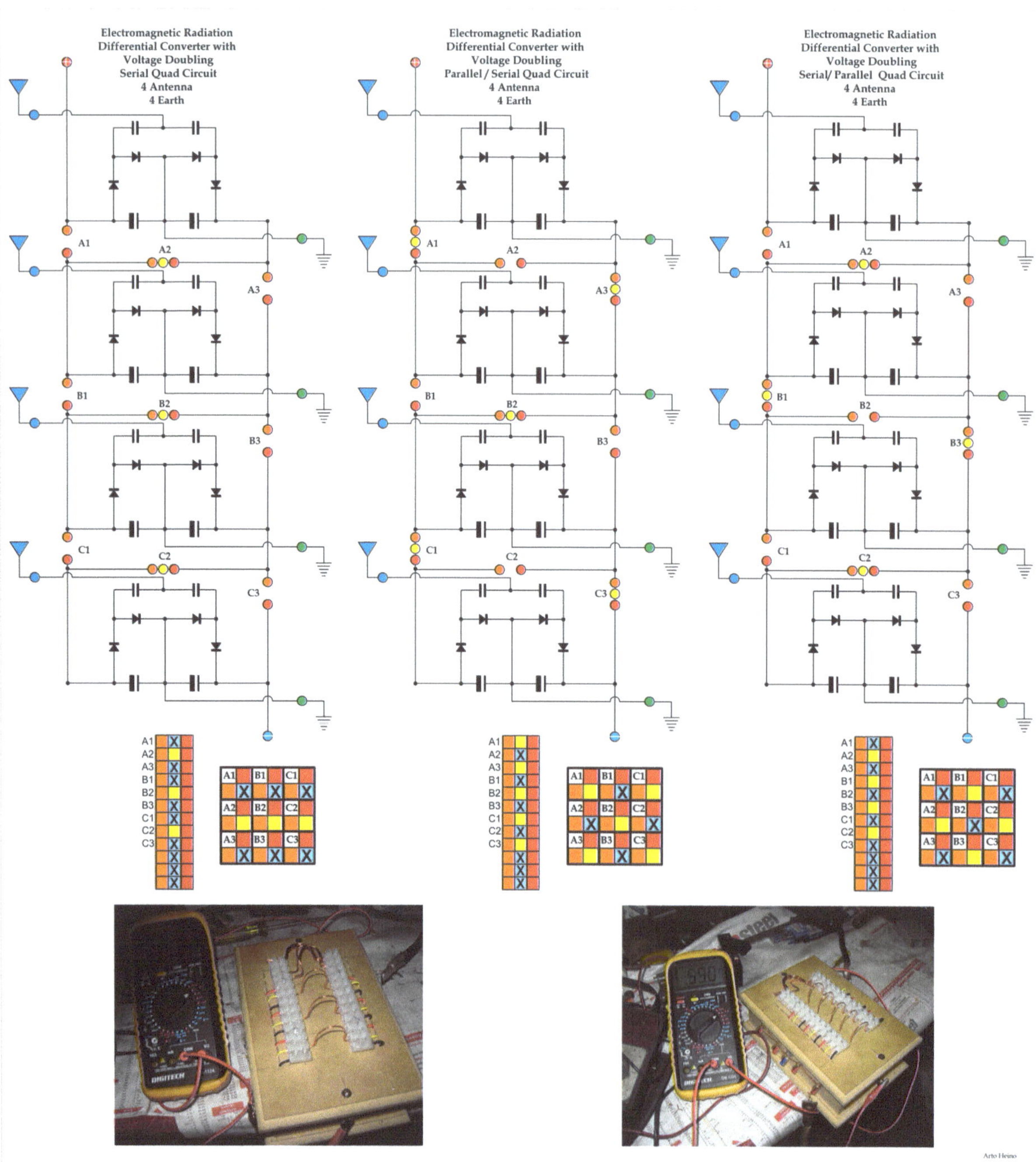

The arrangements for different earth/antenna connections with added coils and capacitors are the vital parameters that must be collected to design a system suitable to the condition to your area. Here are numerous different possibilities that I have put together in a chart to start the process of creative engagement with our electrical environment.

Energy Collector Configurations

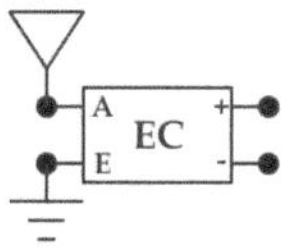

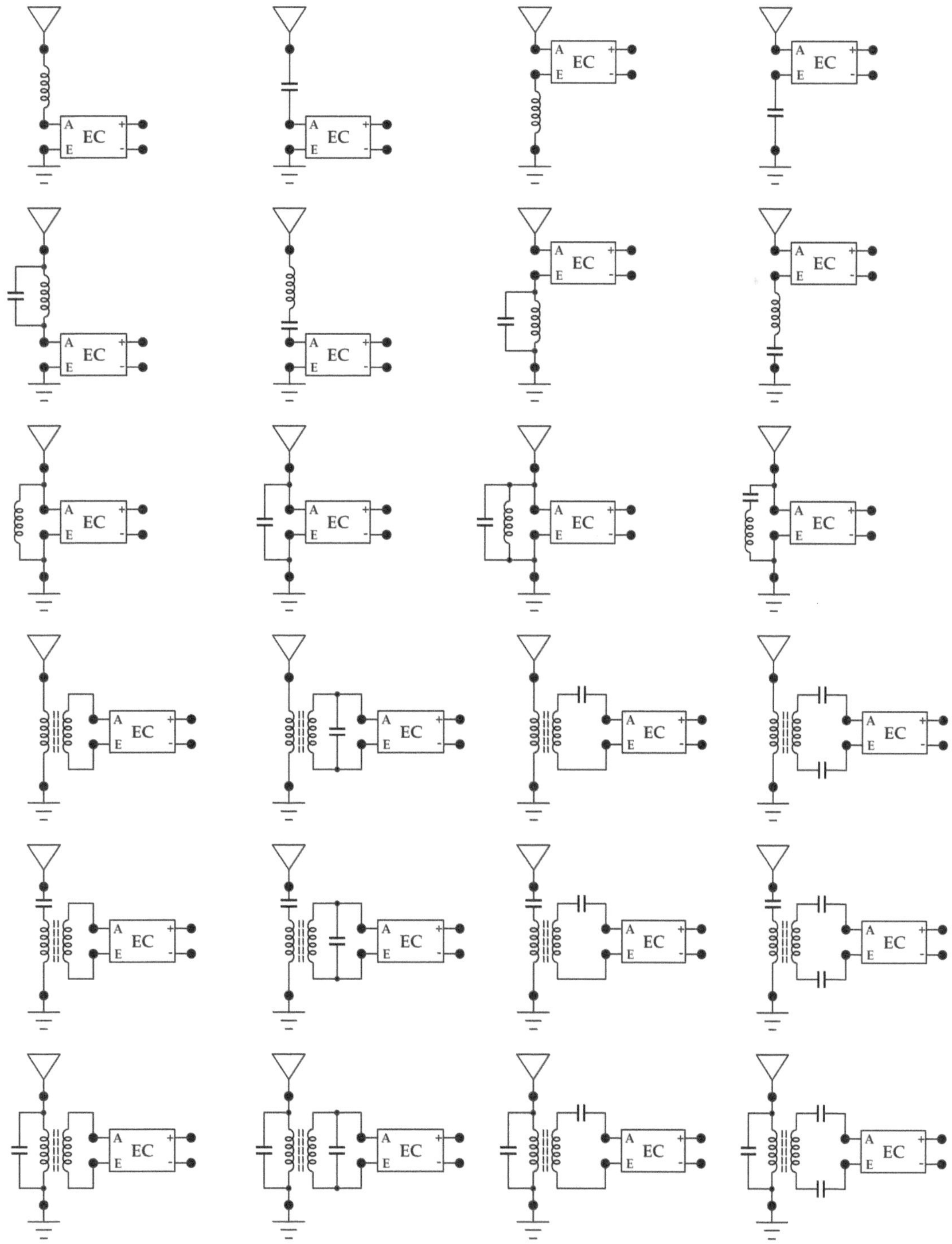

Arto Heino

Chapter Four

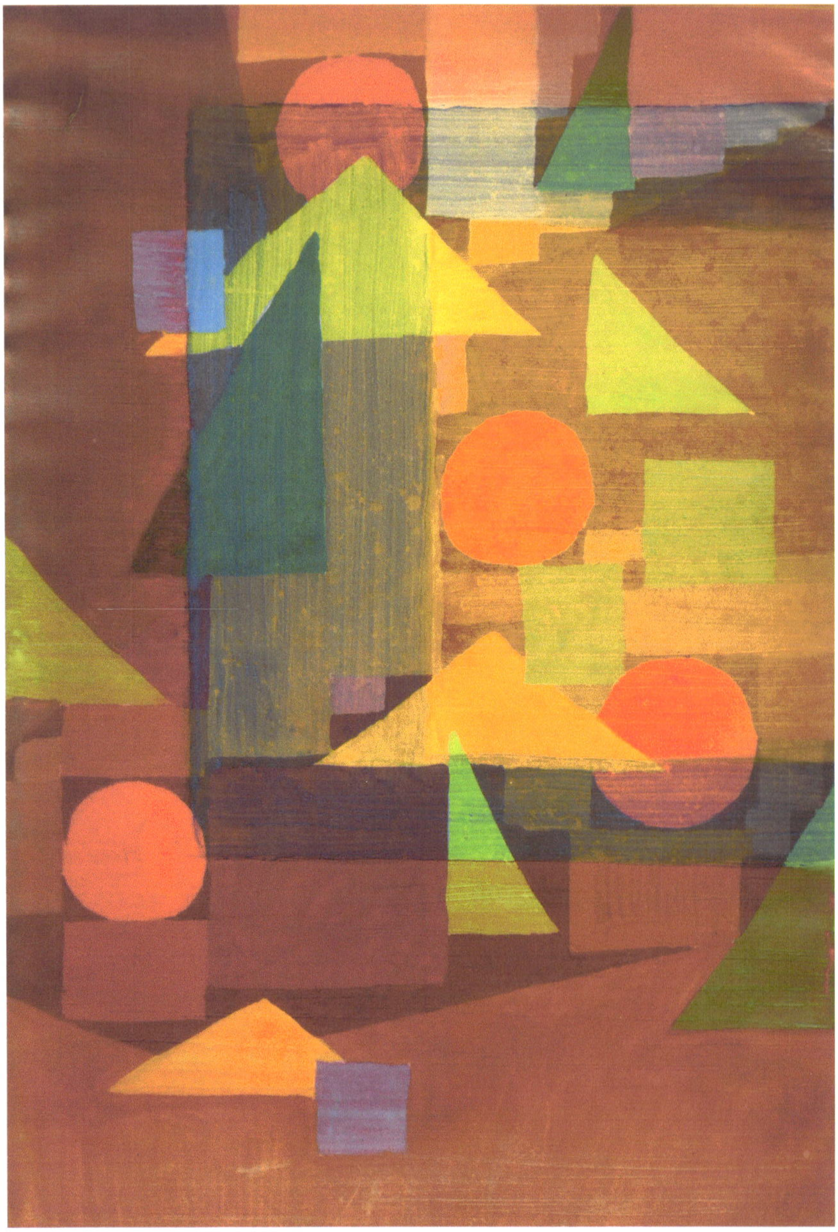

Mathematics

4.1 - Numbers to Pi/2

$\pi/2$

	LD	LN	ACD	RN	RD	
		1		0		
LD = LN / RN	2.00000000	2	0.50000000	1		RN / LN = RD
		3	1.50000000	2	0.66666667	
	1.33333333	4	0.75000000	3		
		5	1.25000000	4	0.80000000	
	1.20000000	6	0.83333333	5		
		7	1.16666667	6	0.85714286	
	1.14285714	8	0.87500000	7		
		9	1.12500000	8	0.88888889	
	1.11111111	10	0.90000000	9		
		11	1.10000000	10	0.90909091	
	1.09090909	12	0.91666667	11		
		13	1.08333333	12	0.92307692	
	1.07692308	14	0.92857143	13		
		15	1.07142857	14	0.93333333	
	1.06666667	16	0.93750000	15		
		17	1.06250000	16	0.94117647	
	1.05882353	18	0.94444444	17		
		19	1.05555556	18	0.94736842	
	1.05263158	20	0.95000000	19		
		21	1.05000000	20	0.95238095	
	1.04761905	22	0.95454545	21		
		23	1.04545455	22	0.95652174	
	1.04347826	24	0.95833333	23		
		25	1.04166667	24	0.96000000	
	1.04000000	26	0.96153846	25		
		27	1.03846154	26	0.96296296	
	1.03703704	28	0.96428571	27		
		29	1.03571429	28	0.96551724	
	1.03448276	30	0.96666667	29		
		31	1.03333333	30	0.96774194	
	1.03225806	32	0.96875000	31		
		33	1.03125000	32	0.96969697	
	1.03030303	34	0.97058824	33		
		35	1.02941176	34	0.97142857	
	1.02857143	36	0.97222222	35		
		37	1.02777778	36	0.97297297	
		CM	0.64528007			
			0.38742867	CD		
AL	**7.57228123**	AL*BR	1.54971468		4.88624218	AR
BL	0.13206060		0.64528007	BL*AR	**0.20465625**	BR
		Pi/2	**1.57079633**			
			0.63661977	2/Pi		
			4.00000000			
			AL*BR/CD			

Arto Heino

The interesting table above is an attempt to describe the use of numbers related to Pi that have some basic geometric and logical steps. The sequence of numbers should extend a lot further to get the accuracy of Pi / 2 to six or more decimal places, as you can see the crossing over of multiples that suggests the actual symbol of the multiplicand.

4.2 - The Hidden Code of Bruce Cathie

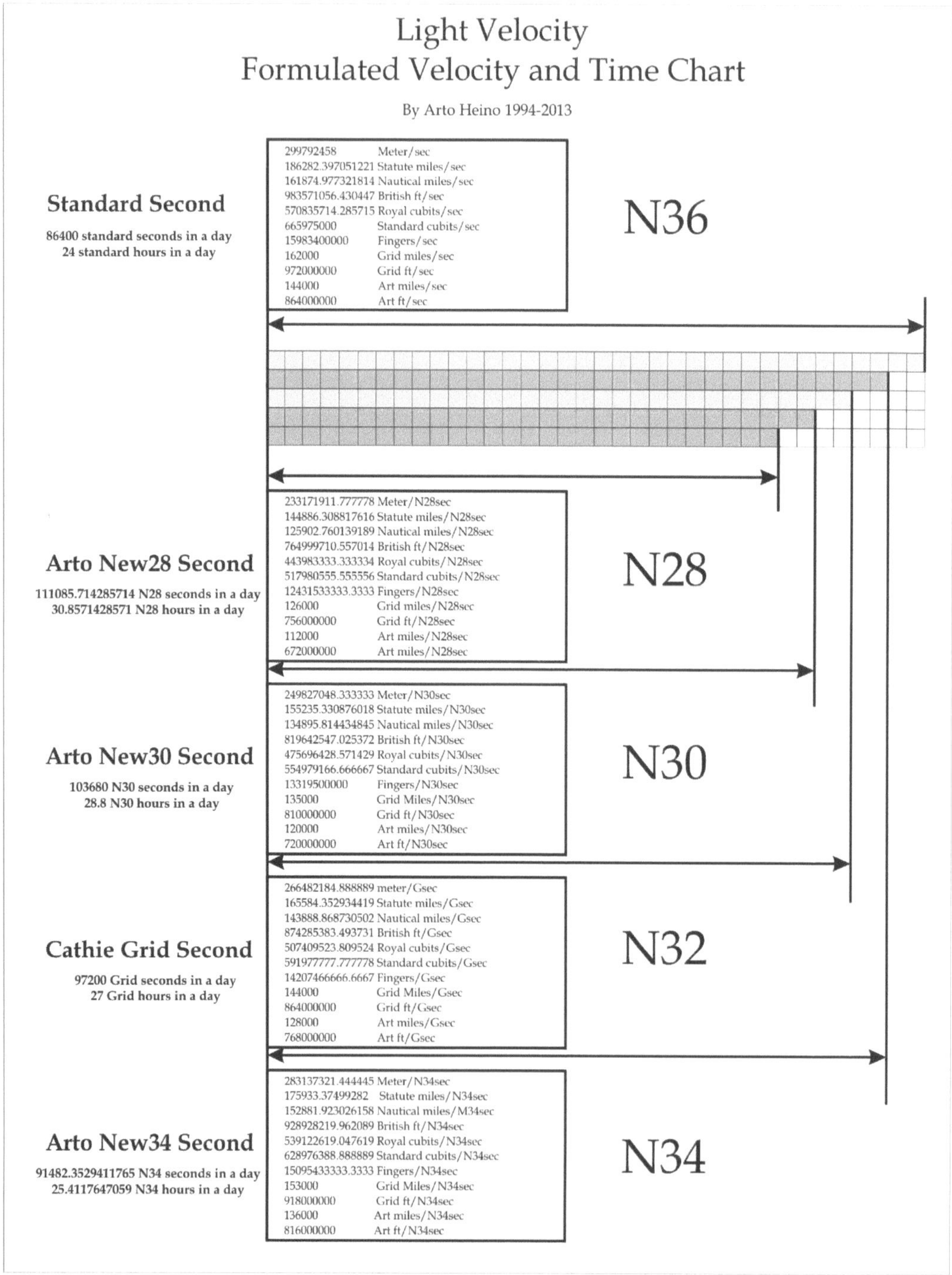

When I first read Bruce Cathies books in the 1980's Harmonic 33, Harmonic 695 and later The Bridge to Infinity and The Harmonic Conquest of Space, I was amazed and fascinated by how Bruce used simple mathematical facility in bringing all his explorations into the 144 light velocity and all the harmonic frame work as well, I loved reading his material and all his exploits.

I could not at first see why he chose 144 as the numbers as the velocity, Bruce had made some assertions about the ancient metric and a lot of biblical and Geomatra references, this made me question the historical reasoning, only later did I find the real reason. You see this was not a common number when measuring anything linear only area, cubits were the rule in the period Bruce points out, besides the Ancients had no need of a decimal version of PHI and its endless decimal fractions or 143.889, 6.9444 decimals fractions either, they are only used in his methods to justify the use of 144 as light speed, the Ancients built and located to natural prime and whole numbers, this is was gives you the PHI ratio. Who had use of a calculator in 3000BC and surely nature never uses parts of anything only whole quantum units for any process. It seems there is something not quite clear in his explanations and has a possible bogus written on this scheme, let us look at some of those that contain vagueness.

Look at the tables and you will see that Standard Cubits and Fingers become rounded off to whole numbers and surely would become more useful that Bruce Cathie's numeric web of harmonic decimal fraction values, why create new measures when the Ancient measure of cubits had so much to offer. The use of a light velocity in ancient times did make for great reading, only its use would have been for astronomy and some form of technology, I could see many ancient places that could have used laser cutting such as Sakkara, Machu Pichu and Puma Punka. Why would the ancients complicate the natural whole number science with complex algorithms with multiple decimal places and complex formulas just to put a monument, where a simple whole number solution would explain more and reveal complex ratios that come from its prime quantum nature.

Only by studying the mathematical and geometric works of Euclid, Archimedes and Pythagoras can you see the numeric tricks that Bruce has employed to result in what may seem like almost magical answers. This sort of manipulation has truly some amazing reverse engineering to give you the results you seek.

The methods are not hard or magical, the term used is the intervals of the harmonic discipline, these are named as Sesquitertian ratios, Sesquiquialter comparisons, Multiple Superparticulars and its attendants. Without writing a formal doctorate on the subject, I will just show you the numbers and how they function. You will recognize an abundance of Geomatra numbers and harmonics that a lot of people write about.

As you can see the top row is pure doubling, nothing very complex, this is how octave work in music, light, radio etc To get the start in next number row you add the first two numbers then place the result in the column under the second number. The process continues for every number pair on the top row, this will now give you the second row,now do the same procedure with this row and this will give the next row, continue with the same process and you will have a table like the one I have made.

Harmonic Tables

Duple - Triple - Quadruple - Quintuple

By Arto Heino 1994-2013

DUPLE

1	2	4	8	16	32	64	128	256	512	1024	2048	4096
	3	6	12	24	48	96	192	384	768	1536	3072	6144
		9	18	36	72	144	288	576	1152	2304	4608	9216
			27	54	108	216	432	864	1728	3456	6912	13824
				81	162	324	648	1296	2592	5184	10368	20736
					243	486	972	1944	3888	7776	15552	31104
						729	1458	2916	5832	11664	23328	46656
							2187	4374	8748	17496	34992	69984
								6561	13122	26244	52488	104976
									19683	39366	78732	157464
										59049	118098	236196
											177147	354294
												531441

TRIPLE

1	3	9	27	81	243	729	2187	6561	19683	59049	177147	531441
	4	12	36	108	324	972	2916	8748	26244	78732	236196	708588
		16	48	144	432	1296	3888	11664	34992	104976	314928	944784
			64	192	576	1728	5184	15552	46656	139968	419904	1259712
				256	768	2304	6912	20736	62208	186624	559872	1679616
					1024	3072	9216	27648	82944	248832	746496	2239488
						4096	12288	36864	110592	331776	995328	2985984
							16384	49152	147456	442368	1327104	3981312
								65536	196608	589824	1769472	5308416
									262144	786432	2359296	7077888
										1048576	3145728	9437184
											4194304	12582912
												16777216

QUADRUPLE

1	4	16	64	256	1024	4096	16384	65536	262144	1048576	4194304	16777216
	5	20	80	320	1280	5120	20480	81920	327680	1310720	5242880	20971520
		25	100	400	1600	6400	25600	102400	409600	1638400	6553600	26214400
			125	500	2000	8000	32000	128000	512000	2048000	8192000	32768000
				625	2500	10000	40000	160000	640000	2560000	10240000	40960000
					3125	12500	50000	200000	800000	3200000	12800000	51200000
						15625	62500	250000	1000000	4000000	16000000	64000000
							78125	312500	1250000	5000000	20000000	80000000
								390625	1562500	6250000	25000000	100000000
									1953125	7812500	31250000	125000000
										9765625	39062500	156250000
											48828125	195312500
												244140625

QUINTUPLE

1	5	25	125	625	3125	15625	78125	390625	1953125	9765625	48828125	244140625
	6	30	150	750	3750	18750	93750	468750	2343750	11718750	58593750	292968750
		36	180	900	4500	22500	112500	562500	2812500	14062500	70312500	351562500
			216	1080	5400	27000	135000	675000	3375000	16875000	84375000	421875000
				1296	6480	32400	162000	810000	4050000	20250000	101250000	506250000
					7776	38880	194400	972000	4860000	24300000	121500000	607500000
						46656	233280	1166400	5832000	29160000	145800000	729000000
							279936	1399680	6998400	34992000	174960000	874800000
								1679616	8398080	41990400	209952000	1049760000
									10077696	50388480	251942400	1259712000
										60466176	302330880	1511654400
											362797056	1813985280
												2176782336

As you can see the second table(Triplet) the numbers are tripled, the same process of addition and carry to the next row the sum is the same as the first table, will give you the table the same as I have made.

A careful study of the first and second table will reveal the tripling is also occurring in the first table, only as the diagonal values. This same view has the diagonals of the triple table now creates quadrature. Many more numeric harmonies are present in these tables. An example of this can be shown, if 256 = C1 in music the its octave is 512 = C2, the next octave in now 1024 = C3, this seems logical and simple, now the 5th of diatonic harmony in music is the most concordant next to the tonic, if the tonic 256= C2 then the fifth is 384 = G2, the 2nd is 288, the ratios are 256 x 3/2 for the fifth = 384, 256 x 9/8 for the second = 288. This is how harmonies function numerically, no mystery here, this information dates back at least 5000 years.

The sextuplet table I am including shows you another Geomatra group that are written about, including the Nineveh Constant. This set is in multiples of 6 , the second row, 7th element is 1959552.

Harmonic Tables

Sextuple - Septuple - Octuple - Novtuple

By Arto Heino 1994-2013

SEXTUPLE

1	6	36	216	1296	7776	46656	279936	1679616	10077696	60466176	362797056	2176782336
	7	42	252	1512	9072	54432	326592	1959552	11757312	70543872	423263232	2539579392
		49	294	1764	10584	63504	381024	2286144	13716864	82301184	493807104	2962842624
			343	2058	12348	74088	444528	2667168	16003008	96018048	576108288	3456649728
				2401	14406	86436	518616	3111696	18670176	112021056	672126336	4032758016
					16807	100842	605052	3630312	21781872	130691232	784147392	4704884352
						117649	705894	4235364	25412184	152473104	914838624	5489031744
							823543	4941258	29647548	177885288	1067311728	6403870368
								5764801	34588806	207532836	1245197016	7471182096
									40353607	242121642	1452729852	8716379112
										282475249	1694851494	10169108964
											1977326743	11863960458
												13841287201

SEPTUPLE

1	7	49	343	2401	16807	117649	823543	5764801	40353607	282475249	1977326743	13841287201
	8	56	392	2744	19208	134456	941192	6588344	46118408	322828856	2259801992	15818613944
		64	448	3136	21952	153664	1075648	7529536	52706752	368947264	2582630848	18078415936
			512	3584	25088	175616	1229312	8605184	60236288	421654016	2951578112	20661046784
				4096	28672	200704	1404928	9834496	68841472	481890304	3373232128	23612624896
					32768	229376	1605632	11239424	78675968	550731776	3855122432	26985857024
						262144	1835008	12845056	89915392	629407744	4405854208	30840979456
							2097152	14680064	102760448	719323136	5035261952	35246833664
								16777216	117440512	822083584	5754585088	40282095616
									134217728	939524096	6576668672	46036680704
										1073741824	7516192768	52613349376
											8589934592	60129542144
												68719476736

OCTUPLE

1	8	64	512	4096	32768	262144	2097152	16777216	134217728	1073741824	8589934592	68719476736
	9	72	576	4608	36864	294912	2359296	18874368	150994944	1207959552	9663676416	77309411328
		81	648	5184	41472	331776	2654208	21233664	169869312	1358954496	10871635968	86973087744
			729	5832	46656	373248	2985984	23887872	191102976	1528823808	12230590464	97844723712
				6561	52488	419904	3359232	26873856	214990848	1719926784	13759414272	110075314176
					59049	472392	3779136	30233088	241864704	1934917632	15479341056	123834728448
						531441	4251528	34012224	272097792	2176782336	17414258688	139314069504
							4782969	38263752	306110016	2448880128	19591041024	156728328192
								43046721	344373768	2754990144	22039921152	176319369216
									387420489	3099363912	24794911296	198359290368
										3486784401	27894275208	223154201664
											31381059609	251048476872
												282429536481

NOVTUPLE

1	9	81	729	6561	59049	531441	4782969	43046721	387420489	3486784401	31381059609	282429536481
	10	90	810	7290	65610	590490	5314410	47829690	430467210	3874204890	34867844010	313810596090
		100	900	8100	72900	656100	5904900	53144100	478296900	4304672100	38742048900	348678440100
			1000	9000	81000	729000	6561000	59049000	531441000	4782969000	43046721000	387420489000
				10000	90000	810000	7290000	65610000	590490000	5314410000	47829690000	430467210000
					100000	900000	8100000	72900000	656100000	5904900000	53144100000	478296900000
						1000000	9000000	81000000	729000000	6561000000	59049000000	531441000000
							10000000	90000000	810000000	7290000000	65610000000	590490000000
								100000000	900000000	8100000000	72900000000	656100000000
									1000000000	9000000000	81000000000	729000000000
										10000000000	90000000000	810000000000
											100000000000	900000000000
												1000000000000

As you can see these are very simple arithmetic calculations, not requiring PI, PHI or Log, this is how nature creates harmonies and natural intervals that exist everywhere not just on your piano. If you require more knowledge to pursue, you must create complete harmonic tables and you must also be willing to understand and make frequency charts covering many octaves.

In the first table and the following table I have created the precise light velocities in:

Meters per second
Miles per second
Feet per second
Nautical Miles per second, Cathie Arc Minutes
Royal Cubits per second
Standard Cubits per second
Fingers per second
Cathie Grid miles per second, Adjusted for maximum
Cathie Grid feet per second, Adjusted for maximum
Art miles per second (an addition to show how it can be done)
Art feet per second

The Royal Cubit = 525.1816774904 mm
The Standard Cubit = 450.1557235632 mm
The Finger = 18.7564884818 mm

These sizes where chosen as the least rounding error.

I have also created a set of new time scales in seconds, N28, N30 and N34 as N36 is the standard second and N32 is Bruce Cathies Grid second. I could extend this time scales in both directions as required. The Grid , and Art lengths are listed here;

Grid inch = 1.011904379 standard British inch
Grid mile = Grid Inch x 12 x 6000 = 72000 Grid Inches = 72857.1152911442 British Inches

The only problem that Bruce had with these tables is lack of justification on the velocity in a different medium, as he shows that maximum is 144000 minutes of arc per grid second, now for this to be correct he must have either the length of the arc grid mile to be different or the grid second to be slightly different. If he uses the grid second as 8/9 of the normal second as he proposes in all his writing then you will get 143888.868730502 minutes of arc per grid second, to correct this he must change the ratio of the grid second to equal 1.124131787 not the theoretical 1.125 as he suggests. Not to labor the issue we will accept a slight variation that allows the value 144000 to be the speed of lights maximum, this will be the adjustment to the grid mile of 1.1498913398, with the Nautical miles as 1.150779448, on the normal minute arc scale, I will include the maximum values in my tables, to be accurate they should be all equivalent.

Light Velocity
Formulated Length and Metrology Chart

By Arto Heino 1994-2013

Light Second

Distance of Light Traveled in One Standard Second

299792458	Meter
186282.397051221	Statute miles
161874.977321814	Nautical miles
983571056.430447	British ft
570835714.285715	Royal cubits
665975000	Standard cubits
15983400000	Fingers
162000	Grid miles
972000000	Grid ft
144000	Art miles
864000000	Art ft

Length Conversion

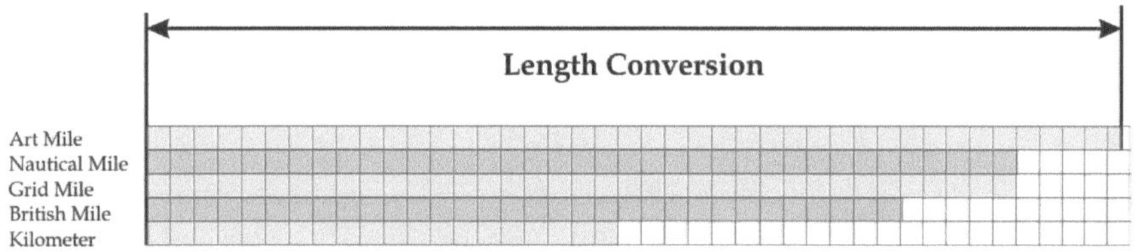

	British Foot	Meter	Grid Foot	Art Foot	Royal Cubit	Standard Cubit
Kilometer	3280.8398950131	1000.0000000000	3319.8962567050	3734.8832887931	1904.1029854251	2221.4534829959
British Mile	5280.0000000000	1609.3440000000	5342.8551213506	6010.7120115194	3064.3567149759	3575.0828341386
Nautical Mile	6076.1154855643	1852.0000000000	6148.4478674176	6917.0038508448	3526.3987290072	4114.1318505084
Grid Mile	6071.4262742620	1850.5707283951	6000.0000000000	6911.6656882123	3523.6772486773	4110.9567901235
Art Mile	6830.3545585448	2081.8920694445	6911.6656882123	6000.0000000000	3964.1369047619	4624.8263888889
Royal Cubit	1.7230369996	0.5251816775	1.7435486852	1.9614922708	1.0000000000	1.1666666667
Standard Cubit	1.4768888568	0.4501557236	1.4944703016	1.6812790893	0.8571428571	1.0000000000
Foot	1.0000000000	0.3048000000	1.0119043790	1.1383924264	0.5803705900	0.6770990216
Grid Foot	0.9882356680	0.3012142316	1.0000000000	0.8888888889	1.7435486852	1.4944703016
Art Foot	0.8784317049	0.2677459837	0.8888888889	1.0000000000	1.9614922708	1.6812790893

We know the velocity of light changes with the medium and this is where he can create a variety of values that nobody can question. The absurd idea that PHI reduction is evident is a slight of hand as you can see that 161874.977321814 Nautical miles per sec is very close to PHI (161803.4), this can be used to make fractional values based on this and rounding errors. The values of PHI from the minutes of arc is a result of ratios that are evident in the size of the earth and the number of degrees in s circle, such as;

In these simple equations you can see some of those harmonic numbers and PHI factors appearing at rounding and simplifying.

360 x 27 = 9720
9720 / 6 = 1620

velocity miles per second x multiple ft per mile / length of N mile in ft = Phi

Here is the equations, the way you can fudge these to fit:

186282.397051221 x 5280 / 6076.1154855643 = 161874.977321814
18.63 x 5.28 / 6.08 = 16.18 Simplify
37.26 x 2.64 / 6.08 = 16.18 Halving
36 x 2.7 / 6 = 16.2 Simplify, Halving and Rounding

Just to note 2.64 is the harmonic that Bruce says is inbuilt into the polar grid structure, which is really another factor of 5280/2 ft in a mile. The other harmonics he mentions are 2545.56, which again is 25.4 ratio mm to inches hiding as this value. Two other numbers Bruce quotes regularly are the magnetic field lines crossing in 1257 and 1850 spacing, firstly 12.57 is a half an inch and 1850(6069.5538 ft) is close to meters in a nautical mile(1852 exact), not a coincidence but as designed so he can correlate his data to fit any situation. As Bruce mentions the double cycle of the universe is 360 x 2 = 720 = 1440/2 as you can see he is eluding to the 144 value again. His use of the pentagon and golden triangle geometries are also used to make it appear to be more of the same harmonics of light generated from these shapes when he maps his grid points to fit the angles of 36, 72, 144 degrees.

One of the main points on Bruce's grid map was his Eltanin Aerial , which was a bogus interpretation as this has been confirmed as a sea sponge(Cladorhiza) that has multiple arms and club like ends , this was revealed by Larry Hatch back in 2006. This and many of Mr Cathie's findings have been exposed by Peter Brookesmith in an article "The Eltanin Enigma". I will not elaborate here so please check his site.

The Art miles and Art feet are also in whole number harmonic structures, very easy to use as these numbers are called perfect,deficient, super-perfect and superabundant numbers, most common in harmonic sequences.

Art ft	= 1.1383924264 British Ft	
Art Mile	= 1.2936277573 British Mile	= 7761.7665438009 B ft

So the Grid foot also comprises of 12 grid inches, so instead of using B/ft and B/inch you create a new unit that divides exactly 6000 times 12 = 72000 grid inches, this simplifies your figures and also creates a base 2,3,4,6,12 etc for harmonics as you can see from the tables 972, 144, 6 are all part of the harmonic structures.

One of the alarm bells that I heard when first viewing his work was his acceptance of Einsteins work as valid science, as we all should know his equations were the borrowings of others mainly Ruđer Josip Bošković and the hard work of his first wife and the blatant plagiarism of the work of David Hilbert, perhaps the most brilliant mathematician of the 20th century, and Felix Klein . Albert was a figure head put their by the power hungry establishment to teach us how to believe a metaphysical universe created from a false set of rules and hide the true knowledge and to accept this false dogma for fear of ridicule and banishment so we can serve the elite. He possess no genius that I can find in any of his writing, because it was not his writing that I was reading it was a patchwork of plagiarism from many sources, great men have a core of clear communication that goes beyond stereotypes and speaks to all men, many of the quotes of Einstein were lifted from other people, original thinking was not the premise of this fascist.

We should also be aware of Max Plank as he had reversed engineered his constant so it would fit with the velocity of light, as noted he was quite disturbed by force fitting this by using Boltzmann work and tried to ignore the implications of his earlier work please refer to the work of Charles William Johnson. Together with Einstein they accomplished an early burial to the science of Walter Russell, Nikola Tesla, Victor Schauberger and others like minded Artists, Scientists and Engineers. It seems Bruce was either ignorant of this side show of the establishment or he was just another unwitting puppet that had been used so his work would give credence to the puppet masters agenda, this way the independent researchers started looking in the wrong direction.

Remember it was not Einstein who created the atomic bomb, it was the Engineers who were disciples of Teslas work that made it work, theory had very little to do with its creation only practical men can build such devices. The Tesla rotating brush was the first step, the disintegration of matter by implosion of spherical reflected force was Tesla's second step, it was this that led men to the atomic era. There are no mysteries on how to create and explode an atomic bomb, only misinformation. The work of Einstein will be long forgotten and relegated to the dust piles of history when the work of Tesla will light the hallways of knowledge for aeons.

As Nikola Tesla quoted in the New York Times (11 July 1935)

"Einstein's relativity work is a magnificent mathematical garb which fascinates, dazzles and makes people blind to the underlying errors. The theory is like a beggar clothed in purple whom ignorant people take for a king, its exponents are brilliant men but they are meta-physicists rather than scientists. "

Harmonic Relations in Energy Equations

By Arto Heino

Name	Intermediate values	As per text book	Units	Simplified
Planck Constant H	6626.068	6.62606957×10−34	Joule sec	6561
Planck Energy E	19560.7871052737	1.956 × 10+9	Joules	19683
Planck Reduced Mass Mr	2176.4310869136	2.17651×10−8	Kg	2187
Planck Momentum P	6524.776252	6.52485	Kg Mtrs/s	6561
Length as Freq = H/P = L	1.0155241719	10.1569336685x 10-35	Mtrs	1
Velocity of Light C	2.99792458	299792458	Mtrs/s	3
Equation	**Intermediate values**	**As per text book**	**Units**	**Simplified**
$E = M \times C^2$	19560.7871052737	1.956 × 10+9	Joules	19683
H = E / (C/L)	6626.068	6.62606957×10−34	Joule sec	6561
E = H x C / L	19560.7871048711	1.956 × 10+9	Joules	19683
Powers of Light Velocity	**Intermediate values**	**Phenomenon**	**Harmonic triplets**	**Simplified**
C^1	2.99792458	Light Velocity	3	3
C^2	8.9875517874		9	9
$C^3 / 4$	6.7360006043	Gravity	27	6.75
$C^4 / 5$	16.1552174261	Length	81	16.2
$C^5 / 5$	48.432123417	Temperature	243	48.6
$C^6 \times 9/8$	645.3149033489	Time	729	648
C^7	2176.4310869136	Mass	2187	2187
C^8	6524.7762521343	Momentum	6561	6561
C^9	19560.7871052737	Energy	19683	19683

Einstein's famous formula

Energy = Mass x Velocity of light squared

This simple relationship was obviously borrowed from Newton's well known equation;

Kinetic Energy + Potential Energy = Mass x Velocity squared

The Bruce Cathie Equation is a unique combination trying to include a version of the Fitzgerald contraction, this being the 1/sqrt(C), this would not help those who are in favor of an Etheric understanding of the Universe. The purpose of the Lorentz-Fitzgerald formula was designed to invalidate Michelson- Morley experiment.

Lorentz Factor = 1 / sqrt(1 - v^2/C^2)

Cathie Equation

E = (C + 1/sqrt(C)) x C^2

This would validate any slight variations due to variable factors, each variable would be unique to the subject he was describing similar to Carl Munck's manipulations. I suspect the reason the authorities were interested in Bruce's work is he was using the powers of the velocity of light in his work, which is what they presented in the scientific community as fundamental constants. They did not want the public to know the level of deception the scientific hierarchy had perpetrated on world regarding the nuclear situation along with the complicity of the military industrial complex and the theoreticians to hide the real truth.

I hope this small snippet can elucidate the unwitting attempts by Mr Cathie over the decades in trying to steer you in unwillingly to a cult-e-sac of misinformation, not much different than the establishment has done for the last 100 years. I am an advocate for what Bruce has tried to do, but his efforts have not been realized by many researchers, scientists or engineers and have not resulted in usage that produces any advantages in designs or concepts. Not to deride his work as I have built many Tesla coils using Bruce's numeric guidance and the results are no different than using, millimeters or inches, as to the advantage I have gained from his work it is primarily by using frequency tables that are clearly of harmonic content. What you gain from Mr Cathie's work is the truly amazing value in understanding harmonic concepts and musical ratios that the Universe is built from

4.3 – Decoding Carl Munck

When I first viewed the work and videos of Mr Munck back in to early 1990's, I thought wow this guy has stumbled on to something, I better check it out. After my initial studies I was not convinced at his methods, so I persevered and allowed time to adjust to his methods so I could begin to find the code just as he had.

At first I was excited and found a few interesting sites I could apply this work on, that is when I decided to write a spread sheet so I could verify the Latitude , Longitude and Geomatra number with accuracy, just as Carl had done. Hidden in his mathematical multipliers was his secret to making you see what he wanted you to see, the numeric decimals he attached and multiplied was nothing but a tricky way to make it look as if harmonics and numeric codes were present, by reverse coding. Here is his method exposed and laid bare for all to see. I hope you can appreciate and understand that nobody in the last 20 years has questioned his means and methods, the main reason I would say is he did a fine job of cryptography, the other reason would be that he filled a wishful thought about the abilities of ancient man, that of his greater knowledge, something which I am still in concordance with, except this knowledge is not based on anything hidden and coded, only on deeper perceptions on what is already given and in front of us.

I will first find a location that has some ancient importance as Carl Munck had decoded and a version that is the exact position, also my back yard as a major grid point on the Earth just to show this fallacy. Each I will give special properties just as Carl did.

Panther Mound

The Panther Mound is in the Everglades of Florida, and by Carl it has a close relationship to The Great Sphinx. The Panther Mound can only be recognized from the air.

Geocode for Panther Mound: Latitude: 25.573164 – Longitude: -80.7589528

Here are Carl P. Muncks latitude/longitude coordinates for the Panther Mound:

80 deg 45 min 46.886 sec W.Greenwich, Longitude ,

this is a close match to the real location

80 deg 45 min 32.23008 sec W.Greenwich, Longitude,

this is the current real location

The difference of 14.65592 Seconds, this equates to 452.668607488 meters(at equator)

25 deg 34 min 3.5 sec North, Latitude,

this is also a close match

25 deg 34 min 23.3904 sec North, Latitude,

this is the current real location

The difference of 19.8904 seconds, this equates to 614.34285056 meters(at equator). This is at least 1/2 of a kilometer from Carl Munck;s calculated position. So something must be wrong with his numbers, his data, his calculations or all of the above.

Now this part is to makes it look like a genuine discovery, but this subject has been talked about and debated for more than a century, ever since they first were deciding where to put the current meridian, the current prime meridian at based at the Royal Observatory, Greenwich which was established by Sir George Airy in 1851, then again in 1884 in the conference of the Meridian in Washington , one of the proposals was the Pyramid of Giza. The need was originally for navigation in the open seas. The ancient Earth prime meridian (as said by Munck) passes through the center of The

Great Pyramid, we add 31 deg 08 min 0.8 sec onto the W.Greenwich longitude, to adjust to the ancient longitude, personally I find this a great location for a meridian, as its location and culture has endured thousands of years, Greenwich will be just a forgotten memory when the pyramids will still be standing, reminding us of their builders.

80 deg 45 min 46.886 sec W.Greenwich

+ 31 deg 08 min 0.8 sec Variance in prime meridians

= 111 deg 53 min 47.686 sec W.Giza

80 deg 45 min 32.23008 sec W.Greenwich, Longitude, (real)

+ 31 deg 08 min 0.8 sec Variance in prime meridians

= 111 deg 53 min 33.03008 sec W. Giza (real)

Now to figure out the Grid Longitude and the Grid Latitude of the Panther Mound. Muncks general rule is to multiply the site's centered (for both longitude and latitude) number of degrees X number of minutes X number of seconds.

Grid Longitude

Grid Longitude 111 deg x 53 min x 47.686 sec = 280536.738

Grid Longitude 111 deg x 53 min x 33.03008 sec = 194315.96064 (Real)

Muncks reverse PI tables told him that this adjusted number was VERY close to the multiplied product of a Gematrian number 'times' (PI to the 4th power).

The Gematrian number	= 2880
The derived number	= 2805.36738
The Actual value	= 1943.1596064
An Invented Geomatra Number	= 6267

So he used the derived number as a base with the PI table, this is the part that makes it look like a discovery.

2880 x (PI 4th power) = 280538.1822 W.Giza

= 111 deg x 53 min x 47.68624548 sec W.Giza

So we will create our own Geomatra number that makes it look real

6267 x (PI to 3rd Power) = 194316.335955438973

= 111 deg x 53 min x33.0301437966 sec W.Giza (created)

Grid Latitude

Grid Latitude 25 deg x 34 min x 3.5 sec = 2975

Grid Latitude 25 deg x 34 min x 23.3904 sec = 19881.84 (real)

Muncks PI table showed this number as being close to the multiplied product of another Gematrian number 'times' the Cube of Pi.

96 x (PI 3rd power) = 2976.602561 North

= 25 deg x 34 min x 3.501885366 sec North

So we will create our own Geomatra number that makes it look real

65 x (PI 5th power) = 19891.2795110432944620

= 25 deg x 34 min x 23.4015053071 sec North (created)

Grid Point Value

To figure his Grid Point Value, take the ratio of Grid Longitude to Grid Latitude , always greater-than-one .

280538.1822 / 2976.602561 = 94.24777961 = 30 x Pi

This result suddenly made many become followers and converted the undecided into the horde of lemmings that become indoctrinated to the numeric religion of the romantic notions of Geomatra. Here is now the door opener for those brave enough to face this truth.

194316.335955438973 / 19891.2795110432944620 = 9.7689208906 (created)

this value approximates

3 x pi = 9.4247779608 or I could choose Pi squared = 9.8696044011

Either one of these values are close enough, now let us work in reverse, I will do both just to prove his method:

187470.89274718436522 / 19891.2795110432944620 = 9.4247779608 (3 x Pi)

195 x (PI sixth power) = 187470.89274718436522

= 111 deg x 53 min x 31.8665464469 sec W.Giza (created)

This will give the result that makes it look brilliant and something unique, also it was not far off from the last value.

Here is the version with Pi squared

196319.0598054914843884/ 19891.2795110432944620 = 9.8696044011 (Pi squared)

65 x (PI seventh power) = 196319.0598054914843884

= 111 deg x 53 min x 33.3705694043 sec W.Giza (created) (very close to exact)

65 is the line number value for the magic square of 5, which they call the square of Mars.

This result will make the crowd dance in the isles, also it was not far off from the last value, this is what makes is look plausible. The Grid Point Value of any site like the Panther Mound that have values PI adjusted in the seconds decimals will always be some multiple of PI, easily designed to match Giza.

Here is my favorite location on the earth, my back yard next to my shed. I have interesting structure I built for my external Tesla coil, a ground system, it has a precise and real location:

Latitude 33.9033671156005 North

Longitude 150.899849953205 West

Using Munck reverse coding and numeric tricks, and not even using a PI table as Munck would have I have found this to be one of the most important grid points on Earth next to the Great Pyramid.

150 deg 54 min = 18.447905898 sec Longitude West

+ 31 deg 08 min 0.8 sec Variance in prime meridians

=181 deg 62 min = 19.247905898 sec Longitude West Giza

Grid Longitude = 181 x 62 x 19.247905898 = 216000 = 60 x 60 x 60

33 deg 54 min 12.1212121212 sec Latitude North

Grid Latitude = 33 x 54 x 12.1212121212 = 21600 = 60 x 60 x 6

Grid Point Value = 216000/21600 = 10 =decahedron is a polyhedron having 10 faces. My Tesla coil has 2160 turns and resonates at 216000hz, the Earth system I made has 8 radial arms 1 vertical down and a vertical up for the coil which makes 10, it also has 10 turns for the primary coil, the top load could be shaped as a decahedron. As you can see I can make fanciful additions, even though all the numbers are correct and accurate, my Tesla coil is also real and its numbers are provable. The only difference is that if I moved it 50 mm in any direction the whole number stack falls apart.

These are based on mathematical tricks that any body with a calculator can create, the problem is the amount of people that have been seduced by the fantasy that there is a hidden code that we must decipher in these ancient sites. The real code is not hidden at all but in plane sight, and you cannot call it a code as much as Carl would like you to believe. The reasoning is a clear headed understanding of the workings of the Universe based on pure natural prime geometry and natures proportions and not religious, ceremonial or Geomatra numbers as most try to make you believe. The very nature of Geomatra numbers act as lures to those who are seeking symbolic meaning and purpose to them. Attaching the velocity of light to 144 is a simple example of this, as this is one of those special numbers that are coded into our history. That does not mean that these numbers do not have some significance to our past, they do, it is just they should not be muddied by superstitious beliefs and foolish notions. Number Theory, Atomic physics, Crystallography and our understanding of electrical phenomena are true pointers for the use of these numbers.

The decimals placed after the whole number seconds are where he hides the PI and Geomatra numbers, when shown on a map this equates to distances under 1 meter,this allows you a great variation of choices to match the particular monument. A good example of this method is to show how numbers actually can do this. I am not a cryptographer as Carl is, but my methods unveiled his number game almost 10 years ago, a similar trick was used by Bruce Cathie, I have sat on this just hoping these guys would fess up and tell the world about their number game. This is when I realized, the purpose behind the deception, they believe their methods would guide any researcher to non-answers and permeate the underground researchers with false leads, 99 percent of the discoveries and conclusions they these types of writers have on the work of people like Schwaller de Lubicz and the engineer Nicola Tesla and his unfinished work have been a total fallacy. Only real Tesla engineers like Eric Dollard have pointed us in the right direction and the work of Christopher Dunn on clues left by ancient engineers proving Schwaller's notions of an enlighten ancient past.

More Tricks

If the number you want in the final answer is 144 then you find what constitutes the parts that can multiply into this number, look at the huge pallet of numeric interplay that can be found in a simple number.

12 x 12

2 x 6 x 2 x 6

2 x 3 x 2 x 2 x 3 x 2

10 x 14.4

2 x 5 x 14.4

8 x 18

2 x 2 x 2 x 6 x 3

6 x 24

5 x 28.8

4 x 36

3 x 48

20 x 7.2

2 x 10 x 7.2

2 x 2 x 5 x 7.2

3 x 4 x 5 x 2.4

(1 x 2 x 3 x 4 x 5 x 6)/5

(1 x 2 x 3 x 4 x 5 x 6 x 7)/ 35

19 x 3 x 2.5263157894736842105263157894737

This last addition lets you see how to hide the intent in the numbers after the decimal point.

19 x 3 x 2	= 114 differential	= 144 - 114	= 30
19 x 3 x 2.5	= 142.5	= 144 - 142.5	= 1.5
19 x 3 x 2.52	= 143.64	= 144 - 143.64	= 0.36
19 x 3 x 2.526	= 143.982	= 144 - 143.982	= 0.018
19 x 3 x 2.5263	= 143.991	= 144 - 143.991	= 0.009

Here is the same number with a hidden PI

2 x 22.918311805232928350719261925642 = 45.836623610465856701438523851284

PI x 45.836623610465856701438523851284 = 144

Now I will hide PI and reveal PHI(1.618034) in the same number

4.8 /PI = 1.5278874536821952233812841283761

(15 x 4.8 x 1.5278874536821952233812841283761)x PHI x PHI / 2 = 144

144 / (15 x 4.8 x 1.5278874536821952233812841283761) = PHI x PHI / 2

If Carl Munck could show how the ancients manipulated ***his*** values without a calculator I would probability would have been more convinced of ***his*** Code and not question his methods. The fact is clear ***his*** methods must be manipulated in inches and many decimal places at the seconds and PI and PHI are always involved, tweaking the bits so it fits in to his code. These two simple facts alone should make you question his results.

I propose the ancients didn't use or need any values after the decimal point, they had no need for the AVERAGE value of PI or PHI, because they used prime numbers as a base property of all of their work, a slow dawning of understanding that only now we are beginning to appreciate. One main clue is the earthquake proof walls that exist in every ancient site on Earth, these were designed using prime number law not by some PI/PHI fudge factor. PI and PHI come out of using whole number values that are manipulated through simple arithmetic, and the use of normal and didactic fractions is paramount to this understanding.

The clear knowledge of Quantum Arithmetic and its true roots in Ancient Egypt, Sumer, Gobekli, Angkor, Mayan and other ancient civilizations are what we should be looking at, not some tricks of code that have no meaning except to promote disinformation and enrich the promoters.

One of the Carl's codes secret is not to use whole prime numbers as they cannot be derived from other numbers, he always uses perfect,deficient, super-perfect and superabundant numbers along side decimal fractions. This keeps the Geomatra numbers easy to transform, hiding PI and PHI in the decimals. This is again will show either the ancients didn't understand prime numbers or Mr Munck avoided using only whole numbers and prime numbers in his code, the later seems to be the case.

4.4 - Quantum Arithmetic

Only a trained mathematician/geometer can recognize the missing elements in the geometric canon, that which has been ignored for the last 60 years, you cannot accidentally or casually find this most unique arithmetic, it is not based just on porisms. This slow discovery has been the work of mainly one man the late Ben Iverson, having worked in this field for at least 40 years. It is not number theory or invented mathematics, it does not need calculus, there are no approximations, it does begin its understanding as the trivial terms that most assume to be understood. Let us begin with the trivial terms so we can dispense with all the mystery.

Quantum Arithmetic uses a natural number system which has a base in all of the prime numbers which occur in the problem you need to solve. All its numbers are interlocked in a geometric arrangement. All numbers are positive integers, no decimal points or irrational numbers. Addition, subtraction, multiplication and division is all that is mainly required, the use of square roots and didactic fractions are also used, they are mainly used to verify the results and find the whole number ratio's. The ordering into Par numbers can also verify and help understand the internal arrangements.

The Quantum Number for one figure is the same for all geometric figures. A single quantum number defines the magnitude of the measurements, as they are related between themselves. Different geometric figures can be connected and their dimensions are calculated once for all of the various shapes. The shapes used are:

1 Right triangles

2 Equilateral triangle

3 Isosceles Triangles

4 Triplet circles (Koenig Series)

5 Triplet Squares (Koenig Series)

6 Ellipses

Each shape is calculated from a single Quantum Number. The one requirement is that the second and third integers of the Quantum Number must be prime to each other. These are in the Fibonacci sequence. They derive from Euclid VII, Proposition 28. The first and fourth integers become, "Sum and Difference" numbers.

A complete unified whole number system should have features based on the interlocking geometry of at least 3 sets of parameters-

1	spacial domain(3 dimensional co-ordinates)	Xs,Ys,Zs
2	time domain(frequency/phase relation/velocity)	Xt,Yt,Zt
3	ether domain(inertia/ether pressure/gravity/radiation)	Xe,Ye,Ze
4	resulting in a 4th(growth/energy/movement)	Xg,Yg,Zg

The parameters are minimum of 12 constants, creating a unity based on 3 numbers. Starting with the smallest set of numbers that result in a whole number unification, the smallest Pythagorean triangle.

a) **3,4,5**

3+4+5	= 12
3x4x5	= 60
3^2,4^2,5^2	= 9,16,25

This gives you a set of eight numbers

3,4,5,9,12,16,25,60

Another type of expansion is the squared numbers as they form a set of three

b) **9,16,25**

9+16+25	= 50
9x16x25	= 3600
9^2,16^2,25^2	= 81,256,625

This gives you another set of three, this sort of expansion can continue with cubed numbers etc.

3^3,4^3,5^3	= 27,64,125
3^4,4^4,5^4	= 81,256,625
3^5,4^5,5^5	= 243,1025,3125

To expand the set you need to create a matrix of areas based on the first three base numbers.

c)	3x3,3x4,3x5	= 9,12,15
	9+12+15	= 36
	9x12x15	= 1620
	9^2,12^2,15^2	= 81,144,225
d)	4x3,4x4,4x5	= 12,16,20
	12+16+20	= 50
	12x16x20	= 3840
	12^2,16^2,20^2	= 144,256,400
e)	5x3,5x4,5x5	= 15,20,25
	15+20+25	= 60
	15x20x25	= 7500
	12^2,20^2,25^2	= 144,400,625

As you can see by these simple and trivial numeric translations, the system of values that can be created that has a unification built into its nature by the virtue of its simplicity. This is the sort of thinking that help build the ancient monuments of the past civilizations.

All this is an application of simple and basic sacred geometric figures and whole number units. I would simplify this by saying , a new set of tools must be created to simplify the hidden complex nature of this understanding, I guess you would call it a new physics? a cosmology? paradigm? at least it should be a game changer.

Structural Fundamentals of Design

3 4 5

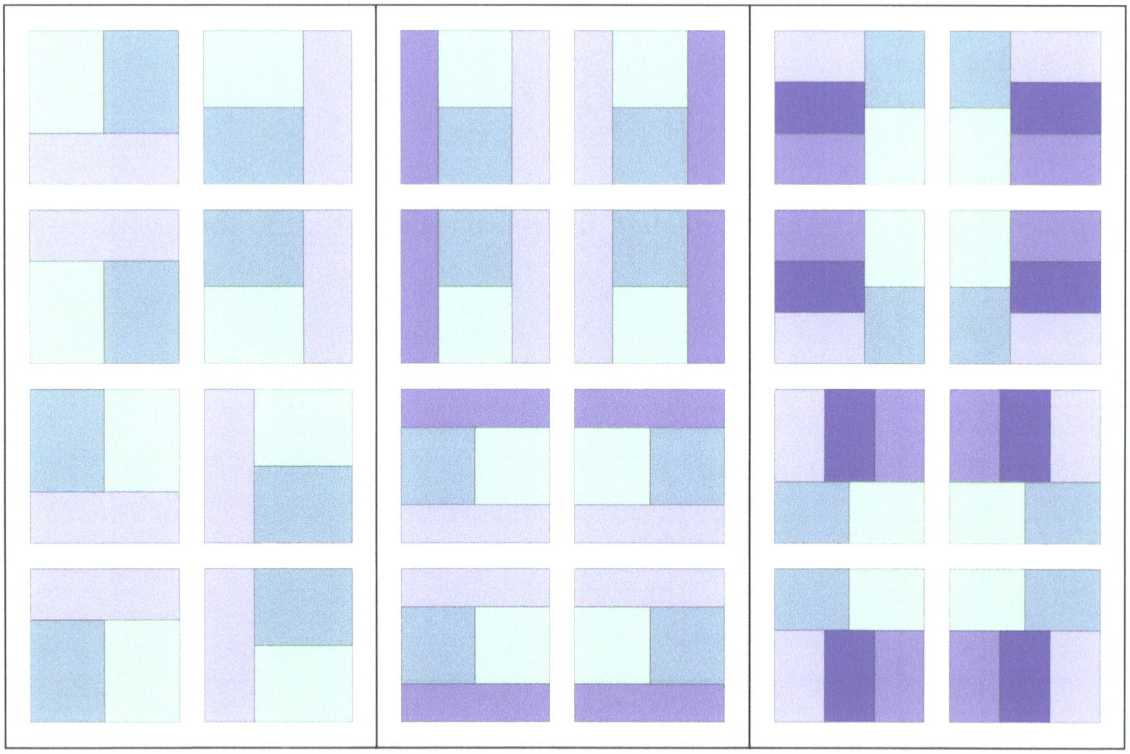

1 Whole Number Linear per unit sections
2 Whole Number Area per unit sections
3 Equal Area Value per unit sections
4 Only 2 Basic Elements per unit
5 Only 8 Permutations per unit
6 Base numbers are 3,4,5
7 Base Linear = 3x4x5 = 60
8 Base Area Total = 60x60 = 3600
9 Area Units are 1200,900,720
10 1200 = 20x60 = 30x40
11 900 = 15x60 = 30x30
12 720 = 20x36 = 24x30

Arto Heino

This diagram above is also an intuitive beginning to unlock the geometry behind the stones walls found at Machu Picchu.

I have found some ways forward and would like to share these findings here, hopefully with those with an open mind. I have been reluctant to bring many of these things forward in the past, mainly because of the personal attacks by those trolls in the late 1990's which led me back underground again. New things get forged in the battle line of ideas, This time around I can see more people are willing to try something different as we have all found out by now, how the methods of the corporate institutions has only given us bankrupt directives with little or no creative thought, designed to lead to their profit motive.

I am pursuing a vision of how to unlock the underlying geometry that make these things work, as an artist I am trying in earnest to manifest its reality. I know with the right environment and support WE ALL will succeed finally and put an end to the conspiracy and secrecy that hinders the truly innovative inventors and rid the system of scammers, profiteers and misinform-ants.

Chapter Five

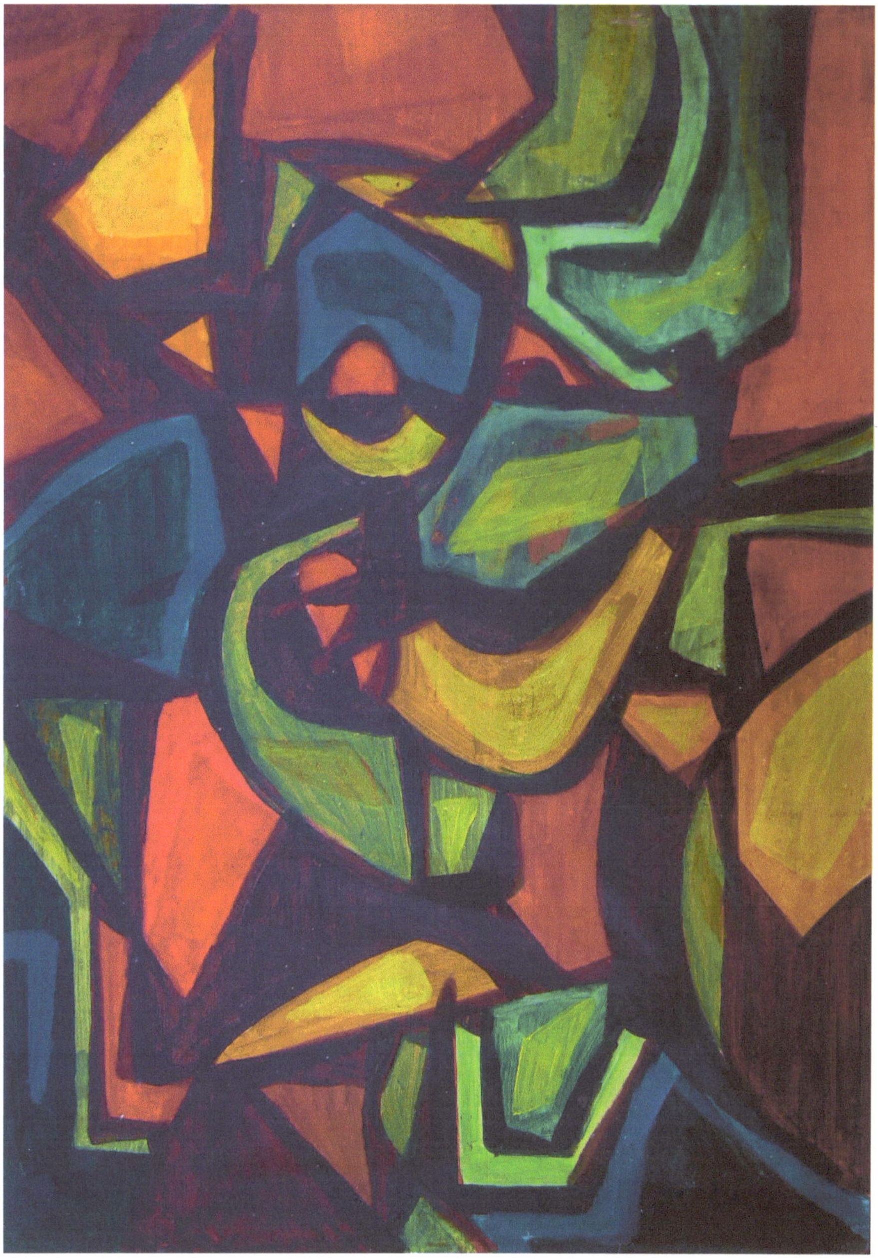

Geometry

5.1 – Variety of Geometry

5.1a - Archimedes

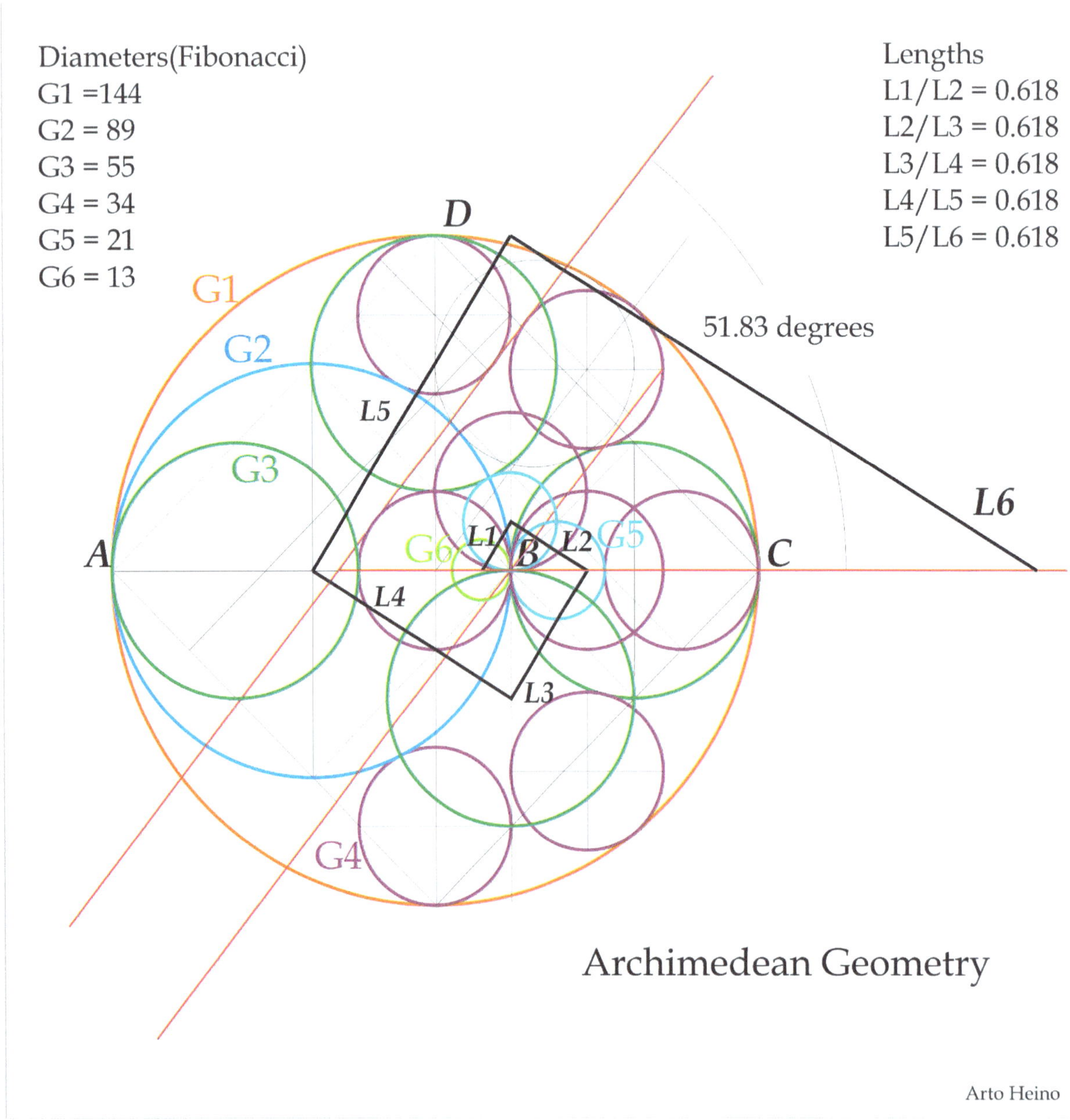

5.1b – The Relations of 2 Shapes

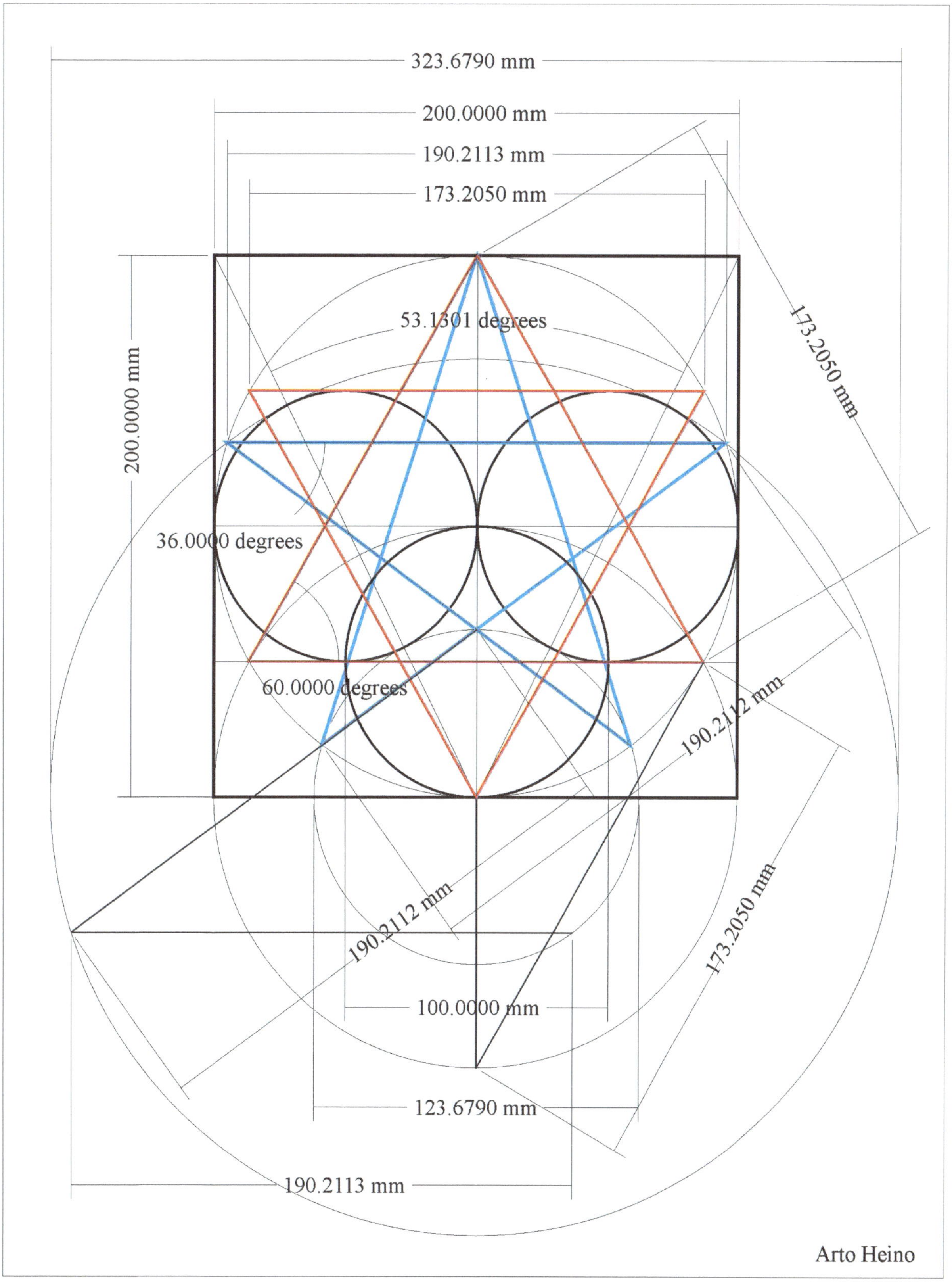

5.1c - The Pythagorean Triangle Series

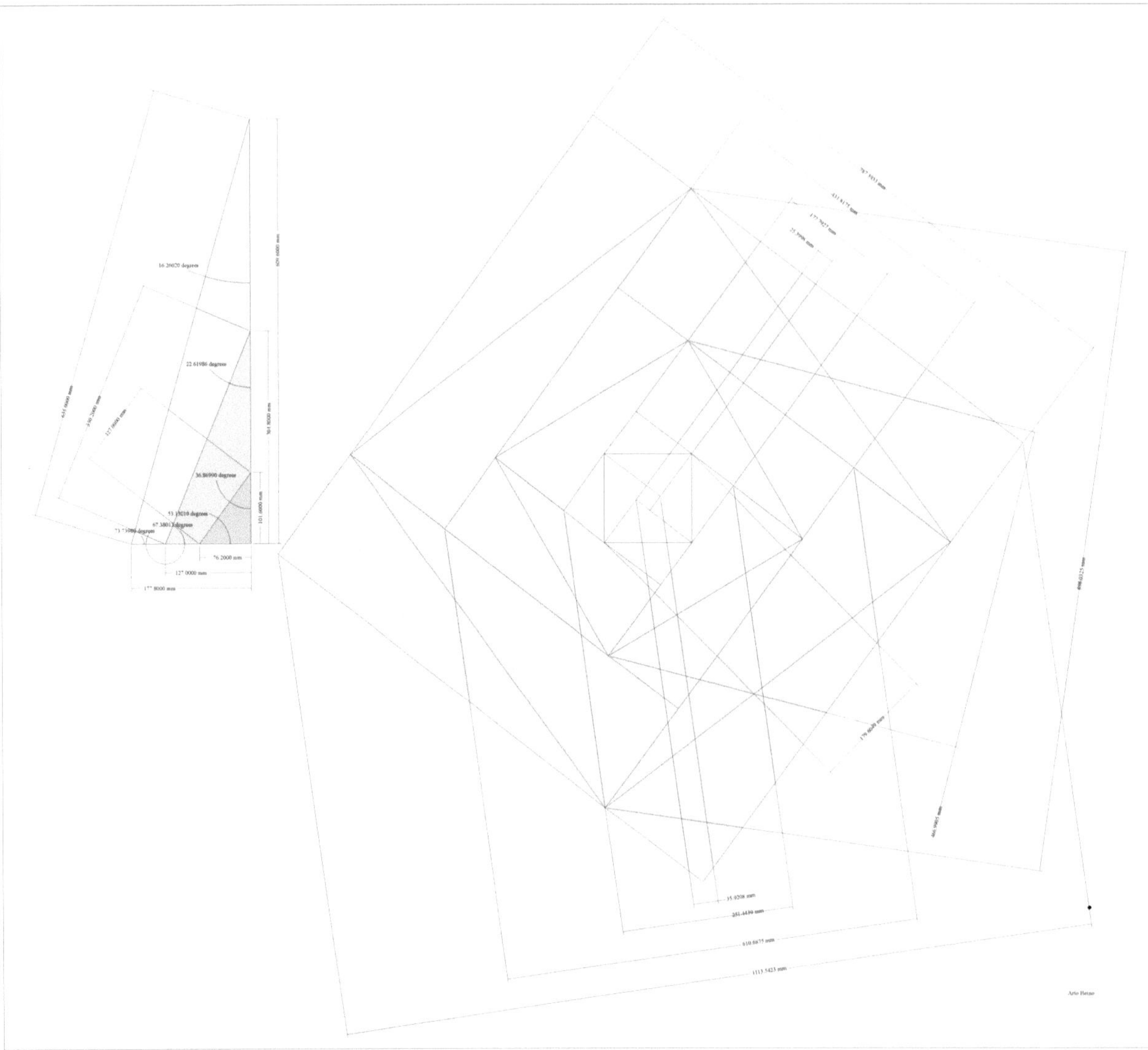

5.2 - Quantum Arithmetic Geometry

There are sixteen primary identities. The first four are given the identity of "a", "b" "e" and "d". These, are the roots of its given problem, and is the base numbers. The next twelve identities are the upper case letters A through L. They are combinations, of the first four base numbers, and are usually considered as one dimension, higher than the base identities. They will denote linear dimension, surface areas, or volumes, for the three standard dimensions above the roots.

There is also one dimension below the "roots" b, e, d, and a. They are called "quaternions", and are the square roots of the root numbers. In conventional mathematics, these are called "Gaussian Integers". They are integers, only when their base number is a perfect square. A problem in QA is well defined and problems solvable when only one of the upper case identities is assigned a value and the name of that identity. All of the values within a problem in QA are intertwined this gives you hundreds of ways to solve any given problem.

The Quantum Number for one figure is the same for all geometric figures as I explained in the last chapter. A ***single quantum number*** defines the magnitude of the measurements, as they are related between themselves. The shapes used are:

1 Right triangles

2 Equilateral triangle

3 Isosceles Triangles

4 Triplet circles (Koenig Series)

5 Triplet Squares (Koenig Series)

6 Ellipses

All Quantum ellipses will give the value ***one*** when you apply the Steiner Inellispe porism. The numeric laws it contain can be geometrically reconstructed with some of the most excellent computer programs written such as GeoGebra, First you need to understand the construction of the "Quantum Ellipse". The usual 2 foci and semi-major length/third point will not suffice only Conic section ellipse will create the correct elliptical form, if you have 5 points that are aligned to the quantum arithmetic standard. I hope some of these diagrams will help clarify all the parameters. The best way to construct an ellipse of this sort is to start with a string with length 3+4+5=12 a Pythagorean triangle. This method was re-discovered by James Clerk Maxwell in the mid 1800's.

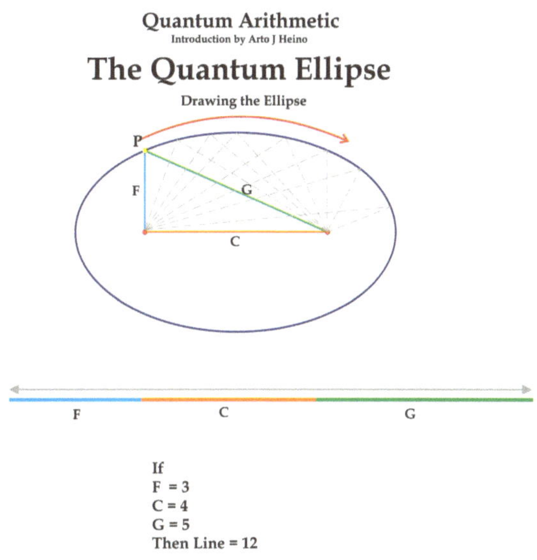

This type of geometry had been developed by Ben Iverson from the 1950's through to to 1990's extended by Dale Pond and myself over the last 20 years. After many calculations and with the application of Quantum Arithmetic. I have created a beginning for the foundation of the Quantum 120 degree Isoceles. The smallest integer units begin at the 4 digit root numbers, which means some calculations are up to 20 numbers long This study has revealed the Quantum Equilateral also within its perimeter, having now multiple **Z** line integers.

Listed below is the some of the raw data for one of the smallest Quantum Isosceles 120 degree Triangle, the added Var1,Var2,Var3,Var4,Var5,Var6 are part of an incomplete data list, you can ignore them or just make each equal zero.

b	= 2646	= d – e
e	= 4343	= a – d
d	= 6989	= b + e
a	= 11332	= d + a
B	= 7001316	= b x b
E	= 18861649	= e x e
D	= 48846121	= d x d
A	= 128414224	= a x a
C	= 60706454	= 2 x d x e
Fe	= 29984472	= d x d – e x e + Var1
G	= 67707770	= d x d + e x e
L	= 151687580848524	= (d x d x d x e – e x e x e x d) / 6
H	= 90690926	= (a x a + 2 x b x a – a x a) / 2
I	= 30721982	= (e x e + 2 x d x e) – d x d
J	= 18492894	= d x d – d x e
K	= 79199348	= d x d + d x e
We	= 109552575	= d x e + d x a
X	= 30353227	= d x e
Ye	= 79568103	= 2 x d x e + d x d + Var2
Ze	= 98060997	= e x e + d x d + d x e + Var3
Wib	= 328657725	= 3 x (d x e + d x a)
Fi	= 109552575	= d x e + d x a + Var4
Yi	= 219105150	= 2 x (d x e + d x a) + Var5
Zi	= 109552575	= d x e + d x a + Var6

Wis = 189750626 = (d x e + d x a) x sq 3

Wh = 94875313 = ((d x e + d x a) / 2) x sq 3

5.2a – Quantum Ellipse

To really come to grips with how these ideas work, follow the simple logic of building these equations without preconceived notions of what are the invented mathematical versions are built from. I have included some data on the right hand side so you can cross check these parameters using other formulas.

Quantum Arithmetic

Introduction by Arto J Heino

The Quantum Ellipse

Quantum set - (b , e , d , a)

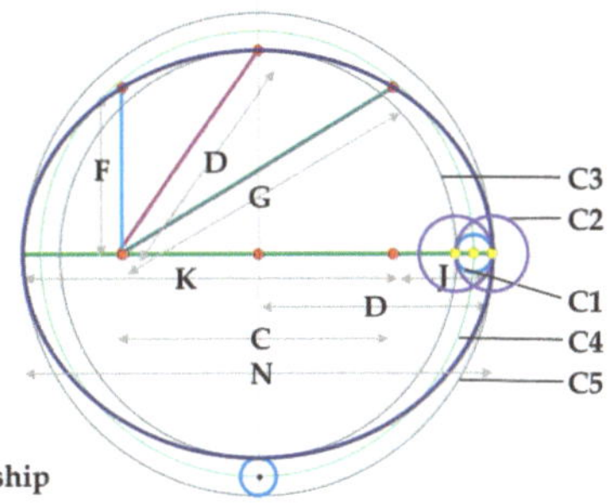

Circle	
Set 1-1	Radius
C1	0.267949192
C2	0.535898385
C3	3.464101615
C4	3.732050808
C5	4
Set 1-2	
C1	1.145898034
C2	2.291796068
C3	6.708203932
C4	7.854101966
C5	9
Set 1-3	
C1	2.708497378
C2	5.416994756
C3	10.58300524
C4	13.29150262
C5	16
Set 1-4	
C1	5
C2	10
C3	15
C4	20
C5	25
Set 3-1	
C1	0.254033308
C2	0.508066615
C3	15.49193338
C4	15.74596669
C5	16

Fibonacci Relationship

b	=	d - e		
e	=	d - b		
d	=	a - e	=	b + e
a	=	d + e		

Quantum Numbers

Set 1-*	b (*Prime*)	e	d	a
1	1	1	2	3
2	1	2	3	5
3	1	3	4	7
4	1	4	5	9
5	1	5	6	11

Circle	*Radius*
C1	(D - SQRT(D^2 - (C/2)^2)) / 2
C2	D - SQRT(D^2 - (C/2)^2)
C3	SQRT(D^2 - (C/2)^2)
C4	SQRT(D^2 - (C/2)^2) + (D - SQRT(D^2 - (C/2)^2)) / 2
C5	D

D	=	d * d	=	(G + F) / 2
C	=	2 * d * e	=	G * G - F * F
F	=	a * b	=	G * G - C * C
G	=	d*d + e*e	=	C * C + F * F
J	=	d * b		
K	=	d * a		

Set 1-*	D	C	F	G	J	K	L	H	I
1	4	4	3	5	2	6	1	7	1
2	9	12	5	13	3	15	5	17	7
3	16	24	7	25	4	28	14	31	17
4	25	40	9	41	5	45	30	49	31
5	36	60	11	61	6	66	55	71	49

Quantum Matrix

b	e	d	a
B	E	D	A
C	F	G	L
H	I	J	K

Set 1-1

1	1	2	3
1	1	4	9
4	3	5	1
7	1	2	6

Set 1-2

1	2	3	5
1	4	9	25
12	5	13	5
17	7	3	15

Set 1-3

1	3	4	7
1	9	16	49
24	7	25	14
31	17	4	28

Set 1-4

1	4	5	9
1	16	25	81
40	9	41	30
49	31	5	45

Set 3-1

3	1	4	5
9	1	16	25
8	15	17	10
23	7	12	20

Quantum Arithmetic

Introduction by Arto J Heino

Quantum Ellipse

b e d a
1, 2, 3, 5

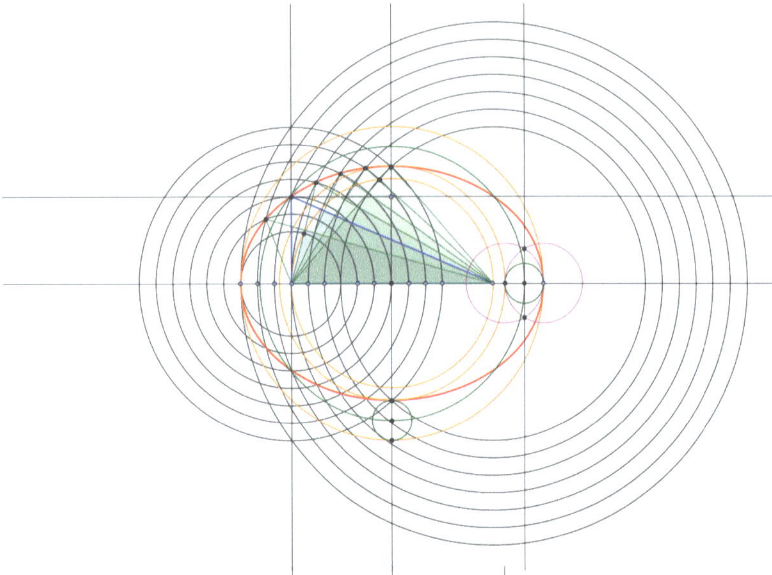

b e d a
1, 3, 4, 7

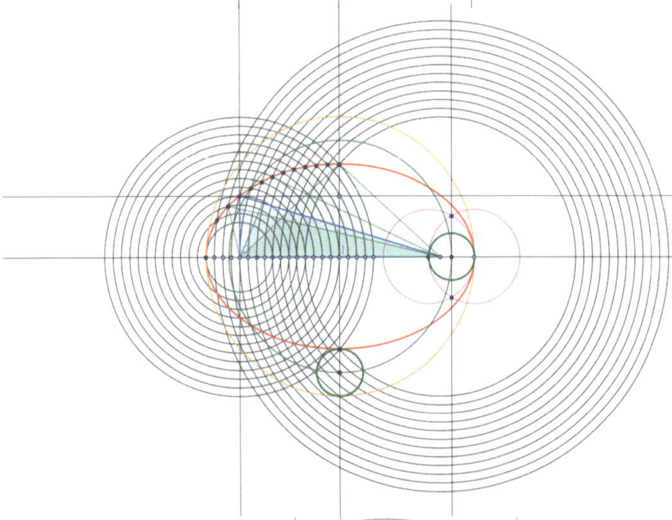

b e d a
3, 1, 4, 5

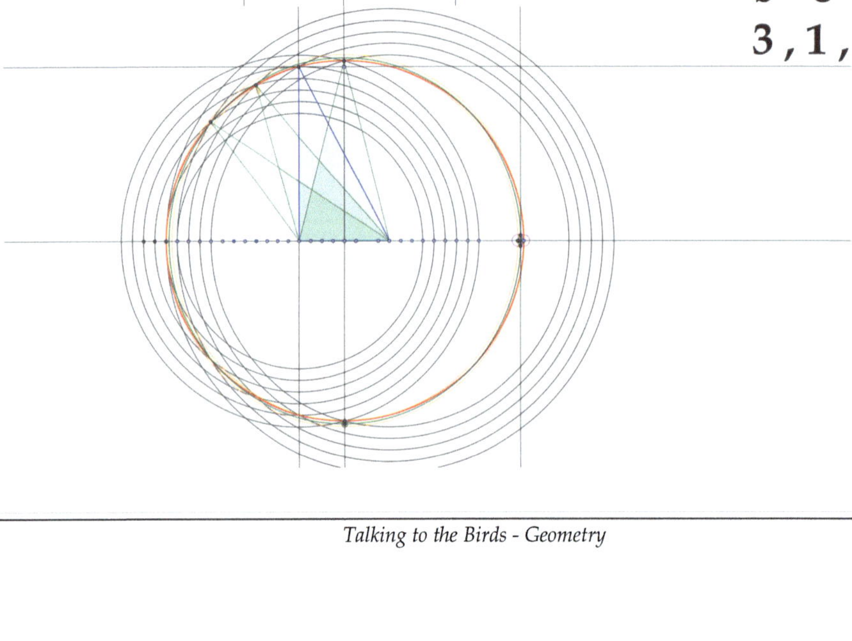

Arto Heino 2012

5.2b – Quantum Isosceles

To push the boundaries of this type of geometry, I decide on an action plan to discover what I could find by new methods.

Quantum Arithmetic

Introduction by Arto J Heino

The Quantum 120 Degree Isosceles

Quantum set - (b , e , d , a)

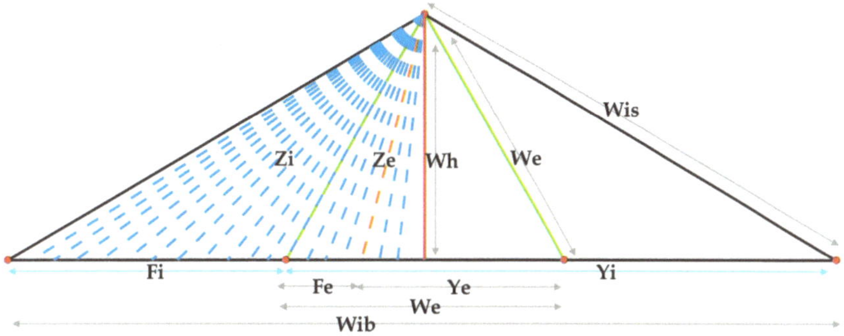

b	e	d	a
B	E	D	A
C	F	G	L
H	I	J	K
We	X	Ye	Ze
Wib	Fi	Yi	Zi
Wis	Wh		

2646

4343

2646	4343	6989	11332
7001316	18861649	48846121	128414224
60706454	29984472	67707770	151687580848524
90690926	30721982	18492894	79199348
109552575	30353227	79568103	98060997
328657725	109552575	219105150	109552575
189750626	94875313		

	Quantum set of 4	D	Semi-Major Diameter of Ellipse
	Pythagorean Triangle	H	Radius of Koenig Circle/Square(outer)
C	Pythagorean Triangle (Base)	I	Radius of Koenig Circle/Square (inner)
C	Distance between Foci of Ellipse	Ze	Equilateral Divisor Line
F	Pythagorean Triangle (Height)	Zi	Isosceles120 Divisor Line
F	Divided Smaller Base of Equilateral Triangle	Fi	Divided Smaller Base of Isosceles120 Triangle
F	Semi-Latus Rectum of Ellipse	X	Half of Pythagorean Triangle Base C
G	Pythagorean Triangle (Hypotenuse)	Ye	Divided Larger Base of Equilateral Triangle
G	Radius of Koenig Circle/Square(middle)	Yi	Divided Larger Base of Isosceles120 Triangle
G	One Foci to End on Semi-Latus of other Foci	We	Side of Equilateral Triangle
L	Area of Pythagorean Triangle / 6	Wib	Base of Isosceles120 Triangle
K	Apogee of Ellipse	Wis	Side of Isosceles120 Triangle
J	Perigee of Ellipse	Wh	Height of Isosceles120 Triangle
D	Divided Larger Line of Pythag Triangle	E	Divided Smaller Line of Pythag Triangle
A		B	Difference between Base and Hypotenuse of Pythag Triangle

Quantum Arithmetic

Introduction by Arto J Heino

The Quantum 120 Degree Isosceles

Quantum set - (b , e , d , a)

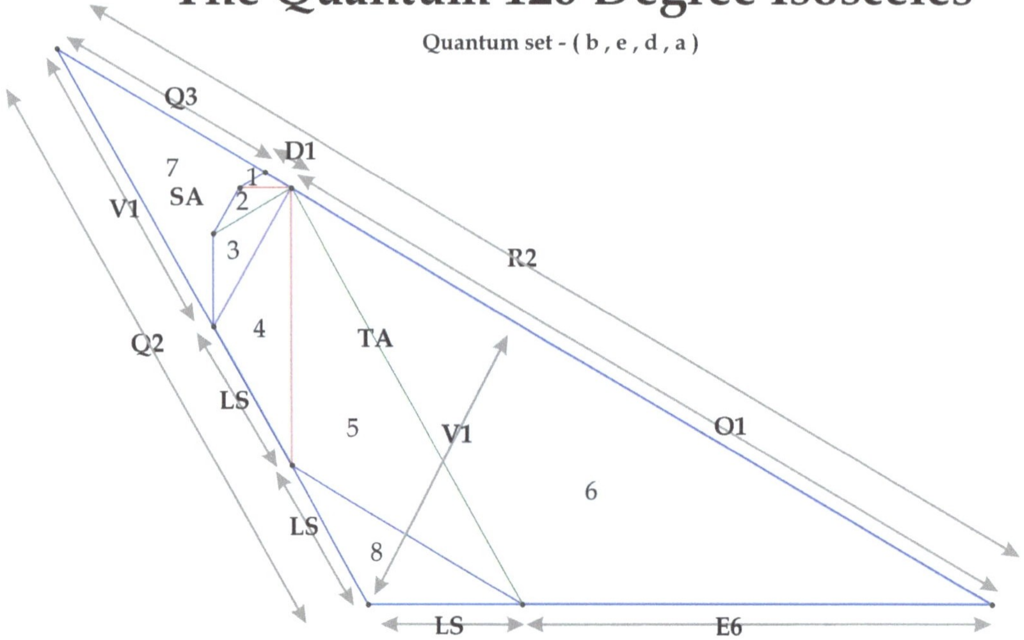

Total Area	**TA**	6735204978316471800
Shape Area	**SA**	639220843418731842

R2	6831022536	**Wib**
O1	5123266902	
Q2	3943892700	**Wis**
E6	2957919525	
V1	1971946350	**Wh**
Q3	1518005008	
LS	985973175	
D1	189750626	

Relationships	**Value**	**Value * Variable**	**Variable**
R2/(O1+D1)	1.2857142857	9	7
O1/D1	27	27	1
Q3/D1	8	8	1
E6/LS	3	3	1
Q2/LS	4	4	1
V1/(LS+LS)	1	1	1
R2/Q3	4.5	9	2

TA/SA	36 / (41 / 12)
Iarea2/Iarea1	3
Iarea3/Iarea1	9
Iarea4/Iarea1	27
Iarea5/Iarea1	81
Iarea6/Iarea1	243

	Wib	**Wis**	**Wh**	**Iarea**	**We**	**Earea**
1	328657725	189750626	94875313	15590752264621462.5	109552575	5196917421540487.5
2	569251878	328657725	164328862.5	46772256793864387.5	189750626	15590752264621562.5
3	985973175	569251878	284625939	140316770381593162.5	328657725	46772256793864387.5
4	1707755634	985973175	492986587.5	420950311144779487.5	569251878	140316770381593162.5
5	2957919525	1707755634	853877817	1262850933434338462.5	985973175	420950311144779487.5
6	5123266902	2957919525	1478959762.5	3788552800303015387.5	1707755634	1262850933434338462.5

Quantum Arithmetic

Introduction by Arto J Heino

The Quantum 120 Degree Isosceles

Quantum set - (b , e , d , a)

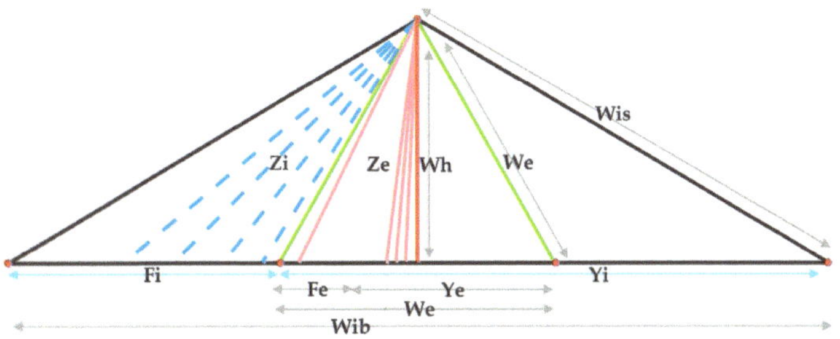

b	e	d	a
B	E	D	A
C	F	G	L
H	I	J	K
We	X	Ye	Ze
Wib	Fi	Yi	Zi
Wis	Wh		

54 6007

54	6007	6061	12068
2916	36084049	36735721	145636624
72816854	651672	72819770	3954392073324
73468526	72165182	327294	73144148
109552575	36408427	108900903	109228197
328657725	109552575	219105150	109552575
189750626	94875313		

2646 4343

2646	4343	6989	11332
7001316	18861649	48846121	128414224
60706454	29984472	67707770	151687580848524
90690926	30721982	18492894	79199348
109552575	30353227	79568103	98060997
328657725	109552575	219105150	109552575
189750626	94875313		

5094 2881

5094	2881	7975	10856
25948836	8300161	63600625	117852736
45951950	55300464	71900786	211763679725400
101252414	9348514	40624650	86576600
109552575	22975975	54252111	94876761
328657725	109552575	219105150	109552575
189750626	94875313		

5042 2911

5042	2911	7953	10864
25421764	8473921	63250209	118026496
46302366	54776288	71724130	211355977924784
101078654	8473922	40099026	86401392
109552575	23151183	54776287	94875313
328657725	109552575	219105150	109552575
189750626	94875313		

5770 2495

5770	2495	8265	10760
33292900	6225025	68310225	115777600
41242350	62085200	74535250	213378295685000
103327550	20842850	47689050	88931400
109552575	20621175	47467375	95156425
328657725	109552575	219105150	109552575
189750626	94875313		

Quantum Arithmetic

Introduction by Arto J Heino

The Quantum 120 Degree Isosceles

Quantum set - (b , e , d , a)

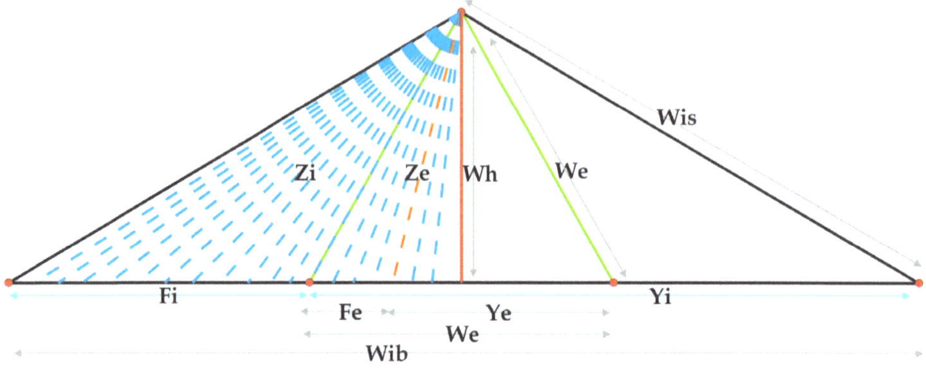

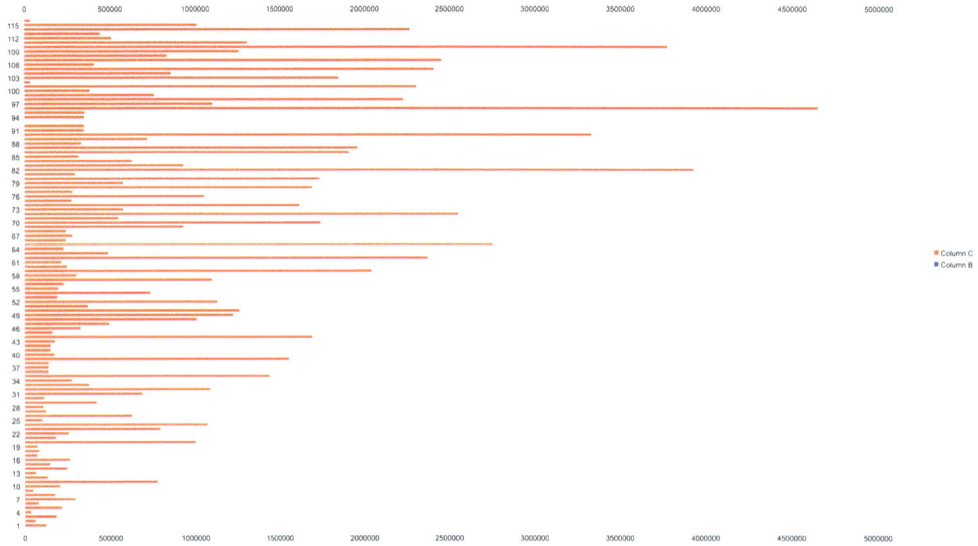

This graph shows the new Zline values for this group

115 Zi values
40 Ze values
117 total Zi if you Add Wis + Wh
42 total Ze if you add We + Wh

Note Element Table 42 + 1 = Technetium
Element Table 117+1 = Completed Periodic Table

5.3 - A Formula of Shapes

General Formula for Areas of Triangles and Quadrilaterals

by Arto Juhani Heino

$$(h * a)+((v * h)/2)+((a * (g-h))/2)+((g * (b-a-v))/2)$$

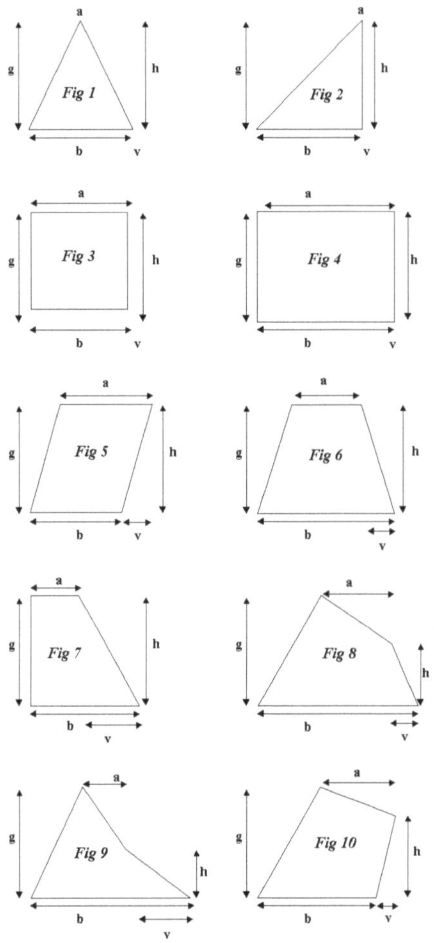

Variables

Figures	a	b	h	g	v	*Area*
1	0	1	1	1	0	0.5
2	0	1	1	1	0	0.5
3	1	1	1	1	0	1
4	1	2	1	1	0	1.5
5	1	1	1	1	0.5	1
6	1	2	1	1	0.5	1.5
7	1	2	1	1	0.5	1.5
8	1	2	0.5	1	0.25	1.1875
9	1	2	0.25	1	0.5	0.9375
10	1	2	0.75	1	0.25	1.34375

Definitions of Variables

Variable	Points/ Planes
a	*x to y*
b	*w to z*
h	*Height y to base*
g	*Total Height*
v	*y to z Edge*

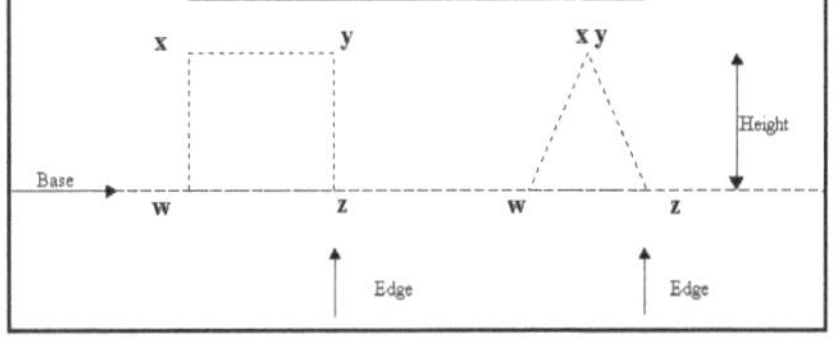

5.4 - The Philo Line

The important points are I,H,G,P,F they make are 5-point circle with the triangle I,J,K. They are what is called concyclic(on the same circle), the two lines of interest are FH and HJ. This interesting geometric construct cannot be created by a compass and straight edge, as this has been tried many times for the last 2000 years.

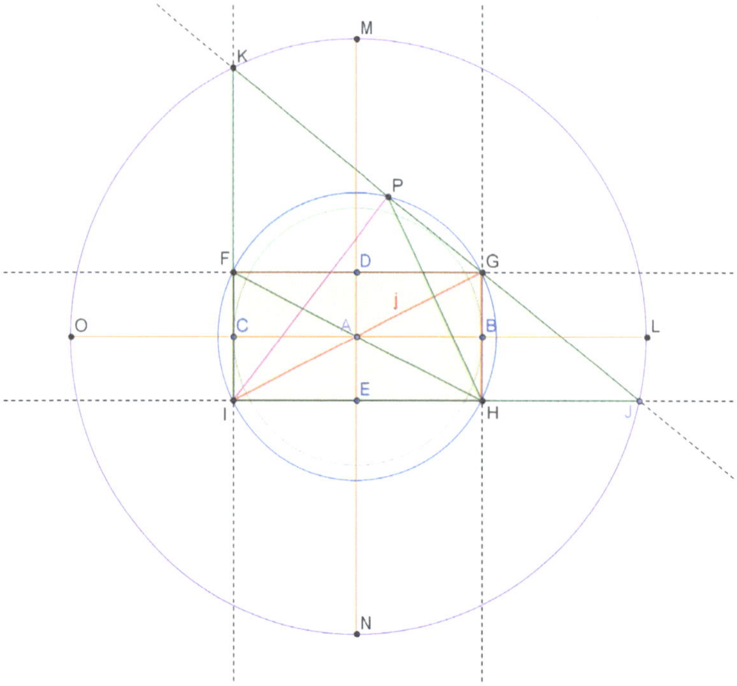

The critical ratio of the triangle is KI:IJ which is the ratio of the two edges of a cube which one is twice the volume of the other, we will designate this value as Z =1.2599210499. Some interesting aspects of this construct is the ratios of its parts, such as KP:KG = JP:JG, which is simple as KP=JG. this ratio is the same as KG:GJ = Z^2. Here is a chart with some of the interesting correlations.

Rectangle	R1	**FGHI**	6.000000000	**Line**	L1	**IF**	1.000000000
Circle radius	C1	**AB**	1.000000000	**Line**	L2	**IH**	2.000000000
Circle	C2	**AG**	1.118033989	**Line**	L3	**FH**	2.236067977
Circle	C3	**AL**	2.314571915	**Line**	L4	**CB**	1.000000000
Triangle	T1	**IJK**	10.009260287	**Line**	L5	**IP**	2.026633453
Circle Area	CA1	**C1**	3.141592654	**Line**	L6	**HP**	1.751444819
Circle Area	CA2	**C2**	3.926990817	**Line**	L7	**IK**	2.587401052
Circle Area	CA3	**C3**	16.830275729	**Line**	L8	**IJ**	3.259921050
Rectangle Area	RA1	**R1**	2.000000000	**Line**	L9	**KJ**	4.161938185
Triangle Area	TA1	**T1**	4.217361577	**Line**	L10	**HJ**	1.259921050

What brought my attention to this geometry was my study of the Kings Chamber which has the floor plan with the same ratios as the rectangle R1, refer to page 273 and my article on this subject. this will prove to any doubter that the Ancient Egyptians were familiar with this geometry 8 thousand years ago.

5.5 - Arto-Parker Pi Series

The Circumference of the Shape Divided by Two Multiplied by the In-Circle Radius Equals the Area of the Shape

AP-Pi Multiplied by the In-Circle Radius Squared Equals the Area of the Shape

The Shape must have All its Sides Tangent to an In-circle

The Circumference of the Shape Divided by Two times the In-Circle Radius Equals the AP-Pi of the Shape

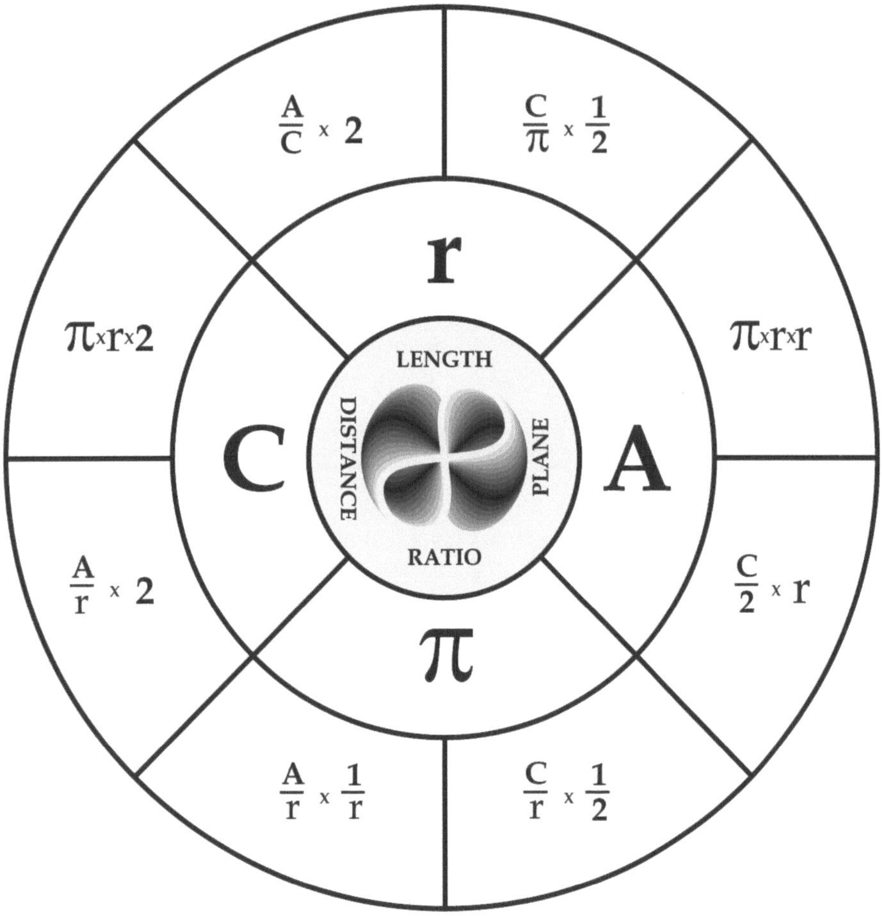

$$\pi_3 \times r^2 = A$$

5.19615242

$$\sqrt{27}$$

π_3 π_3

Π_3

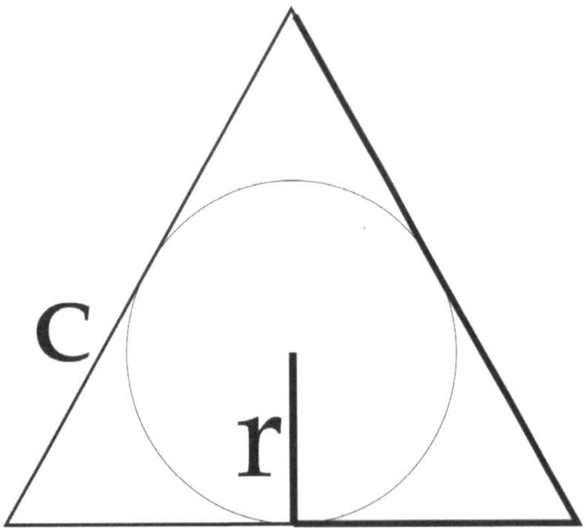

TRIANGLE

$$\frac{C}{2} \times r = A$$

$$\pi_4 \times r^2 = A$$

4

$$\sqrt{16}$$

π_4 π_4

Π_4

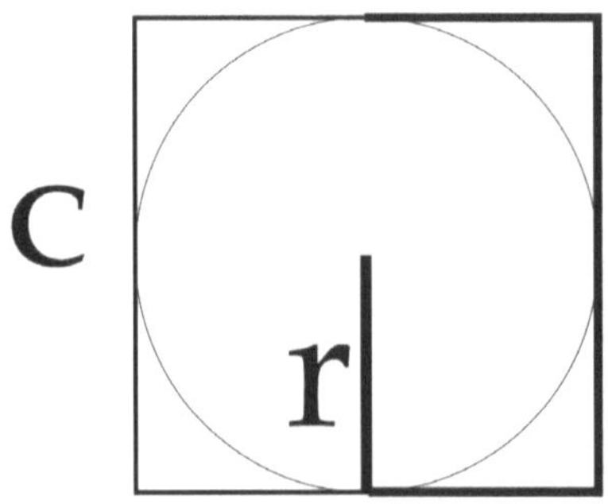

SQUARE

$$\frac{C}{2} \times r = A$$

$$\pi_5 \times r^2 = A$$

3.63271264

46041 / 12674

π_5 π_5

Π_5

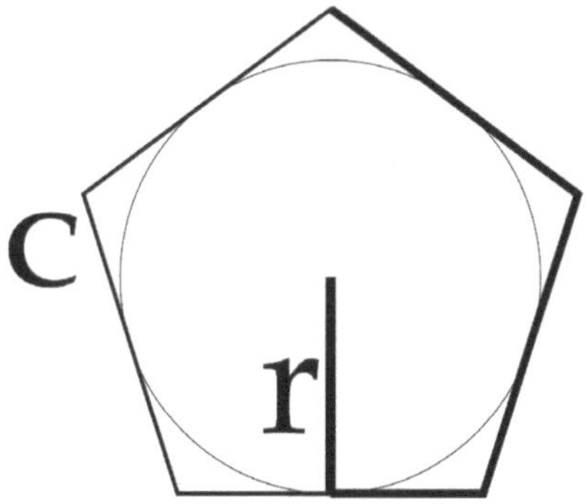

PENTAGON

$$\frac{C}{2} \times r = A$$

$$\pi_6 \times r^2 = A$$

3.464101615

$\sqrt{12}$

π_6 π_6

Π_6

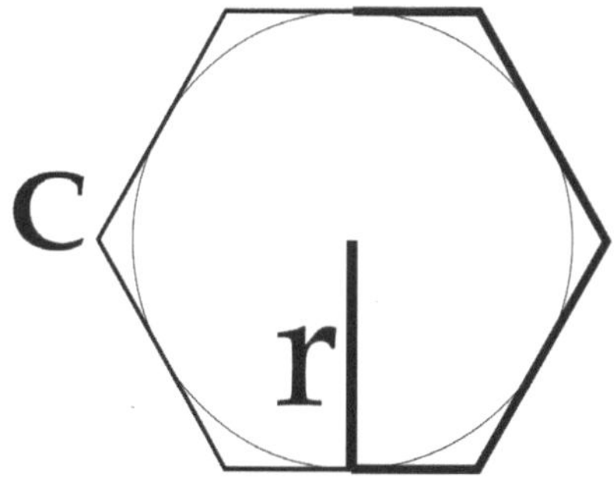

HEXAGON

$$\frac{C}{2} \times r = A$$

$$\pi_8 \times r^2 = A$$

3.31370851

11482 / 3465

π_8 π_8

Π_8

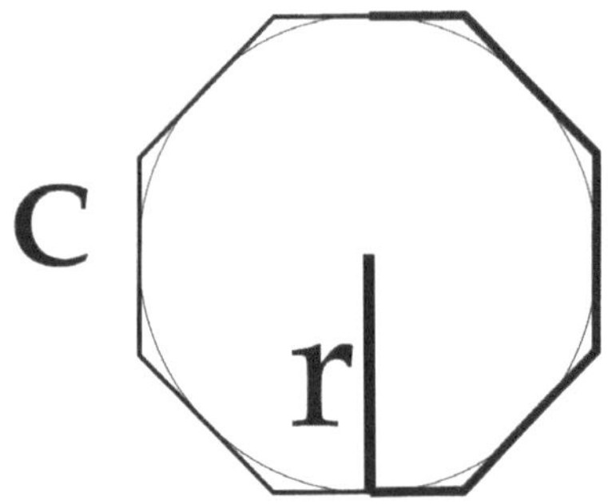

OCTAGON

$$\frac{C}{2} \times r = A$$

$$\pi_9 \times r^2 = A$$

3.25731511

79597 / 24299

π_9 π_9

Π_9

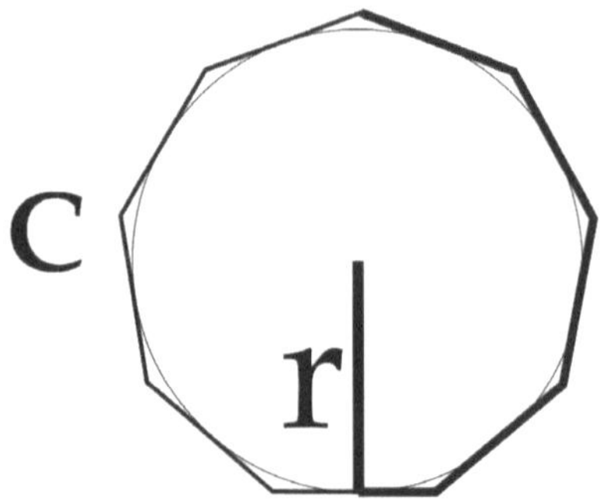

NONAGON

$$\frac{C}{2} \times r = A$$

$$\pi \times r^2 = \mathbf{A}$$

3.141594269

20612 / 6561

$$\pi$$

$$\Pi$$

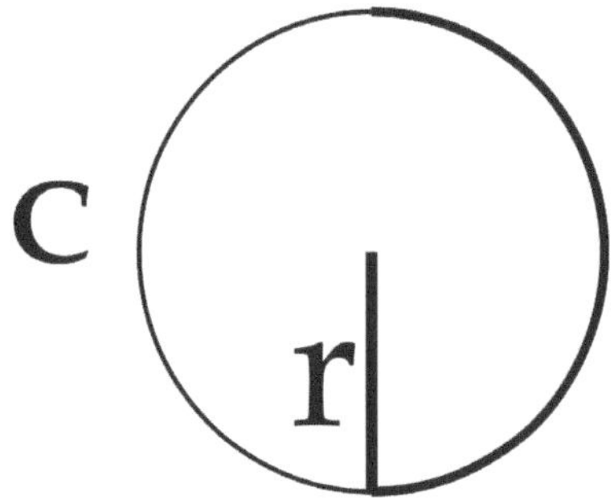

CIRCLE

$$\frac{C}{2} \times r = \mathbf{A}$$

Shape class TRIANGLE Shape name EQUILATERAL Number of sides **3**	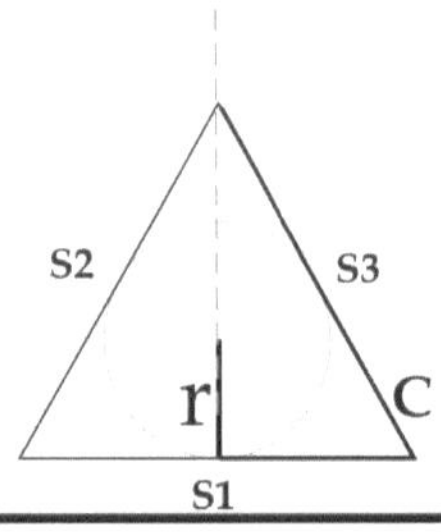	SIDE S1 = S2 = S3 SIDE S2 = S1 = S3 SIDE S3 = S1 = S2 ANGLE S1S2 = 60 ANGLE S2S3 = 60 ANGLE S3S1 = 60	If S1 = 240 , S2 = 240, S3 = 240 Then r = (S1+S2+S3)/(2xAP3EPi) r = 68.282032302755 C = 720 A = 249.5216289918
		SYMBOL | NAME & NUMBER 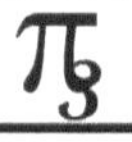 = AP3EPi 5.19615242 SQRT(27)	**Quantum Base Unit** b=1 , e=1 , d=2 , a=3 S1 = 8 , S2 = 8 , S3 =8
Shape class TRIANGLE Shape name ISOCELES Number of sides **3**	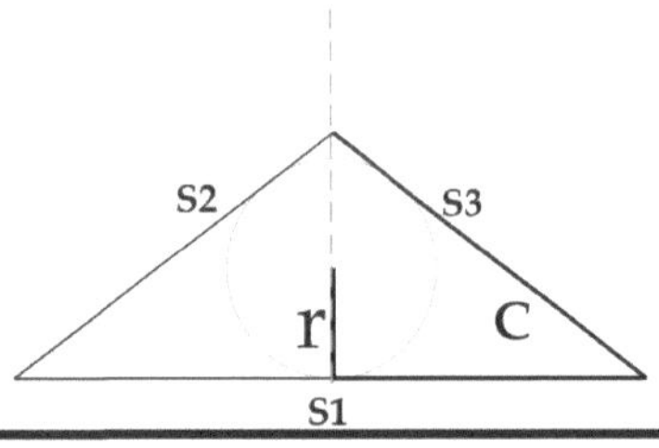	SIDE S1 = S2*1.6 = S3*1.6 SIDE S2 = S1/1.6 = S3 SIDE S3 = S1/1.6 = S2 ANGLE S1S2 = 36.8698976458 ANGLE S2S3 = 106.2602047083 ANGLE S3S1 = 36.8698976458	If S1 = 1152 , S2 = 720, S3 = 720 Then r = (S1+S2+S3)/(2xAP3EPi) r = 192 C = 2592 A = 248832
		SYMBOL | NAME & NUMBER 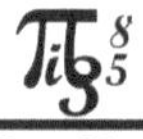 = AP3IPi-BS8/5 6.75 27/4	**Quantum Base Unit** b=1 , e=1 , d=2 , a=3 S1 = 8 , S2 = 5 , S3 =5
Shape class TRIANGLE Shape name ISOSCELES Number of sides **3**	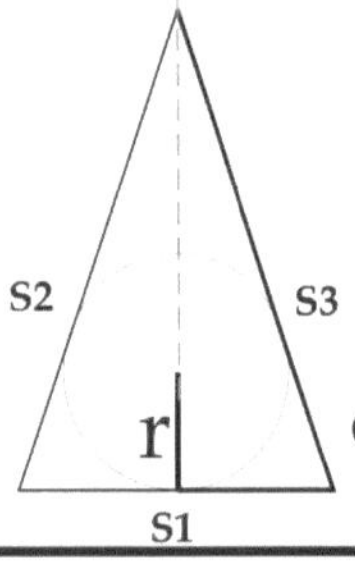	SIDE S1 = S2*24/37 = S3*24/37 SIDE S2 = S1*37/24 = S3 SIDE S3 = S1*37.24 = S2 ANGLE S1S2 = 71.0753555839 ANGLE S2S3 = 37.8492888321 ANGLE S3S1 = 71.0753555839	If S1 = 600 , S2 = 925, S3 = 925 Then r = (S1+S2+S3)/(2xAP3IPi24/37) r = 213.28571429 C = 2450 A = 262500
		SYMBOL | NAME & NUMBER 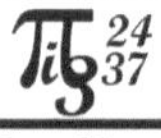 = AP3IPi-BS24/37 5.7166666 1029/180	**Quantum Base Unit** b=5 , e=1 , d=6 , a=7 S1 = 24 , S2 = 37 , S3 =37
Shape class TRIANGLE Shape name RIGHT ANGLE Number of sides **3**	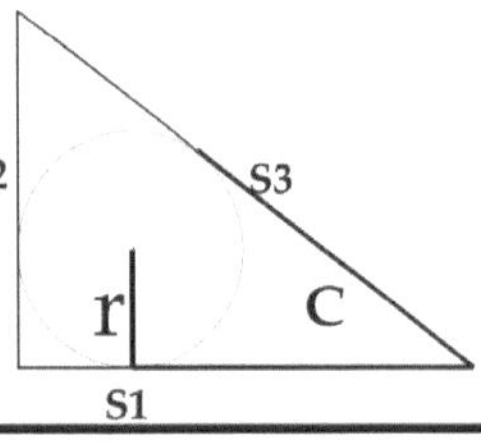	SIDE S1 = S2*4/3 = S3*4/5 SIDE S2 = S1*3/4 = S3*3/5 SIDE S3 = S1*5/4 = S2*5/3 ANGLE S1S2 = 90 ANGLE S2S3 = 53.1301024542 ANGLE S3S1 = 36.6898976458	If S1 = 576 , S2 = 432, S3 = 720 Then r = (S1+S2+S3)/(2xAP3RPi-BS4/3/5 r = 144 C = 1728 A = 124416
		SYMBOL | NAME & NUMBER 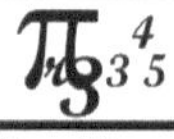 = AP3RPi-BS4/3/5 6 12/2	**Quantum Base Unit** b=1 , e=1 , d=2 , a=3 S1 =4 , S2 = 3 , S3 =5
Shape class TRIANGLE Shape name SCALENE Number of sides **3**	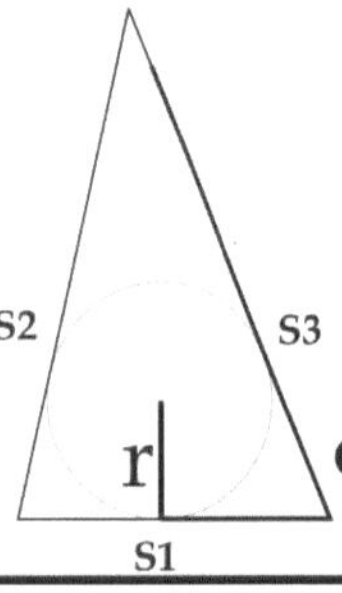	SIDE S1 = S2*17/25 = S3*17/26 SIDE S2 = S1*25/17 = S3*25/26 SIDE S3 = S1*26/17 = S2*26/25 ANGLE S1S2 = 73.7397952917 ANGLE S2S3 = 38.8800696564 **180** ANGLE S3S1 = 67.3801350520	If S1 = 448 , S2 = 1536, S3 = 1600 Then r = (S1+S2+S3)/(2xAP3SPi-BS17/25/26) r = 384 C = 4352 A = 835584
		SYMBOL | NAME & NUMBER 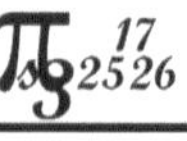 = AP3SPi-BS17/25/26 5.6666666 34/6	**Quantum Base Units** b=4 , e=1 , d=5 , a=6 b=1 , e=3 , d=4 , e=7 S1 =17 , S2 = 25 , S3 =26

Shape class
QUADRANGLE
(TETRAGON)
Shape name
SQUARE

Number of sides

4

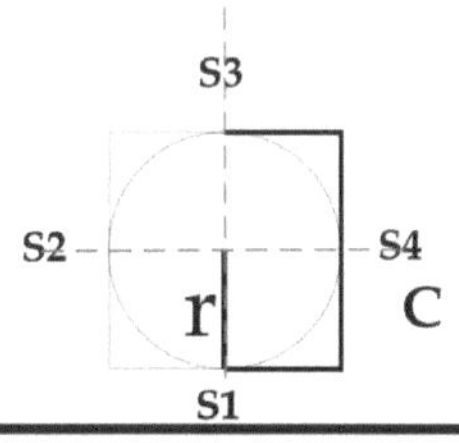

SIDE	S1	=	S2	=	S3	=	S4
SIDE	S2	=	S1	=	S3	=	S4
SIDE	S3	=	S1	=	S2	=	S4
SIDE	S4	=	S1	=	S2	=	S3

ANGLE	Value	Total
ANGLE S1S2 =	90	360
ANGLE S2S3 =	90	
ANGLE S3S4 =	90	
ANGLE S4S1 =	90	

SYMBOL	NAME & NUMBER
π_4 =	AP4SPi 4 8/2

If S1 = 144 , S2 = 144, S3 = 144,S4= 144
Then r = (S1+S2+S3+S4)/(2xAP4SPi)
r = 72
C = 576
A = 20736

Quantum Base Unit

b=2 , e=1 , d=3 , a=4
S1 = 4 , S2 = 4 , S3 = 4 , S4 = 4

Shape class
QUADRANGLE
(TETRAGON)
Shape name
DOUBLE RIGHT
ANGLE TRAPEZOID

Number of sides

4

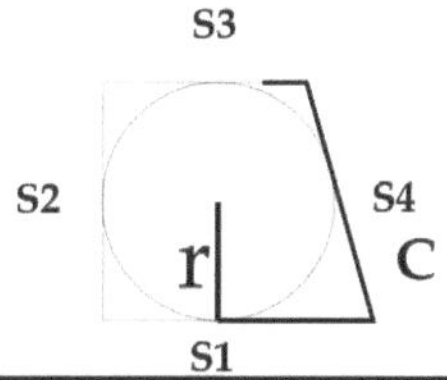

SIDE S1 =	S2*3/2	=	S3/2	=	S4*5/6	
SIDE S2 =	S1*2/3	=	S3	=	S4*3/5	
SIDE S3 =	S1*2/3	=	S2	=	S4*3/5	
SIDE S4 =	S1*5/6	=	S2*5/3	=	S3*5/3	

ANGLE	Value	Total
ANGLE S1S2 =	90	360
ANGLE S2S3 =	90	
ANGLE S3S4 =	126.8698976458	
ANGLE S4S1 =	53.1301023542	

SYMBOL	NAME & NUMBER
$\pi_{4\,drt}{}^{63}_{45}$ =	AP4TPi-BS6/4/3/5 4.5 9/2

If S1 = 576 , S2 = 364, S3 = 288, S4 = 480
Then
r = (S1+S2+S3+S4)/(2xAP4TPi-BS6/4/3/5)
r = 192
C = 1728
A = 165888

Quantum Base Unit

b=2 , e=1 , d=3 , a=4
S1 = 6 , S2 = 4 , S3 = 3 , S4 = 5

Shape class
QUADRANGLE
(TETRAGON)
Shape name
ISOSCELES
TRAPEZOID
Number of sides

4

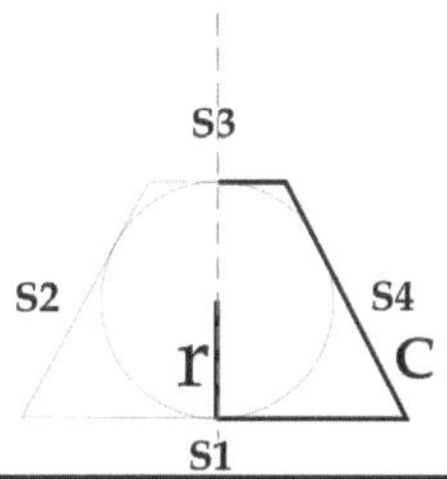

SIDE S1 =	S2*8/5	=	S3*4	=	S4*8/5	
SIDE S2 =	S1*5/8	=	S3*5/2	=	S4	
SIDE S3 =	S1*1/4	=	S2*2/5	=	S4*2/5	
SIDE S4 =	S1*5/8	=	S2	=	S3*5/2	

ANGLE	Value	Total
ANGLE S1S2 =	53.1301023542	360
ANGLE S2S3 =	126.8698976458	
ANGLE S3S4 =	126.8698976458	
ANGLE S4S1 =	53.1301023542	

SYMBOL	NAME & NUMBER
$\pi_{4\,it}{}^{82}_{55}$ =	AP4IPi-BS8/5/2/5 5 10/2

If S1 = 768 , S2 = 480, S3 = 192, S4 = 480
Then
r = (S1+S2+S3+S4)/(2xAP4IPi-BS8/5/2/5)
r = 192
C = 1920
A = 184320

Quantum Base Unit

b=1 , e=1 , d=2 , a=3
S1 = 8 , S2 = 5 , S3 = 2 , S4 = 5

Shape class
QUADRANGLE
(TETRAGON)
Shape name
RHOMBUS

Number of sides

4

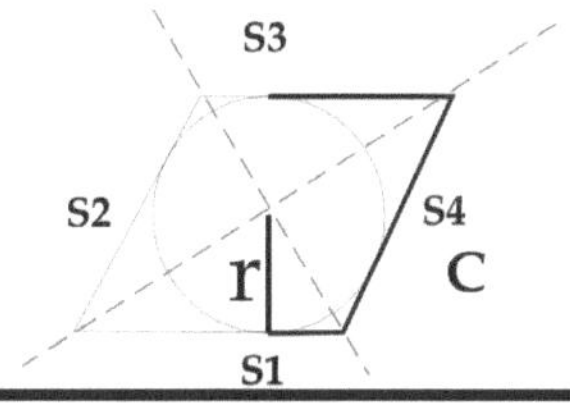

SIDE	S1	=	S2	=	S3	=	S4
SIDE	S2	=	S1	=	S3	=	S4
SIDE	S3	=	S1	=	S2	=	S4
SIDE	S4	=	S1	=	S2	=	S3

ANGLE	Value	Total
ANGLE S1S2 =	53.1301023542	360
ANGLE S2S3 =	126.8698976458	
ANGLE S3S4 =	53.1301023542	
ANGLE S4S1 =	126.8698976458	

SYMBOL	NAME & NUMBER
$\pi_{4\,r}$ =	AP4RPi-BC5/4 5 10/2

If S1 = 480 , S2 = 480, S3 = 480, S4 = 480
Then
r = (S1+S2+S3+S4)/(2xAP4RPi)
r = 192
C = 1920
A = 184320

Quantum Base Unit

b=2 , e=1 , d=3 , a=4
S1 = 5 , S2 = 5 , S3 =5 , S4 =5

Shape class
QUADRANGLE
(TETRAGON)
Shape name
KITE
TRAPEZOID
Number of sides

4

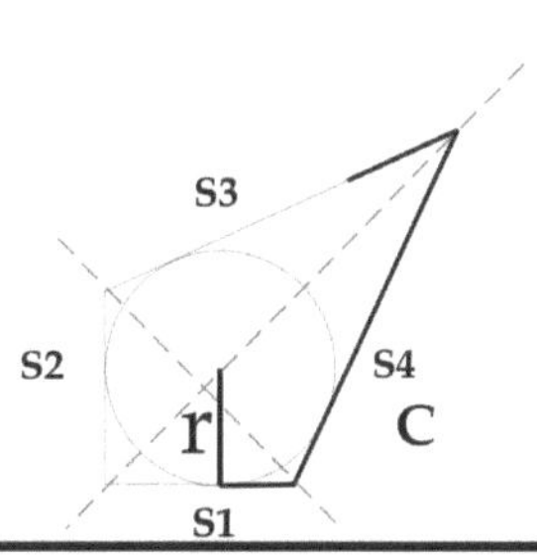

SIDE	S1	=	S2	=	S3/5	=	S4/5
SIDE	S2	=	S1	=	S3/5	=	S4/5
SIDE	S3	=	S1*5	=	S2*5	=	S4
SIDE	S4	=	S1*5	=	S2*5	=	S3

ANGLE	Value	Total
ANGLE S1S2 =	90	360
ANGLE S2S3 =	126.8698976458	
ANGLE S3S4 =	16.2602047083	
ANGLE S4S1 =	126.8698976458	

SYMBOL	NAME & NUMBER
$\pi_{4\,k}{}_{5}$ =	AP4KPi-BS5 9 18/2

If S1 = 192 , S2 = 192, S3 = 960, S4 = 960
Then
r = (S1+S2+S3+S4)/(2xAP4KPi-BS5)
r = 128
C = 2304
A = 147456

Quantum Base Unit

b=2 , e=1 , d=3 , a=4
S1 = 3 , S2 = 3 , S3 =15 , S4 =15

Shape class QUADRANGLE (TETRAGON) Shape name RIGHT ANGLE TRAPEZOID Number of sides **4**	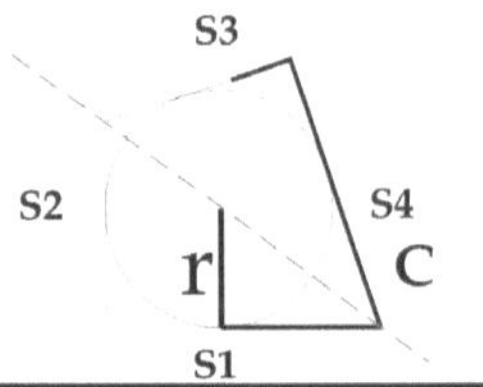	SIDE S1 = S2*2 = S3*2 = S4 SIDE S2 = S1/2 = S3 = S4/2 SIDE S3 = S1/2 = S2 = S4/2 SIDE S4 = S1 = S2*2 = S3*2 ANGLE S1S2 = 90 ANGLE S2S3 = 126.8698976458 ANGLE S3S4 = 90 ANGLE S4S1 = 53.1301023542 **360** SYMBOL — NAME & NUMBER π_4 rt = AP4TPi-BS2 4.5 9/2	If S1 = 384 , S2 = 192, S3 = 192,S4= 384 Then r = (S1+S2+S3+S4)/(2xAP4TPi-BS2) r = 128 C = 1152 A = 73728 **Quantum Base Unit** b=2 , e=1 , d=3 , a=4 S1 = 6 , S2 = 3 , S3 = 3 , S4 = 6
Shape class QUADRANGLE (TETRAGON) Shape name PARALLEL TRAPEZOID Number of sides **4**	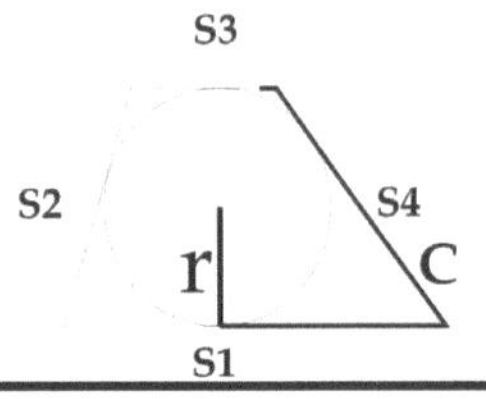	SIDE S1 = S2*21/13 = S3*7/3 = S4*7/5 SIDE S2 = S1*13/21 = S13/7 = S4*15/13 SIDE S3 = S1*1/3 = S2*7/13 = S4*5/7 SIDE S4 = S1*5/7 = S2*15/13 = S3*7/15 ANGLE S1S2 = 67.380135052 ANGLE S2S3 = 112.619864948 ANGLE S3S4 = 126.8698976458 ANGLE S4S1 = 53.1301023542 **360** SYMBOL — NAME & NUMBER π_4 pt 21 7 / 13 15 = AP4TPi-BS21/13/7/15 4.6666666666 14/3	If S1 = 336 , S2 = 208, S3 = 112, S4 = 240 Then r = (S1+S2+S3+S4)/(2xAP4TPi-BS21/13/7/15) r = 96 C = 896 A = 43008 **Quantum Base Unit** b=3 , e=2 , d=5 , a=7 S1 = 21 , S2 =13 , S3 = 7 , S4 = 15
Shape class QUADRANGLE (TETRAGON) Shape name DIVERUS TRAPEZOID Number of sides **4**	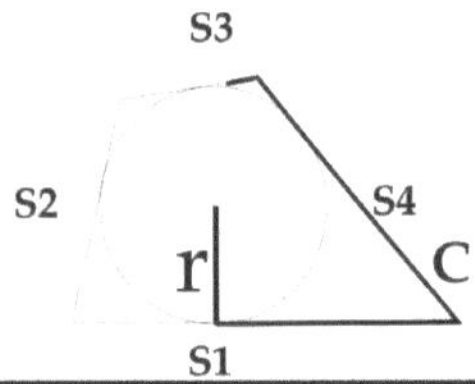	SIDE S1 = S2*21/12 = S3*3 = S4*21/16 SIDE S2 = S1*12/21 = S3*12/7 = S4*21/12 SIDE S3 = S1/3 = S2*12/16= S4*7/16 SIDE S4 = S1*16/21 = S2*16/12 = S3*16/7 ANGLE S1S2 = 67.380135052 ANGLE S2S3 = 126.8698976458 ANGLE S3S4 = 112.6198648477 ANGLE S4S1 = 53.1301023542 **360** SYMBOL — NAME & NUMBER π_4 dt 21 7 / 12 16 = AP4TPi-BS21/12/7/16 4.6666666666 14/3	If S1 = 525 , S2 = 300, S3 = 175, S4 = 400 Then r = (S1+S2+S3+S4)/(2xAP4TPi-BS21/12/7/16) r = 150 C = 1400 A = 10500 **Quantum Base Unit** b=3 , e=2 , d=5 , a=7 S1 = 21 , S2 = 12 , S3 = 7 , S4 = 16
Shape class PENTANGLE (PENTAGON) Shape name Number of sides **5**	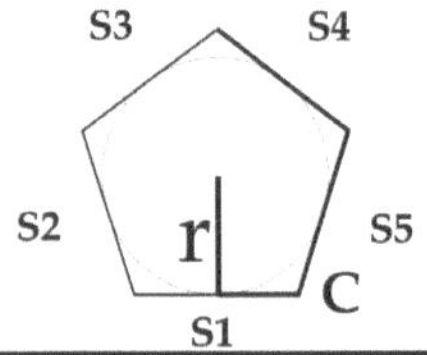	SIDE S1 = S2 = S3 = S4 = S5 SIDE S2 = S1 = S3 = S4 = S5 SIDE S3 = S1 = S2 = S4 = S5 SIDE S4 = S1 = S2 = S3 = S5 SIDE S5 = S1 = S2 = S3 = S4 ANGLE S1S2 = 108 ANGLE S2S3 = 108 ANGLE S3S4 = 108 ANGLE S4S5 = 108 ANGLE S5S1 = 108 **360** SYMBOL — NAME & NUMBER π_5 = AP5PPi 5 10/2	If S1 = 480 , S2 = 480 and S3 = 480, S4 = 480, S5= 480 Then r = (S1+S2+S3+S4+S5)/(2xAP5PPi) r = 192 C = 1920 A = 184320 **Quantum Base Unit** b=2 , e=1 , d=3 , a=4 S1 = 5 , S2 = 5 , S3 =5 , S4 =5

5.6 - Combination Geometry

This is where all the geometric ideas are cross linked and a new discovery can be found lurking behind a simple shape or formula. This particular diagram I have combined Archimedian Circles with magic squares and counter rotational geometry.

Minimum distance
Maximum angle

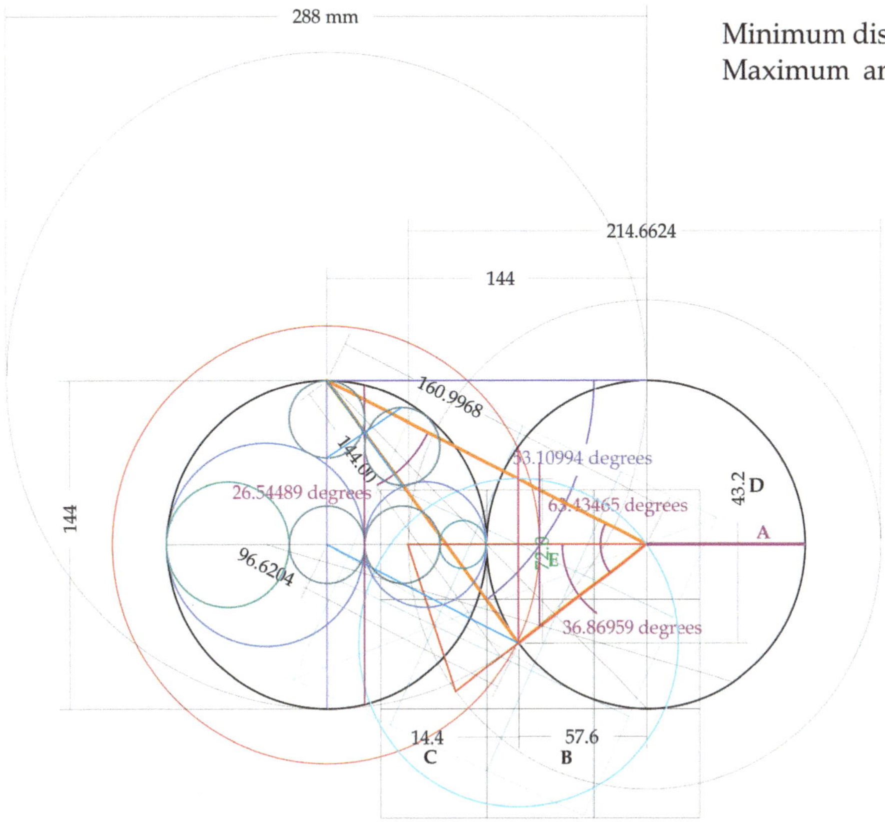

If
A=72
B=A-C
C=A/5
D=B-C
E=A
and
A=B+C
B=D+C
Then
57.6+72=129.6
129.6/14.4=9
43.2/14.4=3
therefore
D=SQR(A+B)/C
Also if
E=D+C+C

8	1	6
3	5	7
4	9	2

Magic square of Three

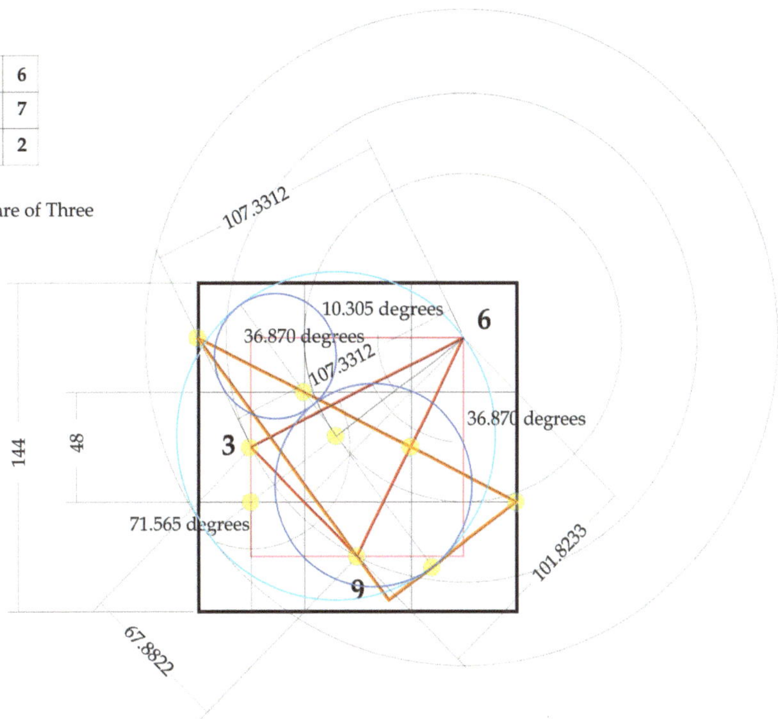

Arto Heino

Chapter Six

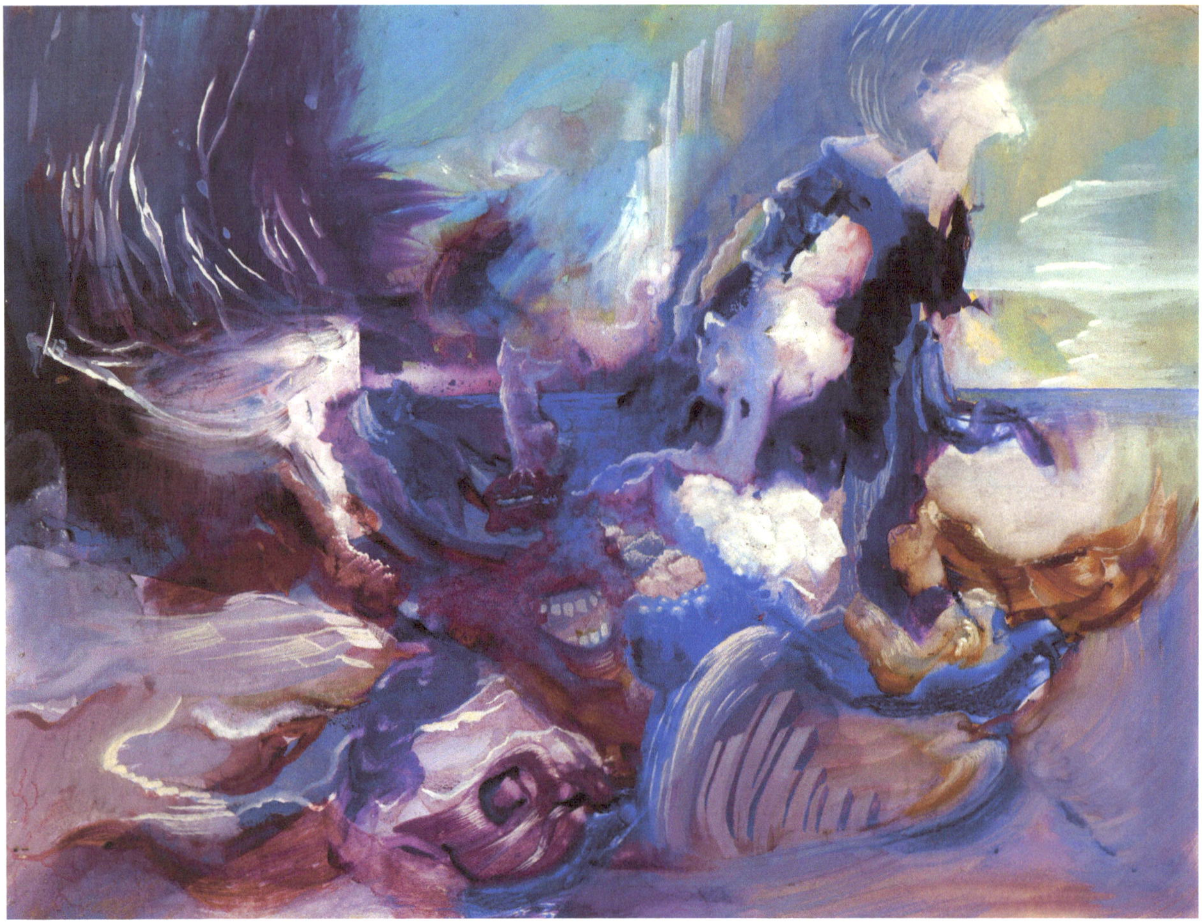

Art

6.1 - Leonardo Da Vinci

6.1a - The Last Supper

The last Supper is one of the most painted scenes from the christian mythology, and Leonardo Da Vinci painted the most controversial and elegant of all the compositions done by artists up to and beyond the Renaissance Period. As I have also flouted in designing this composition, I will reveal some of the hidden stories that must be uncovered before anybody attempts to create such a work.

The first list to be made is who and what are the 12 Apostles, I will show the official Christian List the my derived Ancient Egyptian List and the Zodiac List.

Bartholomaus	Leo	Thoth	Zeus	Lion	Beryl	*10 August – 15 September*
Jacobus Minor	Cancer	Anubis	Hermes	Crab	Topaz	*21 July – 9 August*
Andreas	Gemini	Pheonix	Apollo	Twins	Chysoprase	*20 June – 20 July*
Judas Iscariot	Taurus	Hathor	Aphrodite	Bull	Hyacinth	*14 May – 19 June*
Petrus	Aries	Ra	Athena	Ram	Amethyst	*19 April – 13 May*
Johannes	Pisces	Osiris	Poseidon	Fishes	Jasper	*12 March – 18 April*
Jesus	Orion	Horus	Spring Equinox	Shepard	Diamond	*Easter- March 20*
Thomas	Aquarius	Isis	Hera	Water Bearer	Sapphire	*16 February – 11 March*
Jacobus Major	Capricorn	Shu	Hestia	Sea Goat	Chalcedon	*19 January – 15 February*
Phillipus	Sagittarius	Sphinx	Artemis	Archer	Smaragdus	*18 December – 18 January*
Matthais	Scorpio	Sekhmet	Aries	Scorpion	Sardonyx	*23 November – 29 November*
Jude Thaddeus	Libra	Wadget	Hephaistos	Scales	Sardius	*31 October – 22 November*
Simon Zelotes	Virgo	Horus	Demeter	Maiden	Chrysolite	*16 September – 30 October*

As you can easily see the intention of Leonardo was to inform the viewer of the corollary of the pagan origin of the last supper by reinforcing the salacious drama of the all the figures gossiping and debating about the manner of Jesus announcing a betrayer in the group. Leonardo did not paint his figures worshiping or awe struck by Jesus but more in a natural expression of controversy about who is sitting in the middle of this group and what is his message. The figures actually represent the Zodiac as was known to Leonardo, the ancient interpretation would have only partly been known to Leonardo, so I will add this layer based on the zodiacal figures from Greek knowledge that Leonardo would have known.

I have highlighted the eyes and the hands of each figure, this is to bring your attention to the real story of a group of human beings that do not wholly share the same belief system. So the gestures indicate that there are 12 individuals who have different attitudes and Jesus is the one figure not directing any gesture to the rest of them, only to his personal God. Where Judas has a two geometric lines directed at Jesus, one is the hand hidden behind Peter with a dagger aimed at his right hand and the other hand pointing to the head of Jesus. Phillip adores Jesus and is pleading with him, where Thomas flays his hands out at utter disbelief and is in the heat of argument with Jesus. The gestures of Jacobus minor indicate he either does not want to be part of the debate or he is surprised by what will transpire. This is only a brief of the complexity incorporated in this painting, the Bible should be hand for more details.

Most of the rectangles are golden rectangles and there is a great deal of link geometry between gestures and linear nodes.

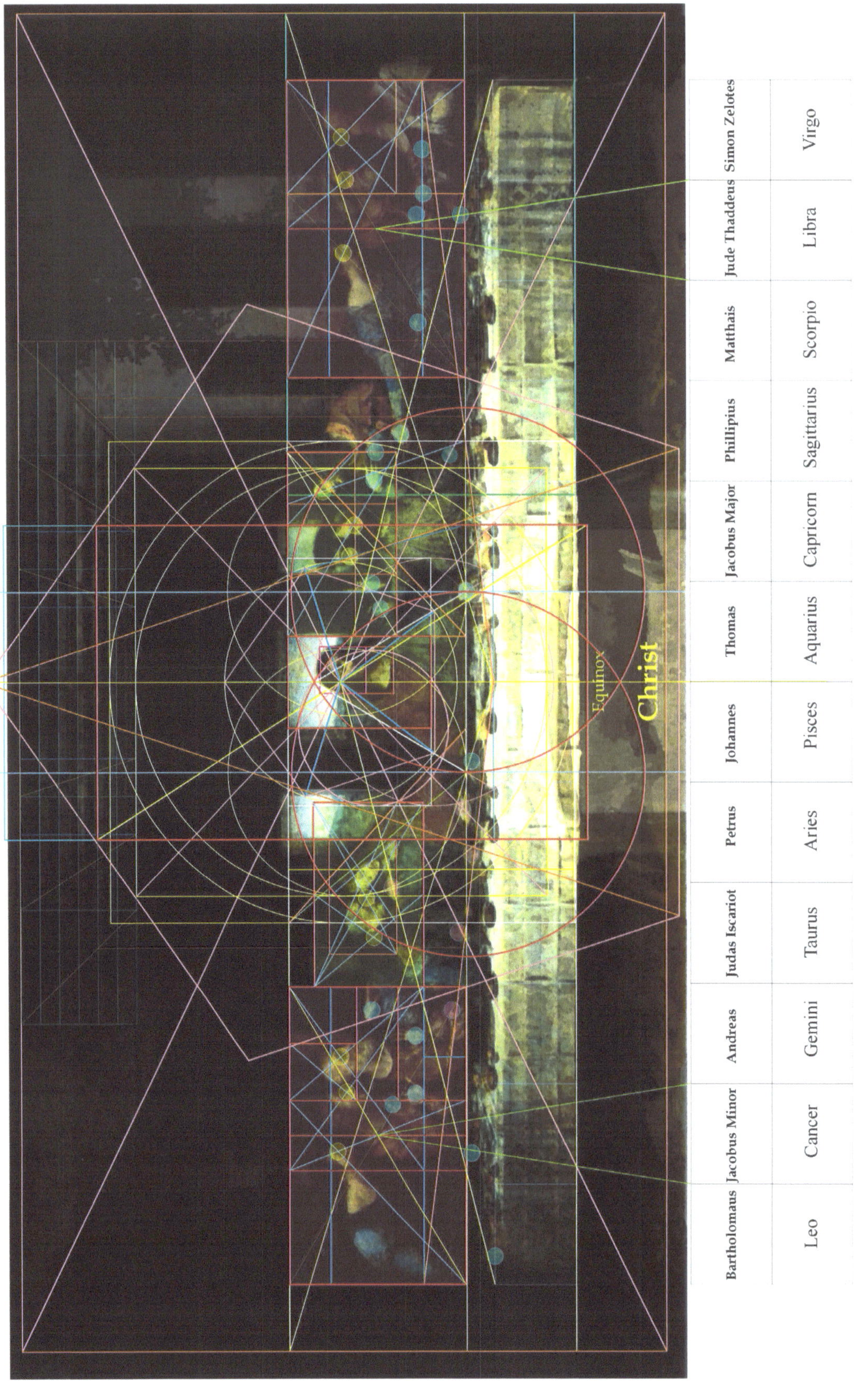
Bartholomaus
Leo
Jacobus Minor
Cancer
Andreas
Gemini
Judas Iscariot
Taurus
Petrus
Aries
Johannes
Pisces
Thomas
Aquarius
Jacobus Major
Capricorn
Phillipius
Sagittarius
Matthais
Scorpio
Jude Thaddeus
Libra
Simon Zelotes
Virgo
Christ
Equinox

6.2 - Hand and Forearm

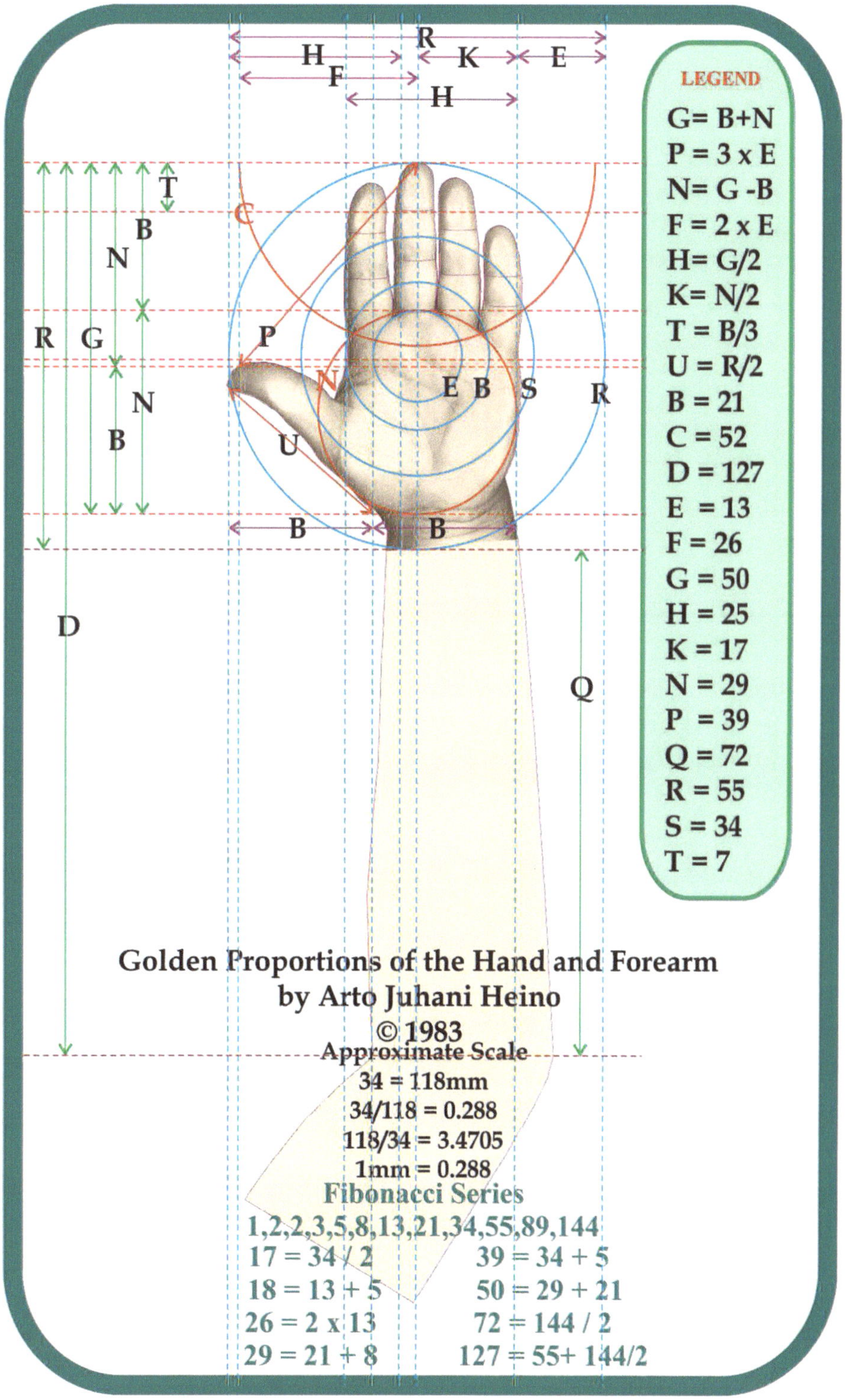

6.6 - Modern Divergence

The start of the 20th century was a time of change for the arts and new technologies were adding to the pallets of sculptors and painters and musicians, while the writers had found other outlets in popular journals, the poets still needed the distance from the modern rush to write meaningful prose. The engineers and architects were becoming the new celebrities that would rebuild the 20th century in the image of the corporate ruler.

While in the past the artist had to pander to the egos of wealthy merchants and the religious fraternity, in the modern era the corporate body became the driving force of commissions to earn a living, where the corporate board would need the skills of the artist to glorify their mantel, this said, it was one of hidden influences that drove the commercialization of the visual arts of the 20th century.

The musicians were adding new sounds to their pallet and expanding the classical formulas, it was composers like Stravinsky that set this whole modernist idea on fire, an art that was forceful and brutal. The painters like Picasso were experimenting and found that the portrayal of the brutality of the modern machine was an art-form in it self , creating a broken and shattered paradigm to match what the bankers, politicians and rulers have created. This form of art was a reflection on the psychopathic nature of those who wielded influence. A disconnection with nature and the dissociation with the deeper philosophies of mankind was to create a new type of artist made in the image of these rulers, the Futurists, Cubist and such were at the spearhead of the breakdown and recreation of the arts. Art was to now serve the corporate body and not just the expansion of any humanist endeavors that have been centuries in the making. This was to drive the artists that were expanding our understanding of ourselves to a smaller and smaller segment of society, relegating them to serve as painters in isolation and not engage the mainstream audiences where the popular dissonant art is being promoted by the self proclaimed owners of the system.

This was a one hundred and eighty degree turn around from the ideals that help start the modern era in the Renaissance period. Those ideals were lost when scarcity was made into a profitable commodity and politics of the narrow idealistic factions made mental slavery to the masses, where choice was reduced to extreme positions and was the now normal for most modern countries. The small number of influential groups co-opted the banking system and with it controlled the politicians and policies, where war was a tool to be used by these groups.

This brings me to the modern artist , a creative demon that resides in him is now only fueled by the torturous nature of the false conventions that were created to control the masses. Where competition is the dominant force and co-operation is considered weak and ineffectual, this engineered cancer keeps the modern artist become a lap dog to the psychopathic rulers of our modern system, the politicians, the bankers and the lawyers. The merry go round of advertising is the only resource the modern artist has , this is also one of the false mind control mechanisms that hound every man at every step in his actions directing him to corporate controlled decisions where choice has a profit motive. These institutions coerce the artists, driving them to follow the rules of the false game laid down by those who control the finances.

Where is the true artists, those born with inherent abilities of insight and dextrous skills to accomplish what they need to express and imagine. They were replaced by commercial copyists and kitsch populist illustrators and graphically trained system monkeys. How can you tell which is what and who is who, easy , those who compete to win, those who are infused with graphic trickery and are disconnected from the real world and live in an imaginary landscape built by the egos of these dominant men, whose choice of genre and iconic output is molded and influenced by the financial inducements created by the institutions ruled by corporate fascists.

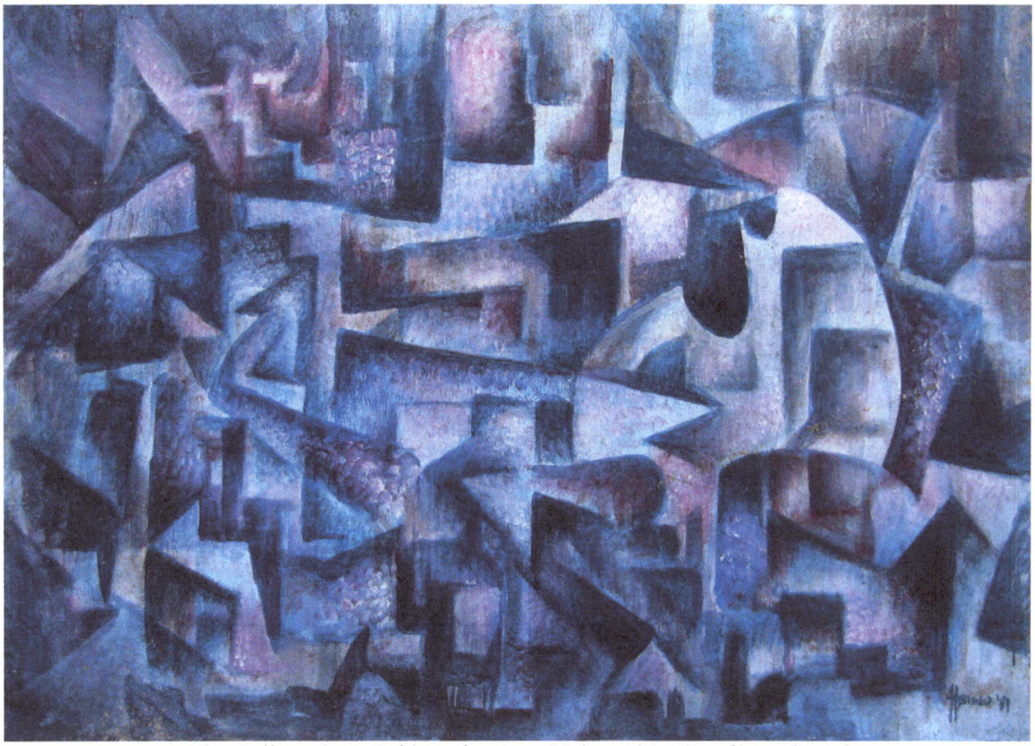

The College Day Cubist – by Arto Heino 1981 Acrylic on board

Chapter Seven

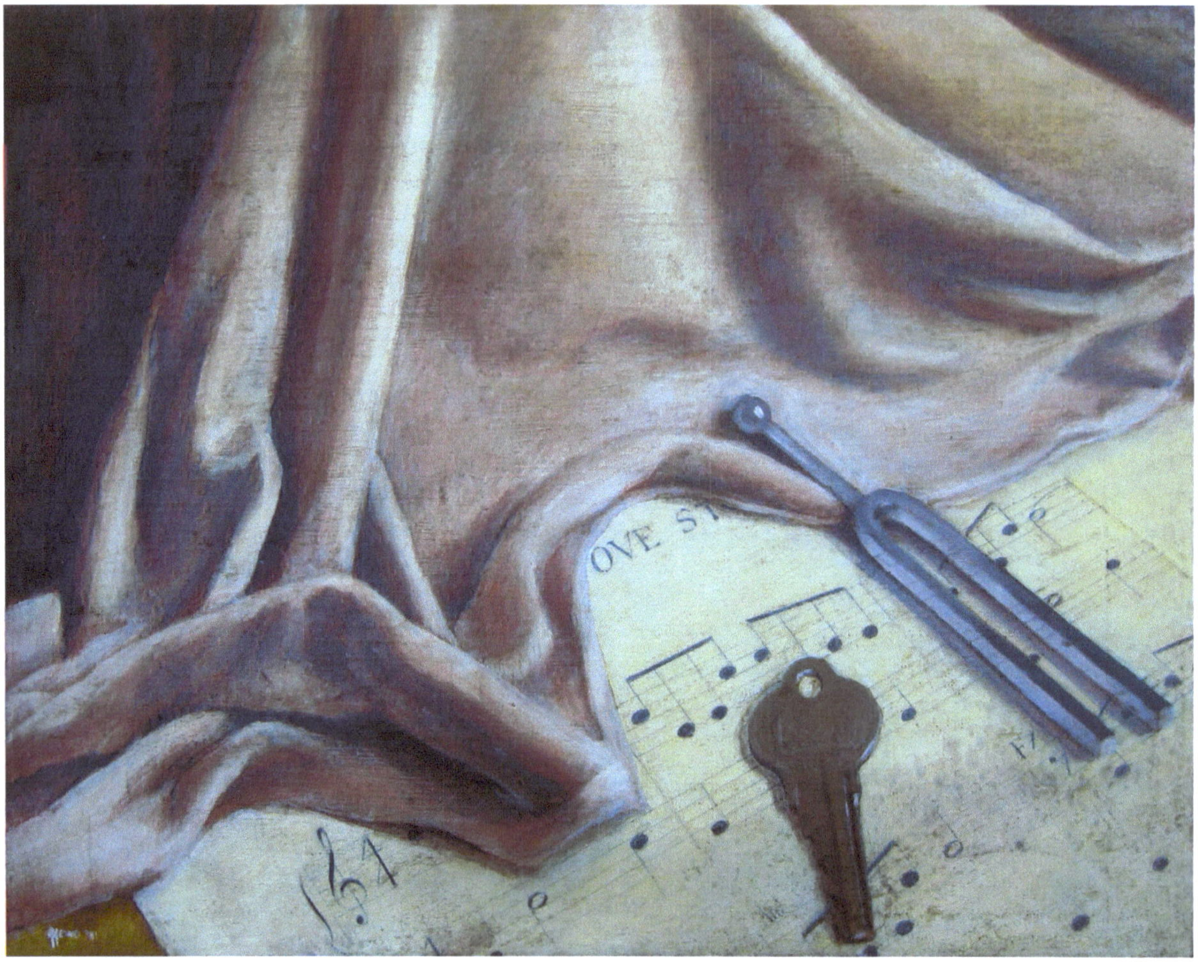

Music

7.1 - Instruments

7.1a - Tutankhamen Trumpet

Howard Carter uncovered many artifact when he discovered the Tomb of Tutankhamen in November 1922, two of these precious objects were musical instruments once played by musicians before 1333 BC. One trumpet is said to be 22 inches long and made of silver and the other is also estimated to be 18 inches long and possibly made of bronze, both had a bell diameter of 4 inches. The Silver Trumpet also had decorative adornments of lotus flowers which look like the horn flare, where both also had figures of Amen-Ra, Horus and Osiris. The other decorative element are the pictorial representation of air flow going out of the end horn, this unit also had a wooden insert that fit inside the complete length.

The sounding mechanism in our modern instruments use a mouth-piece so there is an impedance match to the velocity flow within the trumpet, it seemed these trumpets are not able to produce any tone or harmonic without this piece added. These sizes will produce at least 8 notes, the given sizes are only estimates, I endeavor to make a chart to estimate the tonal scale of the Trumpets of Tutankhamen.

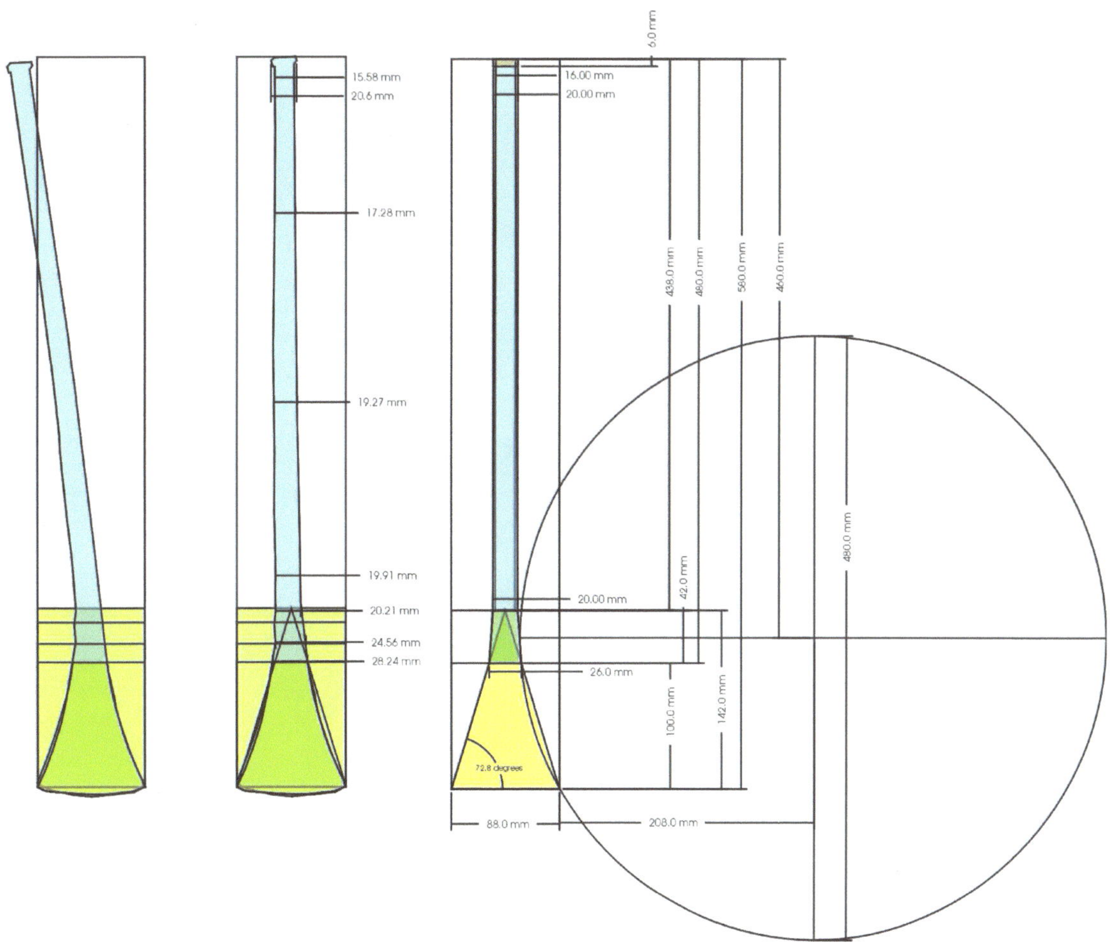

The original dimensions that I had obtained from varied research might have been in error, so I will give different lengths to at least estimate the correct tones.

		Mtrs	Inches
		0.457200	18
Mentioned in Article			
	F#1	188.136483 hz	
	F#2	564.409449 hz	
	F#3	940.682415 hz	

		Mtrs	Inches
		0.523237	20.6
Based on Geo cubits			
	E1	164.392187 hz	
	E2	493.176560 hz	
	E3	821.960933 hz	

		Mtrs	Inches
		0.525000	20.67
Based on Royal Cubits			
	E	163.840000 hz	
	E2	491.520000 hz	
	E3	819.200000 hz	

		Mtrs	Inches
		0.558800	22
Mentioned in Article			
	Eb1	153.929850 hz	
	Eb2	461.789549 hz	
	Eb3	769.649248 hz	

		Mtrs	Inches
		0.580000	22.83
As per Diagram			
	D	148.303448 hz	
	D2	444.910345 hz	
	D3	741.517241 hz	

		Mtrs	Inches
		0.944000	37.17
Mentioned in an Article			
	F#1	91.118644 hz	
	F#2	273.355932 hz	
	F#3	455.593220 hz	

Another set of article;s refer the Silver trumpet was in B natural and the bronze trumpet was 30mm shorter and played in A flat. As you can see the pipes do not match what is written. So without laboring on the estimates I will use my diagram as the final.

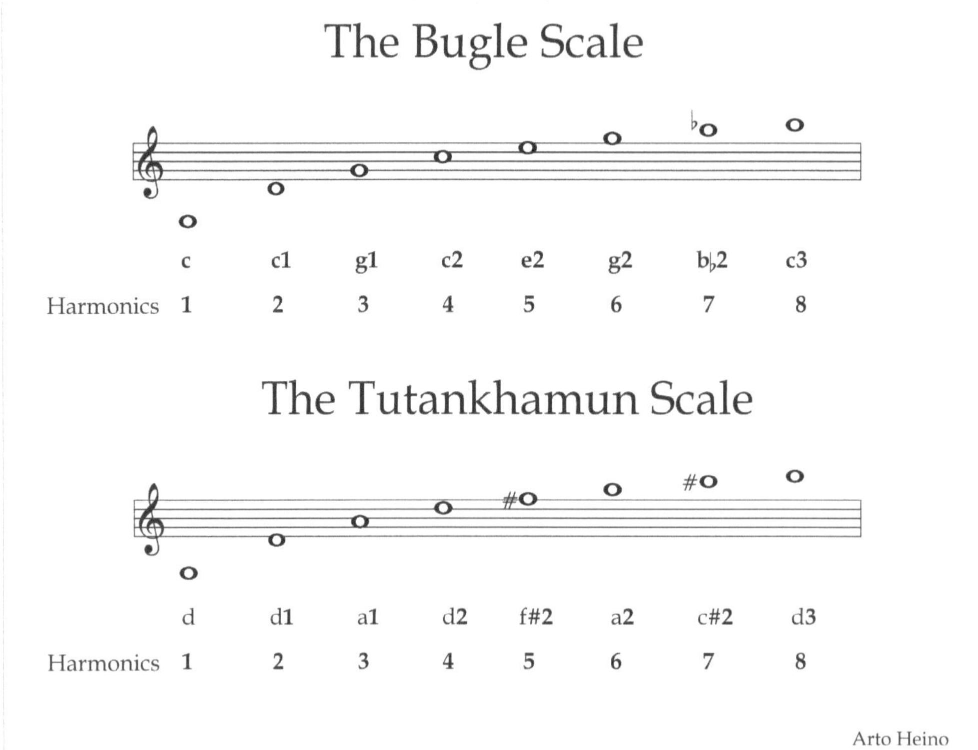

7.1b - Hexaphone

This was developed in 1996 as a electronics instrument that uses touch pads and was designed as a spiral structure, as each section connected in a circular form

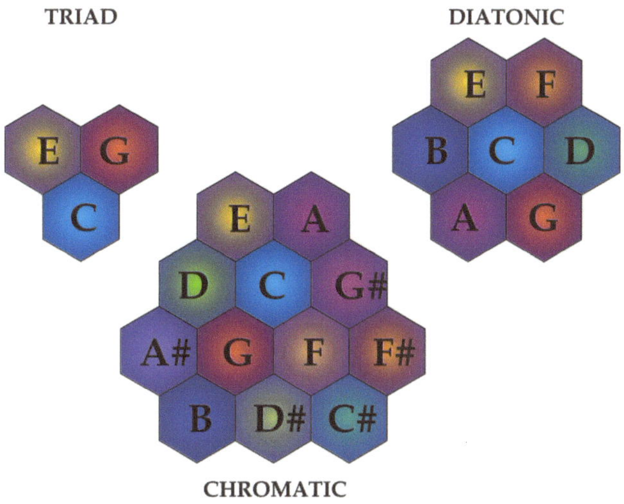

This illustrates the chromatic scale

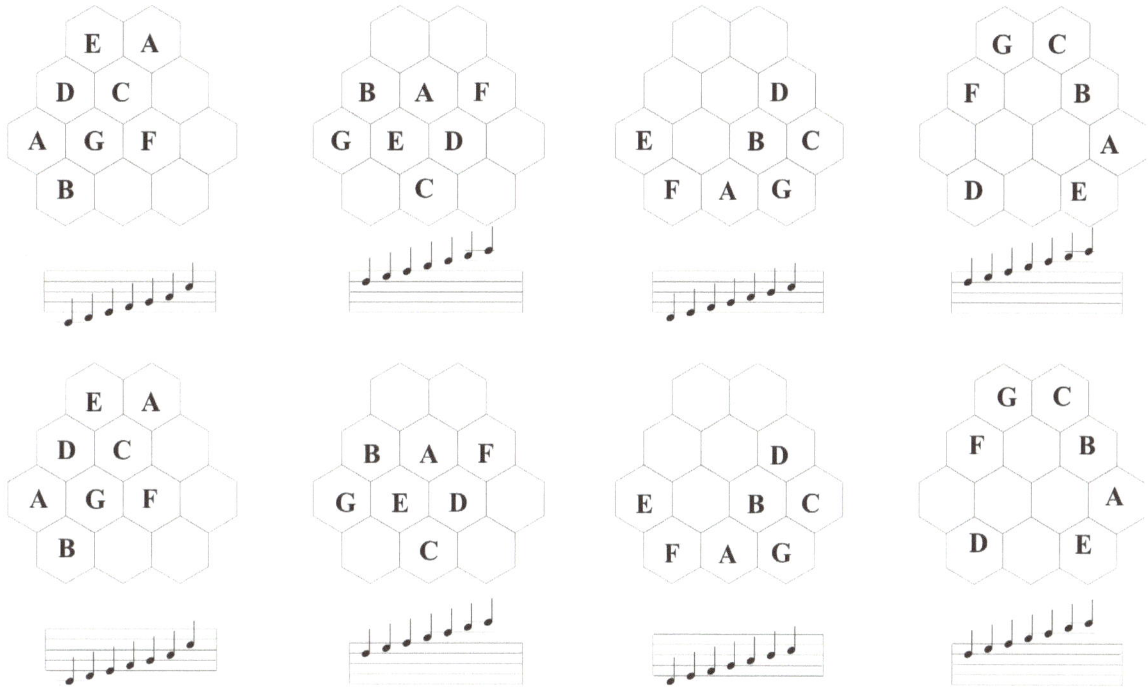

C Major Scale

This are all the scale patterns in all the different segments, as you can see there are only four finger patterns for each scale, which covers 8 octaves.

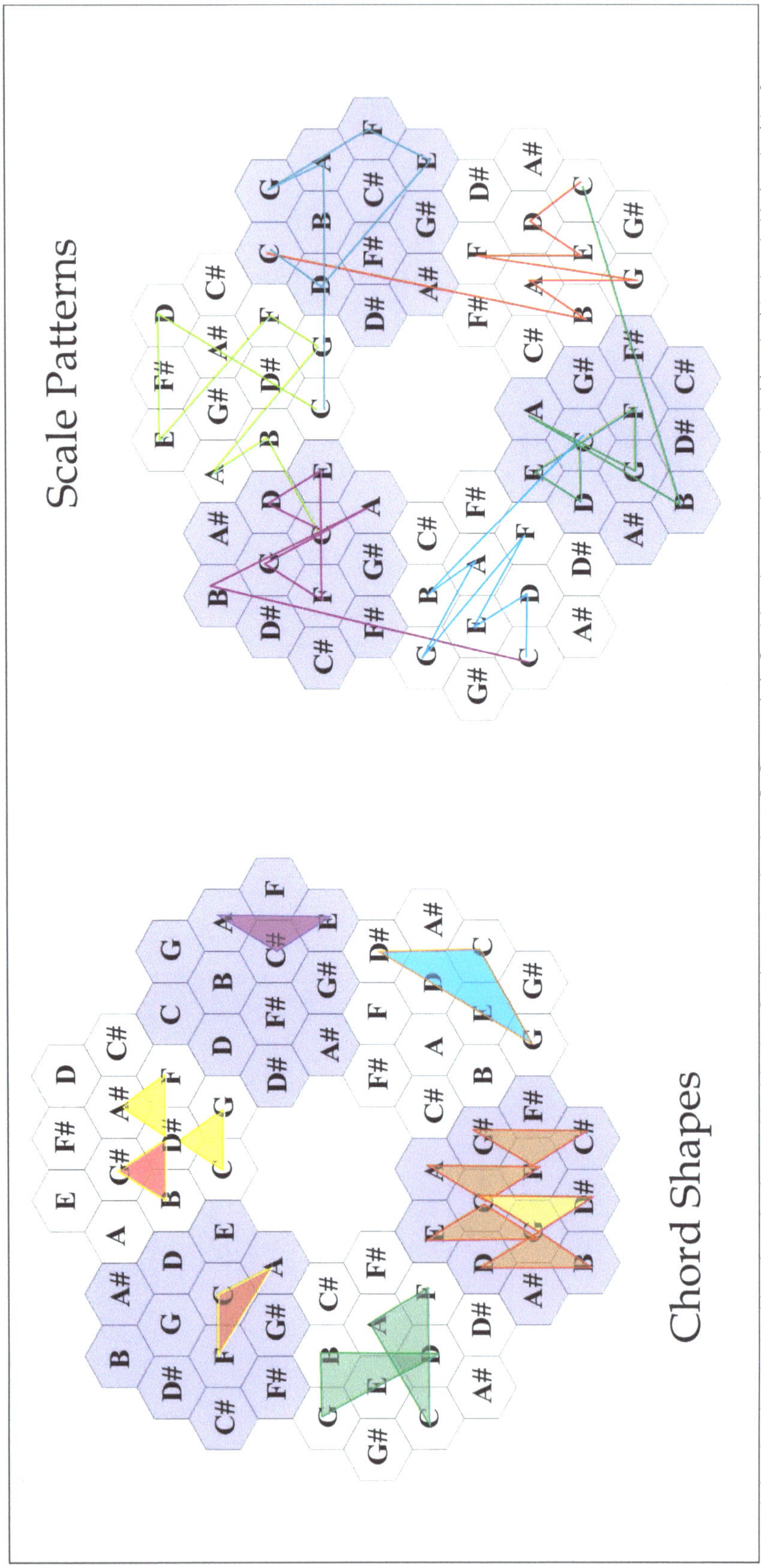

This chart includes all the chord forms and a few others not yet theorized completely. As you can see the Major, Minor, Diminished and Augmented only cover four varieties, so adding the completed units I named them Altered 1,2,3,4 and 5. Later I called three of them the Muror and two as inversions.

These forms are true chords and are used in music as passing chord clusters or tonalities and have some unusual descriptions when using classical or jazz notation. They are not unusual as they are also based on 12 tone chromatic harmony.

The pursuit of the final arrangements for complete chords , scales and technical requirements to play the instrument were fully developed up to 1998, the worked commenced to build the percussive version in 2004, this did not occur as a concurrent development of my "Artone 19" guitar was under way . This led me the the "Hexartom 19" project.

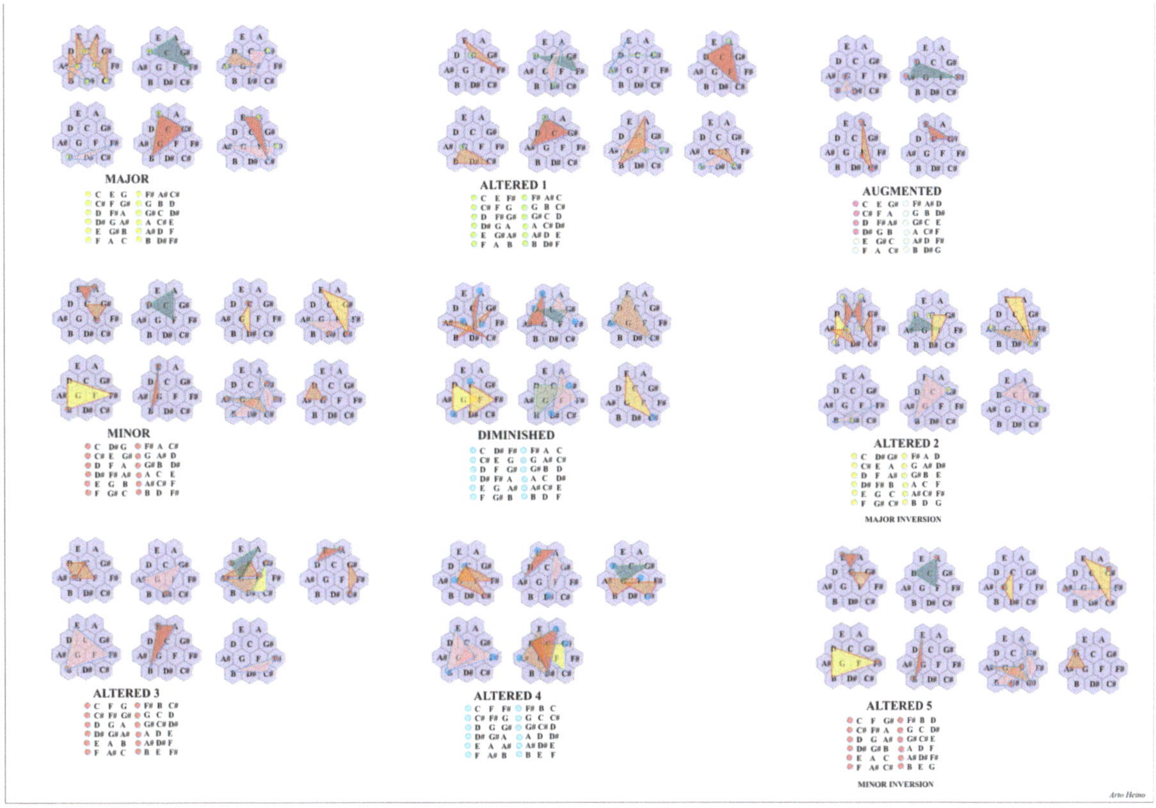

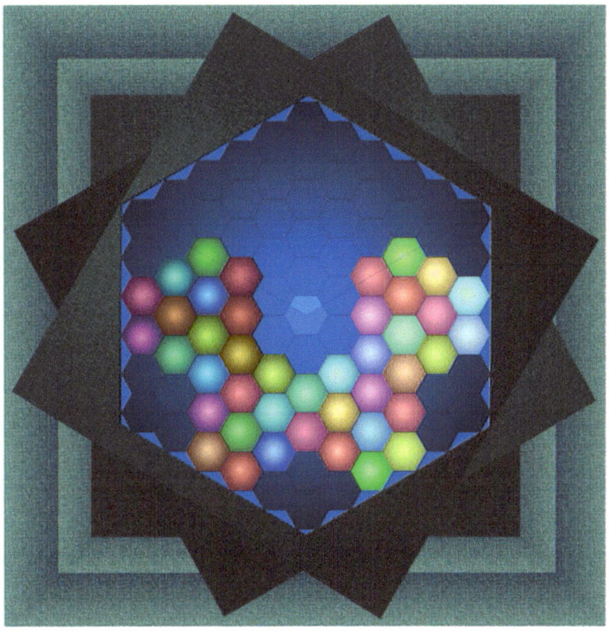

Major = Maj + or = M = Music + a = 1st vowel + j = 10th letter of alphabet

Minor = Min + or = M = Music + i = 3rd vowel + n = 14th letter of alphabet

Muror =Mur + or = M = Music + u = 5th vowel + r = 18th letter of alphabet

As the vowels create a chord 1+3+5 , the middle letters are separated by 4.

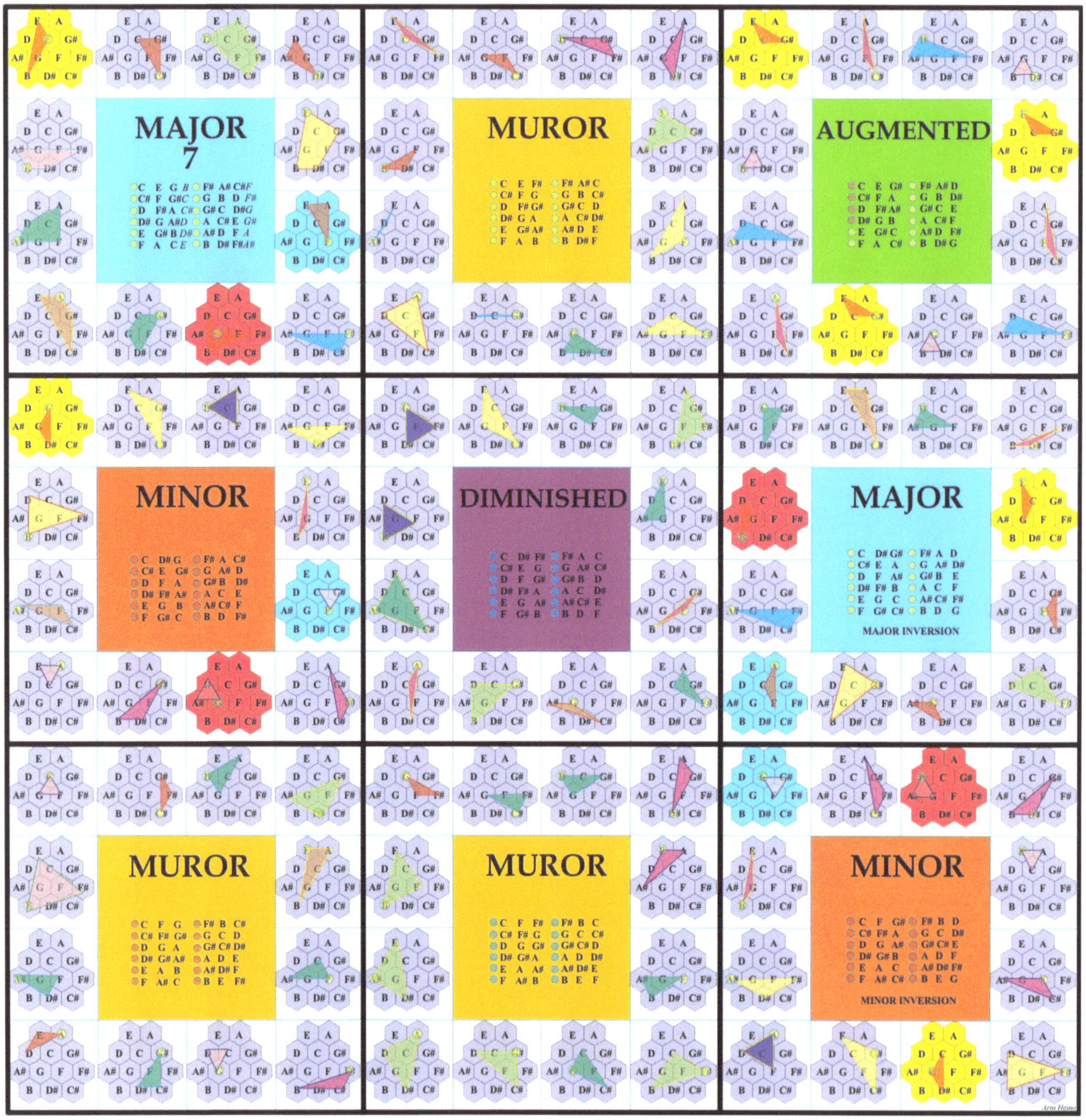

7.1c – Parvatic scale

The development of this instrument having only 12 notes, seemed lacking in tonal scope so the expansion of the standard chromatic 12 note form needed to evolve into a more cohesive and harmonically expansive structure. This research was eventually to introduce a musical scale called the Parvatic scale, which is a 19 interval system of musical notes. Not new in its inception this new instrument called the "**Hexartom 19**" is a 3 octave percussive musical instrument, giving you a total of 57 Parvatic notes, which includes the standard chromatic intervals in its geometry. The Guitar "**Artone 19**" also has this scale at its foundation.

Theoretical

As you know this scale has been known as the 19 tone scale, the justification for adding 7 extra notes to the chromatic scale has been many as history shows. The technical requirements are exacting just as any instruments maker knows. Here are the scale forms as worked out by me.

Parvatic Scale	Chromatic Scale	Diatonic Sale	Pentatonic Scale	Triadic Scale
C	C	C	C	C
Cp				
C#	C#			
D	D	D	D	
Dp				
D#	D#			
E	E	E	E	E
Ep				
F	F	F		
F#	F#			
F#p				
G	G	G	G	G
Gp				
G#	G#			
A	A	A	A	
Ap				
A#	A#			
B	B	B		
Bp				
C	C	C	C	G

As you can see by the simple placement with other basic scale forms the Parvatic Scale is an natural out growth from the 12 tone chromatic scale. When the nineteen note Parvartic scale is employed it would give all instruments a vast musical scope, from ancient to ethnic styles can be played, not missing the most modern sounds that are not available to standard 12 tone instruments.

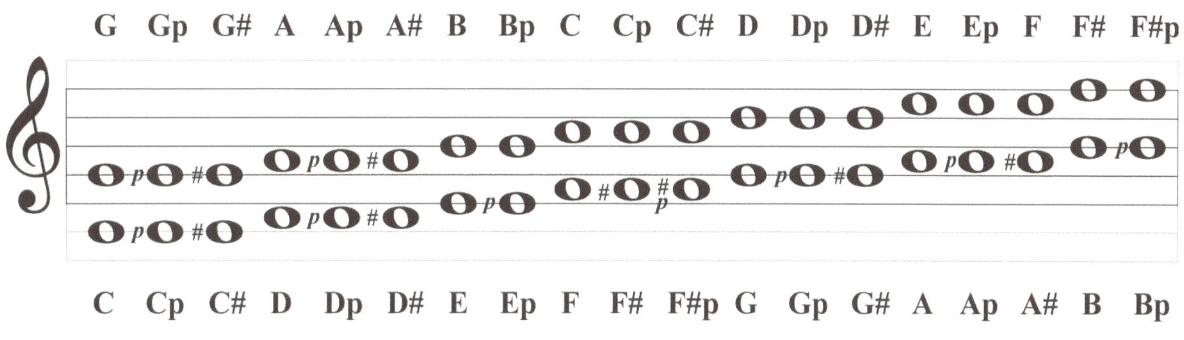

C G D A E B F# Cp Gp Dp Ap Ep Bp F#p C# G# D# A# F

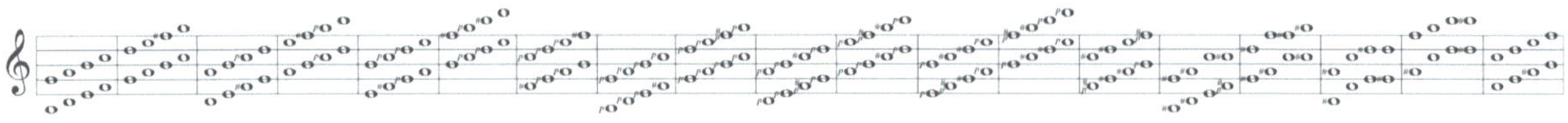

You now have a expanded harmonic language that is new and exciting, a design that allows you to play all your standard 12 tone music and the technique required is not compromised. Historically the musical language needs an upgrade, from Francisco De Salinas in 1577 and Guillaume Costeley in 1588, the musicologist Wesley Woolhouse in 1835 carefully studied all the possibilities and found it the mist agreeable number of tones. The modern era of the 19 tone scale was revived by Joel Mandelbaum in 1961, having looked at alternative ways to express new harmonic content without resorting to electronic sounds, this solution was found.

Parvatic Scale	Chromatic Scale	Chromatic Mode Scales																			
C	C	C																			
Cp			Cp																		
C#	C#	C#		C#																	
D	D	D	D		D																
Dp			Dp	Dp		Dp															
D#	D#	D#		D#	D#		D#														
E	E	E	E		E	E		E													
Ep			Ep	Ep		Ep	Ep		Ep												
F	F	F		F	F		F	F		F											
F#	F#	F#	F#		F#	F#		F#	F#		F#										
F#p			F#p	F#p		F#p	F#p		F#p	F#p		F#p									
G	G	G		G	G		G	G		G	G		G								
Gp			Gp		Gp	Gp		Gp	Gp		Gp	Gp		Gp							
G#	G#	G#		G#		G#	G#		G#	G#		G#	G#		G#						
A	A	A	A		A		A	A		A	A		A	A		A					
Ap			Ap	Ap		Ap		Ap	Ap		Ap	Ap		Ap	Ap		Ap				
A#	A#	A#		A#	A#		A#		A#	A#		A#	A#		A#	A#		A#			
B	B	B	B		B	B		B		B	B		B	B		B	B		B		
Bp			Bp	Bp		Bp	Bp		Bp		Bp	Bp		Bp	Bp		Bp	Bp		Bp	
C	C	C		C	C		C	C		C		C	C		C	C		C	C		C
Cp			Cp		Cp	Cp		Cp	Cp		Cp		Cp	Cp		Cp	Cp		Cp	Cp	
C#	C#			C#		C#	C#		C#	C#		C#		C#	C#		C#	C#		C#	C#
D	D				D		D	D		D	D		D		D	D		D	D		D
Dp						Dp		Dp	Dp		Dp	Dp		Dp		Dp	Dp		Dp	Dp	
D#	D#						D#		D#	D#		D#	D#		D#		D#	D#		D#	D#
E	E							E		E	E		E	E		E		E	E		E
Ep									Ep		Ep	Ep		Ep	Ep		Ep		Ep	Ep	
F	F									F		F	F		F	F		F		F	F
F#	F#										F#		F#	F#		F#	F#		F#		F#
F#p												F#p		F#P	F#p		F#p	F#p		F#p	
G	G												G		G	G		G	G		G
Gp														Gp		Gp	Gp		Gp	Gp	
G#	G#														G#		G#	G#		G#	G#
A	A															A		A	A		A
Ap																	Ap		Ap	Ap	
A#	A#																	A#		A#	A#
B	B																		B		B
Bp																				Bp	
C	C																				C

The chromatic scale will have 19 keys in the Parvatic scale the same as with the diatonic scale has 12 keys in each chromatic scale.

7.1d - Artone 19

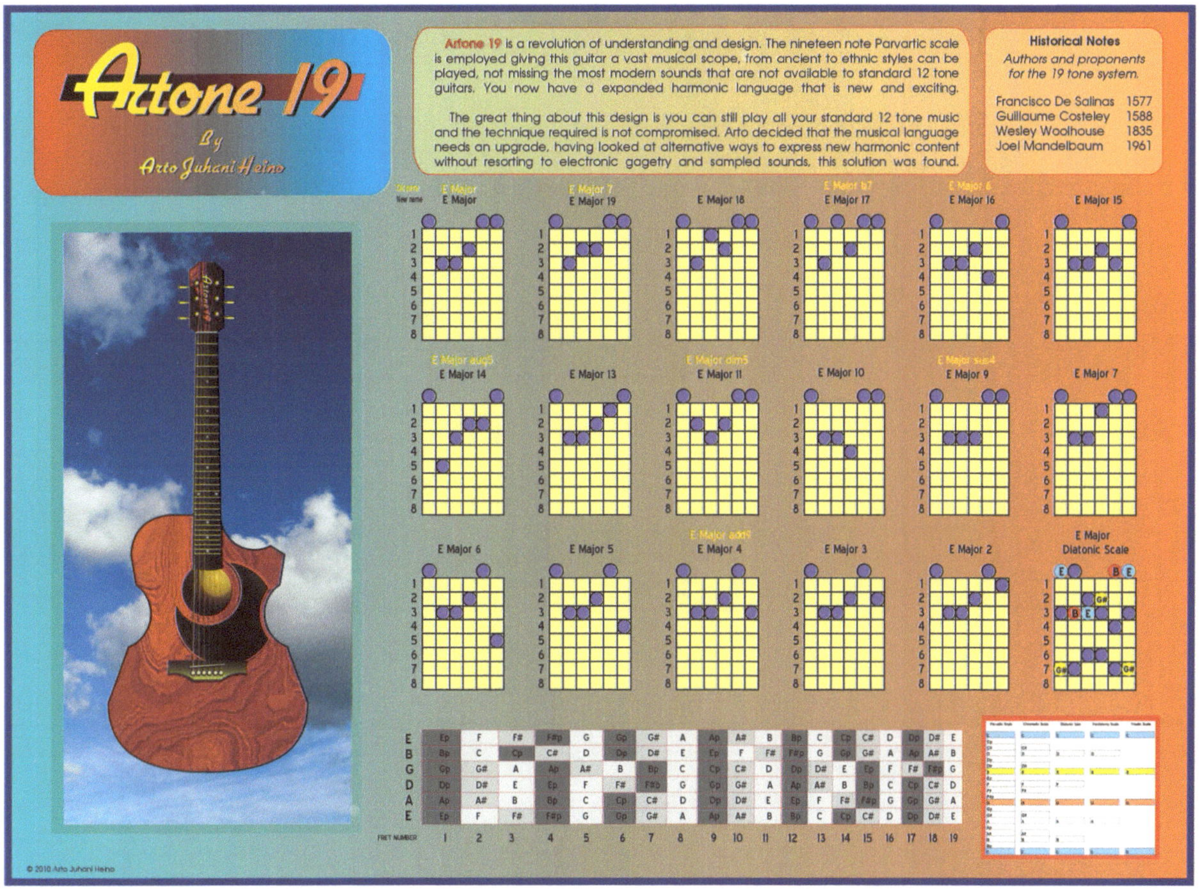

This instrument was designed in 1998 but was never conceptually completed until 2010, this is only the first design. This instrument is quite playable and more will be done in the future, eg electric, bass, etc. I have created the name Parvatic (Latin "parvum" , meaning small steps) so there could be new terms and names of harmonic forms this instrument will introduce to our musical vocabulary. Using "p" instead of "##" or "bb" simplifies the scalar structures, it denotes the new notes that are now usable, the old 3rds are now more in tune than the equal temperament 12 tone system the old 7ths are now sounding good, plus you can now play a lot of other types of music that wasn't available to the 12 tone guitar. This picture is only the preliminary design for the "Artone 19".

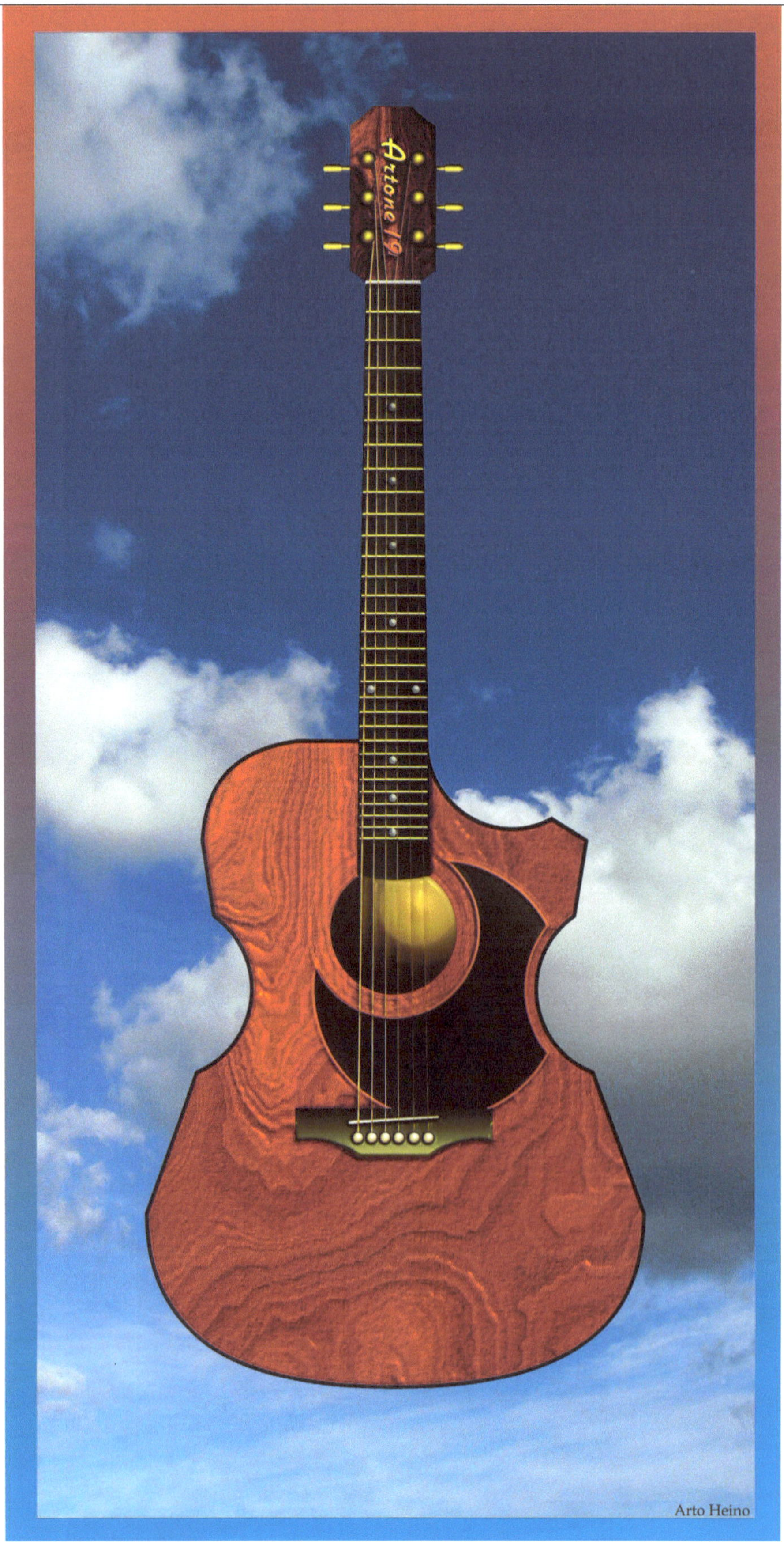
Artone 19
Arto Heino

	1	2	3	4	5	6	7	8	9	10	11	12	13	14	15	16	17	18	19
E	Ep	F	F#	F#p	G	Gp	G#	A	Ap	A#	B	Bp	C	Cp	C#	D	Dp	D#	E
B	Bp	C	Cp	C#	D	Dp	D#	E	Ep	F	F#	F#p	G	Gp	G#	A	Ap	A#	B
G	Gp	G#	A	Ap	A#	B	Bp	C	Cp	C#	D	Dp	D#	E	Ep	F	F#	F#p	G
D	Dp	D#	E	Ep	F	F#	F#p	G	Gp	G#	A	Ap	A#	B	Bp	C	Cp	C#	D
A	Ap	A#	B	Bp	C	Cp	C#	D	Dp	D#	E	Ep	F	F#	F#p	G	Gp	G#	A
E	Ep	F	F#	F#p	G	Gp	G#	A	Ap	A#	B	Bp	C	Cp	C#	D	Dp	D#	E
FRET NUMBER	1	2	3	4	5	6	7	8	9	10	11	12	13	14	15	16	17	18	19

Arto Heino

The fret board skills need to be updated due to a few extra notes between the normal hand positions that most guitarist would use to play a standard scale or chord. A careful study will reveal this to be a minor difficulty as the relative note positions of the diatonic scales are still true the the figure shapes used by the instrumentalists.

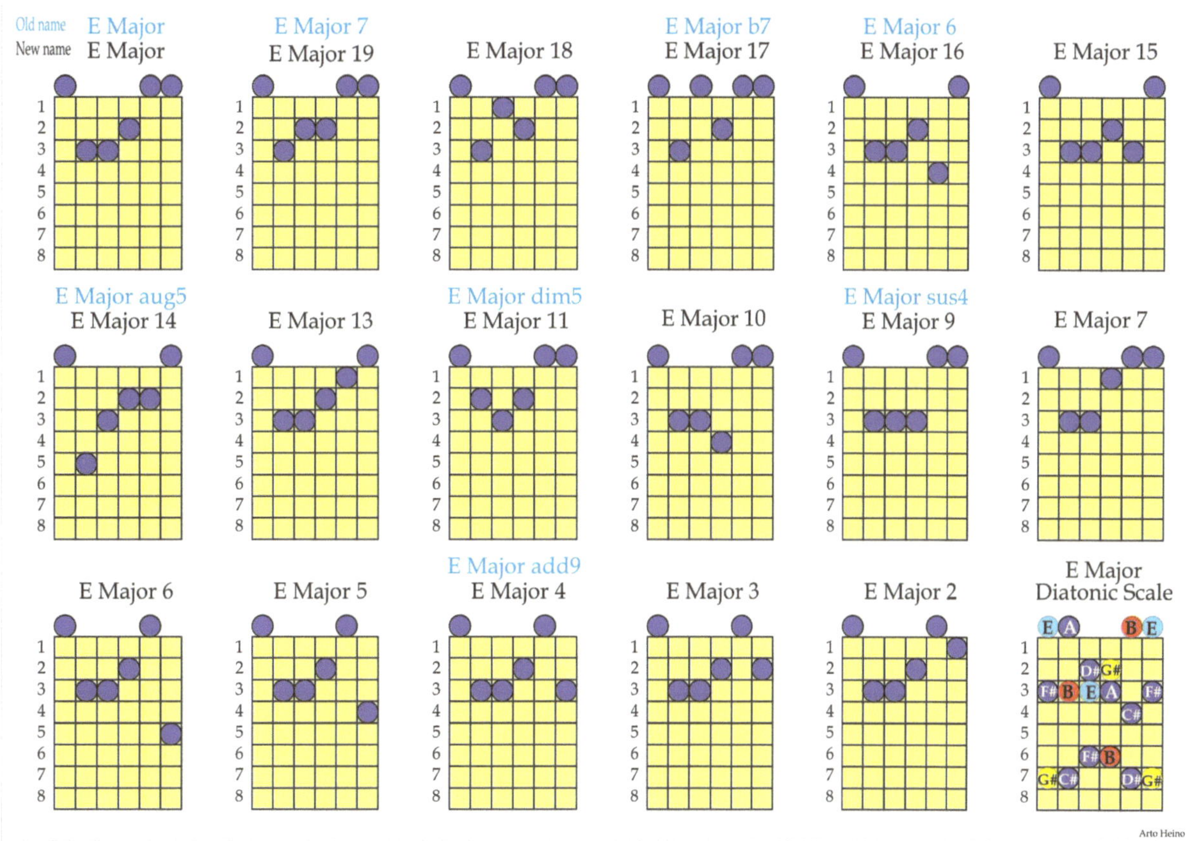

7.1e - Quantum Scale

The basic driver for all these new developments was my discovery of the "Quantum Scale", this was found during my study of Quantum Arithmetic and all the geometric implications, which included frequency scaling and harmonics/overtones that relate to the musical tone. I published my findings on the internet in 1996 and let the online community pick up on innovation. Sadly this did not happen as there were no instruments that could play these notes, excepting pure tone generating electronic synthesizers, as keyboards still only had 12 tones and midi did not allow note variation.

Tempered Scale C=256	Octave 0	Octave 1	Octave 2	Octave 3	Quantum Scale C=252	Octave 0	Octave 1	Octave 2	Octave 3
C	**128.000000**	**256.000000**	**512.000000**	**1024.000000**	C	**126**	**252**	**504**	**1008**
C#	135.611276	271.222552	542.445104	1084.890209	C#	135	270	540	1080
D	**143.675142**	**287.350284**	**574.700569**	**1149.401137**	D	**144**	**288**	**576**	**1152**
D#	152.218511	304.437021	608.874043	1217.748086	D#	151.2	302.4	604.8	1209.6
E	**161.269894**	**322.539789**	**645.079578**	**1290.159155**	E	**159.6**	**319.2**	**638.4**	**1276.8**
F	**170.859501**	**341.719003**	**683.438005**	**1366.876011**	F	**168**	**336**	**672**	**1344**
F#	181.019336	362.038672	724.077344	1448.154688	F#	180	360	720	1440
G	**191.783306**	**383.566612**	**767.133223**	**1534.266447**	G	**189**	**378**	**756**	**1512**
G#	203.187335	406.374669	812.749339	1625.498677	G#	201.6	403.2	806.4	1612.8
A	**215.269482**	**430.538965**	**861.077929**	**1722.155858**	A	**210**	**420**	**840**	**1680**
A#	228.070072	456.140144	912.280287	1824.560575	A#	216	432	864	1728
B	**241.631824**	**483.263648**	**966.527296**	**1933.054592**	B	**240**	**480**	**960**	**1920**
Ratio					**Ratio**				
C#/C	1.059463	1.059463	1.059463	1.059463	**C#/C**	1.071429	1.071429	1.071429	1.071429
D/C#	1.059463	1.059463	1.059463	1.059463	**D/C#**	1.066667	1.066667	1.066667	1.066667
D#/D	1.059463	1.059463	1.059463	1.059463	**D#/D**	1.050000	1.050000	1.050000	1.050000
E/D#	1.059463	1.059463	1.059463	1.059463	**E/D#**	1.055556	1.055556	1.055556	1.055556
F/E	1.059463	1.059463	1.059463	1.059463	**F/E**	1.052632	1.052632	1.052632	1.052632
F#/F	1.059463	1.059463	1.059463	1.059463	**F#/F**	1.071429	1.071429	1.071429	1.071429
G/F#	1.059463	1.059463	1.059463	1.059463	**G/F#**	1.050000	1.050000	1.050000	1.050000
G#/G	1.059463	1.059463	1.059463	1.059463	**G#/G**	1.066667	1.066667	1.066667	1.066667
A/G#	1.059463	1.059463	1.059463	1.059463	**A/G#**	1.041667	1.041667	1.041667	1.041667
A#/A	1.059463	1.059463	1.059463	1.059463	**A#/A**	1.028571	1.028571	1.028571	1.028571
B/A#	1.059463	1.059463	1.059463	1.059463	**B/A#**	1.111111	1.111111	1.111111	1.111111
C/B	1.059463	1.059463	1.059463	1.059463	**C/B**	1.050000	1.050000	1.050000	1.050000
1/(ratio-1)	16.817154	16.817154	16.817154	16.817154	**1/(ratio-1)**	14.000000	14.000000	14.000000	14.000000
	16.817154	16.817154	16.817154	16.817154		15.000000	15.000000	15.000000	15.000000
	16.817154	16.817154	16.817154	16.817154		20.000000	20.000000	20.000000	20.000000
	16.817154	16.817154	16.817154	16.817154		18.000000	18.000000	18.000000	18.000000
	16.817154	16.817154	16.817154	16.817154		19.000000	19.000000	19.000000	19.000000
	16.817154	16.817154	16.817154	16.817154		14.000000	14.000000	14.000000	14.000000
	16.817154	16.817154	16.817154	16.817154		20.000000	20.000000	20.000000	20.000000
	16.817154	16.817154	16.817154	16.817154		15.000000	15.000000	15.000000	15.000000
	16.817154	16.817154	16.817154	16.817154		24.000000	24.000000	24.000000	24.000000
	16.817154	16.817154	16.817154	16.817154		35.000000	35.000000	35.000000	35.000000
	16.817154	16.817154	16.817154	16.817154		9.000000	9.000000	9.000000	9.000000
	16.817154	16.817154	16.817154	16.817154		20.000000	20.000000	20.000000	20.000000

I decided to redesign both units into one design named "Hexartom 19", I fully developed all the technical requirements for note iterations, scales,chords etc in 2009. I realize as many others have for the last 50 years, the need for such an new instrument will help spread a new aesthetic harmony in musical composition and truly usher new beginnings for the 21st century.

7.1f - Hexartom 19

Logo of the Hexartom 19

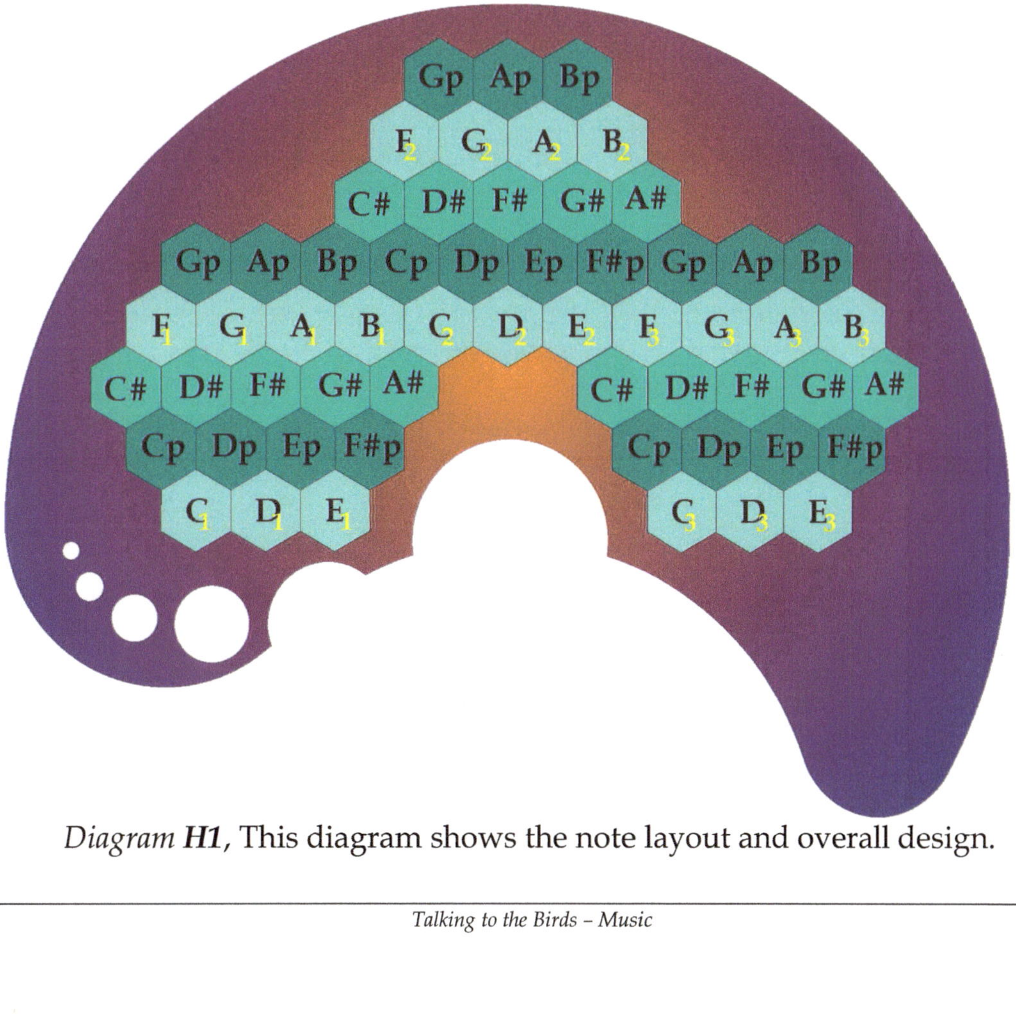

Diagram ***H1***, This diagram shows the note layout and overall design.

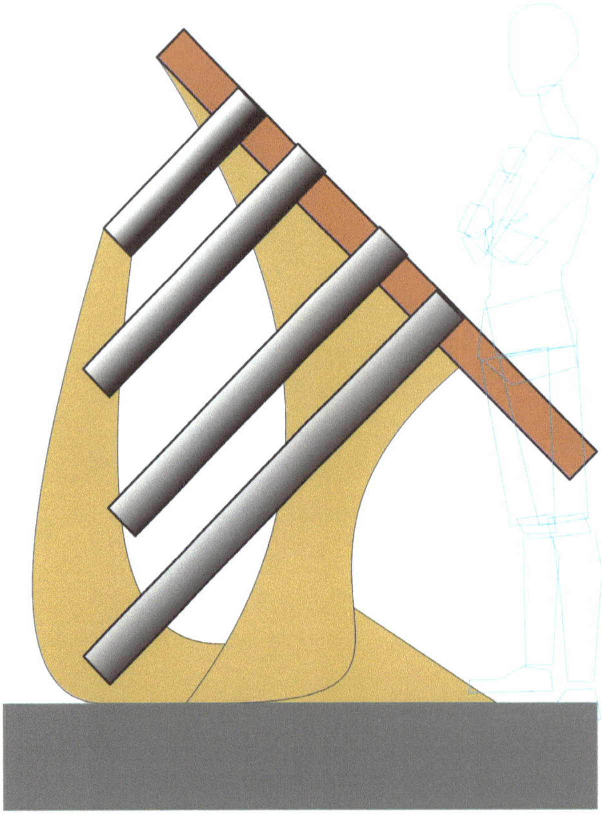

Diagram ***H2***

First stage - building the layout and basic structure

This has 3 parts to complete the basic geometric playing space. This stage will be repeated as the first model will the prototype made from less exotic materials.

1.

Cut and measure 57 unique lengths of PVC/acrylic/plexi-glass tubes with diameters of 100mm and the lengths varying from 155mm to 1297mm. This will give you 57 pieces adding to 31.18 mtrs total, allow 3mm each for cuts giving you at total of 31.351 mtr, rounding this to 32 mtrs.

2.

Create 2 boards from 2000mm x 2000mm square wood oak/pine/marine ply 12mm thick, drill and route holes to fit 57 100mm holes in a 3 form hexagonal shape as per diagram H1. The boards are in pairs, 100mm apart to support the top portions of each tube .

3.

Mount the holding boards onto a custom wood frame made from sturdy 50 to 100 mm timber, all joins will be custom designed without nails or screws, only dowels and tenon joints. All wood surfaces will be rounded and artistically rendered for maximum aesthetic appeal. The prototype frame would be made from pine.

Second stage- Final parts build and tuning

1.

Now we can begin by mounting the latex/leather/plastic drum ends affixed with sliding hollow end caps. each end cap will be colour(tint) coded to indicate the tonal grouping of each row within the Parvatic scale. As indicated by the diagram H1. All fittings are stainless or polished brass.

2.

The instrument is now playable and should be tuned by trimming the prototype pipe and the tension of each drum head. An audio spectrum analyzer, an audio frequency tone generator will be needed to complete the tuning. The final pipe would be glass, plexi-glass , wood or polished metal.

3.

Adding audio pick-ups, mounting toroidal end rings to each pipe, also adding some creative lattice motifs and designing appropriate mallets will complete the final stage.

Conclusions

All the built stages of development should be Photographed, Filmed and a progress report should be published on a popular art blog, these notations will secure an Artistic introduction of the Parvatic scale within the musical culture with a added bonus of the building of a new instrument that can perform with this new scale form.

The whole construction is a repeatable process that can secure employment outcomes and introduce a new and highly professional instrument that can be played in many venues and cultural situations leaving a lasting impression on the creative music and art around Sydney. The 19 interval scale also can be adapted to play the 17 note Pythagorean type scale that is played in some ethnic cultures.

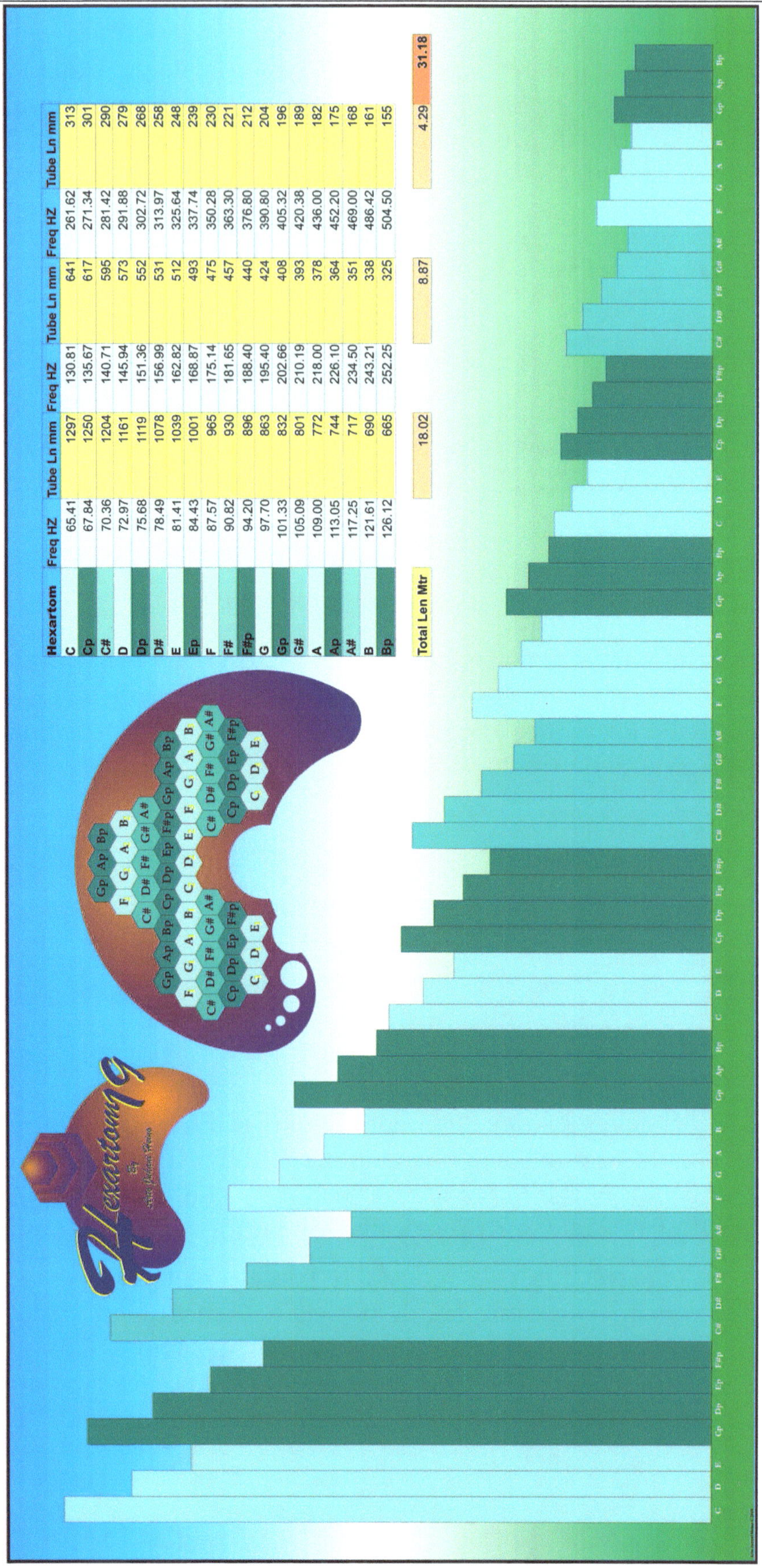

Hexartom	Freq HZ	Tube Ln mm	Freq HZ	Tube Ln mm	Freq HZ	Tube Ln mm
C	65.41	1297	130.81	641	261.62	313
Cp	67.84	1250	135.67	617	271.34	301
C#	70.36	1204	140.71	595	281.42	290
D	72.97	1161	145.94	573	291.88	279
Dp	75.68	1119	151.36	552	302.72	268
D#	78.49	1078	156.99	531	313.97	258
E	81.41	1039	162.82	512	325.64	248
Ep	84.43	1001	168.87	493	337.74	239
F	87.57	965	175.14	475	350.28	230
F#	90.82	930	181.65	457	363.30	221
F#p	94.20	896	188.40	440	376.80	212
G	97.70	863	195.40	424	390.80	204
Gp	101.33	832	202.66	408	405.32	196
G#	105.09	801	210.19	393	420.38	189
A	109.00	772	218.00	378	436.00	182
Ap	113.05	744	226.10	364	452.20	175
A#	117.25	717	234.50	351	469.00	168
B	121.61	690	243.21	338	486.42	161
Bp	126.12	665	252.25	325	504.50	155
Total Len Mtr		18.02		8.87		4.29
						31.18

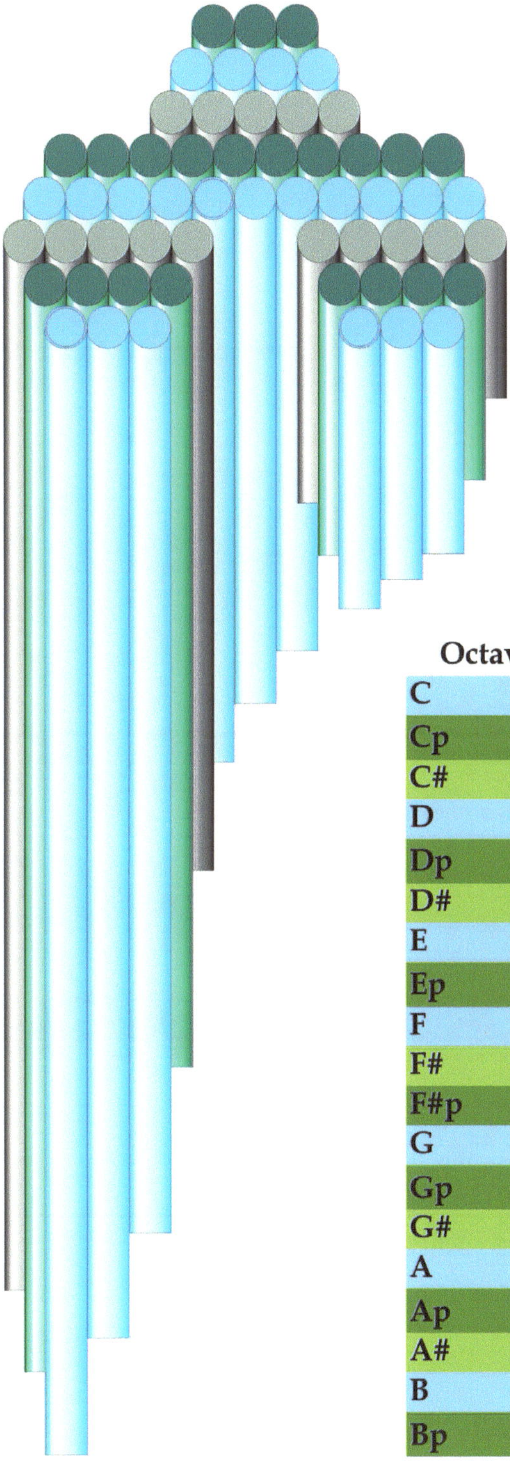

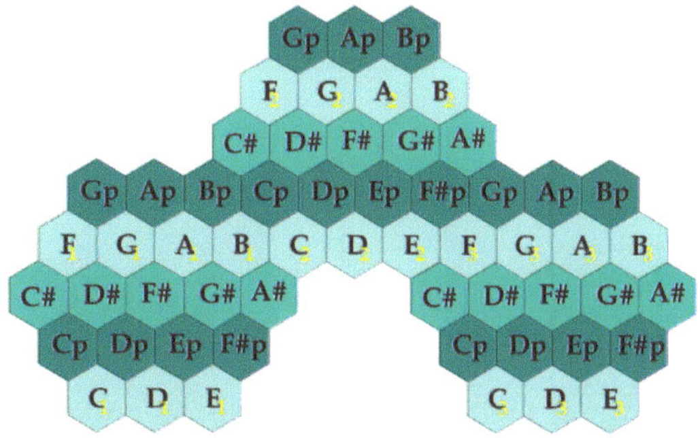

Hexartom Pipe lengths

By Arto Heino

Octave	Length 1	Length 2	Length 3
C	1296.59	640.67	312.71
Cp	1249.71	617.23	300.99
C#	1204.39	594.57	289.66
D	1160.70	572.72	278.74
Dp	1118.57	551.66	268.21
D#	1077.95	531.35	258.05
E	1038.79	511.77	248.26
Ep	1001.03	492.89	238.82
F	964.62	474.69	229.72
F#	929.52	457.13	220.94
F#p	895.67	440.21	212.48
G	863.04	423.90	204.32
Gp	831.58	408.16	196.46
G#	801.24	393.00	188.87
A	771.99	378.37	181.56
Ap	743.79	364.27	174.51
A#	716.60	350.67	167.71
B	690.38	337.56	161.16
Bp	665.10	324.93	154.84

Chapter Eight

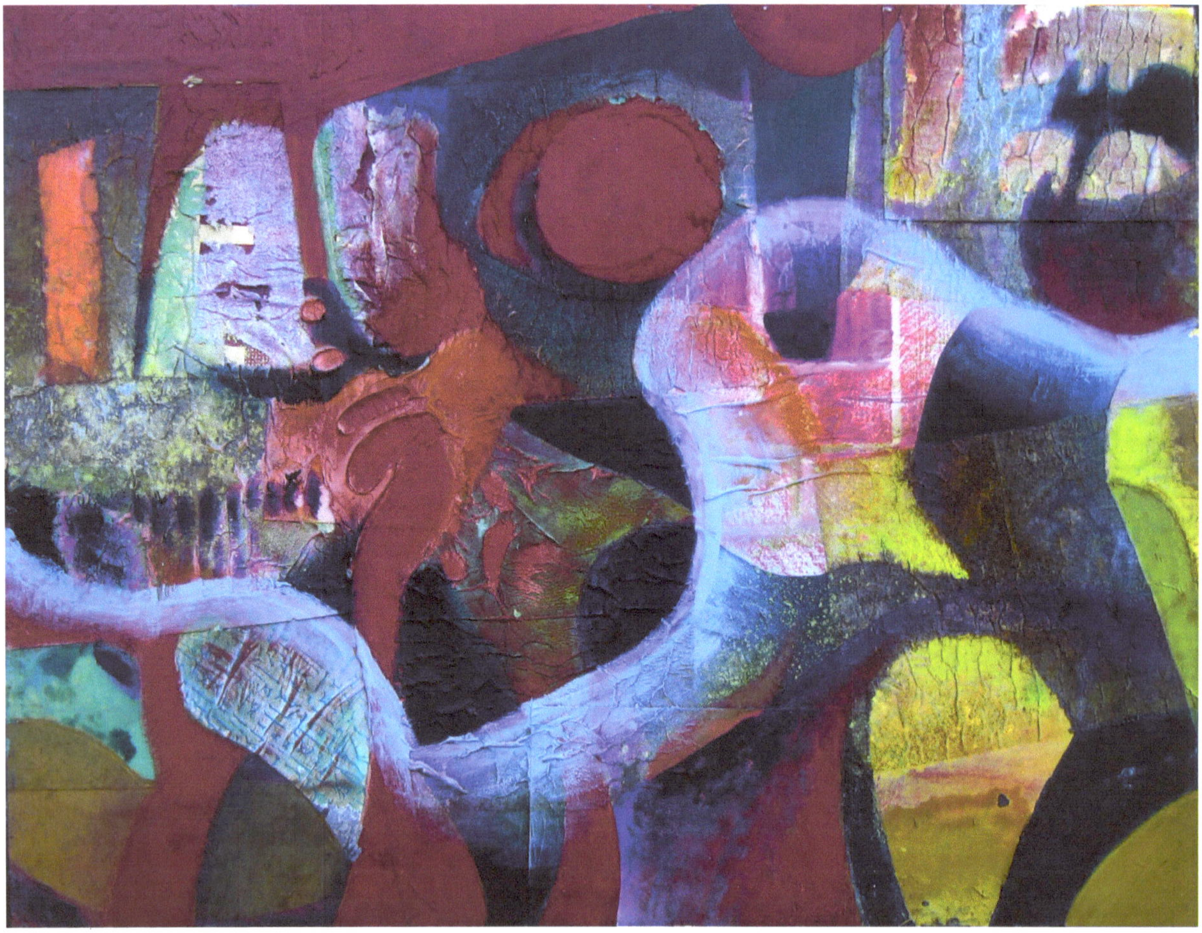

History

8.1 - Modern

8.1a - The Grand Illusion

This dissertation is really about the rampant mental illness of psychopathy that has been allowed to become institutionalized in our modern society led by the most affected and with the cart blanch emergence of their ill conceived control mechanisms, they are those who suffer from a psychosis of hierarchy - Bilderbergers – Illuminati – Masons - Rothschild - Hidden Orders - Banks - Oil Cartels – Law Firms – Security Companies – Power Companies - CFR – DHS – CIA – FBI – NSA, The list is much longer and have three commonalities, it is always fear based, secondly it always involves money and third it steeped in political and legal trickery. In the early part of the 20th century they were called gangsters and dictators and they used all the same modes of operation, easy money, corruption and extortion, the same as the current selection of legalized criminals use today, excepting they are called corporations and governments, sanctioned by the same corruption in the legal and political system.

Edward Mandell House had this to say in a private meeting with Woodrow Wilson (President) [1913-1921]: -

"[Very] soon, every American will be required to register their biological property in a National system designed to keep track of the people and that will operate under the ancient system of pledging. By such methodology, we can compel people to submit to our agenda, which will affect our security as a charge-back for our fiat paper currency. Every American will be forced to register or suffer not being able to work and earn a living. They will be our chattel, and we will hold the security interest over them forever, by operation of the law merchant under the scheme of secured transactions. Americans, by unknowingly or unwittingly delivering the bills of lading to us will be rendered bankrupt and insolvent, forever to remain economic slaves through taxation, secured by their pledges. They will be stripped of their rights and given a commercial value designed to make us a profit and they will be non the wiser, for not one man in a million could ever figure our plans and, if by accident one or two would figure it out, we have in our arsenal plausible deniability. After all, this is the only logical way to fund government, by floating liens and debt to the registrants in the form of benefits and privileges. This will inevitably reap to us huge profits beyond our wildest expectations and leave every American a contributor or to this fraud which we will call "Social Insurance." Without realizing it, every American will insure us for any loss we may incur and in this manner; every American will unknowingly be our servant, however begrudgingly. The people will become helpless and without any hope for their redemption and, we will employ the high office of the President of our dummy corporation to foment this plot against America."

So this will open the free discussion of their motive and execution of this power grab away from the masses. A relevant quote from Alan Bloom:

"Freedom of the mind requires not only, or not even especially, the absence of legal constraints but the presence of alternative thoughts. The most successful tyranny is not the one that uses force to assure uniformity, but the one that removes awareness of other possibilities.", from the book ", The Closing of the American Mind

Distractions, Diversions and Outright lies

History has shown that the simplest method of gaining control of the masses is to keep the public undisciplined and ignorant of basic system principles on the one hand, while keeping them confused, disorganized, and distracted with matters of no real importance on the other., as the king said "Feed them cake", a euphemism for a dietary distraction from their wretched plight while committing generations of his subjects to abject poverty and serfdom for the rest of their lives.

This is achieved by:

1. Disengaging their minds, sabotaging their mental activities, by providing a low quality and highly abstracted program of public education in mathematics, logic, systems design, and economics, and by discouraging core technical creativity.

2. Engaging the focus of their emotions by promoting and increasing selfish, self-indulgent and self centered behavior and their immersion in emotional and physical activities, by:

a. Unrelenting emotional affronts and attacks by way of a constant barrage of sex, violence, sport and wars in the media - especially on the TV, newspapers, Internet, computer games and continuous advertising.

b. Giving them what they desire , in excess ; junk food for thought and by depriving them of what they really need. Junk like; game-shows, sports shows, radio chatter, pornography, mobile phone chat, texting, Internet social networks like twitter, facebook etc. Also the real junk food that will keep them addicted to trans-fats, sugars and starch this will keep them preoccupied with food and health issues and diets etc

3. Rewriting history and law and subjecting the public to the deviant creation, thus being able to shift their thinking from personal needs to highly fabricated outside priorities. Diversions like - war on terror, boat people, global warming, etc . False history by omitting certain events, people and corrupting the true understanding of how a society was built, falsifying ancient historic lineage.

4. Hiding the history and the nature of the psychological control weapons that will be used on them. The main rule is that there is more profit with more confusion. Therefore, the best approach is to create the problems and then offer the a range of unsatisfactory solutions filled with loopholes that detracts from the true needs of the society.

MEDIA:

Keep the adult public attention diverted away from the real social issues, and captivated by matters of no real importance. eg sports events, music concerts, celebrity life styles, blockbuster movies, cartoons, food shows, game shows etc. A continuous blitz on every available TV, Radio and Internet channels, even during shows with scrolling propaganda at the bottom of your screen, sometimes faded into the each of the corners. This will ensure your are primed and filled with enough information so when the Advertising, News or station program break occurs you will be informed with enough to complete the mental picture of the false reality, this will insure you will treat this as truth, repetition becomes indoctrination by stealth. The giant media corporations manipulate and cherry pick news items to sway the public into creating opinions that are loaded in the favor of the particular agenda, human interest stories are used to high-light the humanitarian doctrine of some large corporate groups, like WWF, that are secretly owned and funded by the cabal.

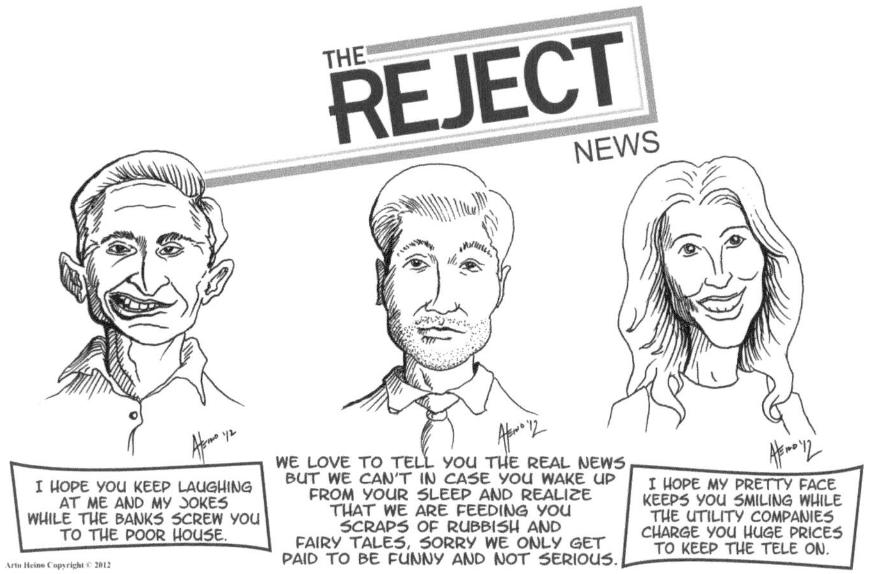

Give complete coverage to the sports world-wide. Allow great profits from any type of sport, this will assure its proliferation into most peoples lives, eg horse racing, football. Highlight the health aspects of precipitation in sport, this will keep the mind diverted from the control mechanism present. The truth is most people who participate in sports suffer greater ailments from their participation in it. The competitive nature of sport will also help keep the society divided, this has the added bonus of making fanatics who will promote the cabals agenda.

SCHOOLS:

Make a good and broad education both too expensive and difficult by creating a menagerie of useless courses that make learning complex, difficult and guided away from crucial primary knowledge. Keep the young public ignorant of real mathematics, real economics, real law, real history and the true nature of psychology and its impact on the human mind. Praise rote learning, ignore and ridicule original thinking. Elevate sports people and public commentators with low IQ's as role models for children. Games shows that reduce historical events to a sound byte. Jingoize false facts and knowledge. Trivia now becomes knowledge, true knowledge is hidden from public view, books will be digitized and e-books will be the norm, an easy method to change the contents to suit the agenda. The rejection and control of information can be manipulated before you read it. School textbooks would omit any references to any knowledge or people that allows alternative solutions to mainstream problems, eg Nikola Tesla, Wilhelm Reich, Victor Schauberger, Eric Dollard etc

Public education indoctrinates students to follow orders and the directives of government, no creativity unless strictly supervised and directed for the proletariat's goals. The curriculum is designed to dumb down the public, no hints on real learning , just follow strict rules and guidelines only. Private education limbers up a layer of society that physically rule the workers and in turn take orders from the governmental bureaucrats which take their directives from the politicians, a well designed hierarchy for the psychopaths.

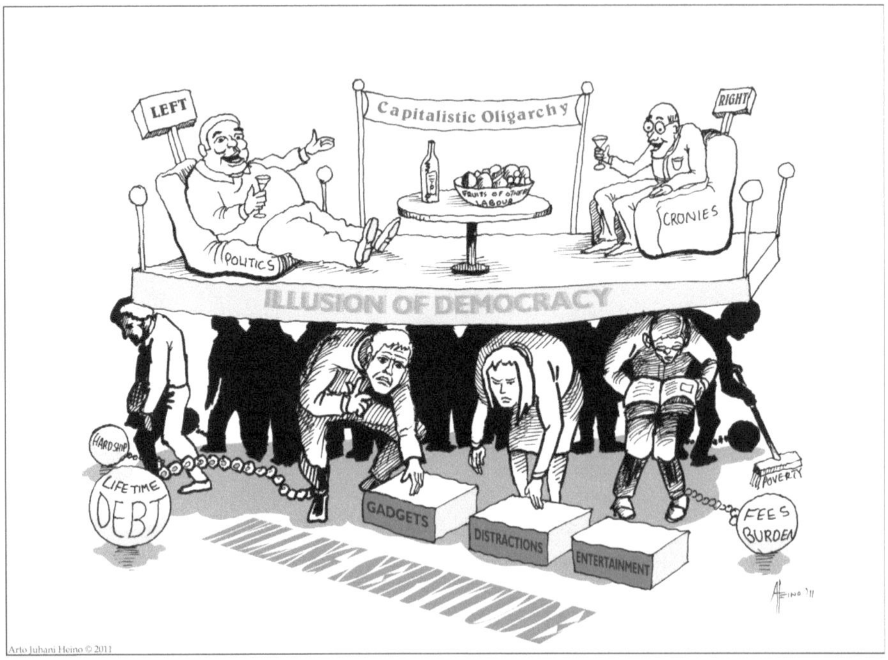

The current toolbox of diversion is the cost of higher education, by charging the students and stacking up their debt, they are now bound to debt repayment servitude for at least a good proportion of their careers. There is no guarantee of jobs after University/College so the extra stress is compounded and made to keep the student preoccupied by these matters on a monthly basis. They will not have enough time to worry about how they are getting screwed by the system, when the standards have dropped and it is very easy for everybody to get a degree, there will be a glut and the degree will be worth very little because of it, this will only be understood many years after graduation.

SCIENCE:

Give stature to compromised academics who can be presenters of extreme abstract concepts such as String Theory, Quantum Physics or Relativity which are totally divorced from our real world, thus keeping the truth of a scientific reality lost in the semantics and word play that fill theoretical physics, theoretical cosmology and all the false mathematical models. The irrational Quantum and Relativity THEORIES take on Religious aspects and are required acceptance or the students fail, this keeps the flame of stupidity alight and true understanding in the dark, thus filling the mind with highly dubious theories and useless information. Thereby blocking understanding of self regulating systems and denying the truth of the underlying nature of reality preventing the true understanding of the wheels of nature that would give us cost effective clean energy machines, anti-gravity, abundance and open clean systems.

Hide the real technology from the masses, such as climate engineering. This technology can be used to effect disasters all over the world, causing havoc to economic situations. This technology along with propaganda can strike fear and exert control over nations, climate change is now a silent weapon with mesmerized followers in all walks of life, they are now willing soldiers ready to defend this false agenda. Using the high frequency radio bandwidths of electronic fog(mobile phones, wireless tech) to keep the brain bombarded with unhealthy EM, and ignore its health implications(cancer, tumor).

The scientific peer review journals is the best place to stop any mention of low cost energy or any breakthroughs that might affect the current status-qua. To co-opt this system, give credence to the science that fits the standard model of creating waste, high profit and feel good science. Any break through technology either dumb it down or ignore it implications.

TECHNOLOGY:

Place this aspect of our current culture as THE MOST important change and direction in our generation, this will assure we all follow like lemmings to the world of iPod, Microsoft, Windows, Google, Facebook, Twitter and we can all live in CPU heaven talking about trivial daily events . This will shape the Job trends and stimulate massive human activity on software and hardware driving our temporary needs for superficiality and mediocrity against mans needs for future long term survival. The games industry will explode into multilevel interaction,virtual reality and on-line gaming competition, dooming the generation who is caught in its wake to computer addiction, maladjustment to reality and mental illness.

Fill the internet with websites that promote the good feel approach to the system facade , fill every corner and page with the same simple messages, allow the adds to follow you from site to site, this will keep everybody busy preoccupied with meaningless trivia without engaging in any research that might lead you to the truth of our current society , its problems and possible solutions.

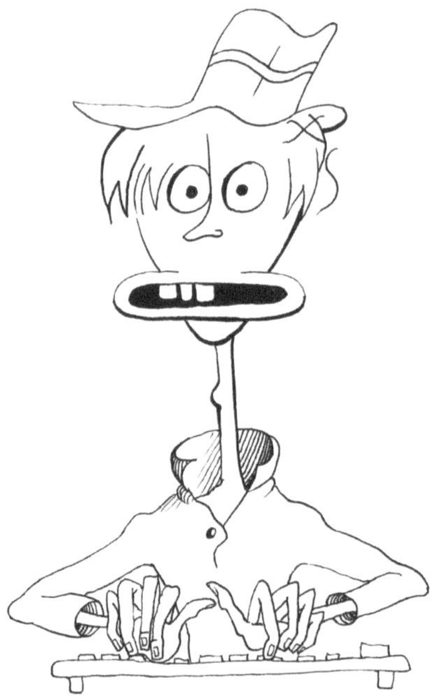

Arto Heino Copyright © 2012

To slow down the positive and useful aspects of the computer revolution and add a huge lump of paranoia into the mix, introduce the idea that some angry techno-mean kid in his room at home is creating a malicious computer virus to take down big companies, and in the process fills your on-line experience with virus created chaos. This is only a ploy, as the technical requirements to create the multitude of computer viruses that fills the Internet in a daily basis would amount to being a corporate structure that employs full time hackers. This IS what they really have done, and if anybody speaks up about this they will aim their cyber attack onto them with a fury. We know this is how they created the computer security industry, the legacy they have given us is firstly a constant reminder(Virus Scanners) and distraction of some lurking virus hiding on the Internet , ready to infect your computer and attack your private information costing you time and money to keep this monster at bay, secondly; is we now have less privacy and anonymity that in the history of this planet. We give our rights away on every program we purchase or use, read your agreement statement at the beginning of every software installation. Linux gave this corporate model a great scare, so they avoid mentioning this type of platform as it promotes the sort of ideas that they are afraid of.

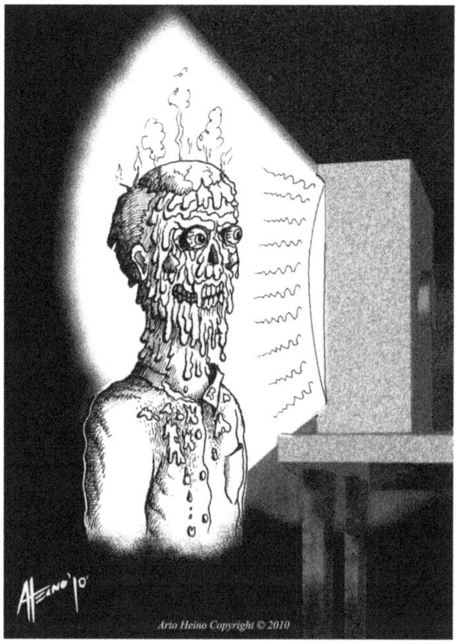

The record use of cell phones will become a health related issue as the number of brain cancers and related tumors start to escalate, this technology has very dangerous outcomes, as the radio frequencies used are the same as your microwave oven. This is another method the elite have reassured population decrease, as the only method of communications will be in this high frequency range.

ENTERTAINMENT:

Keep the public entertainment below a sixth-grade level. Lowering the musical standards so they praise mindless and moronic attitudes and behavior, such as Punk, Rap, Scratch ,Trance, Ambient etc. they own the industry, so crafted manipulation of lyrics and images are infused to keep their message repeated and subconsciously digested. Young artists would do almost anything for recognition, so as they are the owners of the music industry they encourage uninformed, uneducated and undisciplined literary, musical and artistic re-creativity and the copyist mentality, making control over their image and output easy. A good example of this brainwashing format is the TV show "American Idol". Regurgitation of cliches, disengaging the acceptance of anything really new. Standards that are controlled by big media like - cartoons, sample libraries, fashion styles, news segments, reality shows, advertising, etc . Mobile phones will allow you to be diverted where ever you go, applications will keep the distraction level maximized, keep the focus on what THEY have for you. Create followings of mindless babble and foolhardiness on Game shows, You-Tube, Radio shows and Live streaming of these types of events, this will keep the cattle from any real focus on any meaningful and important issues.

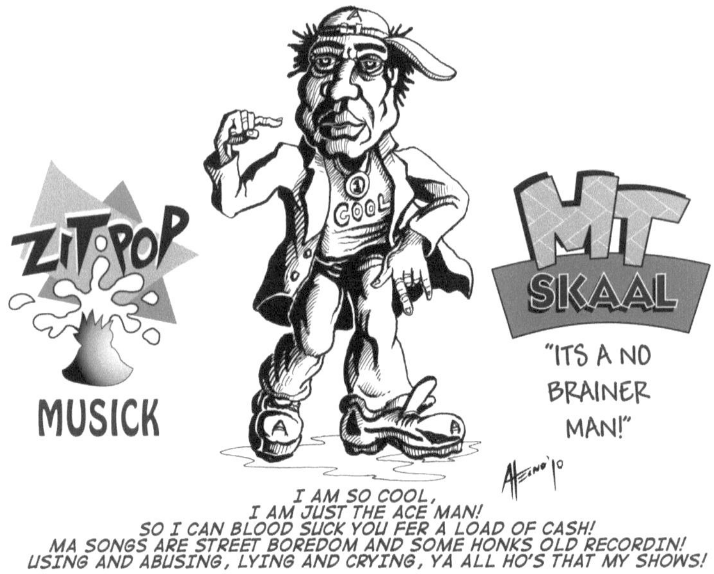

Promote a dysfunctional family lifestyle and abhorrent personal activity as a new norm, while promoting tolerance to deviant behavioral relationships. These shows will be defended as admirable and progressive, as humans in diverse family situations should be naturally tolerated and understood, but with their subversive script manipulation it will only be tolerated in a dysfunctional and deviant setting. This is a facade that hits the very core of human interaction, the freedom of expression and the rights of the individual. This again separates the voices of diversity and nulls the gains made by true humanitarian movements. This type of programming has led to a new wave of shows that promote immersion in the subject of criminality as a daily event, CSI etc

ARTS:

This is one area that the system has trouble trying to contain, you see Artists are not easily manipulated, they generally like to think on their own, without outside coercion or influence, observation is their mantra and stepping outside society is the only way to achieve this.. The control mechanism is the funding issues, competitions and commissions, these must have strict guidelines as the content and the subject matter, you don't want artists making too many political statements that might embarrass the system overlords. To keep artists at bay you either starve them, break their bank account or keep them busy chasing funding, competitions and commissions. This type of programing has the effect of excluding the real artists from contributing to new thinking and creation, marginalize true artists by keeping them as true outsiders, we can't have them telling you how the world is outside the box.

The system will only promote a narrow definition of the Artist,

1 - The warm friendly out door landscape painter(passive art),

2 - The Serious Portrait Painter(need them to paint the leaders and promote the system authority),

3 - The serious Commercial Artist/ Illustrator(tool for the system),

4 - The Genius of Oil painting/Murals(implant a carefully chosen few to keep other artists looking their way) ,

5 - The Genius Child painter(big distraction from real art),

6 - Cartoonist and Comic Artists(could be dangerous, still better to have on the side of the system),

7 - The Abstractionist(the complete fooling of the public on what art truly is),

8 - The Geometric Sculptor(keep the masses amused).

These definitions are narrowly drawn up and are guided by highly biased misinform-ants as these are the only types the system will allow to survive. If you are an artist that tries to truly inform the public on the visual field, you will most likely be ignored. Many real artists have a diverse pallet they use not just paint. The majority of people who enter the field of art, work at menial positions of paste ups, advertising, poster art and sign writing, as tools for the corporation. There is an emergence of a new type of pseudo-artist that has been compromised from the corporate sponsored institutions and led by the lure of money and fame, their usual attractions are public art and competitions, their visual messages are usually commercial and bankrupt of any real humanity, as these have the most appeal to the overlords of the system.

The Novelists, Screen writers, Film Makers, Historians, Biographers, Journalists and the Non fiction writers become the abettors of the knowledge of the age, they must be carefully controlled, and hand picked. The promotion and award system of these writers must be crafted as not to alert the populous of the bias of the system. They will allow many controversial books and films that are paraded as forward thinking and revealing, again this is a sham as they are just pushing the agenda in a direction that will not allow true dissent, only something they can control. They fear true independent groups like Occupy, that could create books and film to truly inform and educate the public.

WORK:

Keep the public busy, busy, busy, with no time to think - back on the farm with the other animals. Complicate roles in the work place with added rules and legislated interactions that hinder smooth work flow and discourage true free enterprise. Make sure all salable products have a limited shelf life, are unreliable and serve limited applications. Then try to fill your spare time with complex forms, contracts, accounts, difficult tax, superannuation and other governmental and private obligations, use legal threats so conformity to any new laws is accomplished, no time to understand the system just follow it or be doomed. This will also cause home life stresses and strains that will alienate the worker from a balanced family life, no time to think of the bigger picture.

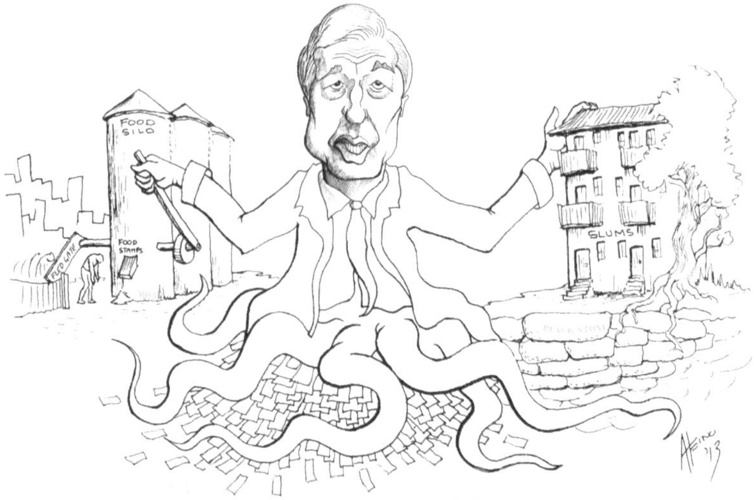

Give corporations tax loopholes and allow them to buy up small business concerns, take the variety out of the equation and replace it with stylized easily mass produced items and designs that have a limited lifetime functional value(cheap molded plastic). Link corporations together to create a seamless identity that follows each others products, designs ideas and crafted identities. This will drive out true competition and avoid quality issues by lowing the standards of acceptable life limits on products and services. Allow corporations to run utilities, health care, prisons,transport and energy, this will assure that you will be working for or directly competing against the transnational giants, no hope of growing a business, so the small players fold and follow them instead. This means you will follow the corporate ethics code and they demand strict obedience when employed by corporate identities, corporate life is not democratic. Corporations lack true human ethics only the ethics of profit and monetary gain at all cost

WAR:

Create many false wars on many fronts, War on Terror, War on Drugs, War on Pollution, War on Cancer this will not only strip our rights as individuals to make personal choices but also encourage the war mentality by creating a culture that seeks the war scenario of conflict, division and confrontation, this also has the added benefit of watering down the real horrific aspects of war. Use statements like "War is Peace", "To have Peace we Need War", this will cloud any arguments against war, the idea of having peace sometime in the near future is the carrot they dangle. The pundits, commentators and ex military service will write books and have lectures about how to achieve peace, meanwhile they move the goal post further and further away from any possible attempts at reconciling differences that actually make war within or between nations.

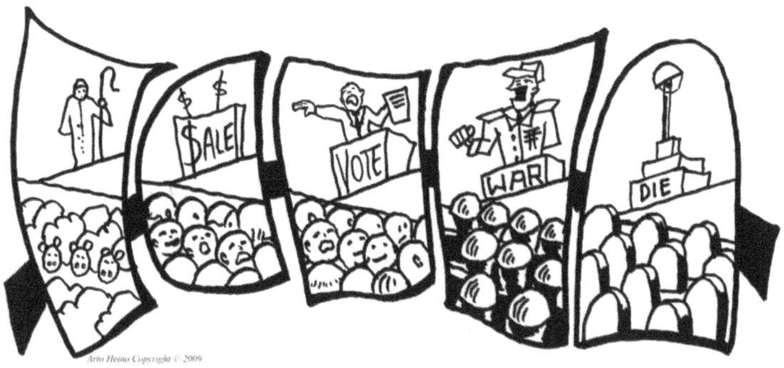

The great bogyman of terror/fascism etc is created either from a single county(Iran, N Korea) or an invisible groups (alqeda = AL CIA DA= of the CIA) that move freely in countries or near your borders(eg USA/Mexico), this is to keep the legislators busy on taking away our liberties while we begin to suspect anybody who is different to the crowd. This maintains the discomfort we feel about people who think differently, this also helps to give rise to racism and division amongst communities and nations, where a policy of a false fronted agenda of multiculturalism will help this divisive agenda along. This has been helped by each third world nation creating a steady stream of refugees to first world counties adding to the prejudicial agenda. No time or energy to really help anybody we are too busy with too many other problems, so we lock them up, this creates a festering hatred within differing communities.

Start wars in third world countries, by infiltration of false social expectations by inserting an invisible anti-group that people can be attracted to, then feed the followers with the idea of revolution as the only alternative to their countries social and economic dilemma. This will create zealots and groups dedicated to unreal causes, preoccupying vast amount of people in theses matters. Supply arms to both sides of these conflicts. When this has been accomplished and the people are weary, reinstate a similar regime and let the people make heroes of the new leaders, after 20 or so years repeat the same. The organizations that promote the politically neutral activities such as the UN Health, Red Cross, Food and Security forces that come into countries of strife are owned by the elite so as to show great philanthropy at work increasing there stature amongst the worlds public, sadly the non-profit/tax free statue allows them to be funded by the donations of the public. The mechanism of taxes funding our wars and by donation the consequences of war, the cabal are making a clear profit from the multiple levels of bureaucracy that are hidden from view. This intentional mind block decimates our own population and resources to a false idea that the cabal impose upon us without our conscious acceptance.

POLITICS:

Keep the number of political parties down to two, this will keep the public polarized. If required to push certain issues, create a radical third party that disguises itself as a supporter of free individuals, only to be really pushing a corporate agenda, that is secretly the issue that is pushed by the other two parties, when presented as a peoples issue the other two parties will adopt it, only to show they want what is best for the people. Preach Liberalism and Freedom to create class warfare. Make the followers of left hate the followers right and visa versa. This will also create a class of pundits and commentators that continuously debate our divided ideologies, this keeps the arguments festering, support emotive announcers and biased presenters, causing much strife and it will further divide people, this will advance their cause.

Set high moral policies and let the opposition water down its application, this will create great public debate and furor, distracting the public eye from any real issues, then the opposition wins government and process repeats until all the peoples rights have been eroded and the required draconian laws have been passed.

Just remember they created the system so there will be two opposites working against each other to keep the division of society alive, while together in collusion they put the hidden agenda into place,and you will be no wiser. We must see past this veil before we can truly act and change this corrupt system. The media play a big role in the public political debate, the corporate agenda and political outcomes can be steered by biased reporting and so called news networks. Giving the impression of a democratic system, but being loaded by hidden meanings and the talking heads of a one sided political debate.

Allow local government to create by-laws that press greater hardship on the community and greater hurdles to jump when activities requires approval. Start revenue raising so as to deplete households of any savings, rates, parking tickets, permits etc. The basic agenda will be the same regardless of the political party in government. The route the the final outcome will be the only difference between any party in power, the objectives will always be the same, keep the public worried about these issues and nothing else. All politicians were trained in the same class, the debate is where they play the game, without the debate game we might figure out the rules ourselves and find out the truth.

The name trading of different political systems of the past are continuously brought up and are tarring there opponents with the stain of communism, fascism, right wing, left wing etc They are there to keep you preoccupied with the differences between the ideologies and not was is being really pushed by current parties of today. This is a dogmatic approach to political debate, where they set the rules for public intercourse and will not allow ideas that might reveal their system of lies that might be exposed. The common method is to change the words of the argument but never change the argument, this way the hidden message can still languish in the public mind.

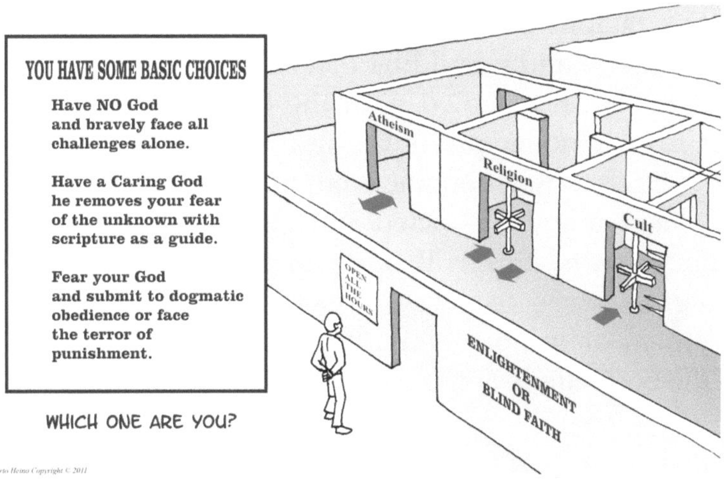

RELIGION:

Allow religions to flourish, under strict guidance using only one book as a guide, make sure there is enough differences in each faith to cause basic conflict. Apply standards like regular scripture reading, prayer meetings, extra services and group activities to keep the followers on track. The supernatural belief system that forbids questioning or defying the leaders should keep most followers away from questioning the true origins and history of their faith. The heads of each church align themselves so as to act in concord with the politics and rule of each nation that houses them, together, the complete control of the masses hearts and minds are assured. Destroy and vilify any belief system that in not sanctified by the ruling class, Waco was a good example of what a can happen when a rogue religion points its finger and accuses the government of being in allegiance with the dark forces of mankind, this sort of open defiance would make people ask questions about our leaders intentions, this would awaken some sleepers.

Religion is the business of sanctifying the most profitable and productive system of indoctrination, it has very little to do with personal spiritual beliefs as the word of a all powerful God need never be written down as it would be understood without the spoken or written word. Only the written word would be under control of the imperfect-ed man who would gain advantage over others for personal gain, which is exactly the way the hierarchy of a religious group has been designed, so each level never questions its structural validity.

ECONOMICS:

The meaning of economics in its inception from the early Greeks was the management of a household administration, this had little to do with money, banks or even politics, it has only been translated in the last 100 years into money based meanings. This way they keep the minds preoccupied with money, both the love of it and the need for it, divide society into rich and poor. Money worries and retirement funding crisis are tools to keep your mind off the real issues. Create a banking or monetary crisis, then bailout the banks and/or industries, shutting down industries and services that are useful and rebuilding those that offer very little but promise a lot, this will keep the public preoccupied with the differing standards of service. Offer financial help then funnel money out of the system into creating a debt problem, then the compromised politicians will demand a cut in spending, causing more stress to the system. This will keep many distracted by trying to find a way to mange their organizations running costs, this will devastate society hurting those in need and ruining any chance of a true recovery. These tactics will require the public to be divided on government debt repayment, this crisis will allow new laws to be passed that will usher in the beginnings of the peoples servitude to a debt ridden society.

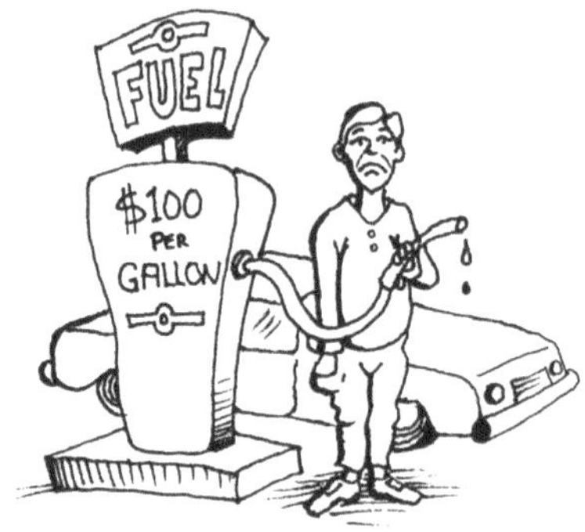

Adding to the economic and banking crisis, whole nations would now be under pressure to pay outstanding debt, to alleviate the crisis an assurance of austerity measures would be promised, if the default continues, repayment by the core wealth of the nation would be handed over eg gold, silver. These austerity measures would impose draconian laws as to increase hardship and stress, again this is to make sure the people will have no resources or will to mount any opposition to these measures. The question never arises, as to who are the faceless money lenders that nations owe their wealth to?

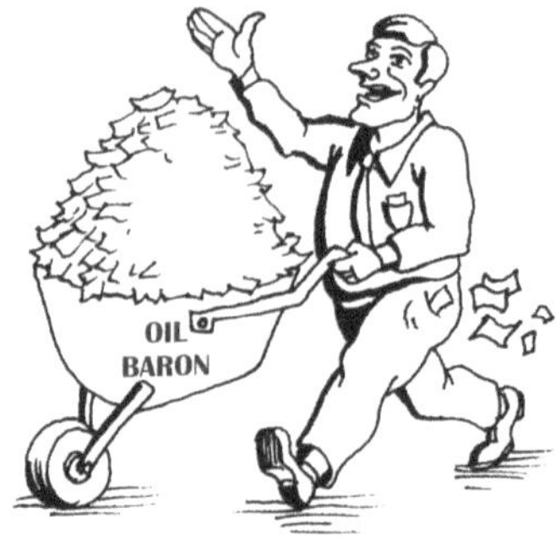

Oil is one of the tools to corrupt government leaders and officials in many nations, offer monetary rewards so oil companies can extract a cheap supply of oil for their local market reaping in large profits. Threaten to destabilize the region, war or even assassination if the game plan is not followed.

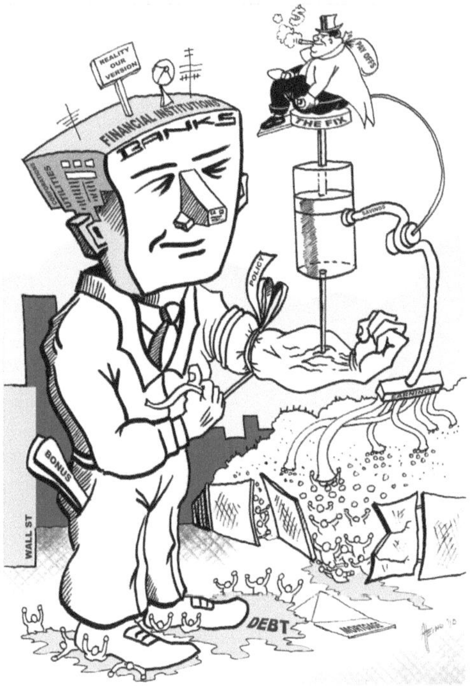

HEALTH:

Putting chlorine, fluorine and aluminum in the water supply and stating that this is to purify the drinking water, when in reality the long term effects are Alzheimer's disease(Al), Fluorosis bone disease(F) and cancer and heart disease(Cl). The use of additives in staples like bread and milk are also another form of slow death. Your life will be a preoccupation with many health issues, no time or energy to expose or fight the system. Keep the references on needed health care plans constantly on show while in reality supplying low quality care at premium prices. Cause a crisis in this industry that will paralyze and divide society, promise to help the poor but in reality force them into private health that will exasperate their debt.

Supply and control pharmaceutical drugs for ailments that can be solved via natural foods, good diet and herbs. Drive the naturopaths into hokey medical categories then make it illegal for them to practice without certification, this will squander the practitioners resources and time, no time to fight them. Through covert means of false standardization of procedural medicine the diversion from benign drugs of application into insidious toxic formulas have been created. Supply dangerous medicines to mild ailments, causing long term damage, too sick to realize what is going on, where Prozac was one of the best known to cause health issues.

Allow the use of technology that will damage our long term health such as mobile phones causing brain cancer.

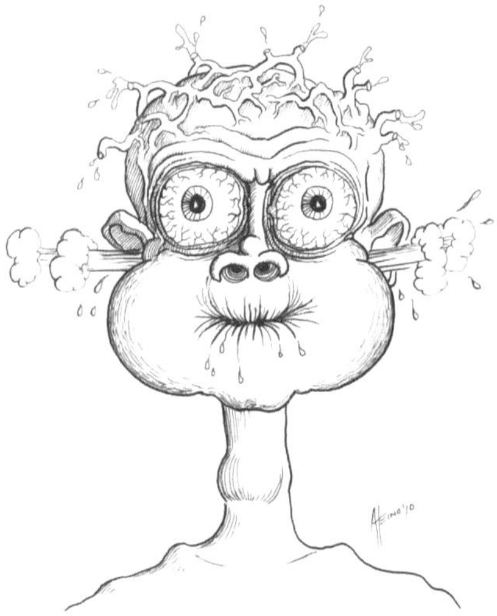

A major controlling element would be the Health Insurance Companies, as these providers supply a ready stream of brain washing to make their audience believe in the impossible task of getting healthy by paying money to a Insurance Company. You health is secured by your life style , your food intake and your chemical and biological environment. A Doctor is only called when you are sick or injured, this is an accidental situation not prevented by having Health Insurance. So how often do you suffer accidents of this nature, I would guess simply as a 5 year event, or as most people even more rare. The main focus they have is creating a fear of not having the funds to pay for a catastrophic life event like cancer, or a car accident. Again this is a form of accident insurance relabeled to fool the unwary.

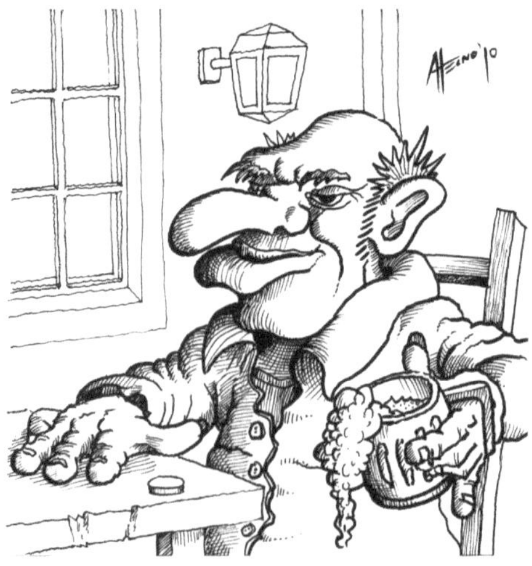

I HAVE ME COIN, SO LET ME DRINK IN PEACE,
SO AS I KIN RUIN ME LIVER AND GET THE HEALTH
INSURANCE TO PAY FER A BLOOMIN NEW ONE!

Give compromised high profile nutritionist the forum to influence the eating habits of the populace, then load the dice in favor of a bad dietary balance, this will enure a generation of unhealthy people that will be ricochet into another round of manipulation to repair their problems. This should be made public as possible, include sports people and trainers to offset the mind and body damaging message(The Fattest Loser etc). The most damaging is the way they divide the nation into body types and degrade their humanity. This loathing and self-conscious approach to the body image affects some of the most vulnerable in society. This ploy will ensure a good proportion of the populous will not have the accoutrement to devote to any other purpose beside their own personal body image plight.

This is what occurred in the 1970's with that carbohydrate loaded diet, so that 30 years later this type of propaganda has caused the obesity epidemic now prevalent in at least 3 generations of people. All these measures will keep the people focused on food in exclusion of everything else, this will create another generation who witness the families preoccupation, again this knock on effect will cause mental disorders surrounding food(Bulimia, Anorexia etc)

Discourage any form of mind altering drugs, it just might break the hold they have on you. Waking from the false reality can be traumatic for some people. Only addictive drugs that dull the mind and senses will be secretly promoted or as in alcohol openly available.

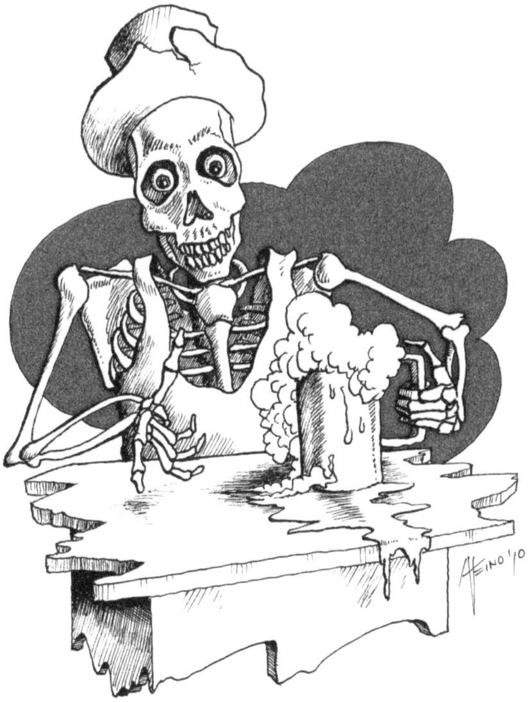

Does any of this describe the world in which we all live?

Of course it does! We live in a society that finds itself continually fascinated with matters of no real importance - such as our obsession with celebrity news, television, computer games, sports and sports stars, super-models and movie stars. A great deal of effort is expended upon these meaningless diversions, such as the trials and tribulations of Oprah, Michael Jackson, Martha Stewart, Paris Hilton, and others who came before and of those who will follow. Entire channels are devoted to this form of tel-cinematic voyeurism, these channels also include shopping and product promotion, feeding the consumerist mentality and offering nothing more than a temporary shopping high.

The problem of disengaging from the melee of thought control, is mainly from the addictive nature of that control. The need to feel like you belong to the mass trend of this inducted way of life is their way of ensuring you promote their agenda without knowing why. The people that fall outside of that control are ridiculed and quite often misaligned by public opinion.

Our so called leaders of society are failures, they have never had the publics interests at heart only the constant building of a system of world wide control. When a true light rises from the trenches of this social mire, he or she is struck down, jailed, misaligned, ignored or even murdered and never to exert their influence over the coming events. A good example of a few such people are Kent Bentkowski , William Cooper, Victor Schauberger, Wilhelm Reich, Stanley Meyers, Nicola Tesla and list grows larger every generation.

Are you ready to ACT?

Are you one of those who have the herd mentality or are one who can see through their deception?

Are you employed in an industry that make items or service support for their agenda?

Are you immersed or engaged in their diversion entertainment?

Are you helping those who are working to free your entrapment or are you condemning them?

If you have a friend, acquaintance or family member that is AWAKE to what has been stated here, do you listen to them and take their experiences seriously or do you just ignore them?

Do you measure up to being an awakened individual or do you prefer to stay as a drone?

Do you feel uncomfortable about all these implications? If you do then please go back to sleep.

If you are a brave soul, then ask yourself all those basic questions, again, will you act upon it?.

What to do?

Here are just a few things you could do:

1. Change your viewing habits.
2. Change your listening habits.
3. Read books, real books not, "So So for Dummies" etc
4. Find alternative view points of all subjects on the net.
5. Understand that big TV, Newspaper, Radio etc NEWS are full of lies.
6. Change your diet, eat clean unprocessed food, do not drink tap water.
7. Wash with natural products.
8. Careful of what you drink, acidic blood is the first sign of a major problem.
9. Get out of debt as soon as you can.
10. Vote true independent, not big party.
11. Find artists, musicians, writers, etc that are no-label and read, look and listen.
12. Ignore the harping of commentators that try to sway public opinion.
13. Ditch the mobile phone and any other wireless system.
14. Get off the electric Grid.
15. Grow your own food, avoid sugars, and any GMO derivatives.
16. Find ways to barter without money.
17. Be brave and speak out about the Emperor who has no clothes.
18. Stand together on these most important issues and unite
19. Study History without bias of winners and losers, just the culture not the wars.
20. Help foster a society that is not run by the psychopathic prone - Lawyers, Bankers, Politicians

Solutions

The main cause of the mufti-faceted ingrained problems our modern society has been enduring for the last 100 years is by those who are inflicted by psychopathy.

The mufti-faceted solutions, must first and foremost catch the arrested development that leads to this problem, early childhood intervention in these matter by normal human contact and empathic guardians, so the mental disorder of psychopathy can be addressed and find solutions before they are damaged beyond solution and become a problem to society. Most of the leaders of the 20th century suffered from this condition, resulting in wars, mass murder, rampant greed and a gangster mentality.

The process of finding the solutions will help society rediscover the complexity of the human condition, neutralizing the hierarchical thinking and redirection to a holistic approach to life's journey. The other great problem will be the simple fact that the control mechanisms of society are under the rule of those who have this disorder, they will not allow any attempt to remove their influence. The solution here would be to nullify the many mechanisms that they need to operate and the non-affected don't need.

1 Monetary System

2 Military System

3 Corporate System

4 Legal System

5 Political System

All theses systems of hierarchical control have a structure , pyramidal in form and easily understood as the base model of the mind set of psychopaths. Even though a captain might be a good soldier and not a psychopath, he is under the direct order of the system and its rules, the leaders know this and exploit the mechanisms created by them so as not to have any form of mutiny of purpose. The legacy of each generation of psychopaths leaves a trail of codes and manifestos that the subsequent generations follow blindly, as they suit the same mental cast, this perpetuates a form of secret society mentality, often Masonic in nature.

The simpler solutions, such as a world wide non-obedience of the forced social control mechanisms, this would be the fastest and most complete method, this would also be called a revolution and all the different systems would fight to regain control.

The long term and secure answers are as I have advocated, are a careful and holistic approach of solving the mental disorders and making it a requirement of society not to allow those with this problem continue to operate without a mental health solution. This will eventually cause our prisons to be empty and crime to disappear, along with the legal mechanisms that incarcerate those as well. The wanton nature of money and how it corrupts those with this mental disorder will be relegated to the basic exchange mechanisms that society originally intended for its use, probably back as far as 2000 BC.

The nature of man is not war like, as war is a disorder and conflict has no solutions, harmony is a natural order of the universe with dissonances resolved though aggregation of group understanding into personal physic resolution.

The corporate model is the pinnacle of this disorder creating a fascist outcome and conformity to false rules as their only model. Co-operatives are capable of delivering the maximum result with the minimum effort, because effort is not dictated by a rule book of profiteering but the conscience of those who co-operate.

For every job lost to a machine, one man must be freed from the shackles of labor. As a machine is essentially a product of the creative mind working with altruistic intention, then this altruistic intent must translate to a man having freedom from labors of servitude.

If this translation is corrupted in anyway we will forever be slaves to an archaic system of forced labor. This will a simple be proof that mankind has not progressed to a higher civilization with the use of technology. The result will be proof that we have been manipulated by a system of psychopathic behavioral mechanisms that creates a class of aberrant humans that require power and fear to keep this mechanism self perpetuating such as a tumor or cancerous growth. The mechanisms of the world monetary system is such a tumor where a growth in assets are multiplied ten fold from those who lack conscience of consequences and force unsustainable interest that is accrued in the process. Inflation(growth) is the result and deflation(austerity) is a money grab from the poor to the rich, either way the loading is always on those at the bottom of the money chain.

The solution here is to decouple the monetary system from the labor system, where labor will now serve to unify the coherence of humanity by creating lasting solutions in terms of technology and housing. The buildings and utility structures will last several hundred generations and be built to a standard that no current corporation would allow. Energy generation must be transformed to serve the public not tax its users, human beings are not just consumers we are also creators, so the terminology must change.

To close this commentary, written with freedom in my own solitude, a quote by Arthur Schopenhauer :

"A man can be himself only so long as he is alone; and if hes does not love solitude, he will not love freedom; for it is only when he is alone that he is really free."

Chapter Nine

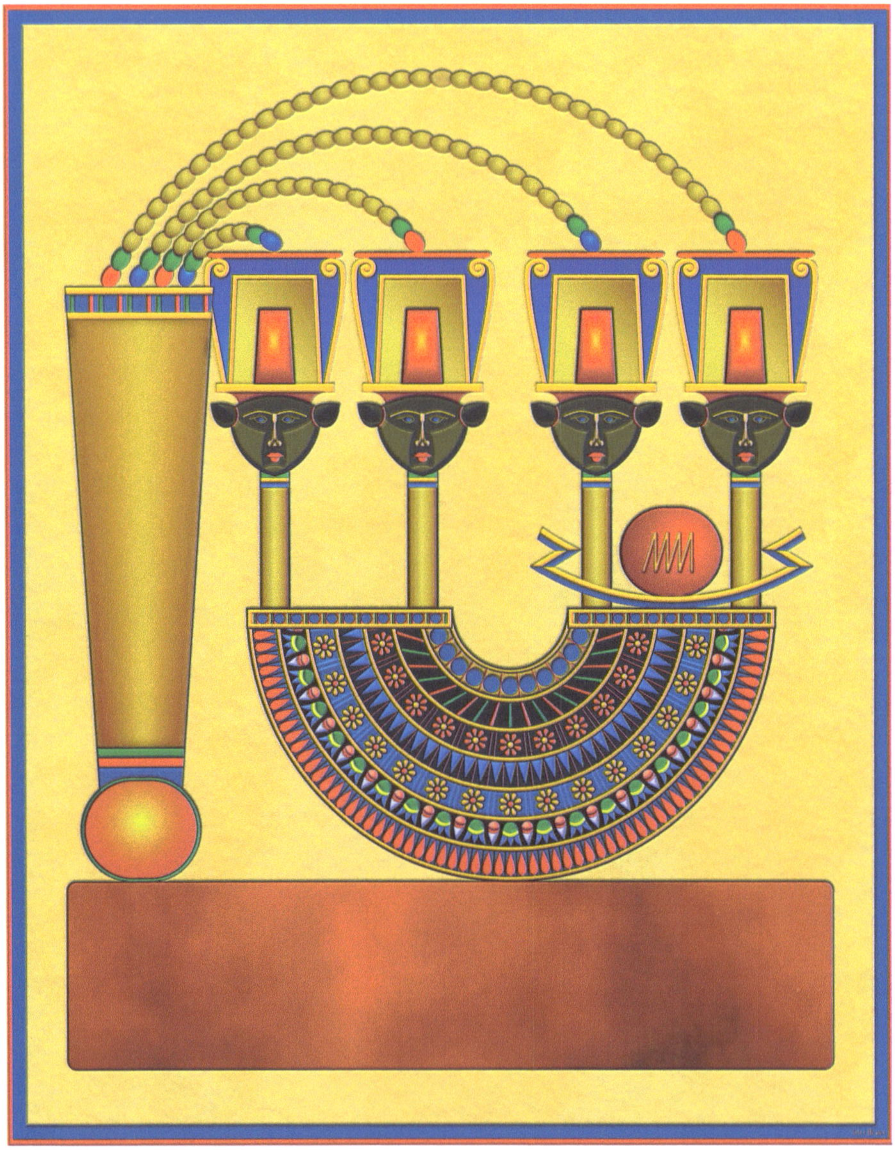

Archo-Astronomy

9.1 - The Hidden Star Charts of the Ancients

9.1a - Conjectures, Alignments and Probabilities

Many scholars and researchers have found that ancient man was obsessed with the heavens and made references to the stars and the planets on a daily basis. As an artist and under that simple premise, I decided to explore what I knew best; which is, the motivations and abstractions of the visual artists, the people who created the historical depictions of all the history and iconography we witness now. Visual artists have inner needs that require pictorial expression, unlike the writer, who uses a system of symbols that represent a language. The visual artist creates a new symbology based on his current exploits while venerating tradition, he uses new visual cues, ideas and science of his generation, thereby extending and modifying the visual and written language.

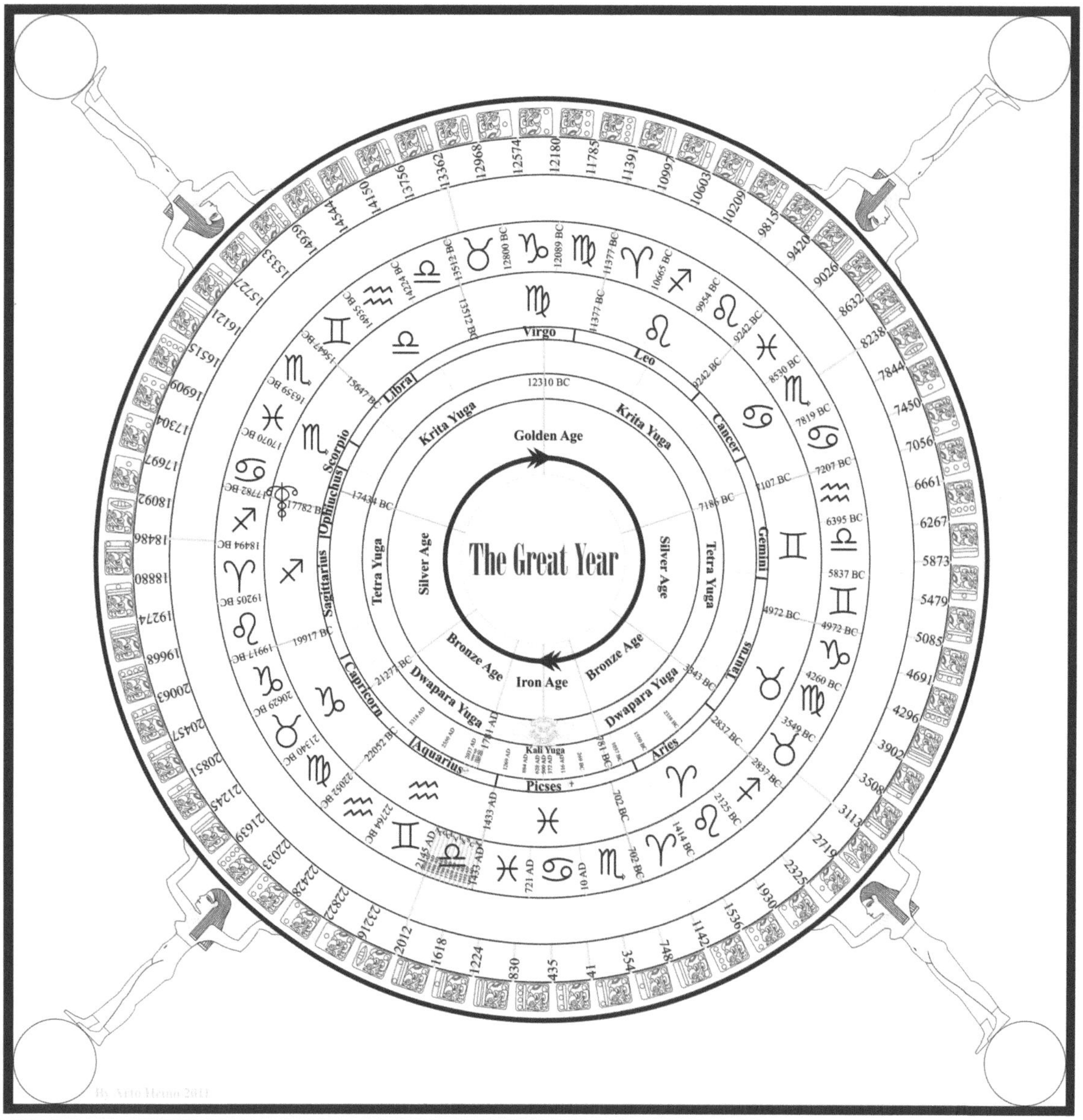

I have found an intellectual and emotional attraction to the beauty of ancient Egyptian art, I thought this might be a good place to start my investigations into the underlying pictorial structure of their art. As the writer Stephen Mehler* has pointed out, the original Egyptians are Khemitians, which are pre 3000BC. This distinction simplifies a lot of the time lines and cultural differences in my discourse.

The understanding of all the cycles would not be clear without a constant reference to the diagram of "The Great Year". I am presenting this astrological and historical picture as a definition of the time cycle covering the last 25920 years, it incorporates the ancient Greek, Hindu, Western and Mayan time periods.

By studying the many derived artworks, translated texts and their building designs, the assumption seems to be that the Khemitians were great astronomers. They understood and expressed their knowledge of the heavens by incorporating it into their artistic layouts and the locations of pyramids and temples. This clarity of purpose has been shown by the work of Wayne Herschel and his book "Hidden Records". You could also include the work of Robert Bauval and Adrian Gilbert in their study of the site of Giza, "The Orion Mystery".

The later Egyptians counted the ages of man with Leo (12000BC) or Sekhmet, as they called her, as the first age of the gods(Zep Tepi). This correlates to Sri Yukteswar's version of the Yuga cycles as the golden age of man. This makes perfect sense, as it is the peak of mans potential and the following ages are examples of the decline of man: like his obsession with death cults; with the distortion of higher knowledge and his past glories, until we reach the age of Pisces (727BC), which became the lowest and most brutal age of mankind.

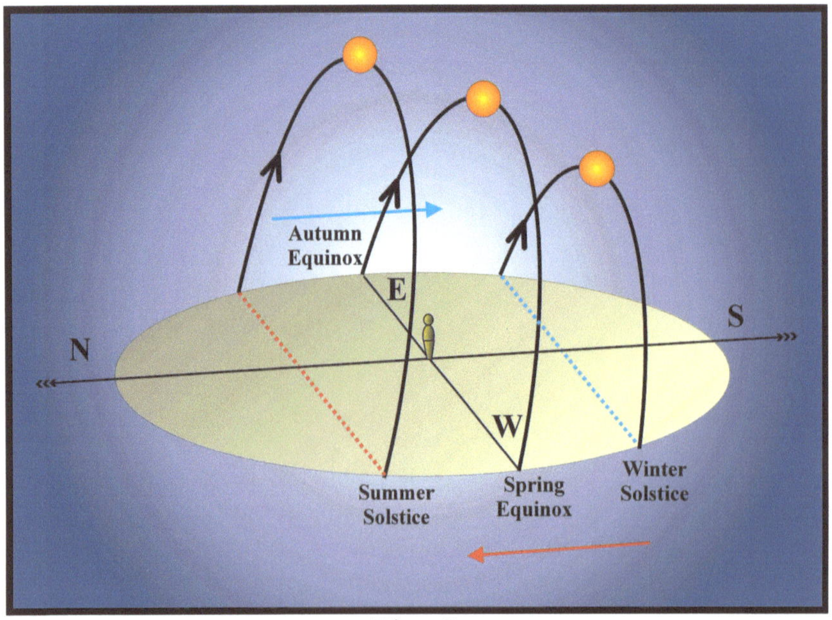

Plate 2

The standard assumption is that Pisces began about 1 AD. After much study, this date seemed to be out of step with the events of history, Terry MacKinnell had also seen these correlations in 1987, with some well founded understanding of why these dates were wrong. His research told him the way ancient man had to observe the constellations. It was not a mathematical exercise, but one of practical and applied knowledge, the one that usually escapes the theorists.

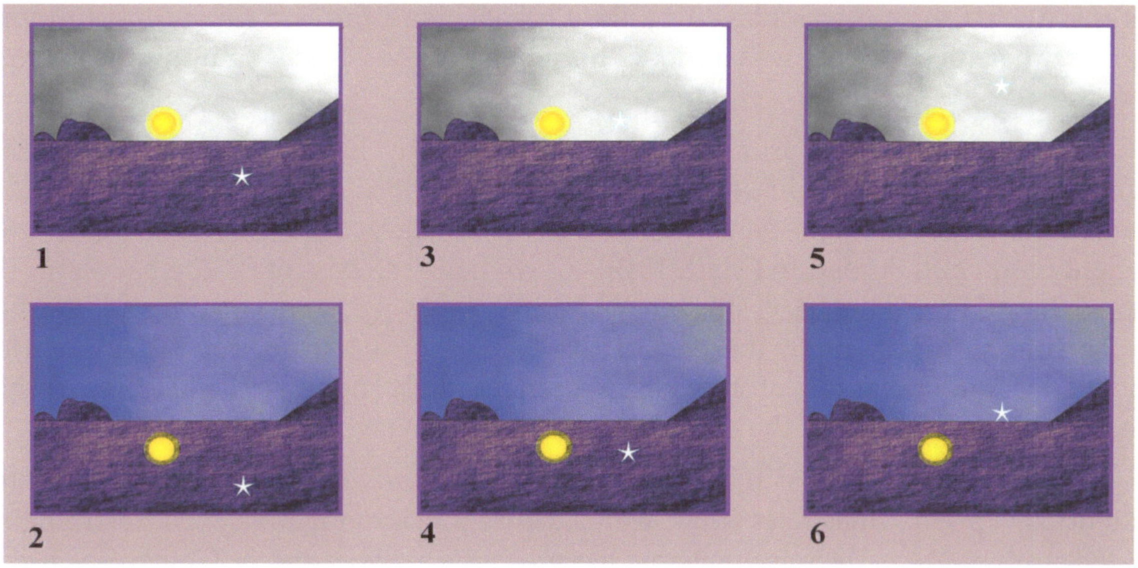

Plate 3

The helical rising of a star or constellation at the vernal equinox graphically displays the astrological age. Instead of written or mathematical calculation, it is based on the last star or constellation to appear above the eastern horizon an hour or so before the sun had risen. This places the constellation earlier than assumed by the Greek and later astrologers by approximately 15 degrees. The helical rising was the observable case before Hipparchus, when his mathematics changed the definition of the occurrence of an age. At the time(150 BC), this did not pose much of a problem, as the definers of the original ages were Egyptians and Babylonians not the Greeks. Hipparchus started the current marker for the way we read the astrological ages, he did not know how they defined the ages previously, therefore he had to codify a definition mathematically that would satisfy the Greek mathematical methods.

Refer to plate 3

Figure 1,3,5 = Sunrise

Figure 2,4,6 = just before Sunrise

Figure 1,2 = the Sun rises before the Star

Figure 3,4 = the Sun rises at the same time as the Star

Figure 6 = the Sun rises after the Star = Helical Rising

By the re-adjusted astrological marker we are in the influence of Pisces, with the Aquarian age in full swing since 1433AD. The borders between astrological ages are not fine lines, the last age influences the new age for at least a few hundred years, not including the influences of the decans of the same sign type (air, water, fire, earth) and are just as strong and last for 720 years per common sign. The division of 12 main signs and 36 decans is how the Egyptians had divided the zodiac, it was to facilitate helical rising in all parts of the year, not just Sirius. The term *"Heliacal Circle"* is used for the ecliptic, which is in the center of the zodiacal circle. The Ancient Egyptian game of Senet has 3 rows of 10 making 30, this is the number of degrees of an age(30), with 3 decans(10). This game was played by the Pharaohs and the last row was split into 5 and 5 with symbols. Senet was a model for an age of the zodiac, befitting the ego of a Pharaoh.

To quote the Bible as a source of understanding in the meanings of the Star signs and the decans, might seem odd, but the Bible has references hoary with age handed down from many generations without change, with possible origins from ancient Egypt and Sumer. In (Matt. 13:39-40) we have the passage ,"The harvest is the end of the AGE, and the harvesters are angels. As the weeds are pulled up and burned in the fire, so it will be at the end of the AGE. ". All the destroying of libraries, knowledge and pillaging of the past were definitely done pre 1400's, when the medieval religious fever of the crusades and inquisitions were in full flight.

Note the year 1433, is the start of the Renaissance,the birth of the USA, and the Gutenberg press, these events were new and innovative, by astrological understanding Pisces was influencing the maritime adventures of China, Spain and England were in full swing, discovering and conquering the world by sea.

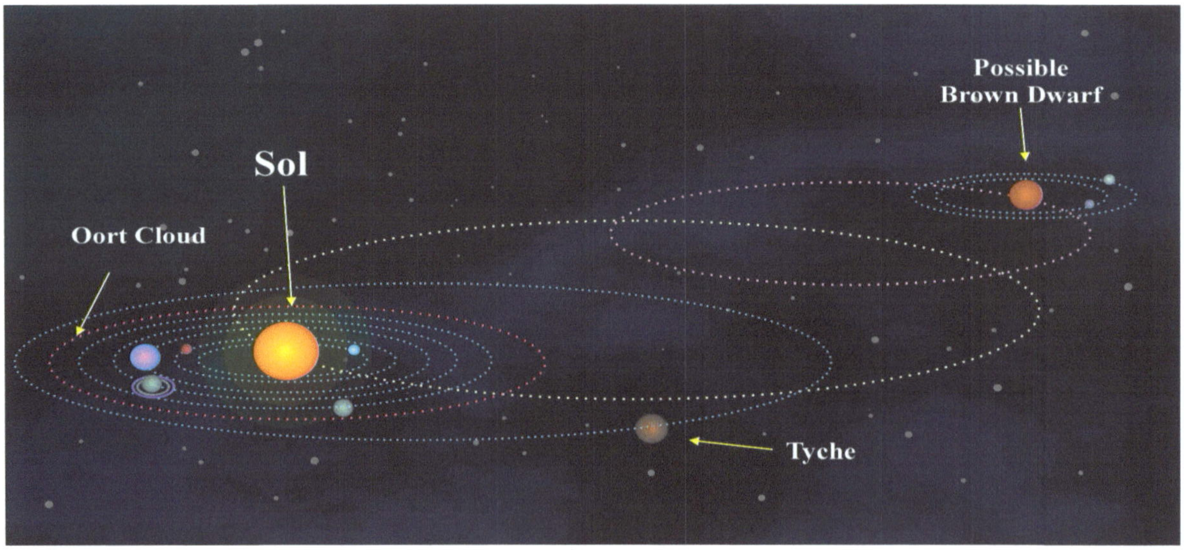

Plate 4

The ancient Greeks(Plato) broke the Great Year into four ages known as the Golden, Silver, Bronze, and Iron Ages both in ascending and descending orders. The same divisions are in the Hindu pantheon of knowledge and according to Sri Yukteswar, the complete Yuga cycle totaled 24000 years. This understanding of the great Yuga cycles has been elucidated by Walter Cruttenden in his documentary, "The Great Year", a must-see video. Ancient mans preoccupation with the Great Year cycle must be looked at with non-jaundiced eyes, as too many astronomers try to scoff at the idea that we have a second sun, as Walter has discovered the precession cycles are but the solar system curving through space, orbiting as a binary pair with a possible brown dwarf, Sirius or another type of star yet unknown. The general and accepted astronomical understanding is that binary star systems are the most probable occurrence in the heavens, while singular stars are rare and often devoid of planets.

While the Khemitians didn't directly say there were two suns, the iconography and pictorial evidence of the culture that followed them, the Egyptians, are overwhelming. Refer to the Dendera Zodiac, a copy of a more ancient document. The symbolism of the sun was not as a worshiped entity but more like having hidden star knowledge that was only revealed to the initiated and educated. In the Pyramid Text , in Utterance 141, it states, "There is no Star-God who has no companion, have you your companion?".

In the diagram(plate 5) the time span between the Sun numbered **1** to Sun numbered **2** would be 26000/4, which is about 6500 years. In current time the Sun location would now be at about at Sun numbered **1**, plus 1500 years, so all our known historical facts are from Sun numbered **4** to **1**. We seem to have forgotten our previous eras, it seems like our consciousness and memories had changed, with this occurring, writing had to be reinvented, possibly to tell our coming future generations about the fall of man by leaving a legacy of laws, buildings and technology so we can re-learn our past in another era.

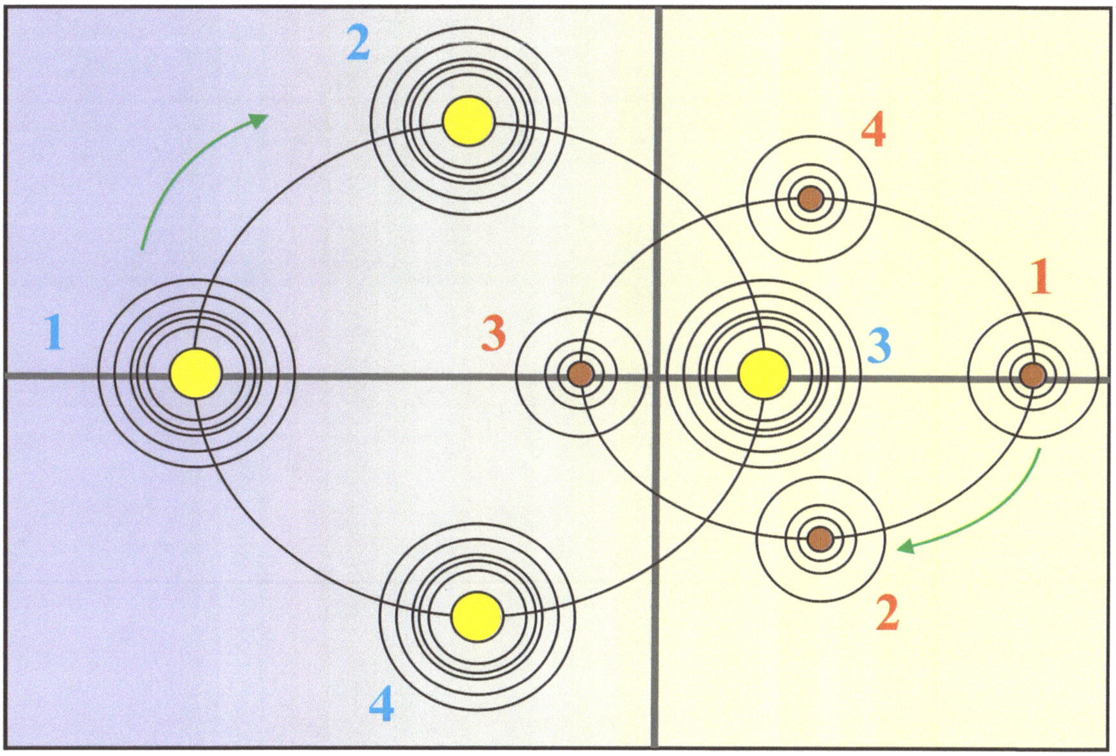

Plate 5

As religion was not the main concern of the Khemitians and early Egyptians but the symbolic representation of a long forgotten knowledge, just like some modern artists do with the abstraction of incorporating hidden symbols in their work. Before 3000BC, the idea of preserving historical facts and maintaining its notations was of the utmost concern, this is the why and the when of the beginnings of writing. As to being religious, this is quite debatable as religion is a group identity for a supernatural belief system, which is more akin to taking sides in a political stance not a spiritual state of being. As most modern scholars make the assumption that everything the very ancient man did was of some religious and ritual meaning, making their great accomplishments look like backwater superstitions. This has distorted the true identity and pursuits of ancient man to such a degree that those who even flout with these ideas are ostracized from the archaeological hierarchy. The dynastic Egyptians turned the vocal traditions of science and knowledge to create a half remembered history to hieroglyphs and symbols creating gods and legends to be venerated , thus turning it into a religion by writing those interpretations on the walls of a preexisting civilization, Khemit, the precursor of what we now know as Egypt. This preservation of the past is a legacy of all the ancient civilizations, for this we will be forever grateful for their efforts.

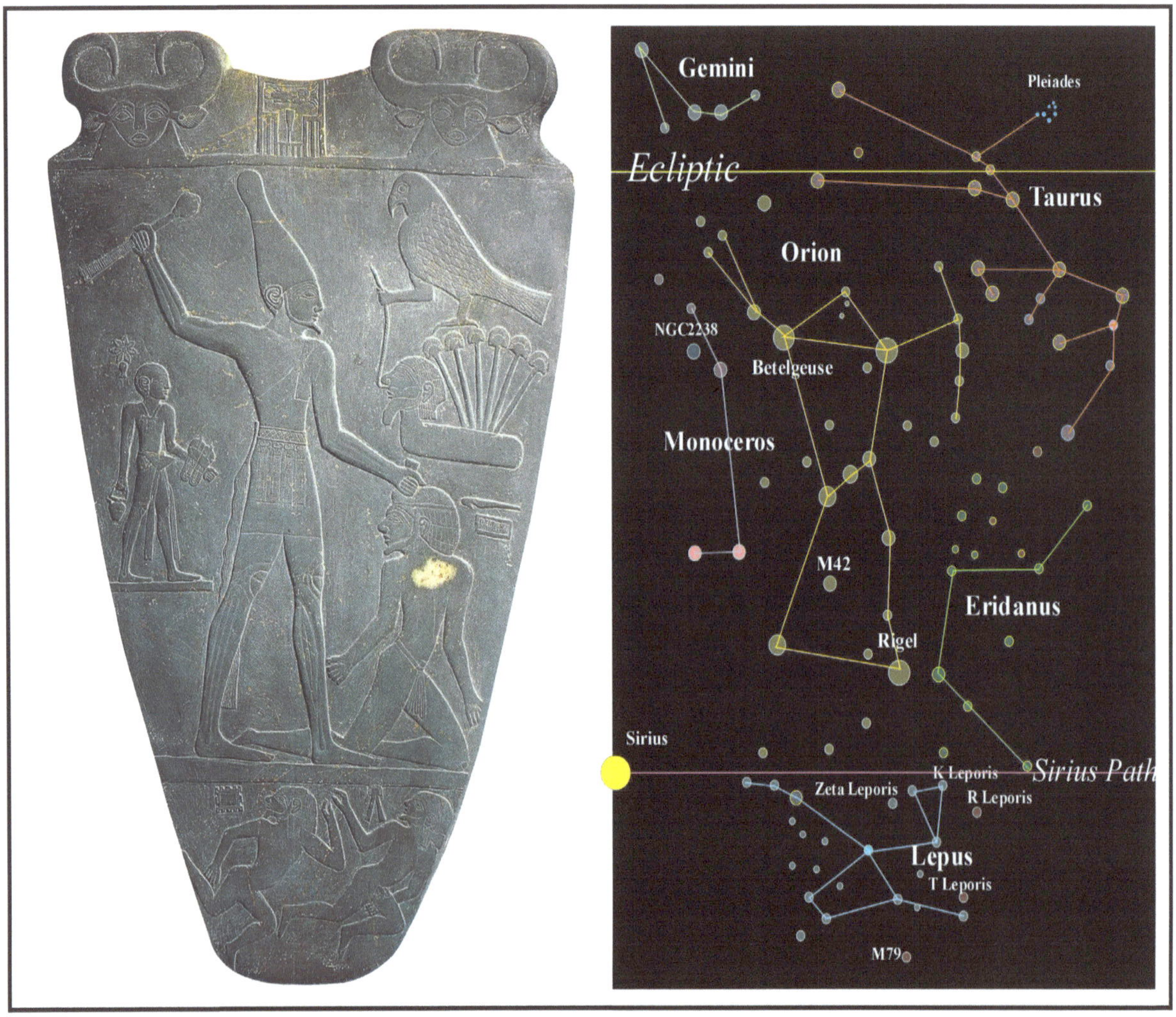

Plate 6

The earliest of the known Egyptian kings was Narmer. He left a lasting legacy that pointed the way for future generations to copy and emulate. Narmer only left a small amount of iconography, still it speaks volumes to those who use their eyes and knowledge to decipher its meaning. The famous and most intensely studied "Narmer Pallet" has pictorial evidence pointing to some of the new revelations of our ancient past. I chose this icon as one of the best examples of the many other star charts I have found in ancient art. This artefact was excavated in 1897 by J.E. Quibell and has more written about it than just about any other ancient Egyptian object, so look it up .

The description most archaeologists give is a make-up palette that commemorates the battle by King Narmer in 3100BC showing him winning against an opponent. The imagery has been described as being bull worship. As you can see, this is an assumption based on prejudice and basic ignorance of the artists inner motivations and his visual culture.

As you can see by my overlay one side of the pallet shows the star chart at the time of Narmer, approx 3100BC. The Egyptologist's interpretation are of two types:

1. Those who believe that the palette is a record of actual events such as the military success over Libya and of Buto.

2. The others who argue that it is an object designed to establish the mythology of united rule over Upper and Lower Egypt by the king. This was shown by Narmer the conqueror delivering a mortal blow to his enemies in the name of the Gods.

These are the ways it has been looked at and are accepted as fact. To my eyes, I see a star chart well defined and understood. The story of Narmer is no more that an figurative overlay onto long existing astrological/astronomical charts, like a few specialized artists do now. The Khemitians understood the precession of the equinoxes and this artifact can also be viewed as a record of the life in the sign of Taurus as the dominant constellation.

The astrological age rendered on this pallet is the third Decan of Taurus, which is Taurus, the reference of the two bulls as shown above his head. Appearing either side, are the early interpretations of the God "Bat", "The Ba of two Faces", who was also represented as having two cow heads. Where "Ba" is the soul, and the god "Bat" who has the ability to see into the past and the future, this idea that the soul can see into both past and future is a basic premise of the teachings of Edgar Cayce, the "sleeping prophet".

In the vertical center line at the top are a few symbols that has been said to be a chisel and a catfish with two chairs opposing each other which supposedly represents two sides of the Nile. As per standard archeology, this is his name and his title. I have my doubts about this standard view, Wayne Herschel states it is possibly a bulls leg pointer showing the location of a star. I deem it is a 'mallet', which would render the story more fulfilling, as now he is shown to be a sculptor of a future empire. Even by conventional Egyptology, the current interpretation of this serekh and his name

"Narmer" is tentative at most. As Wayne Herschel has also pointed out in his book, this could be a bulls leg, which acts like a pointer.

The electric catfish probably represents the view of the Crab Nebulae, which would correlate to its intense activity. It was visually first known to explode in 1054AD, using a small calculation, the real moment it exploded was 1054AD - 6520LY = 5466BC. This means the Khemitians knew what the crab nebulae looked like before it exploded! This alone should make you realize the true implications of our ancient forefathers forgotten star knowledge! If these interpretations are right that would be an astounding revelation!

The Khemitians have shown that in astronomy they have an outstanding record of accuracy with observation and interpretation. Look at the nebulae, NGC2238, called the "Rosetta ", the drawing shows a flower in a vase. The current deep space picture shows such an amazing similarity that you just can not accept as mere coincidence in its design. This might seem like another lucky coincidence, have a look at Orions Ear, 'Lambda Orionis', the modern picture was created from the IRIS version of the IRAS infra-red data. The drawing that is 5000 years old shows a remarkable similarity to the modern version.

Look at Orion's kneecap, the stunning M42 nebula also has a very similar gas clustering as the drawing. After studying many stars and nebulae, I have concluded the original source picture was a lost technical map of the stars and their constructs. The mythological overlay intruded upon its true merits and has distorted the real meaning of this work, or it could have strengthened its hidden story?

The unusual patterned square located as Zeta Leporis, gives you an idea that something is surrounding the center square, it is just like the recent discovery of an asteroid belt around this sun.

There are other nagging questions that also need to be answered, like the identity of the two glyphs on the right(12). The top part is an organ removing knife as used by the Egyptian embalmers, or a harpoon as I have observed and the lower figure represents a pool or lake. It could be the fire lakes of the Egyptian hell. This has been interpreted as "wash" by some archaeologists, possibly a baptismal pool. Could Narmer be crowning or baptising the crouching figure? (He does wear a beard designating royalty). Narmers face looks very calm and the mace is held in the middle, not at the end.

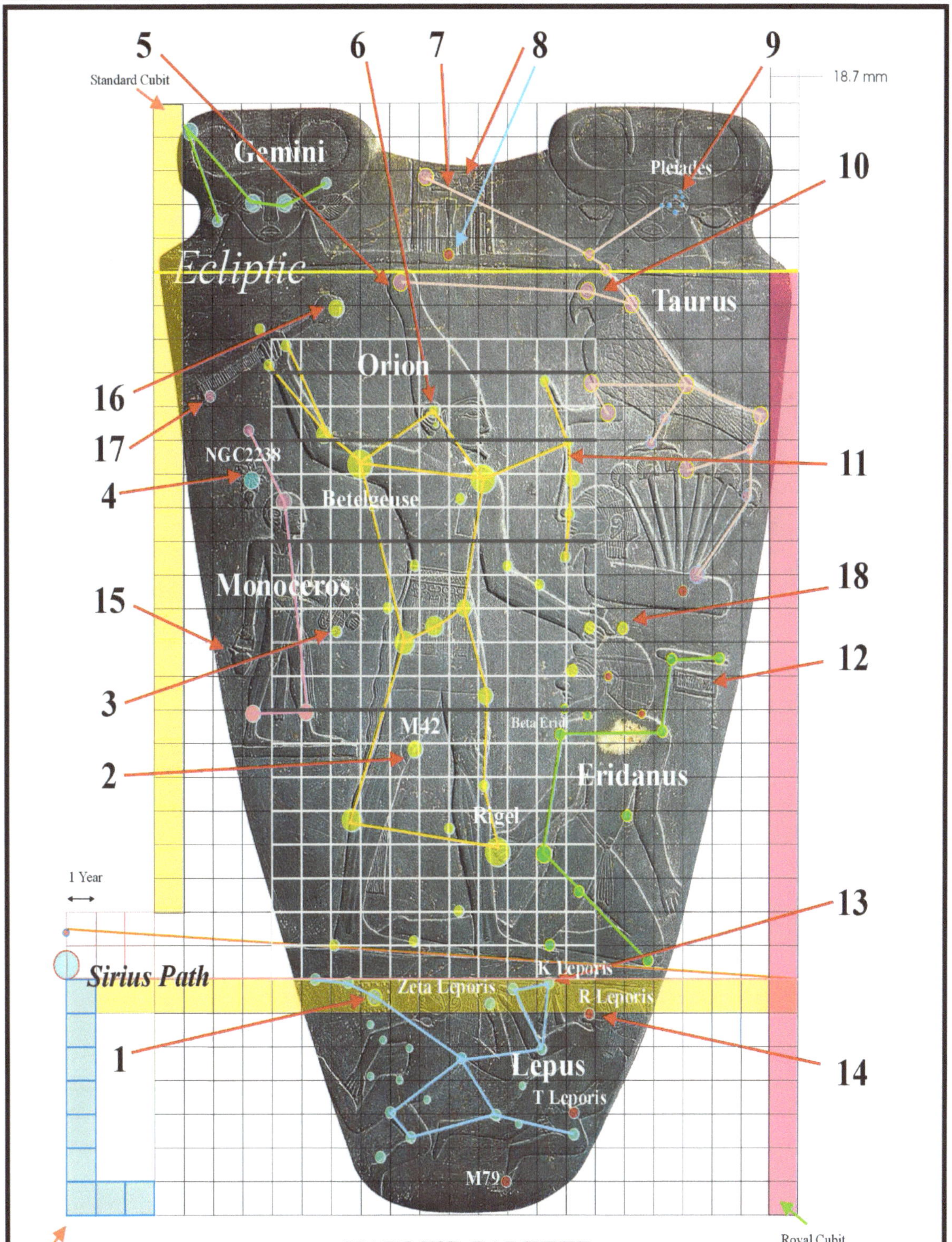

NARMER PALETTE

Plate 7

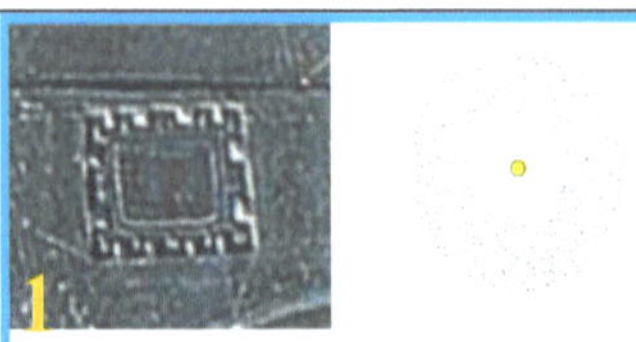

1

At 70 light years Zeta Leporis has shown to have an asteroid belt at about the same distance as our own.

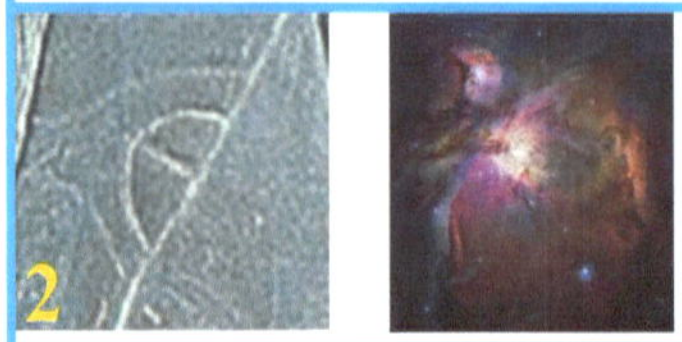

2

The M42 is one of the brightest nebulae, and is visible to the naked eye in the night sky. M42 is located at a distance of 1,344 light years and is the closest region of massive star formation to Earth. it is estimated to be 24 light years across

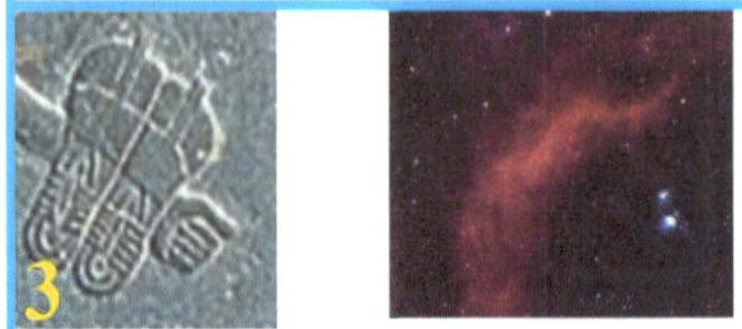

3

Barnard's Loop in Orion is a section of a shell of gas that was blasted out from the Great Orion Nebula.

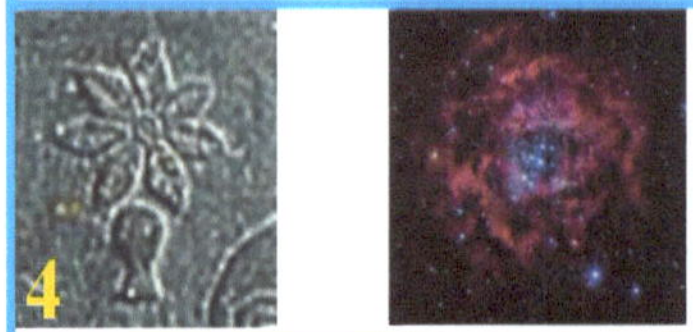

4

The Rosette Nebula is a large, circular H II region located near one end of a giant molecular cloud.

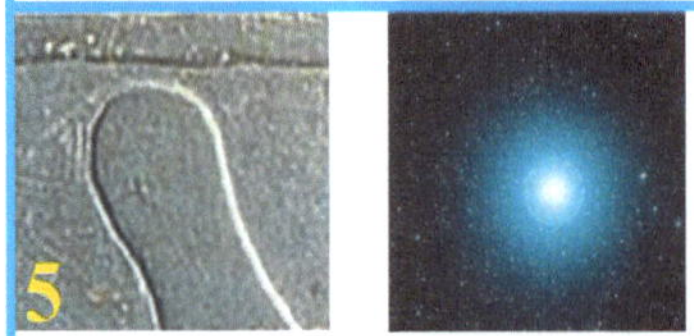

5

Zeta Tauri is a binary star and is a brilliant hot blue-white giant apprx 417LY from Earth. Its brightness varies from mag+2.88 to +3.17 due to variability and an eclipsing binary. The binary system are separated by about one AU and complete an orbit once every 133 days. Its companion star is a yellowG-type star, which has a magnitude of +5.2. A 9 solar ms

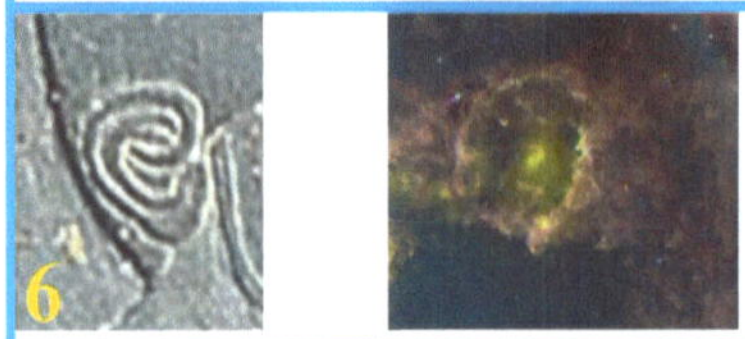

6

Lambda Orionis, the ring can perhaps best be explained by a supernova very close to the O8 III multiple giant
It is about 1 million years old
Sh 2-264 is ionised by L Orionis, and to a lesser extent, the 11 B-class stars near to it, and is part of the Gould Belt.

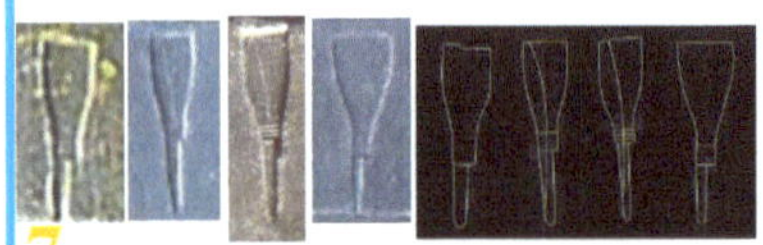

7

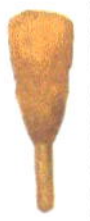

Picture from Turin Museum

This has been denoted as a chisel and could possibly be a mallet with the handle pointing to a location on the star map. When using this type of mallet when carving it is hard to miss with each strike. Hitting anywhere on the mallet is a bull's eye. Using a flat faced hammer you cannot always connect squarely, it will glanced off and cause mistakes or weak strikes.

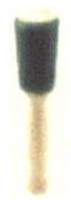

Modern carving mallet

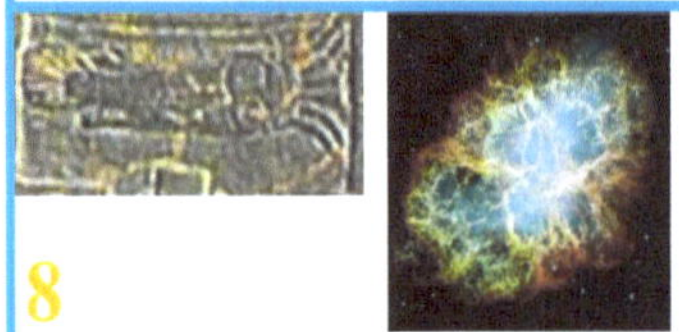

8

The X-ray and gamma-ray energies the Crab is generally the strongest source in the sky. At a distance of 6,500LY from Earth, it has a diameter of 11LY and expands at a rate of about 1,500 kps. At the center of lies the Crab Pulsar, a neutron star , 28-30km across, which radiation from gamma rays to radio waves with a spin rate of 30.2 times per second.

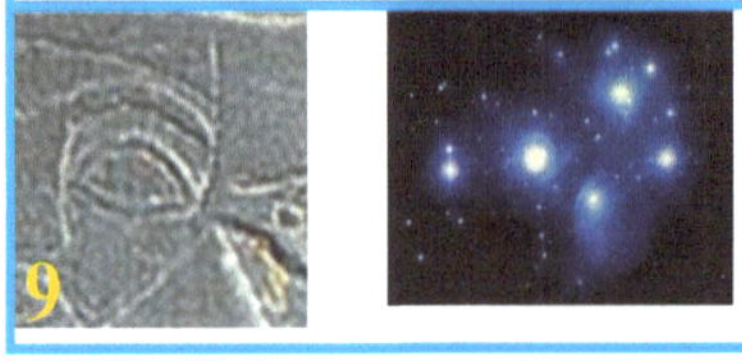

9

The Pleiades star system, sometimes called the seven sisters, it is a group of stars that are not cluster.

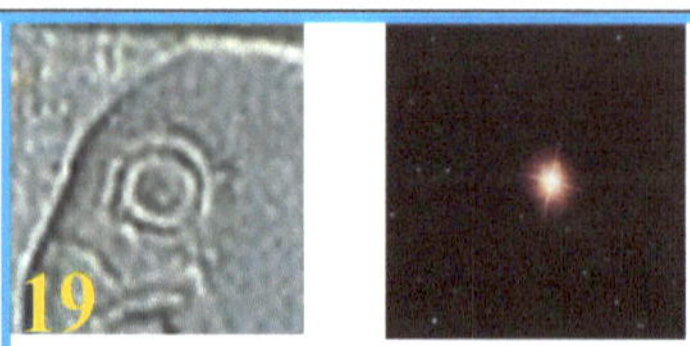

19

Aldebaran (Alpha Tauri) is an orange giant star about 65.1LYs away. With an average app mag of 0.87 it is the brightest star in the constellation and is one of the brightest stars in the sky. You can see why they used this star as Horus's eye.

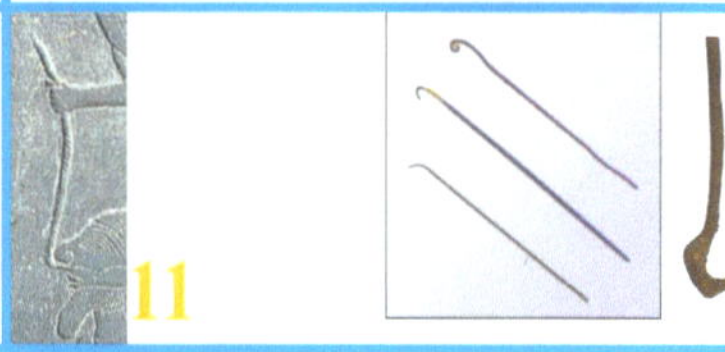

11

It looks like the bird(Horus) is holding a brain extraction hook or an adaze. This could be an early form of the opening of the mouth ceremony, this could relate to *Sahu*

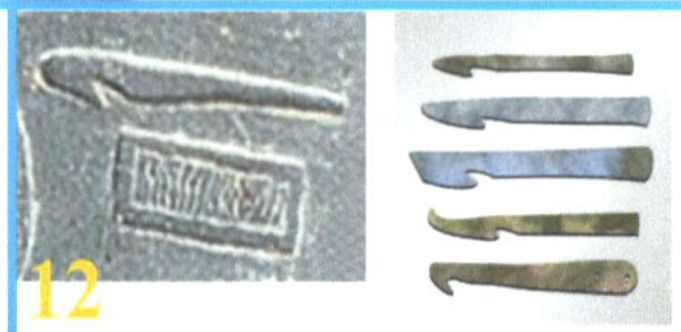

12

The drawing on top has been annotated as an organ removal knife/hook or as I have found it is most likely a harpoon
The drawing below it would be a pond/lake

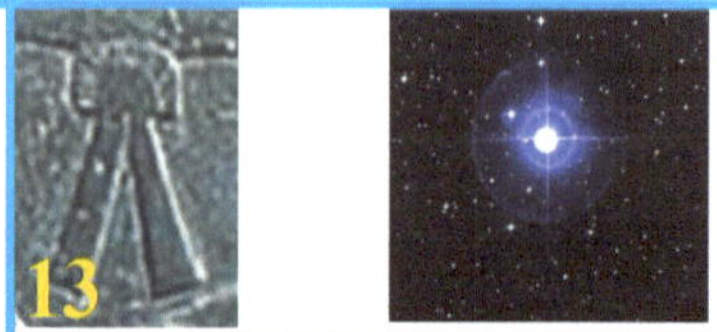

13

Kappa Leporis consists of two blue-white stars, magnitude 4.5 and 7.4. The two stars are separated by 2.6 arc

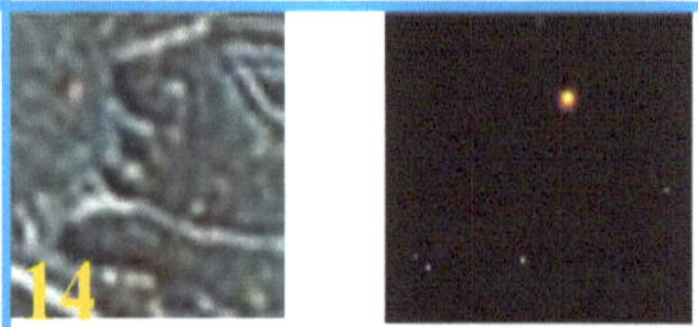

14

R Leporis also known as Hinds Crimson Star, is the reddest bright star in the sky. Its wide variability comes because the star produces enormous amounts of carbon which, through convection, makes its way to the outer layers of its atmosphere.

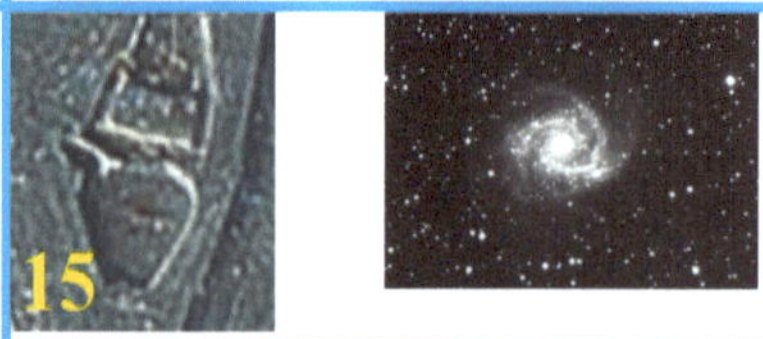

15

NGC2301

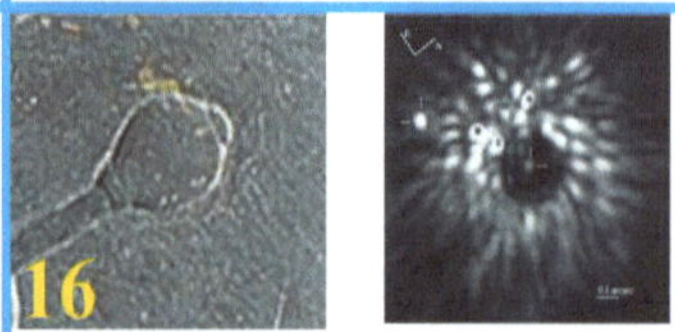

16

Chi-1 Orionis, 28.6 LY away, a solar-type, is hidden at the center (cross-marks) behind "occulting disk" , a dim companion, which is indicated by the cross-marks to the left.
The spots surrounding the bright component are optical artifacts
Chi-1 Ori B is roughly Jupiter's distance from the Sun

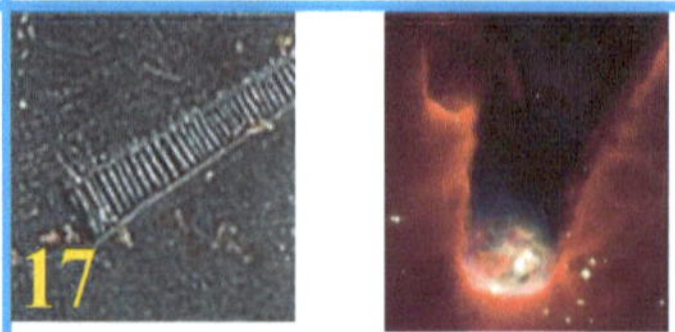

17

NGC2264 , Cone Nebula.

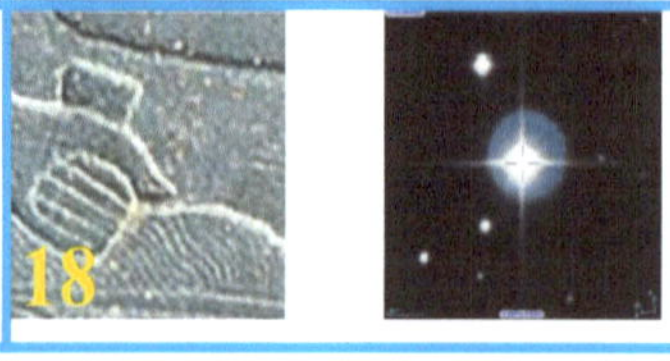

18

Hip22319

Table 1

The image on the other side should be included in any interpretations.

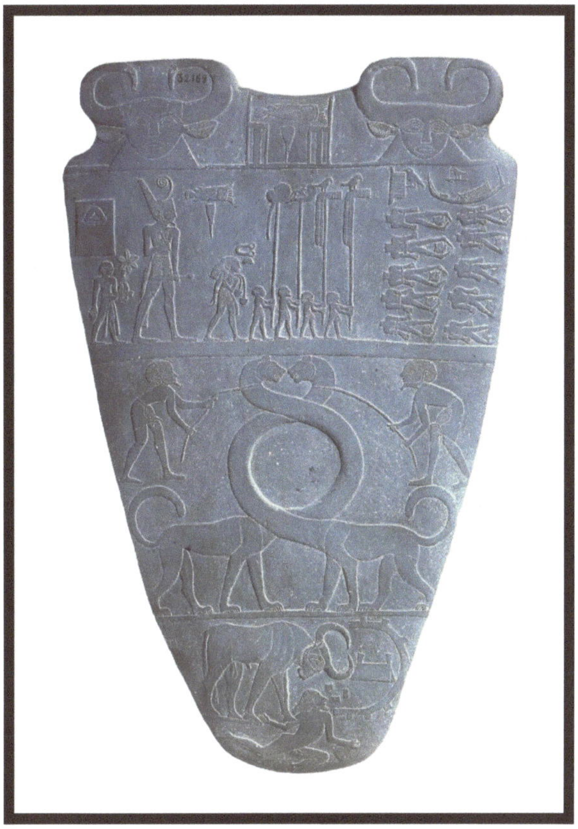

Plate 8

Both these images could also have been re-interpreted as a section from the Bible, from John 1:

Now this was John's testimony when the Jews of Jerusalem sent priests and Levites to ask him who he was. He did not fail to confess, but confessed freely,

"I am not the Christ (the Messiah)." - *Refer this to the smaller figure behind Narmer (plate 1).*

Finally they said, "Who are you? Give us an answer to take back to those who sent us. What do you say about yourself?"

John replied in the words of Isaiah the prophet, - *He is repeating an older source knowledge.*

"I am the voice of one calling in the desert, 'Make straight the way for the Lord.'"... "I baptize with water," - the *Eridanus constellation was considered a river, so the man kneeling was not being punished but could be being baptized. "Make straight the way" could be referring to the direction that Orion travels in the sky, towards Eridanus.*

John replied, "but among you stands one you do not know. He is the one who comes after me, **the thongs of whose sandals** I am not worthy to untie." - *Refer again to the smaller figure holding the sandals.*

This all happened at Bethany(meaning "house of suffering") on the other side of the Jordan, where John was baptizing. - *The other side could refer to the milky way in the sky, which is where these events are occurring, or/and as the Nile and the Necropolis being the Valley of the Kings.*

The next day John saw Jesus coming toward him and said,

"Look, the Lamb of God, who takes away the sin of the world! This is the one I meant when I said, 'A man who comes after me has surpassed me because he was before me.' " - *Refer to Aries who is a sacrificial lamb, the previous age of man, before Pisces, he was the lamb of God and the fisher of men, Aries then Pisces.*

Referring to another Bible reference, from the prophet Isaiah -Chapter 5 Verse 25.

"Therefore, the Lord has become wroth with His people, and **He has stretched forth His hand upon them and has struck them**, and the mountains have quaked, and their corpses were like spittle in the midst of the streets; with all this, His anger has not turned back, and His hand is still outstretched. "- *Narmer has one hand striking and the other outstretched. The other side of the pallet has ten headless corpses, ten degrees per decans in an age.*

Verse 26. And He shall raise an ensign to the nations from afar, and He shall whistle to him from the end of the earth, and behold, quickly, swiftly he shall come. - *the back part of the plate has four marching figures holding banners(ensign), the cow horn was used a wind instrument to herald the King.*

Verse 27. None is weary or stumbles among them, he shall neither slumber nor sleep; the **belt** around his waist shall not open up, nor shall a **thong** of his shoes be torn. - *as you can see he stands straight and only wears his toga with a belt of bulls designs located in the Orion belt and a cord made from the bulls tail, he still does not wear his thongs.*

All subsequent Egyptian dynasties possibly copied the lead of Narmer, by rendering hieroglyphs onto all the ancient buildings as he did in Hierakonopolis, as others also would in their newly discovered structures under the shifting sands. They placed their own versions of what the priests handed down to them about the ancient Gods and their own birth lineage claims on the buildings. The probable outcome here, is that most of the greatest works of the dynastic Egyptians were *not* built by them, they were constructed by an older civilization, the Khemitians, dating back to 12000BC and earlier. The *re*-discovery of their ancient past is what the religious aspects of Dynastic Egypt were all about. Their whole culture was based in age of Aries ,or in the light of Ra (Ram= Aries). The historians try to be convincing about who built what, it just makes each Pharaoh appear to be fraudulent as great builders. They ***Found*** the structures under the sand, they venerated them with all the ancient knowledge they could muster and restored them as well as they could also decorated them with hieroglyphs. Some were never found in dynastic times, like the Osirion, at Abydos, these had no hieroglyphs on them at all. The Great Pyramid and Machu Pichhu also falls into this category.

The building techniques of the Cyclopean type were a mystery to the Egyptians just as it is to us. They made lots of guesses when they tried to restore them and articulated stories on them with hieroglyphic fables, describing the builders as great and venerated them as the first great people of the golden age. These tales were subsequently borrowed from older texts exposed as hieroglyphs to each dynasty and then later translated via the Greeks and Romans to finally be the writing of the Bible, the Torah and other books hundreds of years later.

The modern and ancient artist both uses grids when translating pictorial matter from a scene to another medium, this way he has some accuracy when he articulates the images. The image is usually laid out accordingly to a mathematical plan, thereby facilitating a common point of reference, for both the hidden and symbolic parts of the artists designs and the final placing of the artwork thereby giving the artwork its true meaning.

The grid used for the whole image is 22x33 and measures 411.517mm x 617.2mm, the measurement from the bottom of the pallet to the border just above Narmers hat is exactly one royal cubit = 523.75mm and the measurement from the top of the pallet to the bottom of the inner grid is one standard cubit 448.9mm. I have enclosed Narmer in another grid 11x19, this was the basis for all Egypt's future iconography right there in that simple grid system.

Counting only the lines, they become 12x20 which is the base reduction from 3 x 5 which is part of the Fibonacci series equaling 1.666 where Phi is 1.618. This grid system was most likely used by all the Egyptian dynasties right up to its demise, even Leonardo Da Vinci used a similar grid on the Vertuvian man. The proportions of Dynastic Egyptian figures generally fit in the correct 19x11 grid, it seems Narmers legs were longer by one grid size, the crown would have added to this, also he could have been a real giant, where the standard proportions are different. Like everything on this pallet, nothing is straightforward.

The discovery and hard work by Erik Iversen of the Egyptian canon of proportions has made this understanding a lot easier. We can qualify that this proportional canon was handed to the Greeks, then to the Romans, then through to rediscovery in 15th Century Italy by Renaissance artists such as Leonardo Da Vinci. The fractional golden mean proportions were simplified to the whole number Fibonacci series. This is the way the Khemitians designed almost everything, this is also the core of Quantum Arithmetic.

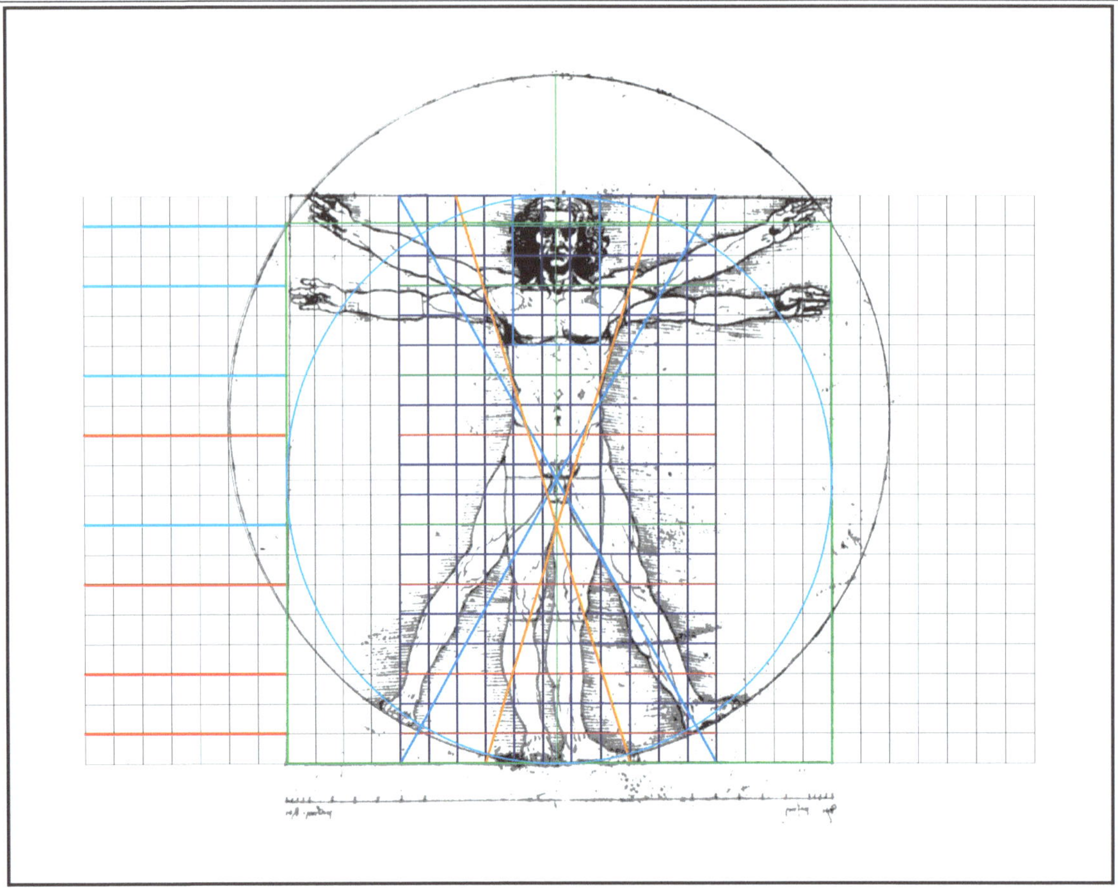

Plate 9

The other side of the pallet has some intriguing figures of two long necked felines that are entwined around each others neck. The ancient name of these mythological beasts were known as Serpopards. These creatures are describing something important, it looks almost technological as this is something not seen in nature. If the Narmer pallet is really describing astronomical data, then these creatures could represent "Aker", the pair of felines that guard sunrise(Sef = yesterday) and sunset(Duau = today) that have a sun between them, or another unnamed pair of guardians that represent the orbiting of two different suns. The full dissertation of this side of the palette will be made in a future article, due to its unique subject matter.

It seems the Narmer pallet has a number of layers of understanding, besides the text book version. I can see it referring to an age of man that had a different set of values and priorities, not as misguided as the 22nd and later Dynasties, with their self serving elite but a practical and evolved society. They had tried to preserve the knowledge of an ancient star map and created allegorical devices so as to preserve the original images and devised a writing form that was used for another 3000 years. A wonderful society that didn't need money. The Egyptians created a empire from what they learned from their Khmetian past, this was also true of other cultures such as the Incas and their past.

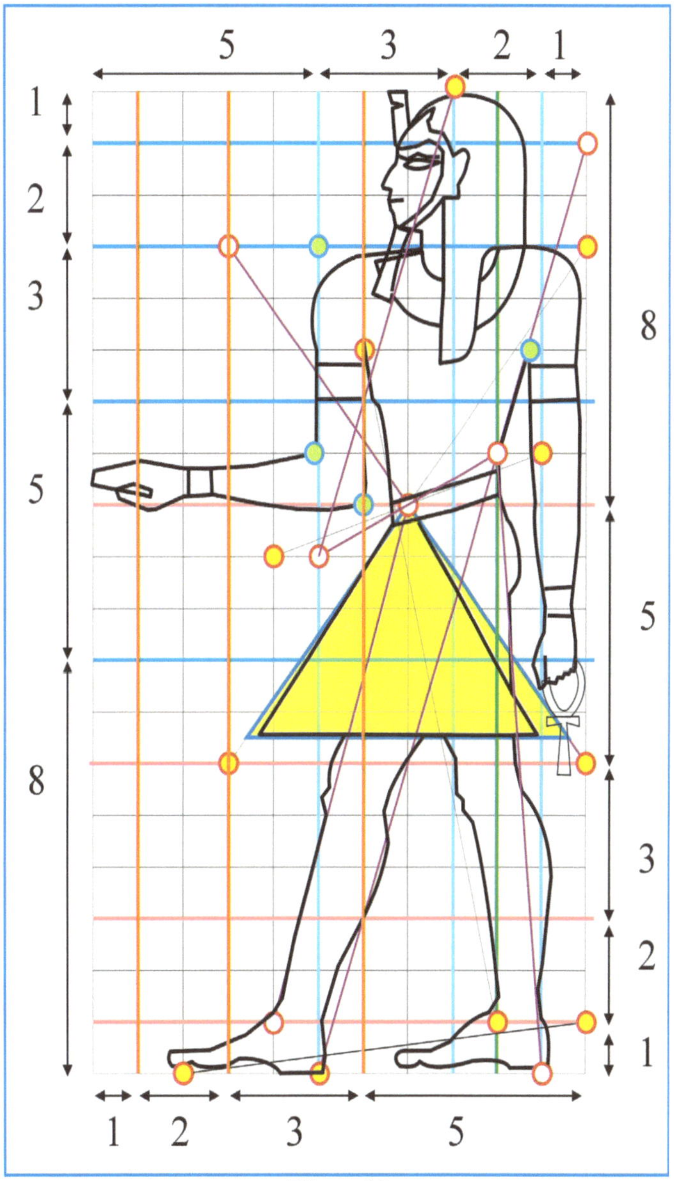

Plate 10

The lines of the ecliptic on the pallet implied a knowledge of precession of the equinoxes. The lines at the lower part are where the decans would be located on the Dendera star map, which demonstrates that this is more than just a zodiac system. It relates to the divisions in the Great Ages of Man, marking different periods where significant evolutionary changes occur in our consciousness and inner spirituality.

The hidden meanings in the myths of man is where the wise Ancients hid as much valuable information as possible, to preserve it. Creating picture writing assured the symbols would never be lost. They knew society was going into a long term decline and much of the important stuff would be destroyed or misinterpreted. The myths were a way of keeping this true knowledge hidden and would be revealed when man was ready to take the next step and realize at the end of the age of Pisces that religion would try to destroy and corrupt the ancient past by conforming man to a supernatural belief system based on fear, rather than a state of enlightened spiritual understanding. This would be the trigger for the true awakening of mankind from the dim lit halls of religion. They wrote these myths into the star maps by creating stories from the mythological figures that were represented in the star charts and only in our current age are we starting to understand these messages.

We have learned from many sources that the myth of Jesus is the same myth as that of Horus and of Mithras. They were created because we needed to carry the ancient stories to the next age. With the Jesus story, you had the Son(sun-visual) of Man and the Son(sun-hidden) of God. With Mithras, you had Helios the Sun(sun-visual) God and Mithras the unconquered Sun(sun-hidden), Horus(sun-visual) and his father Osiris(sun-hidden). In all their different guises they have a common story. Firstly, they were all born on 25th of December(winter solstice), secondly, the story which happens in the sky with the belt of Orion(3 wise men) pointing to the star Sirius(star of East foretelling the event) and relating to the timing of the new Sunrise(birth of a new Son). The only problem is that the star Sirius is not depicted on the Narmer pallet but the path of Sirius was shown, so these revelations are of a different nature. The belt could point in the opposite direction that is, towards the Pleiades. This pallets has a hidden time cycle of 25 years, it is half the orbit of the binary pair that make up Sirius.

The helical rising of the Pleiades in vernal equinox (spring) in the northern hemisphere has been looked as the opening of the fishing (the catfish) and farming (the cow) season, while its dawn autumn equinox, marked the season's end. In the Bible in Job 38:31 the passage “Can you bind the cluster of the Pleiades, Or loose the belt of Orion?” could be a clue.

The belt of Narmer hangs loose on his hip, this could indicate this belt acts as a guide to other events. After all, the emblem of Pleiades is the bull. This is just where it sits on the Narmer pallet. As I have mentioned, Bat as the god that has two cow heads, as all goddesses in the Egyptian Pantheon can be depicted as cows. This could mean that the left head is Isis (Sirius) and Neith the right head (Pleiades), this would make clearer astronomical sense.

Meskhenet is the Egyptian god of birth or rebirth. It is the head attached to the brick (birth-brick) on the right hand side of the pallet. This god also testifies to the newly dead, because it gave them destiny when they were born. The brain hook or adaze held by the bird indicates this is about judgment after death or to convey power over the senses, How this relates to the star chart is only speculative at best, possibly a heavenly body like a brown dwarf (second sun) and it's journey of destiny through the sky. The papyrus stems above the birth brick indicate 1000 x 6 stems = 6000. This number could mean many things. Some archaeologists suggest that it was the number of captives of upper Egypt. I might guess at the 6000 year ending and/or destiny of that age, which would be about 9100BC, the ending of age of Leo, the time of Zep Tepi (The Egyptian age of the Gods). The birth-brick could be the resurrection part of the Osiris myth. All of these themes could indicate that something was occurring in the sky in the same region, in the constellation of Taurus.

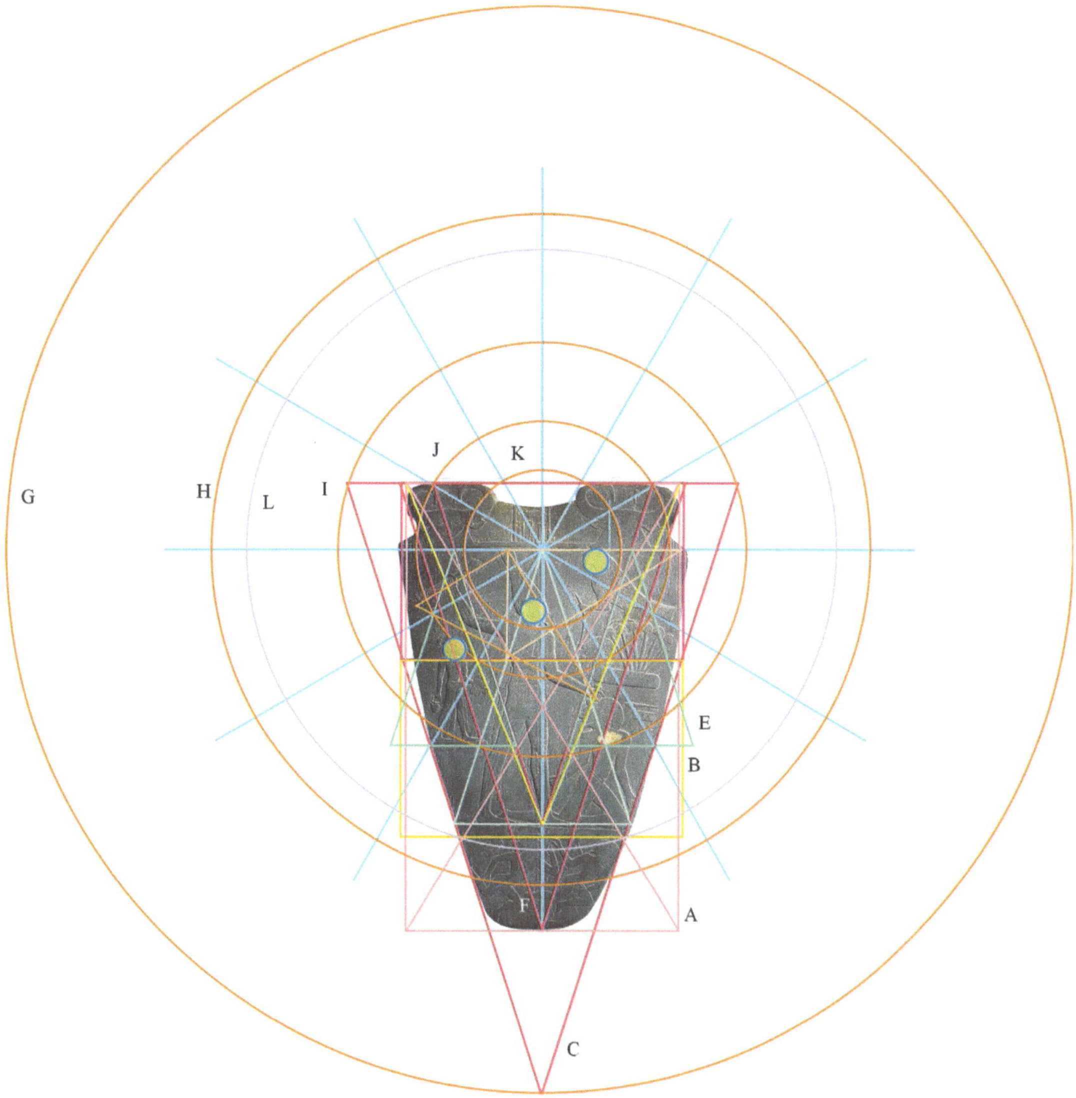

Plate 11

Another diagram will illustrate the true sense of these artists mastery, refer to plate 11. Here are some of the breakdowns:

A	Golden Rectangle	1 x 1.618
B	Golden Rectangle	1 x 0.618
C	Golden Triangle	36 + 72 + 72 degrees

F Golden Triangle 36 + 72 + 72 degrees

G Phi related G:H:I:J:K = G:H = 1:0.618 , H:I = 1:0.618 etc...

A lot of the arm, leg and body positions and directions are related to the golden ratio geometry, as you can see by the added geometric lines. The three circles outlined inside the pallet correspond to the three stars of Orion *exactly*! This culture fused and combined art, science, psychology, theology, history, mathematics and geometry into every piece of iconography as a learning experience. It was a shame that we lost the continuity of its underlying purpose for 1500 years.

The last understanding and writing of Hieroglyphs was about 391AD. It ceased after the closure of all non-Christian temples by the Roman Emperor Theodosius I,.showing the destructive nature of the short lived ignorant men going into the dark ages(500AD). It slowed all aspects of abstract thought, so with the mental blinkers of a brutal mind, the ancient stories were still cleverly disguised in their religious belief system. They read the texts and took them literally. This stopped excess editing and rewriting, so the hidden truths of the stories and their symbols would still emerge. The monks would write verbatim for centuries, unaware of the true content and the truly ancient history of the religious works. Looking at that period of picture making, you can witness the artistry degrading and the ornamentation of elements becoming more pronounced with true symbolism reduced to pattern. This is an historic indicator of a more primitive and degraded society.

In that period, the society would ignore what they couldn't find or see, so a lot of the buried tombs and buildings were safe. They were repositories of the past, which were left to a later era of man that had evolved mentally enough to start to look for them. The Piscean Monastic Idealism of one god Religions, began to poison mans future Aquarian aspirations in this era. The seed had been planted. We have seen through hindsight how this has become an infectious and rampant weed that is hard to remove. Without some of those insights, the Egyptians, Mayans and other old world cultures had established long standing bridges of understanding between the balance of order and chaos. The three great Religions had their birth in the tree Decans of Pisces - Buddhism (702 BC), Christianity (10 AD), Islam (721 AD), the trinity of the Father(Buddha, Statue, Solid, Decan of Scorpio), the Son(Jesus, Image, Liquid, Decan of Cancer) and the Holy Ghost(Allah, Symbol, Gas, Decan of Pisces).

Truly the apocalypse has already happened. Look at the years 200AD to 800AD, the darkest point of the Yuga cycle(500AD) and the third Piscean Decan(727AD). These were the most disease-ridden and bloodthirsty times of man. Nothing but destruction. The passages in the most spoken Psalm in the Bible mentions this transitional period.

Psalm 23:4, "Yea, though I walk through the valley of the shadow of death, I will fear no evil: for thou art with me; thy rod and thy staff they comfort me. ", *I go through the Constellation of Pisces, I am brave, the Sun shines with me, your Flail(Wheat thresher) and Staff (Sheppard staff) they are tools that will feed and protect me*. The actual shape of Pisces constellation is a large "V", denoting a valley. This refers to the Valley of the Kings in Egypt, the reference of death is the necropolis.

The next line reads, "Thou preparest a table before me in the presence of mine enemies: thou anointest my head with oil; my cup runneth over.". This surely mentions the next constellation Aquarius, the water bearer, the cup spilled, in preparation of the table means it will be happening, future tense, after he has been through Pisces. The anointing of oil could be a hint at mummification.

The modern version of the pending apocalypse is just a complete misunderstanding of ancient texts and bible reading. As I have said previously, they were never meant to be taken completely literal. The unimaginative mind will see only what has been stated, not what has been implied or abstracted. Revelations read like someone trying to interpret and read an ancient Egyptian mural, with passages that remind me of the different Gods, their actions and their hieroglyphs.

We have a least half the world who have a belief in the medieval inspired apocalypse. They have created a mind set that wants this to occur, slowing true human progress in all aspects of life and allowing our life to be run by a cabal of religious, political and merchant zealots. The Aquarian ideas of technology and mastery of some of natures forces, have been co-opted by generations of misguided people who have forgotten their ancient beginnings. This has us re-inventing all the tools that we need to restore balance to society, back to a life of abundance. We are at the cusp of creating a fair and abundant society in our own time, if the cabal will allow us!

The next boost in inventive behavior that mankind experienced was in the 1400's, it was due to the first astrological decan in Aquarius (1433-2145AD, decan of Libra) then then again in the 1700's, "*the age of reason*", by shifting into the Dwapara Yuga (1700AD, silver age), as well as slowly losing the Piscean influence. By the astrological view the *Sun of Good* (our second hidden sun), would guide our subconscious intentions through the trinity of the Decans, when only at the end of the age is when we truly understand our journey, in our last Decan in Aquarius, which is Aquarius.

The historical ages of the different civilizations were not staggered like they show in the history books, but concordant and ran in parallel . The renaissance in Italy, Germany, England, China, Japan and India also happened in South America. The Inca were the result of the Renaissance in the South American continent. They were re-discovering their ancient past. It was a pity the Spaniards arrived and changed the course of the Inca culture. The knowledge they were re-claiming had only just begun, this could be the main reason the understanding of their language was difficult to decipher. They were still in the process of learning from a chronology of their own past, Aztec, Maya and the Toltec, just as the Renaissance artists and writers learned from the Romans, Greeks and the Egyptians. The Incas who were versed in the writing were killed, this stopped any progress of decipherment and delayed it until the late 20th Century. The re-discovery of the Egyptian Hieroglyphic writing and its understanding occurred in 1824AD, because of references to living Coptic Egyptians and their language, the true expansion of their legacy is still going on.

A similar situation happened in Thailand and Cambodia or as they called it in the 12th century, Siam. They only re-claimed the ancient buildings that were already there, because of the declining dark age. They originally abandoned their cities and they let nature destroy most of their historical records and also lost their earlier hierarchy, giving them the inability to form a new forward looking civilization. You can still see the twisted banyan trees covering many old sites. Some of these trees are said to be over 1500 years old. This age alone would suggest the Khmer never built them in the 1300's, there is absolutely no evidence ever found that they built any buildings from the Angkor(Ankh+Horus) region. The reclamation of nature occurred in most of the world, where the Dark Age(500AD) was a recipe for wholesale destruction of man and society. The decline of mankind was to annihilate a lot of the verbal history of most of the ancient cultures and as much written history was destroyed as the madness and superstition increased. There was only a very small amount of true documentation of those times, which were mostly religiously swayed, in all cultures, with no exceptions.

In Plate 12 there is a perfect example of the ancient documents in stone is the Temple of Angkor Prassat Preah Vihear. This represents the undamaged stories of the ancient past, codified in stone with each compass face elucidated by an ancient Hindu deity, expressing the stories and nature if the builders. The Layout and underlying harmony of the Temple on the border of Cambodia and Thailand is very finely proportioned, and has caused a war between those countries. Said to be much older that the archeologists prescribe as 900AD, more like 3000 BC.

The Plate 13 shows the Sumerians have a great deal of ancient artifacts that have not been truly understood. Here is an example of a clay impression taken from a seal found on a site said to be from 4000BC. The Grid here is 25 x 11, it is not necessarily the correct size but a beginning of some sort must preclude a clearer understanding. As can be noted the modern names of the constellations are not of the same character as the ancient Sumerian versions.

One last excursion into unknown esoteric understanding, this side of the Narmer plate could have another hidden level to express, refer Plate 14, and Plate 15. To conclude some of my thoughts about mans journey through the ages, I can truly say that I am only touching the tip of a very large iceberg. There are many ancient artifacts that need fresh interpretation. Some will show the hidden star charts that we know were made, some will guide us to new interpretations of our past and some will just be pretty pictures to admire.

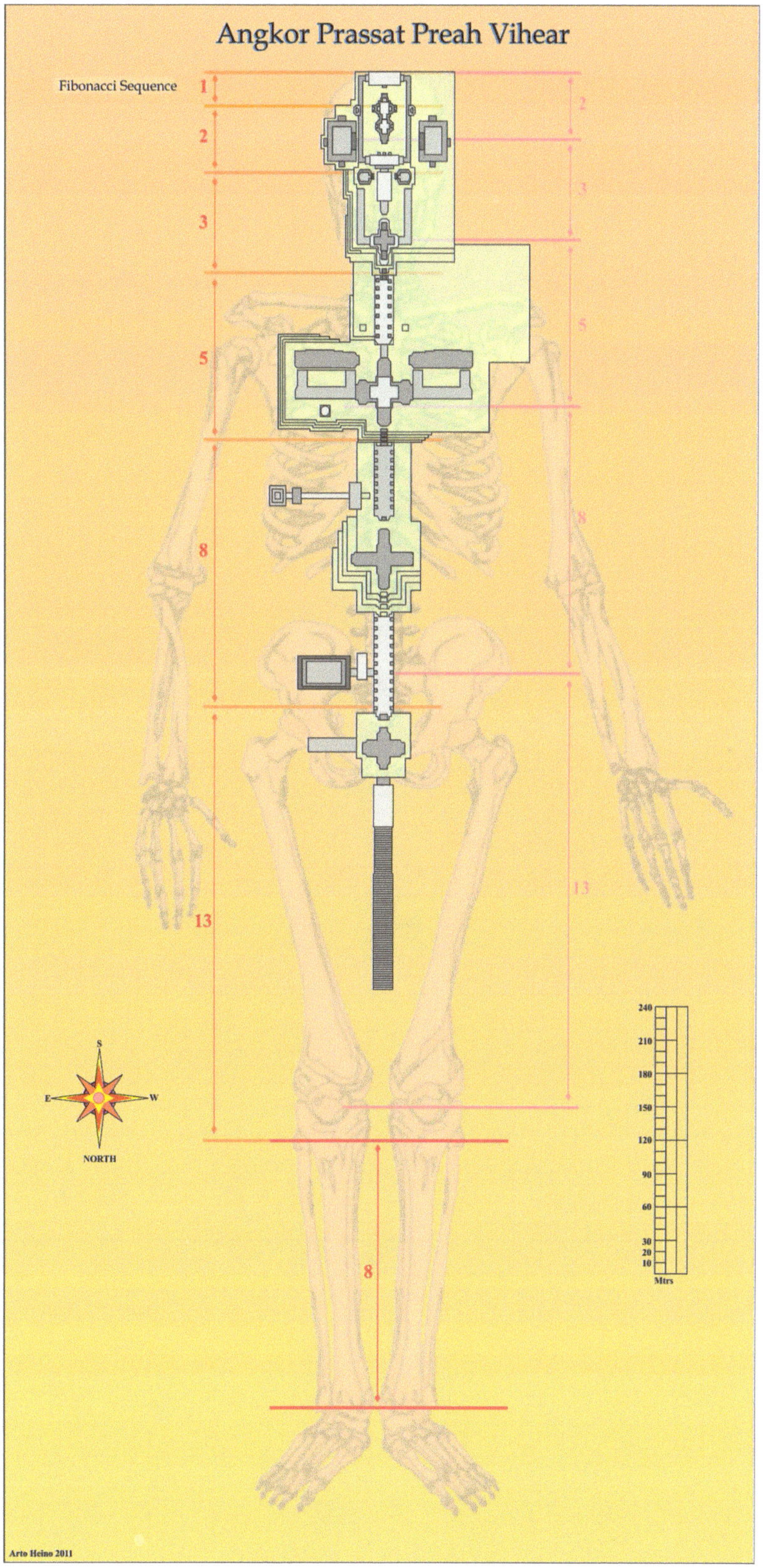
Angkor Prassat Preah Vihear
Fibonacci Sequence
1
2
3
5
8
13
8
2
3
5
8
13
S
E
W
NORTH
240
210
180
150
120
90
60
30
20
10
Mtrs
Arto Heino 2011

Plate 12

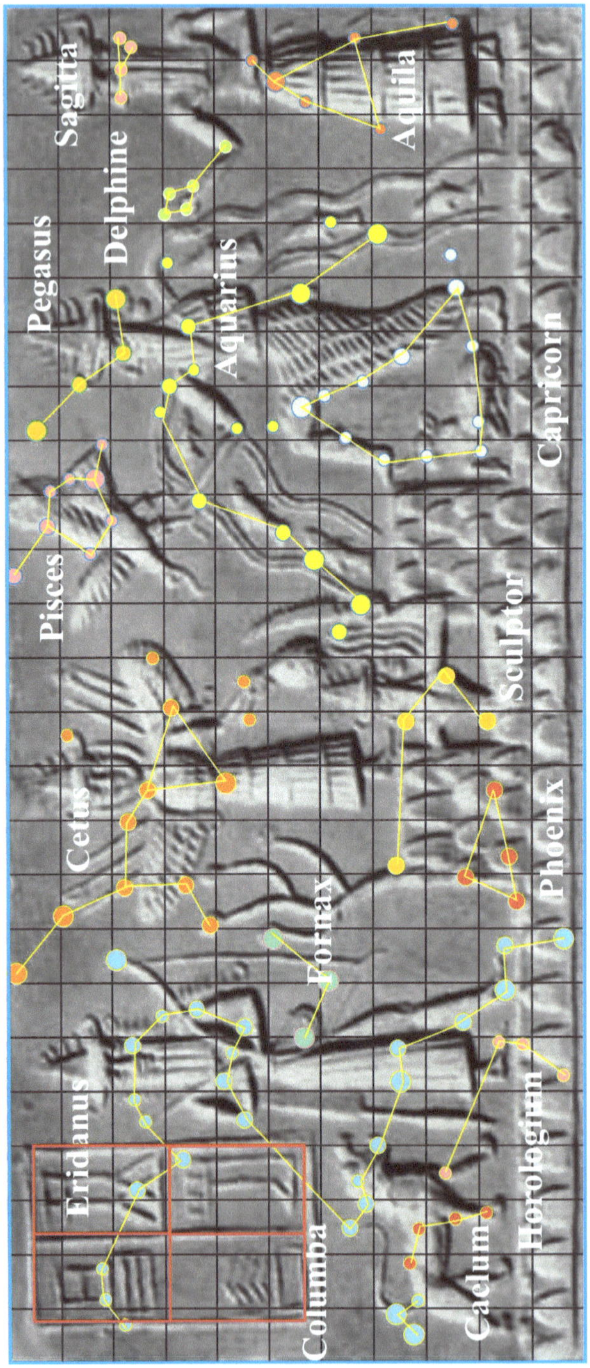

Plate 13

The stories of ancient man living 800 years and having great abilities, has validity in all aspects as long as you do not close your mind to the true possibility that we are part of a binary star system, which incorporates a cyclic change in human consciousness, that we have an ancient past that understood more than we know now and they have left a legacy for us to study, learn and follow.

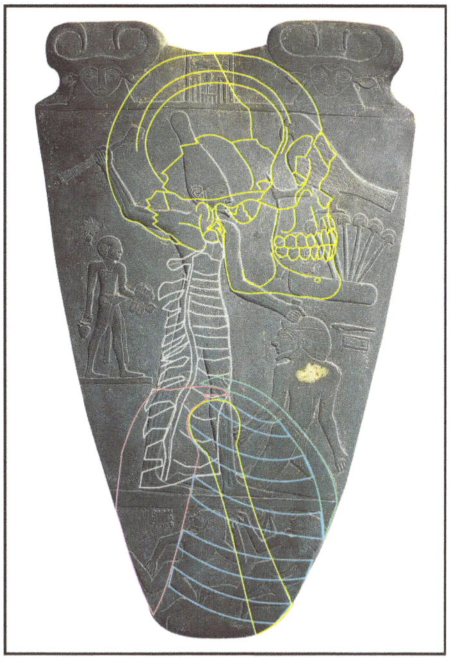

Plate 14

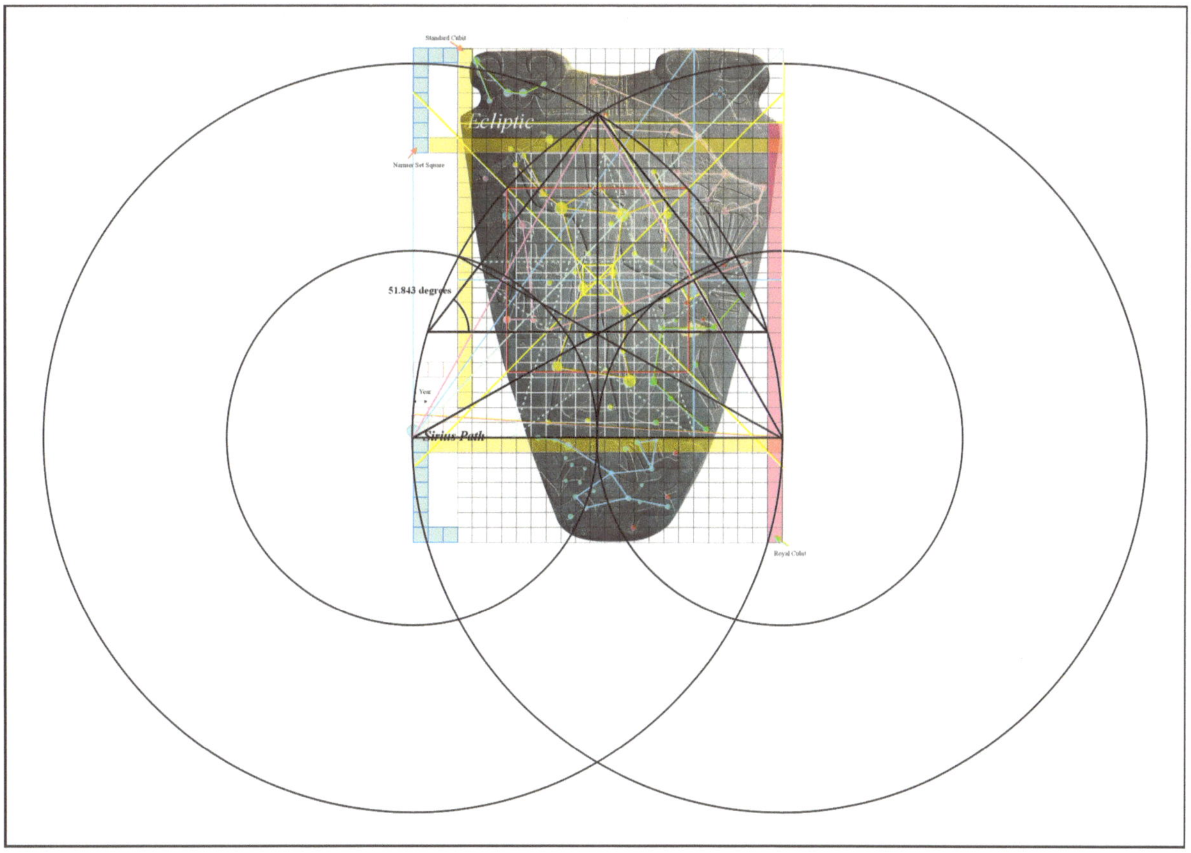

Plate 15

9.2 – Egyptian Crowns

The mystery of the ancient Egyptian crowns have always made for some great reading, where the least understood is some of the ideas by R. A. Schwaller de Lubicz. His interpretations are an enlightening read and very well versed in the symbolic interpretations. As I have my own enlighten ideas based on some possible world cultures that existed at least 12000 years ago. The Easter island Statues, the Olmec and the buried per-dynastic Egypt that has never been written about, these are tenuous connections that need to talked about not ignored like most left field theories. Too many new ancient cultures like Gobekli Tepe have been unearthed in recent times that point to a world culture of at least 15000 years ago.

Here is a small addition to the debate, a possible origin of the Crowns of Egypt.

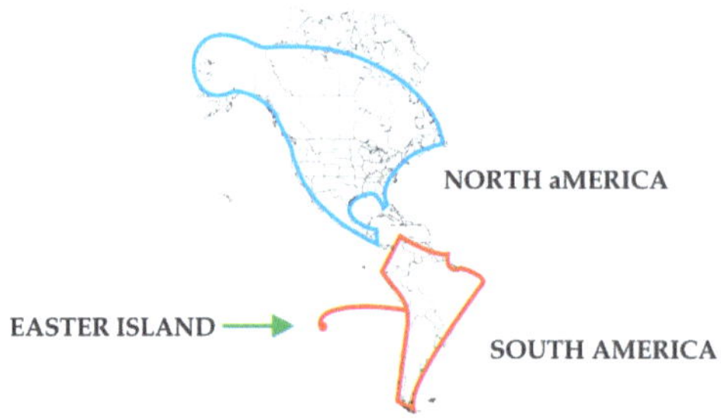

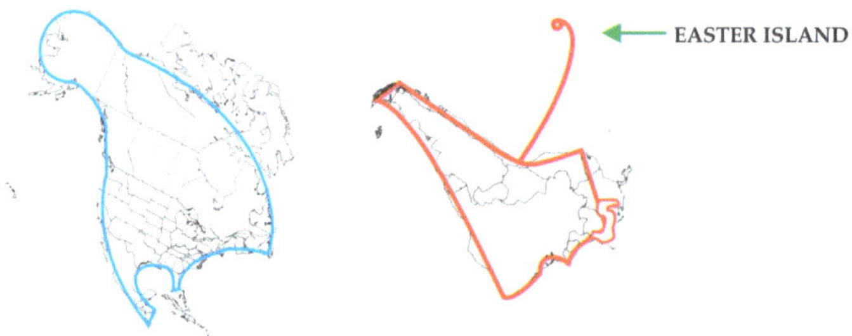

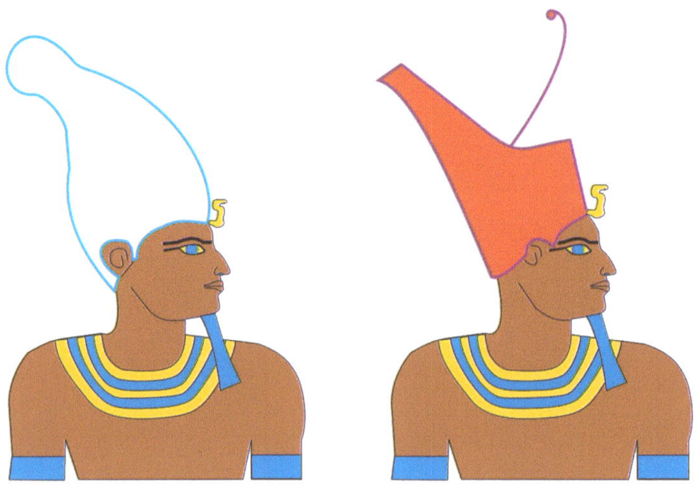

WHITE CROWN RED CROWN

Arto Heino

9.3 -Ancient Metric

Region	Light velocity	Unit	mm	Units/sec	Unit	mm	Units/sec	Unit	mm	Units/sec
Mtrs	**299792458**	**Miles**	1609344.0006 mm	186282.396980070000	**Feet**	304.8000 mm	983571057.925474000000	**Yards**	914.4000 mm	327857019.308491000000
Egypt	299792458	Djeser	300.0000 mm	999308193.333333000000	Ancient Egyptian Ft	308.0000 mm	973352136.363636000000	Remen	369.6000 mm	811126780.303030000000
	299792458	Geographic Cubit	462.0000 mm	648901424.242424000000	Royal Cubit	525.0000 mm	571033253.333333000000	Pyk Belady	577.5000 mm	519121139.393939000000
	299792458	Megalithic Yard	840.0000 mm	356895783.333333000000	Fathom	1848.0000 mm	162225356.060606000000	Remen 2	375.0000 mm	799446554.666667000000
	299792458	Large Span	250.0000 mm	1199169832.000000000000	Small Span	225.0000 mm	1332410924.444440000000	Fist	107.5000 mm	2788767051.162790000000
	299792458	Hand	93.8000 mm	3196081641.791040000000	Palm	75.0000 mm	3997232773.333330000000	Finger	18.8000 mm	15946407340.425500000000
	299792458	Khet	52500.0000 mm	5710332.533333330000	Iteru	10500000.0000 mm	28551.662666666700	Cubit 2	450.0000 mm	666205462.222222000000
Sumer	299792458	Grain	2.5000 mm	119916983200.000000000000	Finger	15.0000 mm	19986163866.666700000000	Foot	333.0000 mm	900277651.651652000000
	299792458	Cubit	497.0000 mm	603204140.845071000000	Step	1000.0000 mm	299792458.000000000000	Reed	3000.0000 mm	99930819.333333300000
	299792458	Rod	6000.0000 mm	49965409.666666700000	Cord	60000.0000 mm	4996540.966666670000	Cable	360000.0000 mm	832756.827777778000
	299792458	League	10800000.0000 mm	27758.560925925900	MY 2	829.6600 mm	361343752.862618000000	Cable	360000.0000 mm	832756.827777778000
Arto	299792458	Hand	93.4351 mm	3208563652.732610000000	Palm	74.7481 mm	4010704565.915770000000	Finger	18.6870 mm	16042818263.663100000000
	299792458	Standard Cubit	448.4885 mm	668450760.985961000000	Royal Cubit	523.2365 mm	572957795.130824000000	Pyk Belady	560.6106 mm	534760608.788769000000
	299792458	Hand PI/2	59.4826 mm	5040000000.000000000000	Palm PI/2	47.5861 mm	6300000000.000010000000	Finger PI/2	11.8965 mm	25200000000.000000000000
	299792458	Standard Cubit PI/2	285.5166 mm	1050000000.000000000000	Royal Cubit PI/2	333.1027 mm	900000000.000001000000	Pyk Belady PI/2	356.8958 mm	840000000.000001000000

These are a selection of ancient measure with the millimeter measure and the light velocity in units per second.

9.4 - Archeogeometry

9.4a -Saqqara Ostrakon a Different and Exact Solution

Having solved many numeric puzzles using Quantum Arithmetic, I decided to attack this interesting artifact with relish. It seems everybody had investigated this item including Graham Hancock and Samuel Laboy. I am glad to express my delight in the solution that I have found, revealing possibly the true principles of Ancient Khemitian(Egyptian) geometry, this being part of the foundation of building and pyramid design and the reason for their construction principles. As my investigations into Quantum Arithmetic show, I have found some of the the vital keys to knowledge of the ancient Pyramidic geometric cipher, no weird magic number or symbol required, this is truly exciting.

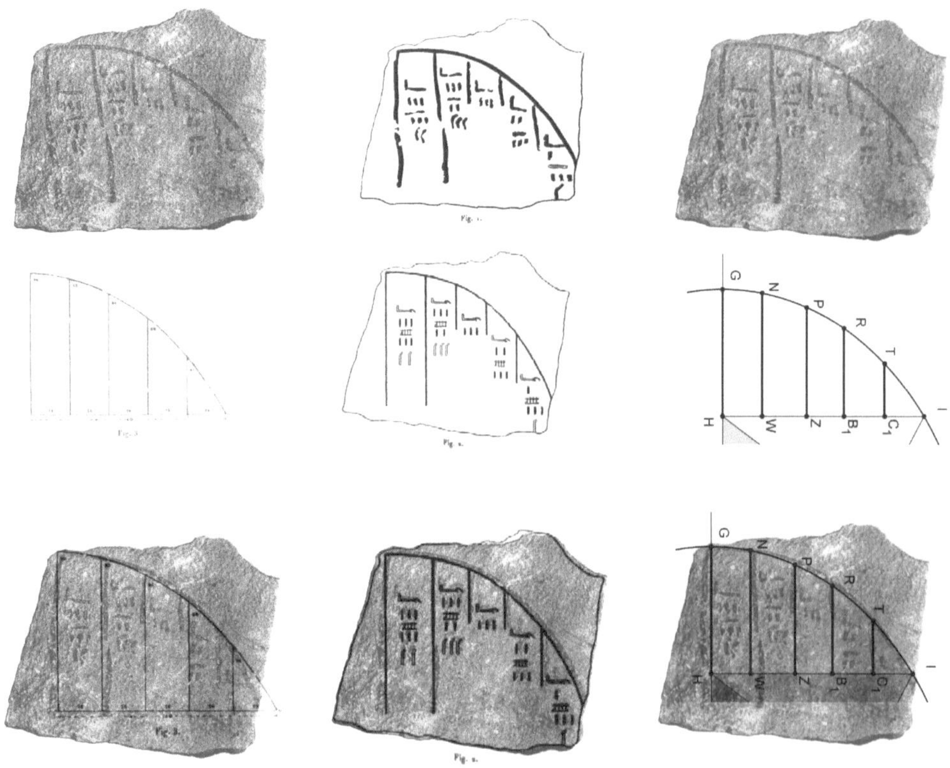

The Saqqara Ostrakon was found at the excavations in Saqqara in 1925, the writing reveals this object has a geometric value. The inscription is a form of shorthand to some larger idea.

3 Cubits, 3 Palms, 2 Fingers = 98 Fingers = 1837.5 mm

3 Cubits, 2 Palms, 3 Fingers = 95 Fingers = 1781.25 mm

3 Cubits, 0 Palms, 0 Fingers	= 84 Fingers	= 1575	mm
2 Cubits, 3 Palms, 0 Fingers	= 68 Fingers	= 1275	mm
1 Cubit , 3 Palms, 1 Finger	= 41 Fingers	= 768.5	mm

1 Cubit	= 7 Palms	= 28 Fingers = 525 mm *(using this conversion size in millimeters)*	
1 Palm	= 4 Fingers	= 75 mm	
I Finger	= 18.75 mm		

The small curved line and a few strokes and the numbers above is all that was needed. The 98 Fingers was the first clue, this would be the Perihelion or in QA the J factor. The plane number 98 didn't fit with any Quantum Ellipse so a multiplication or conversion number must have been used, after a few attempts it looks like the prime number 5 was the cipher.

G _ H	= 490	= 5 x 98
N _ W	= 475	= 5 x 95
D _ Z	= 420	= 5 x 84
R _ B1	= 340	= 5 x 68
T _ C1	= 205	= 5 x 41

Now that I know J= 98 then as in all QA you can work backwoods to find the roots of the ellipse. These are the original root numbers that I found:

b = 2 e = 3 d = 5 a = 8 these will give you J = 10 and the eccentricity of 0.6

b = 14 e 21 d = 35 a = 56 these will give you J = 490 and the eccentricity of 0.6

The prime factor underlying the x 7 is a way of finding the other derived numbers in a simple translation as J=98 is not a Quantum root size it is a derivation from a physical measuring system from a Quantum number set. This diagram should explain the geometry.

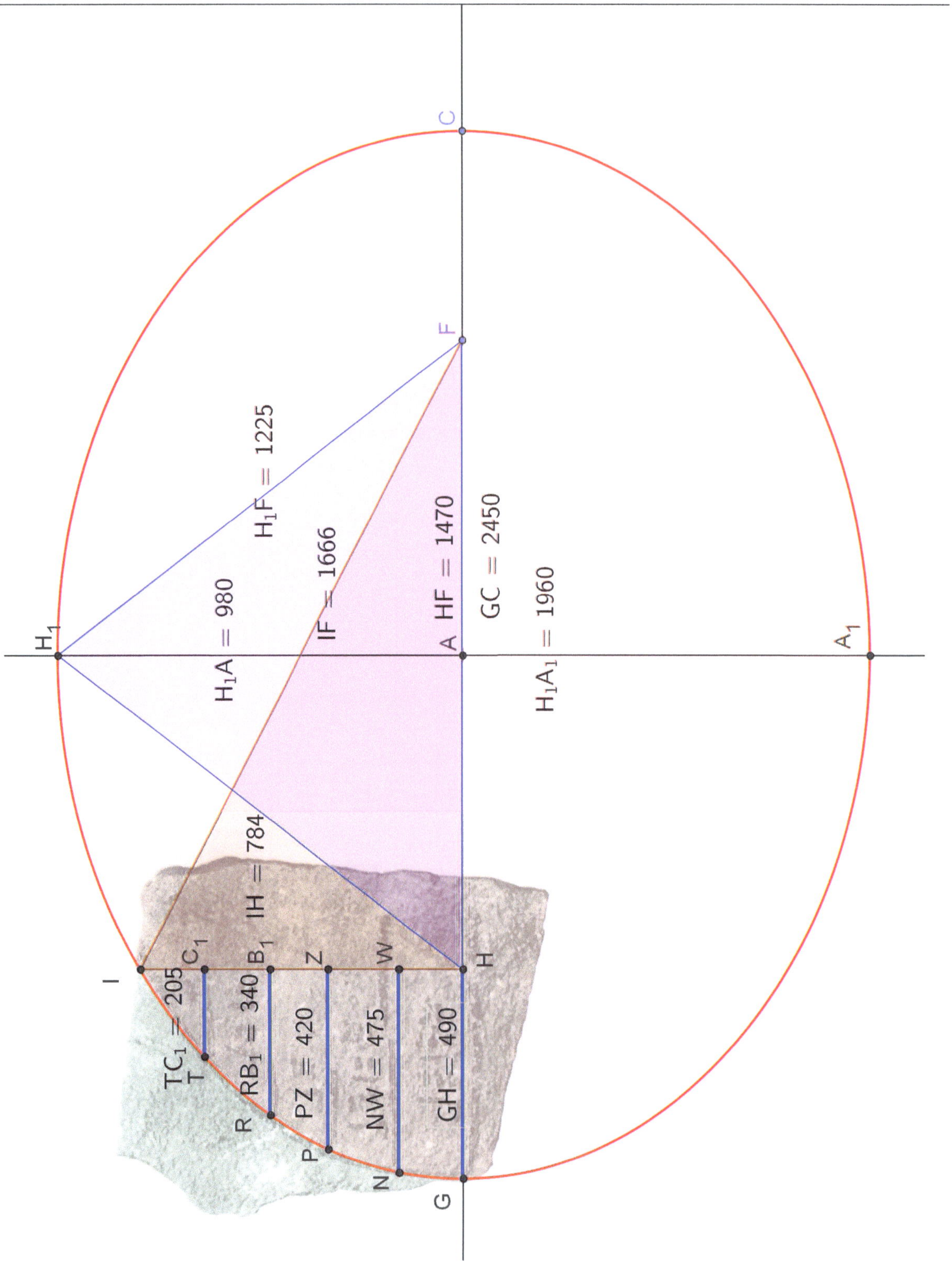

The reason I exclaimed the true solution is because of the following hidden variables. The redrawn diagram everybody followed by Battiscombe Gunn is incorrect, looks like his pet theory of 28 fingers between lines was wrong, as you can see by the original carving the lines are not evenly spaced, it was only a crude diagram drawn without a rule, you cannot base the intention of the geometer on a rough sketch, only on the description he gives and the approximate position of all the lines. All QA requires is some real numbers and an approximate shape to define the correct form.

These are exact values with no decimal fractions:

H _W = sqrt(23376)
H _Z = sqrt(106624)
H _B1 = sqrt(220800)
H _C1 = sqrt(394896)
H _ I = Sqrt(614656)

These are the values when J=490, they reveal the correctness of the QA precision You will never arrive at this sort of numeric clarity using any mathematical tricks, the Quantum Ellipse is a unperturbed form and non empirical. The values when J=98 re listed below:

H _ W = sqrt(935.04)
H _ Z = sqrt(4264.96)
H _ B1 = sqrt(8832)
H _ C1 = sqrt(15795.84)
H _ I = sqrt(24586.24)

What we are looking for is the gap sizes, here is J=490

H _ W = sqrt(23376)
W _ Z = sqrt(106624) – sqrt(23376)
Z _ B1 = sqrt(220800) – sqrt(106624)
B1_ C1 = sqrt(394896) – sqrt(220800)
C1_ I = Sqrt(614656) – sqrt(394896)

Here are the final values decoded and back to empirical finger and metric measure

H _ W = sqrt(935.04) = 30.5784237 Fingers = 573.3454456 mm

W _ Z = sqrt(4264.96) - sqrt(935.04) = 34.7282374 Fingers = 651.1544523 mm

Z _ B1 = sqrt(8832) - sqrt(4264.96) = 28.6720597 Fingers = 537.6011207 mm

B1_C1 = sqrt(15795.84) - sqrt(8832) = 31.7027812 Fingers = 594.4271479 mm

C1_ I = Sqrt(24586.24) - sqrt(15795.84) = 31.1184877 Fingers = 583.4718333 mm

These sizes were not needed by the Khemitian geometer to their requirements only the right angle edge and the curve was needed, only modern calculus require these values as to plot X * Y curves. The Khemitians had no need of such crude mathematical devices such as calculus, they treated the curve as a different type of measure than the straight line. Now we can extrapolate the Cubit/Finger and metric size of the whole ellipse.

Length of the ellipse is = 2450 = 490 x 5 in Fingers

= 490 = 17 Cubits 3 Palms 2 Fingers = 9.1875 mtrs

Width of the ellipse is = 1960 = 392 x 5 in Fingers

= 392 = 14 Cubits = 7.35 Mtrs

The values are correct as no extra bit of a finger is required, this one of the reasons the Khemitians used these types of measure. The translation into quantum units were simple and there were no fudge factors that required PI, PHI or Log, no fractional decimals, only whole numbers So the final outcome underlies how truly powerful Quantum Arithmetic is when fully understood and utilized in the right hands, when looking for a accurate solution without errors from empirical measurements and data from approximation using calculus.

The Major Resonant Cavity of the Pyramid of Alnitak

(The Kings Chamber of the Great Pyramid)

By Arto Juhani Heino (12-12-2012)

Quantum Number Set

26	13	39	52
676	169	1521	2704
1014	1352	1690	114244
2366	338	1014	2028

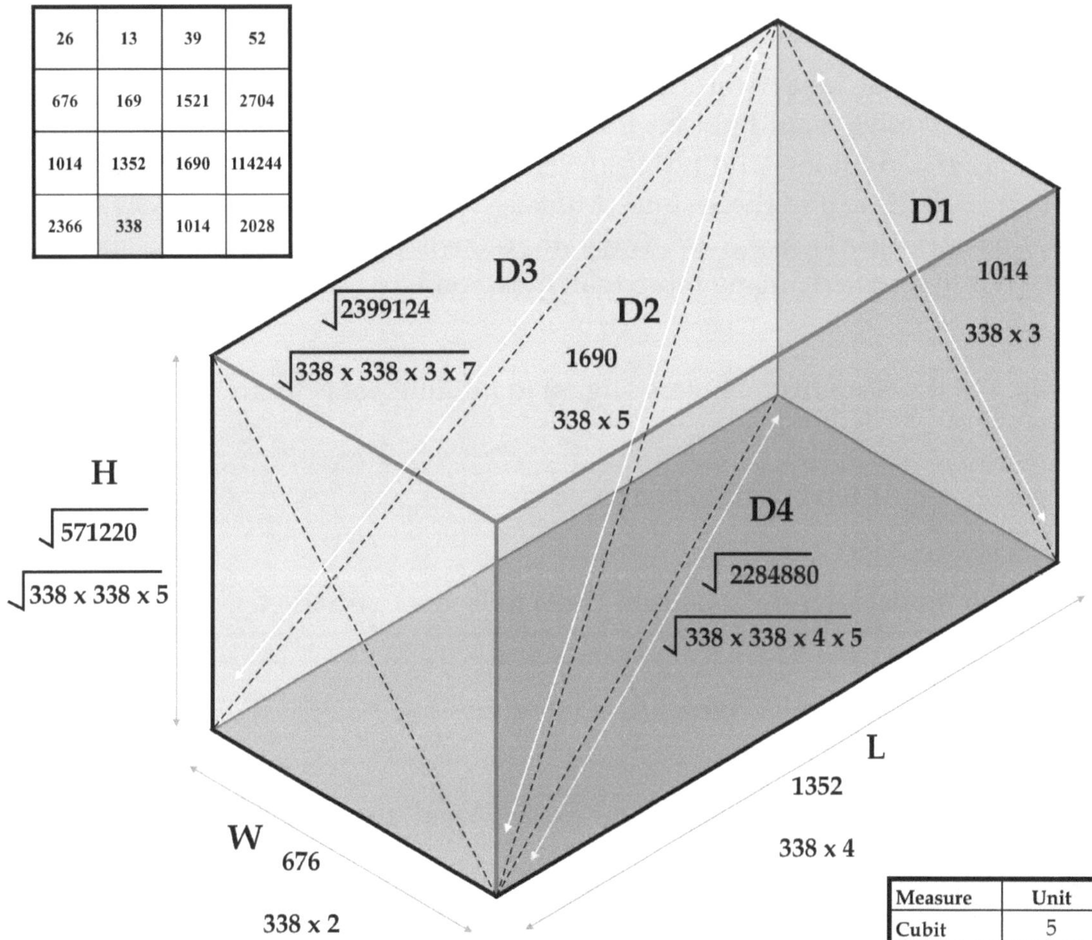

Measure	Unit
Cubit	5
Palm	35
Fingers	140
Inches	103
Feet	8.58333
Meters	2.6162

Quantum Volumetric Cavity

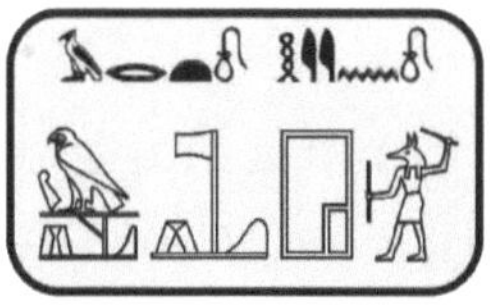

Quantum	338
x2	676
x3	1014
x4	1352
x5	1690
x7	2366
x8	2704
squared	114244
/2	169
/13	26
/26	13

Arto Juhani Heino 2012

Empirical Measures			
Measure	H	D3	D4
Standard Cubits	13	26	26 217/338

Unit Multiplier						
W	D1	L	D2	H	D3	D4
2	3	4	5	$\sqrt{5}$	$\sqrt{20}$	$\sqrt{21}$

Empirical Measures							
Measure	W	D1	L	D2	H	D3	D4
R Cubits	10	15	20	25	11.1803398	22.3606797	22.9128784
Palms	70	105	140	175	78.2623792	156.524758	160.390149
Fingers	280	420	560	700	313.04951	626.099034	641.560597
Inches	206	309	412	515	230.315001	460.630003	472.005296
Feet	17.16666	25.75	34.33333	42.916666	19.1929168	38.385833	39.3337747
Meters	5.2324	7.8486	10.4648	13.081	5.85000104	11.700002	11.9889345

Arto Heino 12-12-12

9.4b - The Kings Chamber

Here is the complete numeric code of the Kings Chamber(Resonant Cavity) from the interior of the Great Pyramid(Alnitak), the numbers cannot be fudged and are precise , no allusion to the value PI or the Phi. Using transcendental numbers only gives you approximations, not true values, only Quantum Arithmetic can reveal the underlying order behind these types of structures.

The original ideas were not coded from computers and approximated with calculus, these tools would not have been needed 10000 years ago. The dawning of a new paradigm in understanding of ancient monuments are written in the quantum code that underlies all of these stone artifacts, if they knew how and why, we can also know, this was their legacy to us. The path to the enlightenment of our ancestors does involve many sacrifices, many more than most modern people would want to endure.

The value to translate from Fingers(Digits) to millimeters is to divide by:

1 Finger = 18.6871428571429 mm

The value to translate from Quantum Units to feet is to divide by:

1 Foot = 39.378640776699 QU (Quantum Unit = 338)

By the research I have here it seems the Cubit(Royal) used for the Kings Chamber is:

1 Cubit (CR) = 523.24 mm = 20.6 inches *(523.236546546474681 refer Earth Chp 11)*

A quantum unit in not a empirical measure, it is an absolute value that can be derived from number law and prime numbers it does not have a distance measure like meters and feet. This is what the ancient Khemitians (Egyptians) express as divine , sometimes giving it names like Osiris, Horus, Thoth, Atum etc to explain the amazing qualities of pure numbers. I suspect the nine gods of the Ennead are also known as the geometers Nine point circle and all the discipline and understanding that comes from its knowledge and deep connections with the absolute and pure numbers of the universe.

As you can see the use of cubits, hands and fingers are very easy to translate from quantum units, the people from Khemit knew that these measures would not be absolute, only as a translational device. This is the reasons our current society cannot pinpoint why cubits had so many variants, we think that our measures are correct and absolute and the ancients were vague and approximate, I agree this is the case, only if you look at it with a narrow idea of what absolute measure really is, meters or feet are just convenient rulers for our utilitarian needs, just as cubits and palms were to the ancients.

There is a another level of hidden measure here, as the Khemitians did not use decimal fractions like we do, we must look at another schema that does not rely on our modern approach. It looks like the unique quantum values have been used again, this time to create another cubit measure that can be used for the height and one of the diagonals(D3). This was has been called the standard cubit:

1 Standard Cubit (CS) = 17.7165385909599 inches

This length reduces the problem required by making the height now 13 CS and the diagonal D3 become double value of 26 CS. These numbers are the first 2 numbers of the Quantum set that I have shown. This alone will tell you the Khemits knew the Quantum code used and that the rediscovery of their methods is complete, no more esoteric geomatra that only confuses the numeric issues by introducing some strange masonic ritual into the understanding of numbers, not that this did not serve the purpose of showing these values as sacred to the hidden code and help perpetuate the use of them through history, thanks to the people who showed their continued reverence and usage in architecture, art and geometry. Here is some more figures that confirm this:

13 x 26	= 169 x 2	= 338
338 X 338	= 114244	

A right angled triangle created from a base of 26 CS(D3) and a height of sqrt(33.8) has a hypotenuse that is equal to the floor diagonal(D4), this value is Sqrt(709.8)

7098	= 239912/338	= 1014 x 7	= 338 x 21
	= 338 x 7 x 3	= 26 x 13 x 7 x 3	= 169 x 42
	= 169 x 7 x 6	= 13 x 13 x 7 x 6	

There seems to be some strange idea that the ancient builders used some sort of special ruler to measure and build the structures and maintain the PI/PHI properties, this is just fantasy as you do not need to use these numeric or ruler type devices to measure its proportions, they arrive out of using Quantum numbers as PI/PHI are built from Quantum proportions.

The D4 measure is most interesting it looks like the possibility to translate into whole number measure the Khemits may have used is 26 cubit CS and a didactic fraction 217/338 this is 0.00006015717542851 from being exact, in millimeters this equates to 0.000041959721260mm which is about the thickness of a human hair. This would seem be an acceptable clearance allowing for temperature differentials and seismic activity. As for that number 217, this is another interesting quantity as 216 is 6 cubed, further investigations would seem fruitful in this direction, even though we are only dealing with such a small fractional part, the Khemitians gave us the questions, so we are obliged to answer.

The QA numbers seem almost trivial compared to the modern edifice of mathematics, millions of volumes full of the most complex ideas of mankind, still it seems not to speak as simply and directly as Quantum Arithmetic. The Quantum number set used in the cavity is the part of a series of numbers that have particular properties, such as:

2 x 39 x 13 = 1014 = 39 x 39 - 39 x 13

the common divisor is 13

as 39 = 3 x 13, this only works by multiplying 2 prime numbers and having a common divisor(13).

2a x b = a^2 - ab

if b/a is an integer and is a divisor of the other

what if the QA set was 26,12,38,50 then

2 x 38 x 13 = 912 and 38 x 38 - 38 x 12 = 988

The number 38 does not divide by 12 without remainders, this cannot be used in this type of construction, but maybe in another shape or form these numbers are suitable as part of its required building blocks. This particular rectangular cube has harmonic properties just like the Quantum number set has.

My methods are original and the answers speak for themselves. if you doubt the veracity of my approach please by all means find fault with the numbers not me. I am not an Engineer, I am not a Mathematician, I do not hold a degree in any subject; BUT: I am an Artist and this is part of my art, something that I have worked a lifetime to accomplish, I hope what I reveal through my work, we can all gain a better understanding of the world around us.

Quantum Numbers for the Pyramid of Alnitak

I am One that transforms into Two
I am Two that transforms into Four
I am Four that transforms into Eight
After this I am One

(Coffin of Petamon, Cairo Museum no: 1160)

Finally I have unearthed the Quantum Arithmetic set of numbers for the Great Pyramid. I have been working on this for a number of years(1988), only now have I discovered some of the methods of the underlying geometry to build such a complex structure from only two numbers. The deep understanding of geometry by the Ancient Khemitians(Egyptians) should make our current system of calculus of approximates seem poorly suited for simplification and shorthand. All the pyramids can be constructed from this understanding, and if these insights become common knowledge then we have entered a new paradigm of Art, Architecture and Science.

The first two numbers and a ratio translation number are all the information required to reconstruct the Great Pyramid's elusive design. There has been many theories about what constituted the design and the actual numbers used to build the structure, each one has failed to show the true beginning of the numeric journey. I found many ideas quite amazing and interesting how they weave their numeric code into the design.

There are many amazing and promising designs I have encountered, great work using pure geometry to crunch those angles and make it fit, sadly there are many devices used that are artificial and eludes to squaring of the circle, a pursuit rooted in the misunderstanding between a curved line and a straight line, two different forms of measure. The constructs used are the same devices used by generations of geometers who construct perimeters of different types, rooted in Phi and Pi. Just remember PI and Phi are not quantum units they cannot be used without resorting to approximations, you cannot build wholeness from endless fractional decimals, this is a modern illusion. I have included the Phi relations in Diagram 1 so you can crunch your own numbers and design a Golden Pyramid, based of PHI. The short list is a PI Pyramid, a PHI Pyramid and the Quantum Pyramid, all three are valid and related to each other, only the Quantum Pyramid is exact solution the other two are approximations.

The Quantum Aspects of the Pyramid of Alnitak

(The Holistic Relationships in Nature as Expressed in The Great Pyramid)

By Arto Juhani Heino

Quantum Number Set

70	37	107	144
4900	1369	11449	20736
7918	10080	12818	6651120
17998	2162	7490	15408

I am One that transforms into Two
I am Two that transforms into Four
I am Four that transforms into Eight
After this I am One

(Coffin of Petamon, Cairo Museum no. 1160)

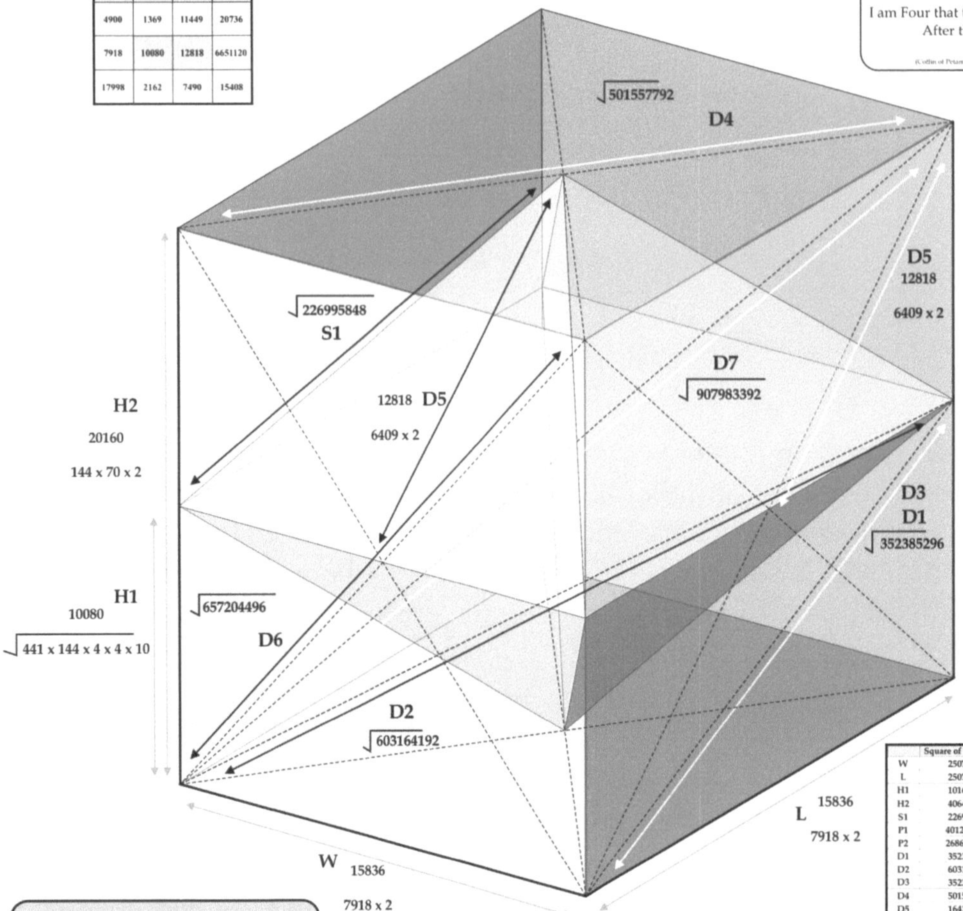

Measure	Unit
Royal Cubit	220
Standard Cubit	256.6666
Palm	1540
Fingers	6160
Inches	4560
Feet	380
Meters	115.824
Quantum	7918

	Square of QA measures	Divide by 107	Remainder x/107
W	250778896	2343728	0
L	250778896	2343728	0
H1	101606400	949592	56
H2	406425600	3798370	10
S1	226995848	2121456	56
P1	4012462336	37499648	0
P2	2686556224	25108002	10
D1	352385296	3293320	56
D2	603164192	5637048	56
D3	352385296	3293320	56
D4	501557792	4687456	0
D5	164301124	1535524	56
D6	657204496	6142098	10
D7	907983392	8485826	10

Quantum Volumetric Division

Decimal Quantum Unit Multiplier

W	L	H1	H2	S1	P1	P2	D1	D2	D3	D4	D5	D6	D7
2	2	1.273048749684	2.546097499369	1.902801387185	8	6.546097499369	2.370791665100	3.101717769100	2.370791665100	2.828427124746	1.618843142208	3.237686284415	3.805602774370

Decimal Phi Unit Multiplier

W	L	H1	H2	S1	P1	P2	D1	D2	D3	D4	D5	D6	D7
2	2	1.272019649514	2.544039299028	1.902113032590	8	6.544039299028	2.370239226059	3.101295533926	2.370239226059	2.828427124746	1.618033988750	3.236067977500	3.804226065181

	Measure	W	L	H1	H2	S1	P1	P2	D1	D2	D3	D4	D5	D6	D7
	Royal Cubits	440	440	280	560	418	1760	1440	521	682	521	622	356	712	837
	+ Fingers	0	0	1	3	17	0	3	16	10	16	7	4	8	6
31672 = 7918 x 4	+ Parts of x/7918	0	0	560	1120	4880	0	1120	..18185/31672	...11969/31672	..18185/31672	2011	1152	2304	1842
P1 = Nautical mile/2	Measure	W	L	H1	H2	S1	P1	P2	D1	D2	D3	D4	D5	D6	D7
Royal Cubit =	Inches	9120	9120	5805.102299	11610.204597	18240	36480	29850.204597	10810.809993	14143.833027	10810.809993	12897.627689	7381.924728	14763.849457	17353.548651
20.7272727273 Inches	Feet	760	760	483.758525	967.51705	723.064527	3040	2487.51705	900.900833	1178.652752	900.900833	1074.802307	615.160394	1230.320788	1446.129054
Multiplier Unit = 220	Meters	231.648	231.648	147.449598	294.899197	220.390068	926.592	758.195197	274.594574	359.253359	274.594574	327.599743	187.500888	375.001776	440.780136
	Measure	W	L	H1	H2	S1	P1	P2	D1	D2	D3	D4	D5	D6	D7
Royal Cubit =	Inches	9077.2	9077.2	5777.859055	11555.718111	18154.4	36308.8	29710.118111	10760.075051	14077.456267	10760.075051	12837.099348	7347.281485	14694.56297	17272.108752
20.63 Inches	Feet	756.433333	756.433333	481.488255	962.976509	719.671198	3025.733333	2475.843176	896.672921	1173.121356	896.672921	1069.758279	612.273457	1224.546914	1439.342396
Unit Sum	Meters	230.56088	230.56088	146.75762	293.51524	219.355781	922.24352	754.637	273.305906	357.567389	273.305906	326.062323	186.62095	373.241899	438.711562
W = 755.750692 ft	Measure	W	L	H1	H2	S1	P1	P2	D1	D2	D3	D4	D5	D6	D7
Royal Cubit =	Inches	9069.008304	9069.008304	5772.644841	11545.289682	18138.016608	36276.033216	29683.30629	10750.364649	14064.752102	10750.364649	12825.514541	7340.65095	14681.301899	17256.521581
20.6113825088722 inches	Feet	755.750692	755.750692	481.053737	962.107474	719.021733	3023.002768	2473.608857	895.863721	1172.062675	895.863721	1068.792878	611.720912	1223.441825	1438.043465
As Per Diagrams	Meters	230.352811	230.352811	146.625179	293.250358	219.157824	921.411244	753.95598	273.059262	357.244703	273.059262	325.768069	186.452534	372.905068	438.315648
W = 755.7488 ft	Measure	W	L	H1	H2	S1	P1	P2	D1	D2	D3	D4	D5	D6	D7
Royal Cubit =	Inches	9068.9856	9068.9856	5772.630389	11545.260779	18137.9712	36275.9424	29683.231979	10750.337735	14064.716892	10750.337735	12825.482432	7340.632573	14681.265145	17256.47838
20.6113309090917 inches	Feet	755.7488	755.7488	481.052532	962.105065	719.019933	3022.9952	2473.602665	895.861478	1172.059741	895.861478	1068.790203	611.719381	1223.438762	1438.039865
	Meters	230.352234	230.352234	146.624812	293.249624	219.157275	921.408937	753.954092	273.058578	357.243809	273.058578	325.767254	186.452067	372.904135	438.314551
Absolute Whole Unit	Quantum Unit	15836	15836	10080	20160	√226995848	63344	51832	√352385296	√603164192	√352385296	√501557792	12818	25636	√907983392

Arto Juhani Heino 2012

Quantum	7918
x2	15836
x3	23754
x4	31672
x5	39590
x7	55426
x8	63344
x840	6651120
squared	250778896
/2	3959
/37	214
/74	107
/107	74

Here Listed here are some of the Pyramids investigators and a brief of their work.

Leon Cooper	-The Architects Plan
Ole Jergen Bryn Apex	- Retracing the Egyptian Pyramids
John Cadman	- Engineering Science and the Great Pyramid

David Furlong	-The Key to the Temple
Samuel Laboy	-The Perfect Symbol
Tim G. Hunkler	- Symbolism and Coincidences of the Great Pyramid
Christopher Dunn	-Giza Power Plant
Graham Hancock	-The Orion Mystery
Mark Lehner	- The Complete Pyramids
I.E.S. Edwards	-The Pyramids of Egypt

Now that you can see some of the Writers, Archaeologists, Engineers and Investigators that have have walked this well worn ground and have not found the first principals, the reasons of structural placement and the functional meaning matching the form. They are all right in their conclusions , as this Pyramid includes ALL of these views and ideas, only their conclusions are but a different view point on the same thing. The main problem is the teaching of maths and geometry in schools, treated like a adjunct to abstract notions in finance and computer modeling. Only very few investigators in pure mathematics bothers to find out why numbers works in nature this way and why some solutions have multiple equations and some need calculus to try and figure it all out.

The only problem with invented mathematics are the results are not an arrow to the question what the Pyramid originally imposed upon mankind, only some numeric abstraction that has no basis in the natural world or order of the universe. Empirical results cannot give mankind much headway in answering our future dilemma. only the enlightened human being who understands the inner needs on the living system can know the path we must make to change our current perceptions on the nature of our reality and survival.

One question is why does the circle have only one center and the triangle have more than 6000 and the list grows. These things seem trivial, remember it is not the answer that matters but the question that is important. from questions arise the underlying principles that makes these geometric constructs work.

Here Listed is the basic Quantum Arithmetic equations that you need to design the correct pyramid. I recommend all researches to use these numbers to correct their models, as this is the pure whole number set that you cannot fudge to fit, they are absolute and cannot be corrected by any small decimals using Phi, Pi or Log. They are a constant and the roots to these the higher dimensions can be translated to any unit of measure by using a conversion ratio. We are dealing with whole number solutions and the ratio Phi is only a derived number, in the natural system you can refer this to the root numbers in Fibonacci series, where the oscillation about the Phi point generate amazing geometry. Pi is also a ratio and cannot have an exact solution so we will not use it as the value 3.1415727.....but as a series of ratios like the Phi point. Here is a list between 1.25 and 1.75 that you should recognise as common usage by Engineers, Geometers and Mathematicians.

5/4	= 1.25000000000000000	
9/7	= 1.28571428571428571	
4/3	= 1.33333333333333333	
3*HDp/WDp	= 1.350272105524200	
e/2	= 1.35914091422952261	= Natural logarithm/2
7/5	= 1.40000000000000000	
SQR(2)	= 1.41421356237309504	
10/7	= 1.42857142857142857	
3/2	= 1.50000000000000000	
163/107	= 1.52336448598130841	= Alnitak fraction 56+107 = 163
Fg/3	= 1.55640053636766355	= Feigenbaum constant /3
168/107	= 1.57009345794392523	
PI/2	= 1.57079632679486619	= PI/2
Wp/Hp	= 1.571031746031750	= Pyramid Width/Height
11/7	= 1.57142857142857142	= Refer Plato also 11! / 7! ^2
8/5	= 1.60000000000000000	
PHI	= 1.61803398874989484	= PHI
2*HFp/WFp	= 1.6188431422076200	= Quantum Pyramid Face Height/Width
5/3	= 1.66666666666666666	
12/7	= 1.71428571428571428	
265/153	= 1.73202614379084967	
SQR(3)	= 1.73205080756887729	
1/V	= 1.73245471460063347	= 1/ Euler-Mascheroni constant
100/Rad	= 1.74532925199432957	= Radian
3/4	= 1.75000000000000000	

Fg	= 4.6692016091029906718532038204662	= Feigenbaum constant
e	=2.7182818284590452353602874713526	= Natural logarithm
V	=0.5772156649015328606065120900824	= Euler-Mascheroni constant

Rad = 57.295779513082320876798154814105 = Radian = 180/Pi

A series of Quantum versions, some are exact, most are approximates:

5/4	= 9/7, 35/28, 180/144, 6300/5040, 8201/6561
4/3	= 9/7, 37/28, 192/144, 6720/5040, 8748/6561
Sqr(2)	= 10/7, 40/28, 204/144, 7128/5040, 9279/6561
3/2	= 11/7, 42/28, 216/144, 7560/5040, 9842/6561
Fg/3	= 11/7, 44/28, 224/144, 7844/5040, 10212/6561
PI/2	= 11/7, 44/28, 226/144, 7917/5040, 10306/6561
11!/(7!)^2	= 11/7, 44/28, 226/144. 7920/5040, 10310/6561
PHI	= 11/7, 45/28, 233/144, 8155/5040, 10616/6561
5/3	= 11/7, 46/28, 240/144, 8400/5040, 10935/6561
Sqr(3)	= 12/7, 48/28, 249/144, 8730/5040, 11364/6561
1/V	= 12/7, 49/28, 249/144, 8732/5040, 11367/6561
100/Rad	= 12/7, 49/28, 251/144, 8796/5040, 11451/6561
7/4	= 12/7, 49/28, 252/144, 8820/5040, 11482/6561

PHI		= 1.6180339887498948482
PHI	= 11/7	= 1.5714285714285714285 below
PHI	= 45/28	= 1.6071428571428571428 below
PHI	= 173/107	= 1.6168224299065420560 below
PHI	= 233/144	= 1.6180555555555555555 above
PHI	= 8155/5040	= 1.6180555555555555555 above
PHI	= 10616/6561	= 1.6180460295686633135 above

As you can see these ratios converge to the irrational value PHI, as a abstract. This cannot be used a a building block only the two convergent whole numbers are usable as true constructs of concrete reality.

PI		= 3.1415926535897932384
PI	= 22/7	= 3.1428571428571428571 above
PI	= 88/28	= 3.1428571428571428571 above

PI = 336/107 = 3.1401869158878504672 below

PI = 453/144 = 3.1458333333333333333 above

PI = 7918/2520 = 3.1420634920634920634 above

PI = 15834/5040 = 3.1416666666666666666 above

PI = 20612/6561 = 3.1415942691662856271 above

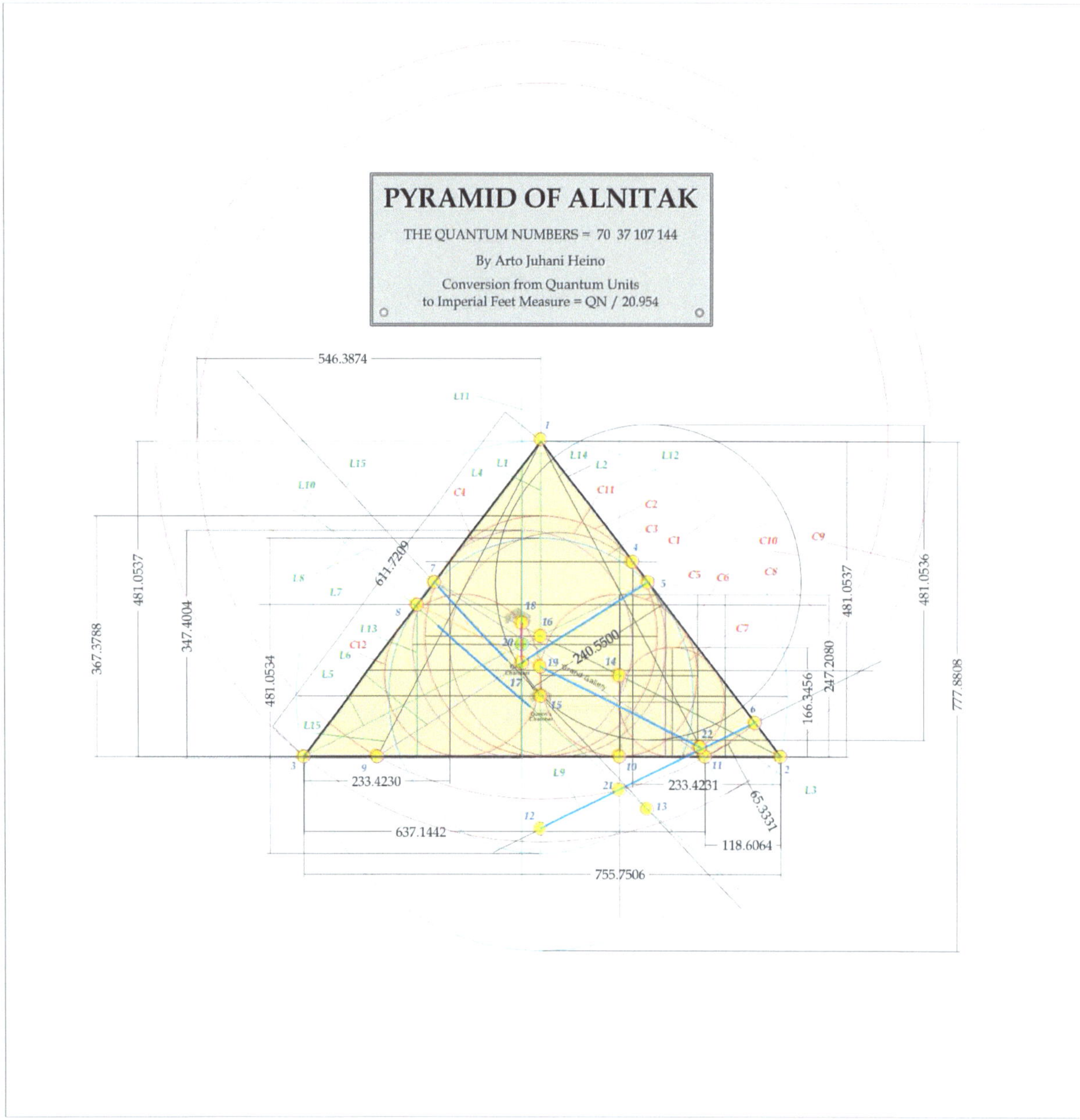

Diagram 2

As you can see these ratios converge to the value PI, again it is an irrational abstract. This cannot be used a a building block as the value is not a whole number, only the two convergent whole numbers are usable as true constructs. These lists are guide lines to investigators to look for whole number relations that can be used to create whole forms without PI/PHI, only this type of knowledge can help you realize the hidden potential of Quantum Arithmetic.

The diagram illustrates the Pyramids design is based only on the properties of the Quantum Triangle Unit and the centers that are generated from such a triangle and the circular distances(radii) from points. There is no esoteric magical number or Pi / PHI related ratio grid and the use of any design constructs are removed and only the pure triangle geometry is allowed. This constraint is to eliminate the modern propensity to try and own and control the geometric space by artificial means, this allows the clean use of the quantum numbers that you have begun the construction with and not to wander into transcendental number schemes, this would only cloud the underlying integer structure and clear definitions that are created by staying to the whole number constraint.

The numbers contain the constants of the geometry of the triangle and all the units derived from these base units for the construction of circles squares and many other forms.

b	e	d	a
70	37	107	144

B	E	D	A
4900	1369	11449	20736

C	F	G	L
7918	10080	12818	6651120

H	I	J	K
17998	2162	7490	15408

W	X	Y	Z
19367	3959	9287	16777

Some of these numbers you might recognize if you are a student of Geomatra and the hidden meaning behind numbers. I am sure the number 37, will ring a bell to those mathematicians who will also recognize 107 + 56 = 163, these numbers are called "Almost Integer", these types of numbers are found in the Quantum Pyramid in abundance. Some of these number are 25, 37, 43, 58, 67, 74, 148 etc. Why are these numbers important as a large class of these irrational numbers can be found using the theory of modular functions, a few examples are given by Ramanujan (1913-14). Such approximations were also studied by Hermite (1859), Kronecker (1863), and Smith (1965). They can be generated using some properties of the j-function. How this relates to to the Quantum Pyramid is fascinating as the mathematicians where searching for an underlying structure to prime numbers and number theory in general, these crossings can only guide our understanding.

There are a minimum of three basic triangles in the pyramid and a minimum of one square :

1	BH	Base width x Height
2	DH	Base Diagonal x Height
3	BS	Base Width x Face slope

Each Triangle has a minimum of three quantum parts

1	BHQ1	C+C,G,G
2	BHQ2	Q,G,D
3	BHQ3	G,D,M

Each Triangle Has a minimum of 5 Centers, depending on the type of triangle , some centers coincide.

1	CEcent	Centroid
2	CEinct	Incenter
3	CEcrct	Circumcenter
4	CEorth	Orthocenter
5	CE9pnt	9 Point Center

Each Triangle Center Has at least a Circle associated with it.

1	CRcent	Centroid**
2	CRinct	Incircle
3	CRcrct	Circumcircle
4	CRorth	Orthocenter**
5	CR9pnt	9 Point Circle

Each center and circle will have lines either connecting each other, tangent from centers to all circles, from each point in each triangle. from each point in each triangle and each circle center and from each point of each triangle to the tangent of each circle. Quite a lot of these values generated by the line point/tangent co-ordinates will have direct relationships to the quantum set.

BH

BHQ1

BHQ1CEcent

BHQ1CEinct

BHQ1CEcrct

BHQ1CEorth

BHQ1CE9pnt

BHQ2

BHQ2CEcent

BHQ2CEinct

BHG2CEcrct

BHQ2CEorth

BHQ2CE9pnt

BHQ3

BHQ3CEcent

BHQ3CEinct

BHQ3CEcrct

BHQ3CEorth

BHQ3CE9pnt

DH

DHQ1

DHQ1CEcent

DHQ1CEinct

DHQ1CEcrct

DHQ1CEorth

DHQ1CE9pnt

DHQ2

DHQ2CEcent

DHQ2CEinct

DHS2CEcrct

DHQ2CEorth

DHQ2CE9pnt

DHQ3

DHQ3CEcent

DHQ3CEinct

DHQ3CEcrct

DHQ3CEorth

DHQ3CE9pnt

BS

BSQ1

BSQ1CEcent

BSQ1CEinct

BSQ1CEcrct

BSQ1CEorth

BSQ1CE9pnt

BSQ2

BSQ2CEcent

BSQ2CEinct

BSQ2CEcrct

BSQ2CEorth

BSQ2CE9pnt

BSQ3

BSQ3CEcent

BSQ3CEinct

BSQ3CEcrct

BSQ3CEorth

BSQ3CE9pnt

The list above is only a short version of the centers and points that could be used to build these structures, these are only basic guides that will be in harmony with the quantum numbers that you started with. These guides are allocation design parameters and the purposes of the shafts and cavities that you build, each will reflect the meaning of the geometric center that is used and the numeric harmony it has with the quantum number set. As mankind has forgotten the inner meaning and names of these hidden geometries, we will give names to each as we find them, just as the ancients did.

This List is relative to the *Diagram 2* showing the pure geometric reasoning for placements of internal section of the design.

Circles

C1 Circumcircle

C2 Incircle

C3 Centroid Circle

C4 Nine Point Cicrle Ztr

C5 Incircle 2nd

C6 Incircle 3rd

C7 Tan L4/L2/ Diam H

C8 Apex circle

C9 Quantum Zline Circle

C10 Diam = H

C11 Incircle Ztr

C12 Incircle 2nd

Lines

L1 Central axis

L2 Triangle slope

L3 Triangle Base Line

L4 Triangle slope

L5 3 point Angle 1/2 division Main tr

L6 Angle 1/2 division Ztr

L7 Circumcircle center/L4/

L8	2 point Angle 1/2 division Main tr
L9	Point 21/Point 6/Point 11/Zline Circle/L1/L12
L10	Point 8/point 13
L11	Point 18/Point 17/ Point 20
L12	Point 10/ Incircle 2nd center/point 21
L13	Point 8/
L14	Quantum Zline
L15	Point 7/Point 13/Point 15/C7

Points

1	Cross L2/L4/L1 Apex
2	Cross L2/L3/L8 Main Tr corner
3	Cross L2/L3/L5 Main Tr corner
4	Tan C3/L2
5	Cross L5/L2 Circle center Diam=H
6	Cross L2/C9/L9
7	Cross L4/L8
8	Cross L4/L7/tan C7
9	Cross L3/L16/C9
10	Cross C5/tan L3/L12
11	Cross C9/L14/
12	Cross L1/L9
13	Cross L10/C7
14	Cross C5 center/L12/L8
15	Cross L1/Center C1
16	Cross L1/Center C2/C10/Centroid of Main Tr
17	Cross L6/L11
18	Edge Incircle C5/L11
19	Cross L1/L7
20	Center 9 Point circle C12/L11
21	Cross L12/L9/C9
22	Cross L14/C6/L7

The early Egyptian deities were all on a equal footing and each reflected a particular type of action and philosophy, this will have to be rediscovered again before we can realize the subtle reasoning that the ancients had with the structural form allocations. The clear truth starts to dawn that the Khemitians NEVER used the pyramids as a tomb for a Pharaoh, because at the time they were built there were no Pharaohs only an enlightened mankind that left a legacy to the future. The legacy of pure geometry and number that is immutable and is aligned with the actions and harmony of the universe.

The Khemitians left a structure that was filled by the early Egyptians and slowly converted in to the religious worship that was never part of the original designs of the Khemitians , they knew their legacy would be followed if they stuck strictly to geometric rules and numeric constraints, eventually the true knowledge would rise from the observations of the art and architecture, even if the the civilization that embraces them have lost the original meanings. An example is the sun sign on the heads of many deities, these are never shown as circular only as ellipses, and if observed without preconceived notions they would be understood as Quantum ellipses.

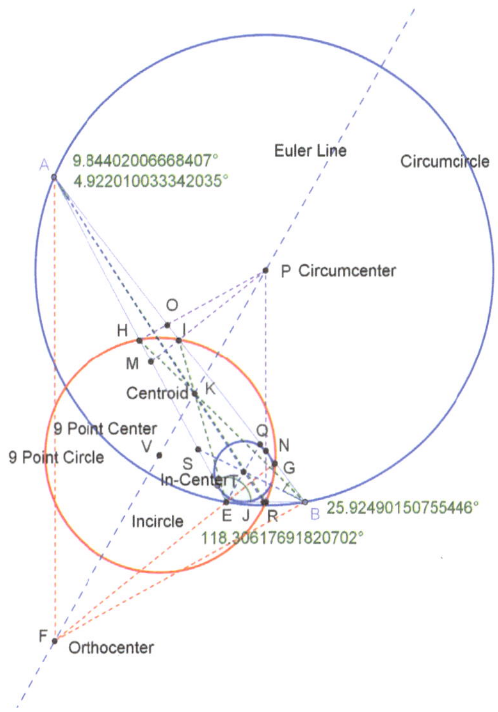

The diagram above is the BHQ3 = G,D,M

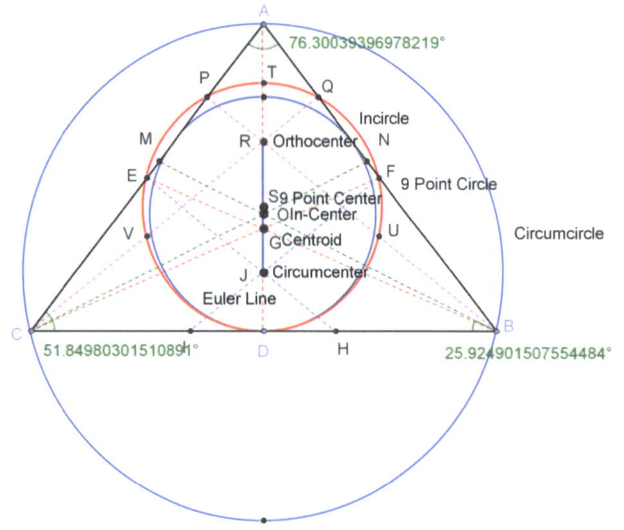

The diagram above is the BHQ1 = C+C,G,G

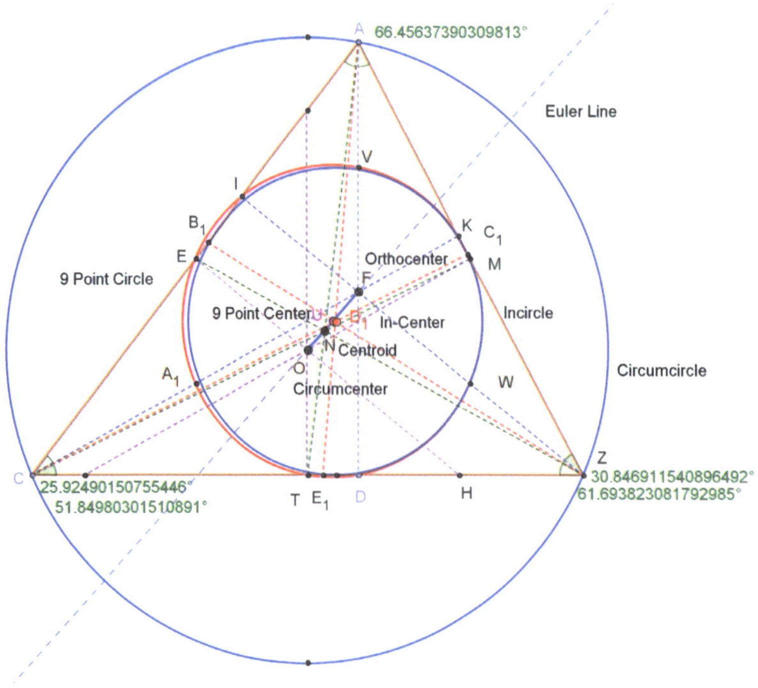

The Diagram above is the BHQ2 = Q,G,D

The Origins of Geomatra Numbers

As you are aware there has been a New Age numeric enlightenment towards ancient monuments and the associated location and the dimensions of them. These ideas are from numerous sources old and new and they place importance to different numbers and quantities, as you know if the values are PHI related then you can be sure they have a common background in Quantum Arithmetic. The list is very long and the variety of scared numbers is very long and grow longer with each generation, it looks like we should find the source to this attachment to these special numbers. Here is a short list and some of the background of its attachment.

7 = Days

8 = Eightfold Path

9 = Egyptian Gods

11 = Master numbers in modern occult numerology , it denotes hidden dangers and great trials and difficulties.

12 = 12 Links of Dependent Origination, 12 Powers of Buddha, 12 solar rays.

13 = Natures number, Mersenne prime

32 = Parts of the Body. 32 Physical Marks, 32 Characteristics Which a Great Man can be recognized found on Buddha. 32 parts to build a Vimana.

37 = Factors of Enlightenment, 37 things that are conducive to awakening. Mark the perfect man, and behold the upright: for the end of that man is peace. That's Psalms Chapter 37, Verse 37.

40 = days and nights of darkness,

52 = Mental Formations. These range from greed and hatred to discretion and balance of mind. Weeks in the year.

70 = spheric number, Pell number, heptagonal number, smallest weird number

107 = 28th prime number, Mersenne prime

108 = two angles of pentagon, beads in Buddhist rosary

144 = Light

153 = Bible, miraculous catch of fishes

216 = 6 x 6 x 6

288 = Double light, the Kingdom of Heaven.

352 = Mary

360 = Degrees of the circle

432 = Consecration (also the square root of the classical speed of light, 186624 miles per second). Luke 8:15, Revelation 2:17, etc.

396 = Classical earth radius (3960 miles).

576 = Prophecy and Gospel.

666 = Sign of the beast, Solar, Chinese procreative dragon,total of magic square 6

864 = Time , the source of light and life, Most Holy. Diameter of the sun=864000 miles

888 = Jesus

1080 = the Holy Ghost, Lunar, fountain of wisdom

1081 = the abyss

1152 = Witness (576 x 2). Biblical references include Luke 14:26, Revelation 3:12, 12:11 and 19:9.

1164 = Son of God

1176 = only begotten son

1285 = Babylon, Jerusalem

1296 = Circle of space (360 x 60 x 60). 1296 was also Plato's favorite number.

1440 = minutes in 24hours. Light

1548 = Priest of God.

1665 = spirit of the Earth

1728 = 12x12x12 , A-flat in music.

1746 = mustard seed, fusion, pearl of Mary,

2160 = Diameter of the moon: 2160 miles

2260 = the image of the beast, the man of sin, the son of man

2304 = False Christ's and False prophets (1 Cor. 14:22, Mark 13:12).

2385 = the son of perdition, power of Christ, baptism of John

3168 = Lord Jesus Christ.

3888 = New Jerusalem (Rev. 21:2, 1 Cor. 12:27, Luke 8:21, etc.).

5040 = 7! , Platos mystical number.

5184 = 72 x 72, Victory over the beast (Rev. 15:2).

25920 = The Great Year

There are many more than I could list, a common sequence is those numbers that divide by 3, and 7. The mental attachment to these and other numbers are a symptom of hundreds of years of subtle religious reinforcement, financial methodologies and forceful masonic application each using the influence as a form of magical control over the general public, if you are one of those who have this affliction then you are in good company.

The main source of the western Geomatra is the Hebrew alphabet and Greek alphabet, as the Hebrew has its origins in Egyptian hieroglyphs, I would say they are related by basic forms and meanings. Then as a consequence of the common heritage the lineage of Geomatra numbers go back to the Khemitians who wrote the first codes of understanding into the monuments and the writings on the early civilizations of that region. Other parts of the world used the same ideas and basic forms , not necessarily the exact same glyphs but similar enough to recognize the same origin, as the local culture was global NOT regional as the establishment want us to believe.

The first sign of enlightenment and awareness is to recognize these numbers, the second is to be see to who is wielding the influence, third is in the clear knowledge that these things can no longer control your actions. At the moment of this , we will have entered the new paradigm and have gone past the hierarchical control from the leaders who wield agendas of enslavement. This was the purpose behind the Egyptian Horus, the complete and enlightened human, a model for all to follow, NOT a God to worship.

Opening the door to this is accepting a new numeric organization, one that has not been taught since the Ancient Khemitians built their Pyramids. Only Pythagoras had tried to pass some of his methodologies and ideas that was taught to him from the Egyptians who kept the knowledge through the ages, only to be misunderstood by many. Archimedes and others who read the manuscripts from the Alexandrian library and passed some of his interpretations back to us.

At the entrance of the new awareness you will have to leave the coded tricks of invented mathematics and walk a narrow path that will lead you to answers that are simpler and full of meaning. It is only truly free men that are not trapped by a wall of numbers built by accountants, lawyers and politicians and the rest of the masonic orders that wield power and influence hidden to most of mankind. Are we ready for such a new journey or are you still comfortable in the entrapment of occultists numeric policies that run our system..

Remember all the valued listed are derived from the b and e roots by simple arithmetic. The first 4 terms are the Roots, the next 12 are derived from the the 4.

b = 70 = d – e

e = 37 = a – d

d = 107 = b + e

a = 144 = d + a

B = 4900 = b x b

E = 1369 = e x e

D = 11449 = d x d

A	= 20736	= a x a
C	= 7918	= 2 x d x e
F	= 10080	= d x d - e x e
G	= 12818	= d x d + e x e
L	= 6651120	= (d x d x d x e - e x e x e x d) / 6
H	= 17998	= (a x a + 2 x b x a - a x a) / 2
I	= 2162	= (e x e + 2 x d x e) - d x d
J	= 7490	= d x d - d x e
K	= 15408	= d x d + d x e
W	= 19367	= d x e + d x a
X	= 3959	= d x e
Y	= 9287	= 2 x d x e + d x d
Z	= 16777	= e x e + d x d + d x e

These two are non whole number terms that are used to help solve the Circular and Triangular Equations.

AP-Pi = 5.387334464 = (2 x d x e + d x d + e x e) / 2 x ((b x e x d) / a)
= Arto-Parker Pi Ratio for Quantum Triangles

ICR = 3849.027777 = 2 x ((b x e x d) / a)
= Incircle Radius
= 138565 / 36

Where 138565 has whole number correlations in the QA set of numbers

138656 = (b x X)/ 2
= e x 3745
= d x 1295

9.5 - Egyptian Star

Amongst the many interesting geometries that I have encountered are the unusual stars that adorn the ancient Egyptians walls. These have a specific dimension that give a distinctive look, it was not based on a symmetrical pentagon and not just freehand artistry. Here are the numeric parameters:

Angle	AN1	**73.73979529**	degrees	4.88203145	Ratio to 360
Angle	AN2	**53.13010235**	degrees	6.77581981	Ratio to 360
	K1	104		10816	Squared value
	L1	78		6084	Squared value
Triangle A1 D1 A1	TR1	8112	Area	416	Perimeter

Angle	AN3	**64.44003483**	degrees	5.58658916	Ratio to 360
Angle	AN4	**57.77998259**	degrees	6.23053147	Ratio to 360
	M1	111.04581223		12331.17241379	Squared value
	P1	72.42118189		5244.82758621	Squared value
	S1	132.57450735		17576	Squared value
Triangle S1 C1 A1	TR1	9601.20251089	Area	392.57450735	Perimeter

Angle	AN5	**78.69006753**	degrees	4.57491029	Ratio to 360
Angle	AN6	**50.65496624**	degrees	7.10690435	Ratio to 360
	N1	101.51294523		10304.87804878	Squared value
	Q1	85.27087399		7271.12195122	Squared value
Triangle S1 B1 A1	TR1	8656.09756098	Area	429.05573753	Perimeter

Angle	AN7	**11.42118627**	degrees	31.52036849	Ratio to 360
Angle	AN8	**84.28940686**	degrees	4.27099933	Ratio to 360
	W1	26		676	Squared value
	T1	132.57450735		17576	Squared value
Triangle T1 W1 T1	TR1	3446.93719119	Area	291.14901471	Perimeter

Angle Ratio	AN1/AN3	**1.14431650**	1 1/7	Fraction approx
Angle Ratio	AN5/AN1	**1.06713162**	1 1/9	Fraction approx
Angle Ratio	AN5/AN3	**1.22113633**	1 2/9	Fraction approx

Line Lengths	A1	130	16900	Squared value
	B1	166.48123017	27716	
	C1	130	16900	
	D1	156	24336	
	E1	114	12996	
	F1	146	21316	
	G1	52	2704	
	H1	160.68883707	25821	
	J1	130	16900	
	K1	104	10816	
	L1	78	6084	
	M1	111.04581223	12331.17241379	
	N1	101.51294523	10304.87804878	
	P1	72.42118189	5244.82758621	
	Q1	85.27087399	7271.12195122	
	S1	132.57450735	17576	
	T1	132.57450735	17576	
	V1	26	676	
	W1	26	676	
	X1	286	81796	
	Y1	234	54756	
	Y2	260	67600	
	Y3	286	81796	

Perimeter Lengths	BP	803.44418537	Base Pentagon
	P	768.99103032	Star Pentagon
	BS	1144	Base Square Y3xX1
	S	1040	Square Y2xX2
	R	1092	Star Rectangle Y1xX1
	Circle Main	816.81408993	Large Circle A1=Radius
	Inner Circle	81.68140899	Small Circle W1 = Diam

Area	BP	44424.27897407	Base Pentagon
	P	40560	Star Pentagon
	BS	81796	Base Square Y3xX1
	S	67600	Square Y2xX2
	R	74360	Star Rectangle Y1xX1
	Circle Main	53092.91584567	Large Circle A1=Radius
	Inner Circle	530.92915846	Small Circle W1 = Diam

The Egyptian Star Form

A practical five sided star geometry

Roof inside the Unas Pyramid
Wall of Seti Tomb

By Arto Heino

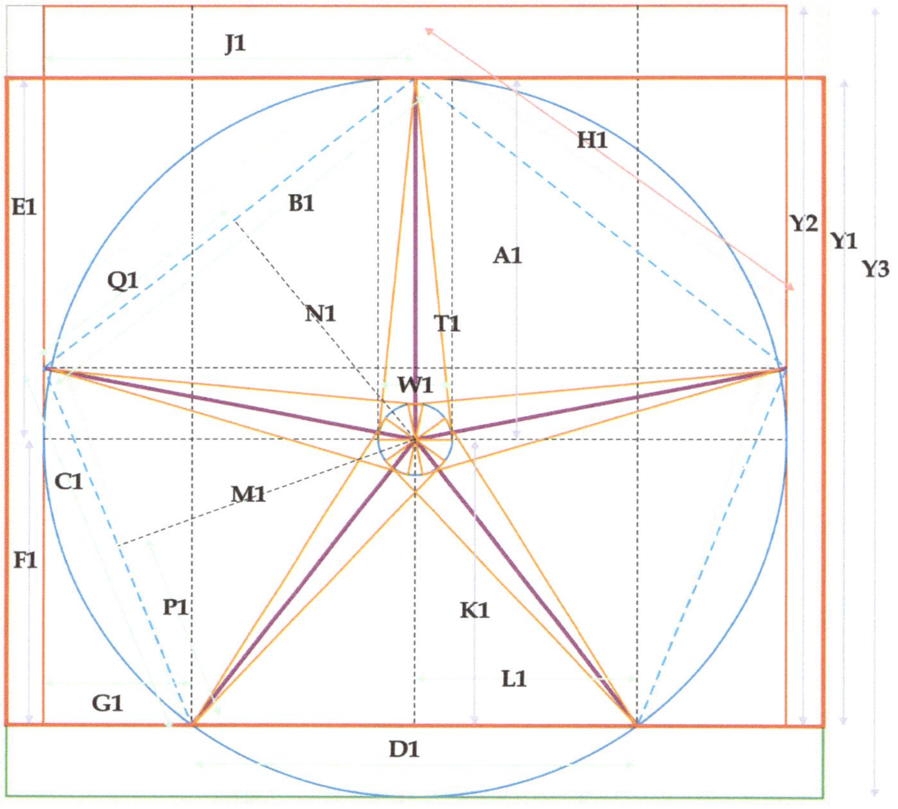

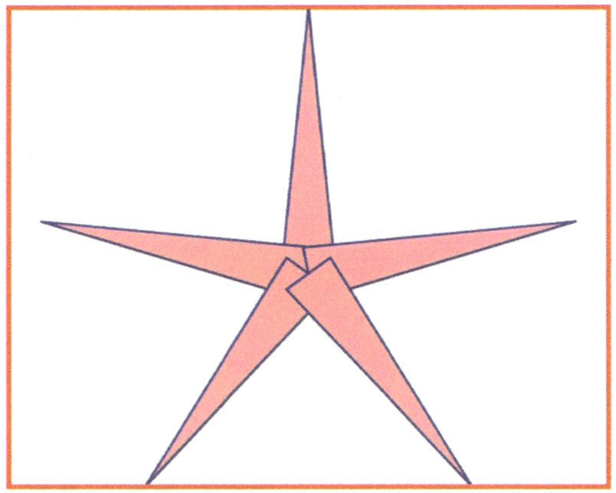

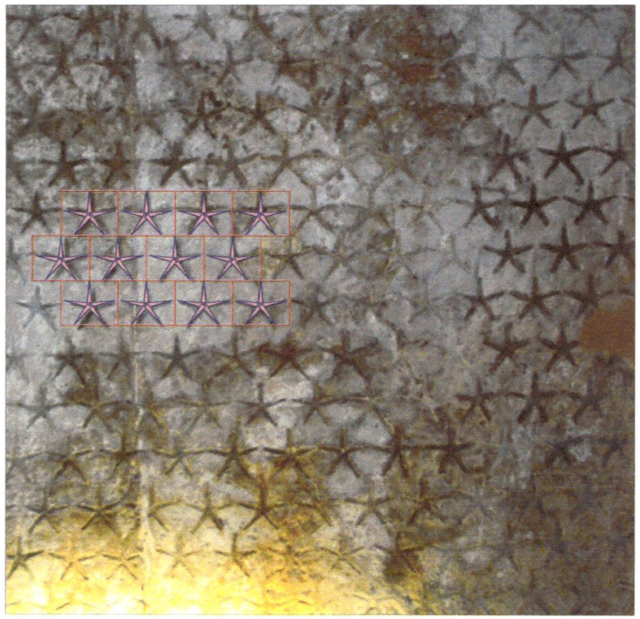

These stars have a uniform appearance and would be based on a simple open rectangular box that would have some sort of stencil attached to its base. The utility of a whole number form is internal to its design. These would be simple finger lengths based on 24, 28 lengths from cubits. These added become 52, which is the basis for my geometrical designs, where 52= 13x4 .

234	= 18 x 13	= 2 x 9 x 13	= 2 x 3 x 3 x 13
260	= 20 x 13	= 2 x 10 x 13	= 2 x 2 x 5 x 13
286	= 22 x 13	= 2 x 11 x 13	

With many values presented here, the next logical step is to find the Quantum set of 4 numbers. As this was an exercise in finding a Egyptian foundational geometry, I have not completed the whole system of numbers required to arrive at a Quantum set, here are some tentative sets.

b	e	d	a
52	78	130	208

b	e	d	a
78	104	182	286

Chapter Ten

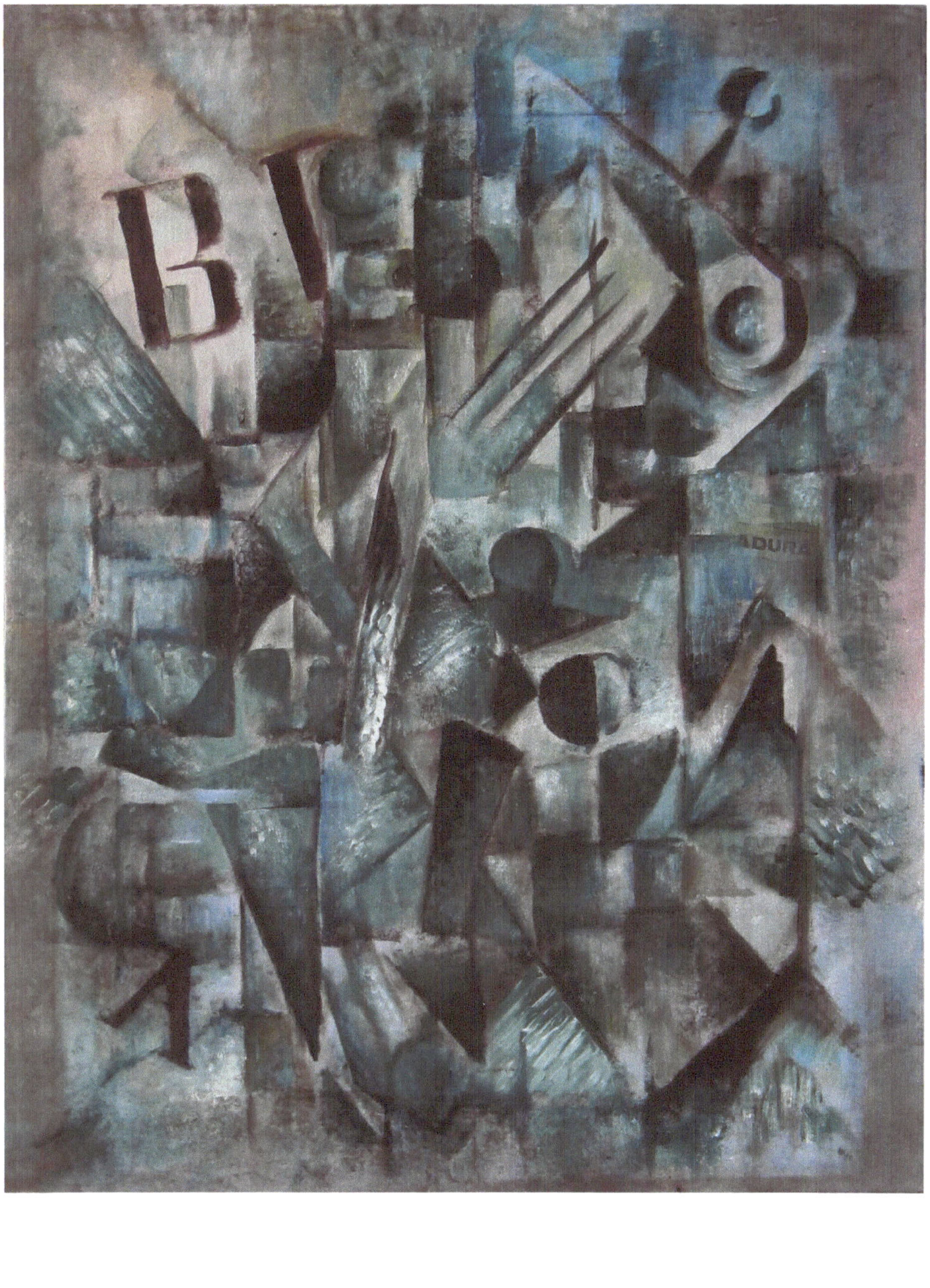

Magic Squares

10.1 - The Energy Square of Three

The Magic Square of Three has many interpretations, the first being a number pattern that has each row and column and diagonals, equal the same number. The special geometry that is internal when you start to follow the numbers in order is the Magic component, something nature has given us to discover.

As with the Ether these squares displays inherent balance, having other properties that belay its simple arrangement, this indicates a chaotic core. There are **9** numbers **8** of which surround a central **5** (Phi, Fire, Pyre) the outer pair to **10**(decan). The Magic Square is telling you that it is **5** Lengths average on either side. Note the outer ring 618,34, and 729 = 3^6 .

If Time = 1/ Possibilities x Types

Therefore Frequency = Possibilities x Types

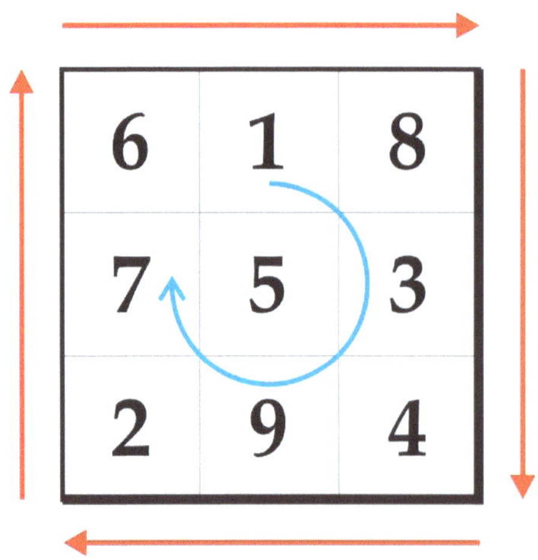

When circulating around the central 5 axle, all the line lengths would be the same ***WAVELENGTH*** giving you **8** Magic Square Units, refer De Broglie Equation

1	**2**	**3**	**4**	**5**	**6**	**7**	**8**	**9**
9	**8**	**7**	**6**	**5**	**4**	**3**	**2**	**1**
10	**10**	**10**	**10**	**10**	**10**	**10**	**10**	**10**

Start this line pattern by traveling from 1 to 9 and closing by joining Alpha and Omega.(1 and 9).This pattern is the most familiar and well used Ancient design.

This type of pattern doesn't give us reliable results, this is due to fact that other arrangements of numbers will give the same result, here are some other possibilities;

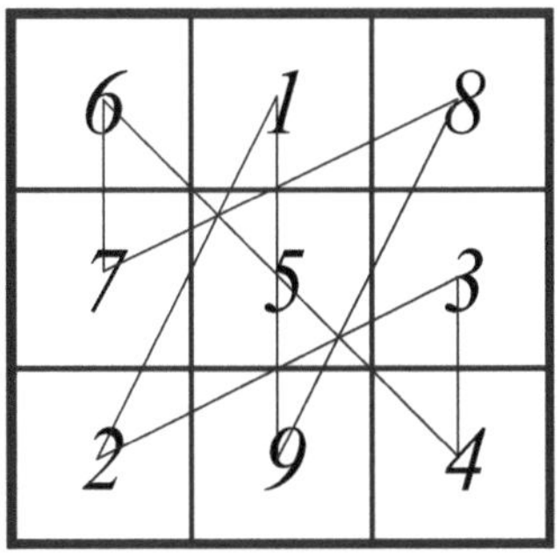

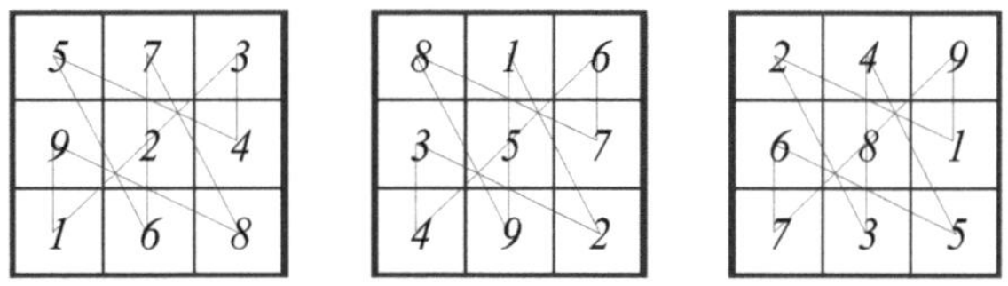

One Diagonal = 6 All = 15 One Diagonal = 24

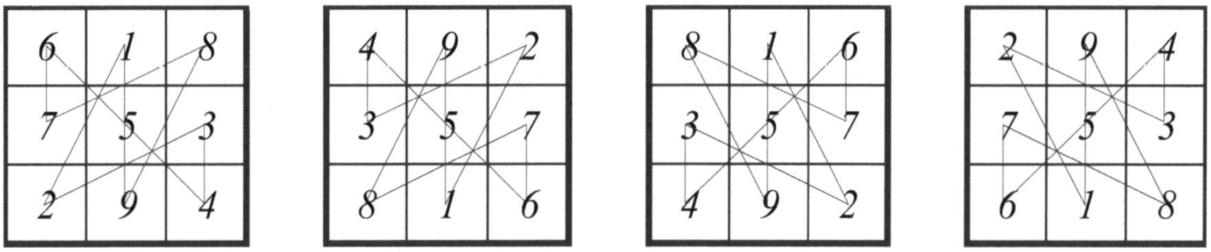

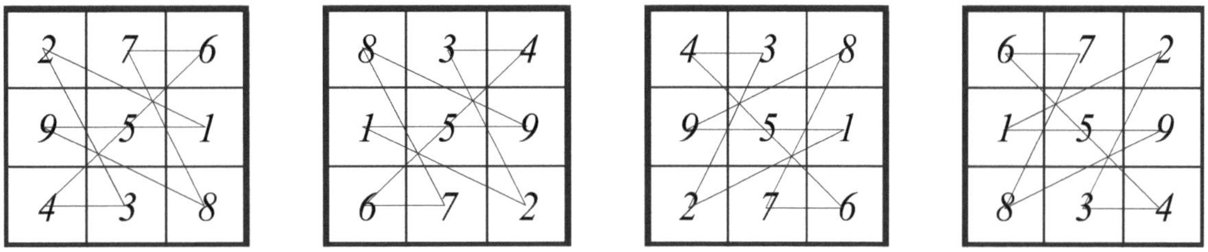

These diagrams illustrate the patterns of all **8** of the reflections and rotations possible. The **8** different ways to represent ***1*** type of arrangement for a magic square of three is called its basic ***FREQUENCY.*** Because a Sqr 3 has only **1** type and a Sqr 4 has **880** type, so they will have different frequencies. To understand their frequency, Sqr 3 has 1 type and 8 permutations = 8 where Sqr 4 has 880 types and 8 permutations = 7040.

The size of the beginning number in the square must be considered as the ***LENGTH*** because it now becomes a multiple from the base square numbers e.g., Sqr 3 has a start at 1 making a total of 45, 2 = 54, 3 = 63, 4 = 72 .So mass increases when a Sqr 3 get larger and its increase is by 9 to each higher level. These abstracts must be seen before understanding the relationship of energy to a mass is made clear.

The numbers related to size is the Magic Line Number Values and its Total as in this case the line value is **15** and the total is **45** from a Square 3 which has **9** Main Cells.. The total of 45 spread over 9 cells is 45/9 = 5.

				Diff Start Value
LENGTH	= Start	= 1	= 1	= 2
AREA	= Height * Width	= 3 * 3	= 9	= 9
VOLUME	= Area * Length	= 9 * 1	= 9	= 18
DENSITY	= Mass / Volume	= 45 / 9	= 5	= 6
MASS	= Density * Volume	= 5 * 9	= 45	= 54

We have now successfully translated a Magic Square into a solid mass.

MASS-ENERGY SQUARES

1	*4*	*7*
2	*5*	*8*
3	*6*	*9*

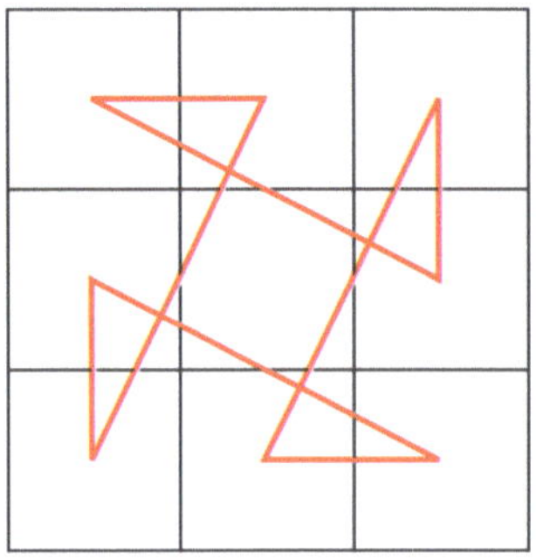

A1

1	4	7
2	5	8
3	6	9

B1

8	1	6
3	5	7
4	9	2

A2

1	4	7
2	5	8
3	6	9

B2

8	1	6
3	5	7
4	9	2

The Sequence is A1,B1,A2,B2....etc

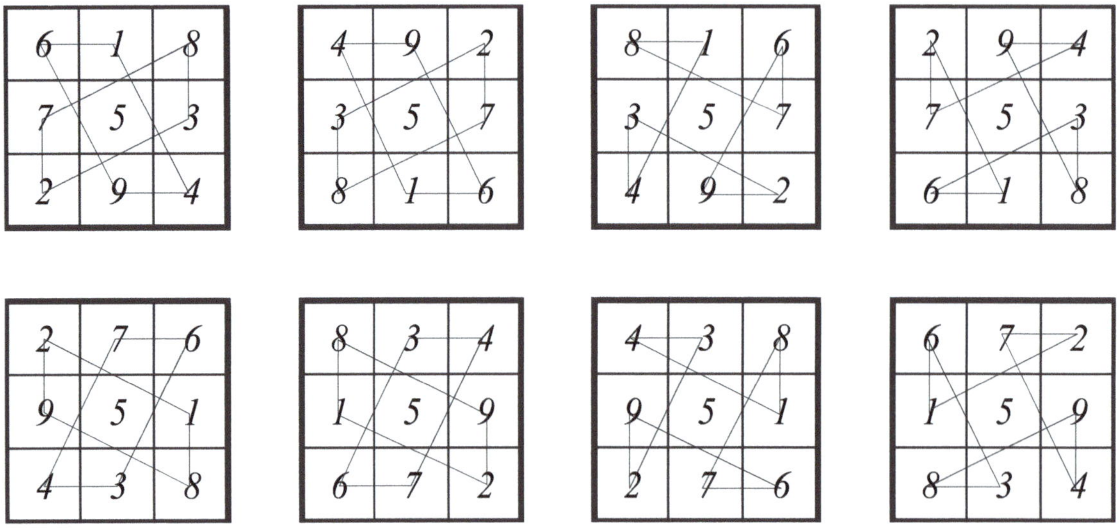

These lines show the switching between a **GRAVITY** mass stacked square and an **ANTI GRAVITY** energy square arrangement This arrangement is the only set of patterns to use because no other numeric arrangements can give this pattern unless all of its optional line numbers equal each other.

$$E = m\,v^2$$

m = *MASS = DENSITY X VOLUME*

v = *VELOCITY = FREQUENCY X WAVELENGTH*

E = *ENERGY = MASS X (FREQUENCY X WAVELENGTH)*^{2}

Energy Square = Mass Square x (Frequency of Square x Wavelength of Line)^2

$$\mathbf{184320 = 45 \times (8 \times 8)^2}$$

$$4096 = 8^4$$

So there are 4096 **M**agic **S**qr **E**nergy **U**nits in each of the 45 Cell Mass Number Groups lets call them MSEU, so the whole 45 separate components each has the same value in size

1 0
2 00
3 000
4 0000
5 00000
6 000000
7 0000000
8 00000000
9 000000000

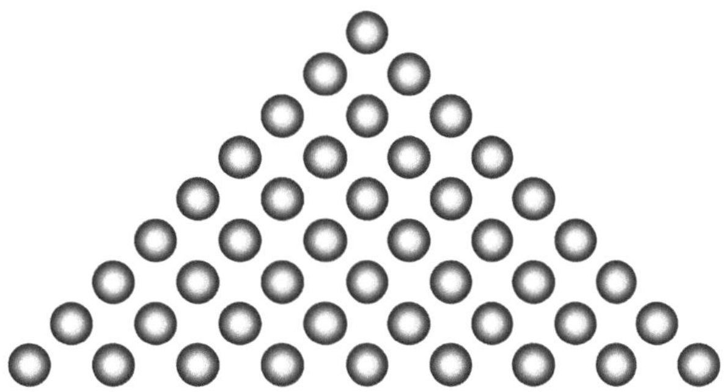

The MSEU groups become redistributed when in a Magic condition

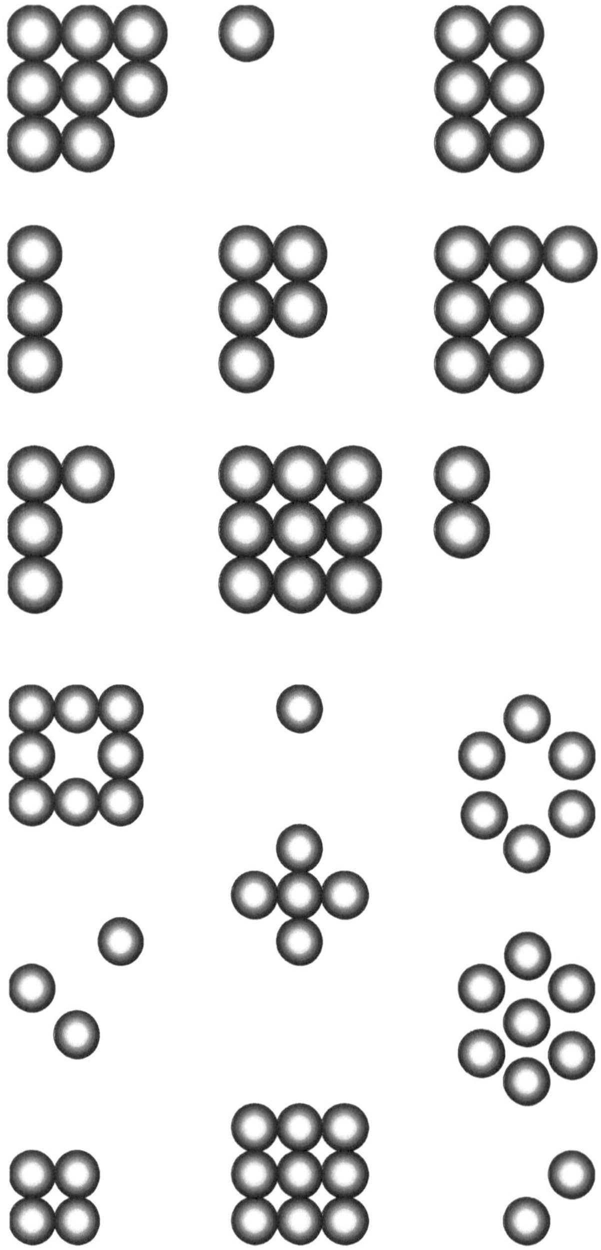

These are the some Geometric and mass packing grouping of these numbers, others can be deduced.

The Energy per Group /32

1	=	4096	128
2	=	8192	256
3	=	12288	384
4	=	16384	512
5	=	20480	640
6	=	24576	768
7	=	28672	896
8	=	32768	1024
9	=	36864	1152

$5 * 20480 = 102400 = 320^2$

The other component not mentioned in the equations is the

GRAVITY of a Magic Square.

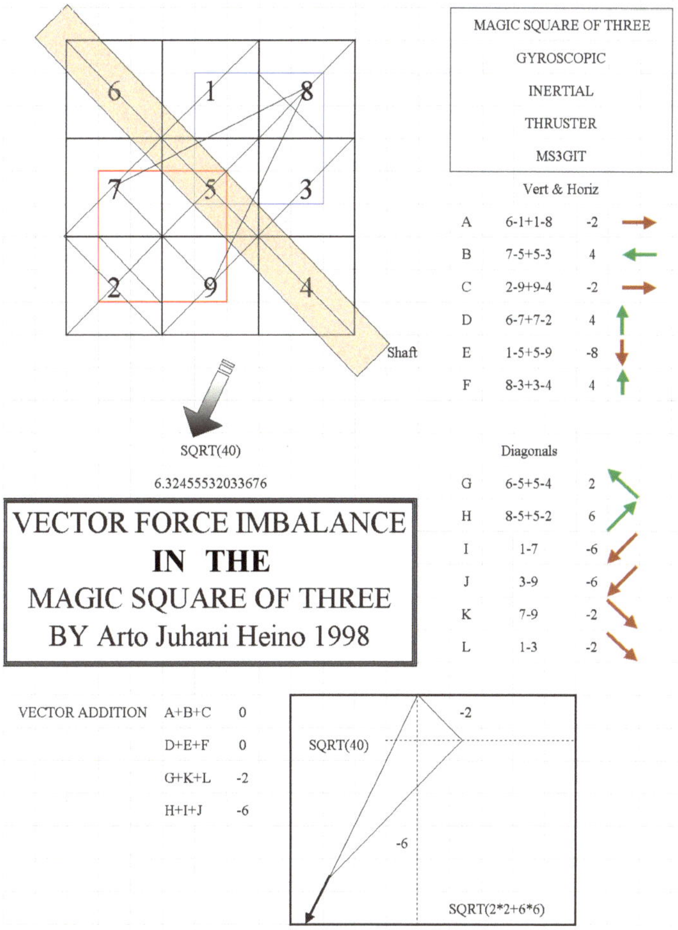

Each rotation will reorient the gravity vector of the square in different directions and each different type will also have a different vector sum.

PAIRING GROUPS

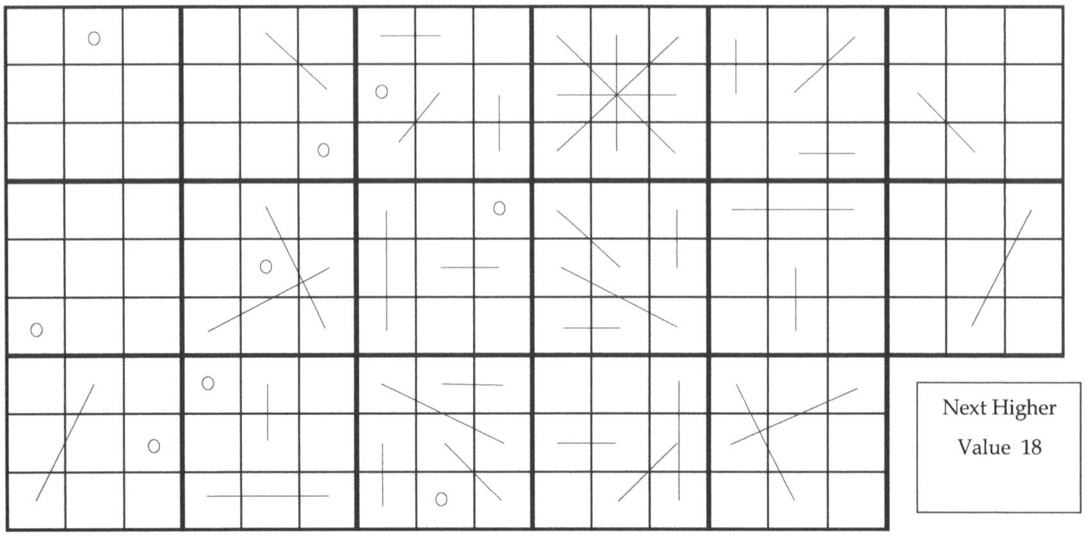

1 4 7 10 13 16

2 5 8 11 14 17

3 6 9 12 15 18

Breaking down the square in to these types helps you understand that these numeric arrangements are not arbitrarily placed. Note the number 10 displaying symmetry evenly placed around the center. Number 5 and 15 have mirrored placement. The number sets 3,17 and 4,16 and 7,13 and 9,11 another 6,8,14 are also mirrored or rotated. Observe how nature has some chaos in the groups.

	Sqrt(5)	Sqrt(2)	L	WL	E	E/45	F	F/45
1	0	0	0	0	0	0	0	0
2	0	0	1	1	2880	64	18214.71932	404.7715405
3	0	1	0	1.414213562	5760	128	36429.43865	809.543081
4	0	0	2	2	11520	256	72858.87729	1619.086162
5	1	0	0	2.236067977	14400	320	91073.59661	2023.857703
6	0	1	1	2.414213562	16785.87012	373.019336	106163.1642	2359.181426
7	0	2	0	2.828427125	23040	512	145717.7546	3238.172324
8	0	0	3	3	25920	576	163932.4739	3642.943865
9	1	0	1	3.236067977	30159.75155	670.2167011	190747.0171	4238.822603
10	0	1	2	3.414213562	33571.74024	746.038672	212326.3283	4718.362852
11	1	1	0	3.65028154	38374.71932	852.7715405	242703.0353	5393.400784
12	0	2	1	3.828427125	42211.74024	938.038672	266970.4863	5932.677473
13	1	0	2	4.236067977	51679.5031	1148.433402	326849.8763	7263.330584
14	0	3	0	4.242640687	51840	1152	327864.9478	7285.887729
15	0	1	3	4.414213562	56117.61036	1247.058008	354918.9312	7887.087359
16	2	0	0	4.472135955	57600	1280	364294.3865	8095.43081
17	1	1	1	4.65028154	62280.34099	1384.007578	393895.462	8753.232488
18	0	2	2	4.828427125	67143.48048	1492.077344	424652.6567	9436.725704
19	1	2	0	5.064495102	73869.43865	1641.543081	467191.3512	10382.03003
20	1	0	3	5.236067977	78959.25465	1754.650103	499382.1741	11097.38165
21	0	3	1	5.242640687	79157.61036	1759.058008	500636.6857	11125.25968
22	2	0	1	5.472135955	86239.5031	1916.433402	545426.5082	12120.58907
23	1	1	2	5.65028154	91945.96266	2043.243615	581517.3273	12922.60727
24	0	2	3	5.828427125	97835.22072	2174.116016	618764.2657	13750.31702
25	2	1	0	5.886349517	99789.43865	2217.543081	631123.8251	14024.97389
26	1	2	1	6.064495102	105920.9304	2353.798454	669902.7841	14886.72854
27	0	3	2	6.242640687	112235.2207	2494.116016	709837.8623	15774.17472
28	2	0	2	6.472135955	120639.0062	2680.866804	762988.0685	16955.29041
29	1	3	0	6.478708665	120884.158	2686.314622	764538.5444	16989.74543
30	1	1	3	6.65028154	127371.5843	2830.479652	805568.6313	17901.52514
31	3	0	0	6.708203932	129600	2880	819662.3695	18214.71932
32	2	1	1	6.886349517	136574.8119	3034.995819	863774.953	19194.99896
33	1	2	2	7.064495102	143732.4222	3194.053827	909043.6557	20200.97013
34	2	2	0	7.30056308	153498.8773	3411.086162	970812.141	21573.60313
35	2	0	3	7.472135955	160798.5093	3573.300207	1016979.068	22599.53483
36	1	3	1	7.478708665	161081.5199	3579.589331	1018768.984	22639.31075
37	1	3	1	7.478708665	161081.5199	3579.589331	1018768.984	22639.31075
38	3	0	1	7.708203932	171119.2547	3802.650103	1082253.192	24050.07094
39	2	1	2	7.886349517	179120.1851	3980.448557	1132855.52	25174.5671
40	1	2	3	8.064495102	187303.914	4162.3092	1184613.966	26324.7548
41	3	1	0	8.122417495	190004.158	4222.314622	1201691.808	26704.2624
42	2	2	1	8.30056308	198430.1206	4409.558236	1254982.275	27888.495
43	1	3	2	8.478708665	207038.8818	4600.86404	1309428.861	29098.41914
44	3	0	2	8.708203932	218398.5093	4853.300207	1381273.454	30694.96564
45	2	3	0	8.714776642	218728.3159	4860.629243	1383359.334	30741.31854
46	2	1	3	8.886349517	227425.5583	5053.901296	1438365.525	31963.67833
47	3	1	1	9.122417495	239669.2827	5325.984061	1515801.637	33684.48083
48	2	2	2	9.30056308	249121.364	5536.03031	1575581.848	35012.92995
49	1	3	3	9.478708665	258756.2437	5750.138749	1636518.178	36367.07062
50	3	2	0	9.536631057	261928.3159	5820.629243	1656580.124	36812.89165
51	2	3	1	9.714776642	271805.4294	6040.120653	1719048.475	38201.07721
52	3	1	2	10.12241749	295094.4075	6557.6535	1866340.905	41474.24233
53	2	2	3	10.30056308	305572.6073	6790.502385	1932610.859	42946.90798
54	3	2	1	10.53663106	319739.3108	7105.318018	2022208.959	44937.97688
55	2	3	2	10.71477664	330642.5429	7347.612063	2091167.054	46470.37897
56	3	3	0	10.95084462	345372.4739	7674.943865	2184327.317	48540.60705
57	3	1	3	11.12241749	356279.5323	7917.32294	2253309.611	50073.54692
58	3	2	2	11.53663106	383310.3057	8518.006794	2424267.233	53872.60519
59	3	2	2	11.53663106	383310.3057	8518.006794	2424267.233	53872.60519
60	2	3	3	11.71477664	395239.6563	8783.103474	2499715.071	55549.2238
61	3	3	1	11.95084462	411329.3389	9140.651976	2601475.159	57810.55909
62	3	2	3	12.53663106	452641.3006	10058.69557	2862754.946	63616.77658
63	3	3	2	12.95084462	483046.2039	10734.36009	3055052.439	67890.0542
64	3	3	3	13.95084462	560523.0689	12456.0682	3545059.158	78779.09239

Groups			1 Unit	Sqrt(2)	Sqrt(5)	Energy
3	17		0	0	1	14400
4	16		0	1	0	5760
5	15		0	0	2	57600
6	14	8	3	0	0	25920
7	13		2	1	0	33571.74024
8	6	14	3	0	0	25920
9	11		2	1	1	91945.96266
10			4	4	0	57600
11	9		2	1	1	91945.96266
12			3	1	0	56117.61036
13	7		2	1	0	33571.74024
14	8	6	3	0	0	25920
15	5		0	0	2	57600
16	4		3	0	0	25920
17	3		2	1	0	14400

These groups also make clear that this geometry begins at 3. Recapping this table with the numbers are 3 to 17 inclusive which are 15 in all, the same as the line number value. The next table gives you the Energy values of 64 combination of 0,1,2,3 line lengths. The table also gives you the Gravity Force values.

Polynomial expansion

The sum of two numbers is called a binomial. Examples

$a + b$ $\qquad$ $x + 5$ $\qquad$ $4 + 6$

When a binomial is raised to a power it becomes a Polynomial. Examples

$(a + b)^2$ $\qquad$ $(x + 5)^3$ $\qquad$ $(4 + 6)^4$

Looking at this, the numbers expand internally in the brackets

		Coefficients	
$(a+b)^2$	$= (a+b) * (a+b)$		
$(1+1)^1$	$= (1+1)$	$= 1$	$= 1$
$(1+1)^2$	$= (1+1) * (1+1)$	$= 2 * 2$	$= 1 + 2 + 1$
$(1+1)^3$	$= (1+1) * (1+1) * (1+1)$	$= 2 * 2 * 2$	$= 1 + 3 + 3 + 1$
$(1+1)^4$	$= (1+1) * (1+1) * (1+1) * (1+1)$	$= 2 * 2 * 2 * 2$	$= 1 + 4 + 6 + 4 + 1$
$(1+1+1)^2$	$= (1+1+1) * (1+1+1)$	$= 3 * 3$	$= 1 + 2 + 2 + 1$
			+ 2 +
			1
			= 1+2+2+1+2+1

A B C D E F

1 + **1** x 1 + 1 x 1 + 1

ACE	1 * 1 * 1	= 1
ADE	1 * 1 * 1	= 1
ACF	1 * 1 * 1	= 1
ADF	1 * 1 * 1	= 1
BDF	1 * 1 * 1	= 1
BCF	1 * 1 * 1	= 1
BDE	1 * 1 * 1	= 1
BCE	1 * 1 * 1	= 1
		= 8

1 + 2 x 1 + 2 x 1 + 2

By grouping in this manner you can easily see how the Pascals triangle could be built.

ACE	1 * 1 * 1	= 1
ADE	1 * 2 * 1	= 2
ACF	1 * 1 * 2	= 2
ADF	1 * 2 * 2	= 4
BDF	2 * 2 * 2	= 8
BCF	2 * 1 * 2	= 4
BDE	2 * 2 * 1	= 4
BCE	2 * 1 * 1	= 2
		= **27**

1 222 444 8

1 3 3 1

**Note these groupings as they appear everywhere*

Refer to Pascals Triangle below for the Expansion Numbers

Formula	*n*	*Number of terms*	*Coefficient*	*Sum*									
$(a+b)^n$	0	2	1	1	1								
	1	2	2	2	1	1							
	2	2	4	4	1	2	1						
	3	2	8	8	1	3	3	1					
	4	2	16	16	1	4	6	4	1				
	5	2	32	32	1	5	10	10	5	1			
	6	2	64	64	1	6	15	20	15	6	1		
	7	2	128	128	1	7	21	35	35	21	7	1	
	8	2	256	256	1	8	28	56	70	56	28	8	1

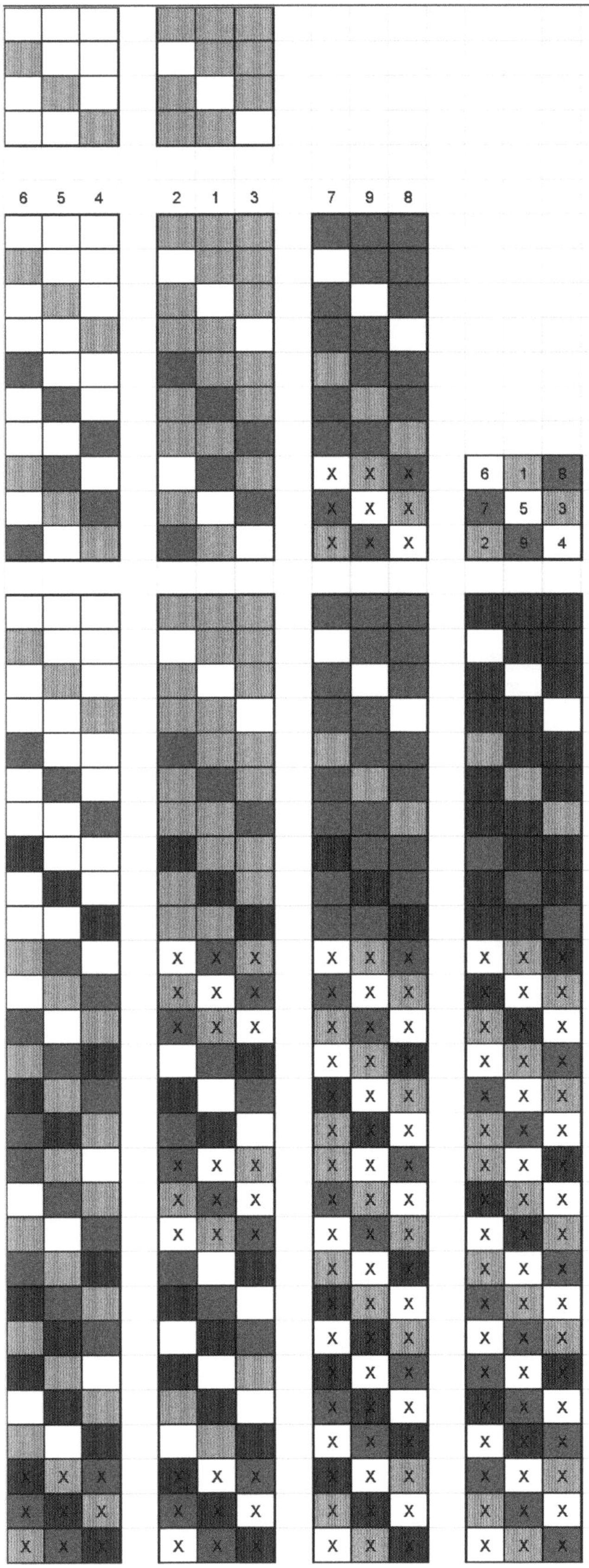

Binomial 8 $(a+b)^3$

Trinomial 27 $(a+b+c)^3$

*Note that the redundant "**X**" set has created a Magic Square of Three.*

Quadnomial 64 $(a+b+c+d)^3$

*Note that there are 16 redundant "**X**" sets of Square Three.*

Energy of a Magic Square Four (*Fr Index 176)*

Pattern of 1 to 16 Line.

LENGTH	= 1	=1
AREA	= 4 * 4	=16
VOLUME	= 1 * 16	=16
MASS	= !16	=34 * 4=136
DENSITY	= 136/ 16	= 8
GRAVITY	= Sqrt($15^2 + 9^2$)	= Sqrt(306)
	= 17.49285568454	
	= Sqrt((Proton Mass / Electron Mass)/ 6)	

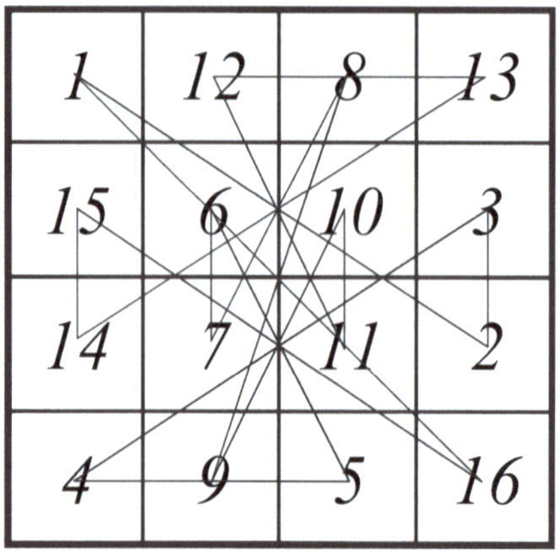

10.2 - The Circular Geometric of Magic squares

This simple arrangement has numerous applications in Art, Engineering and Science. The exploration of these forms are open to the uses of circular rotational machinery or even molecular ideas can be expressed using a similar type of arrangement.

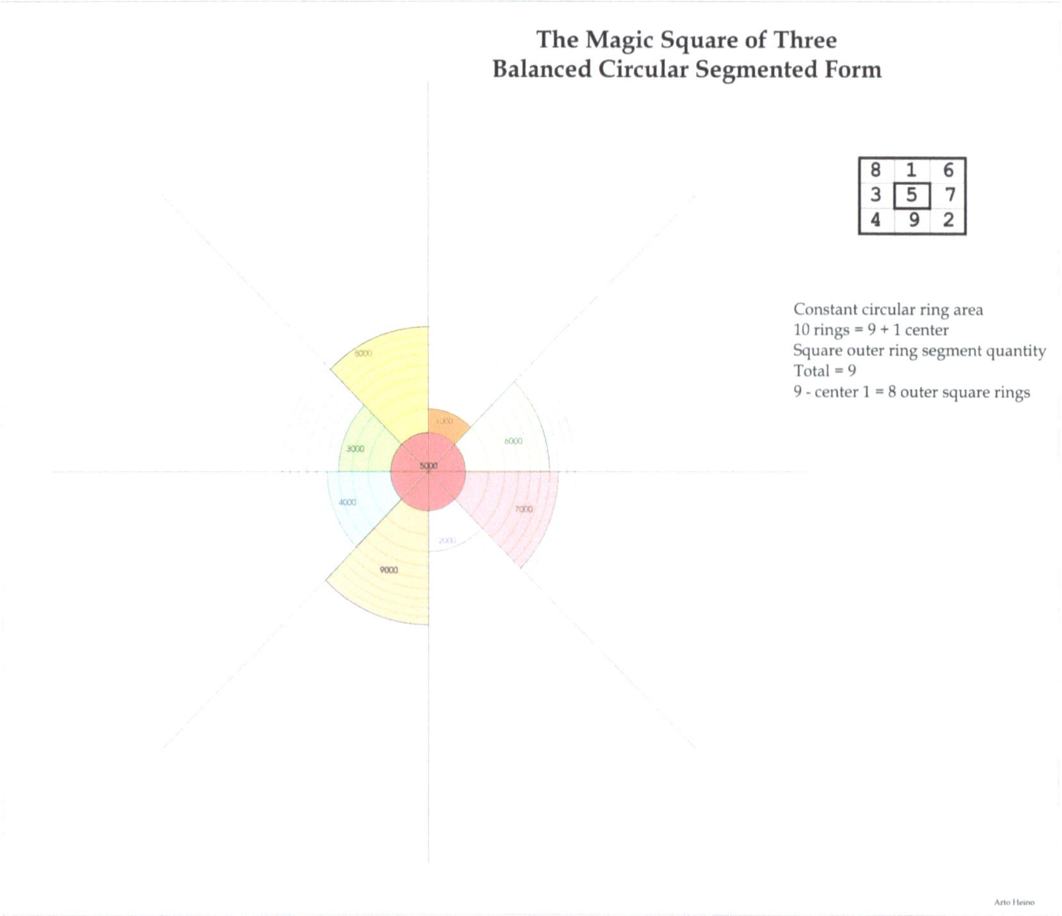

The Magic Square of Four
Balanced Circular Segmented Form

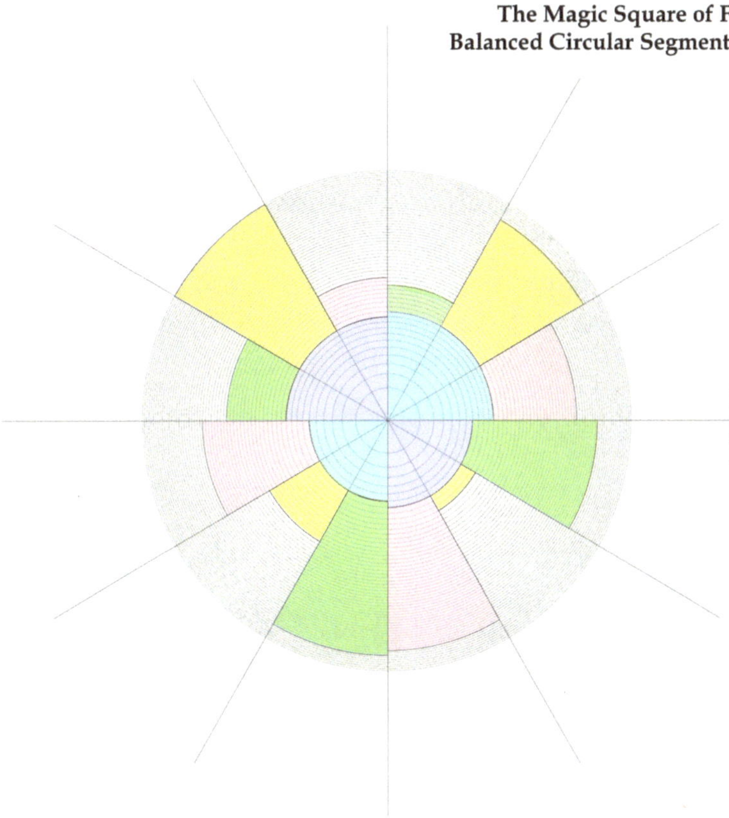

16	3	2	13
5	10	11	8
9	6	7	12
4	15	14	1

48	9	6	39
15	30	33	24
27	18	21	36
12	45	42	3

Constant circular ring area
78 rings = 3 x 26 = (16+10)x 3
Square outer ring segment quantity
Total = 16
16 - center 4 = 12 outer square rings

Arto Heino

The Magic Square of Five
Balanced Circular Segmented Form

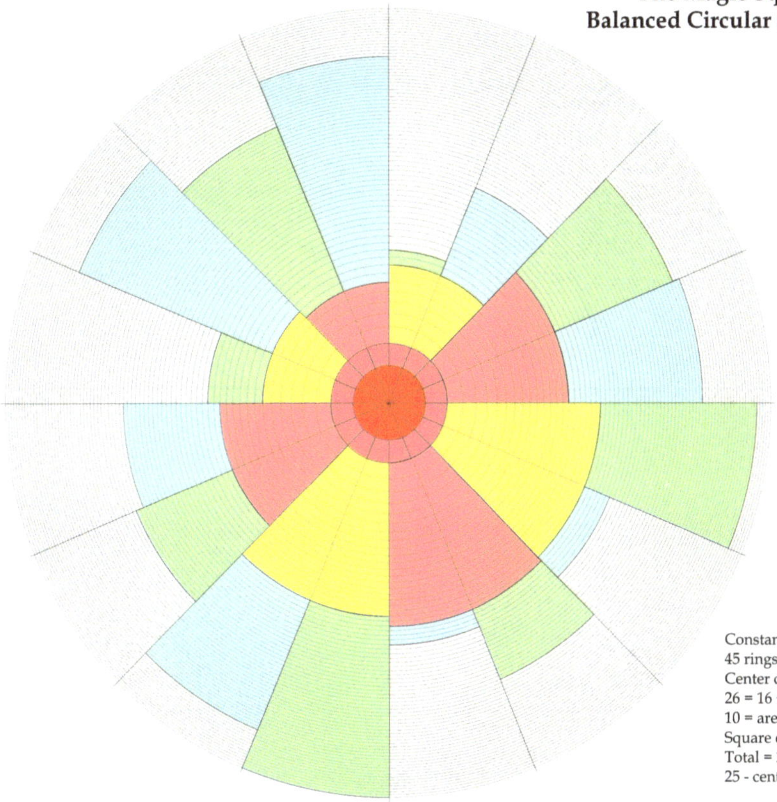

17	24	1	8	15
23	5	7	14	16
4	6	13	20	22
10	12	19	21	3
11	18	25	2	9

34	48	2	16	30
46	10	14	28	32
8	12	26	40	44
20	24	38	42	6
22	36	50	4	18

Constant circular ring area
45 rings = 1 + 19 + 25
Center circular ring 2 parts
26 = 16 + 10
10 = area/16 = segment areas
Square outer ring segment quantity
Total = 25
25 - center 9 = 16 outer square rings

Arto Heino

10.3 – Nested Square 15

This is one of my early(1996) investigations into different interpretations of the magic square sequences. This has led me to redefine particular geometries that could be understood in terms of the changes in dynamics that can create rotational causality, such as imbalanced wheels etc. The placement of the sequential string of numbers towards the center, offsetting one number after two turns and on the third creating a completed section before returning back to the first compliments a vortex action that can been seen not just imagined, actually 4 vortices. I am including this here so to further the understanding of a new way of to interpret the magic squares, numbers patterns that has always existed and never created by man.

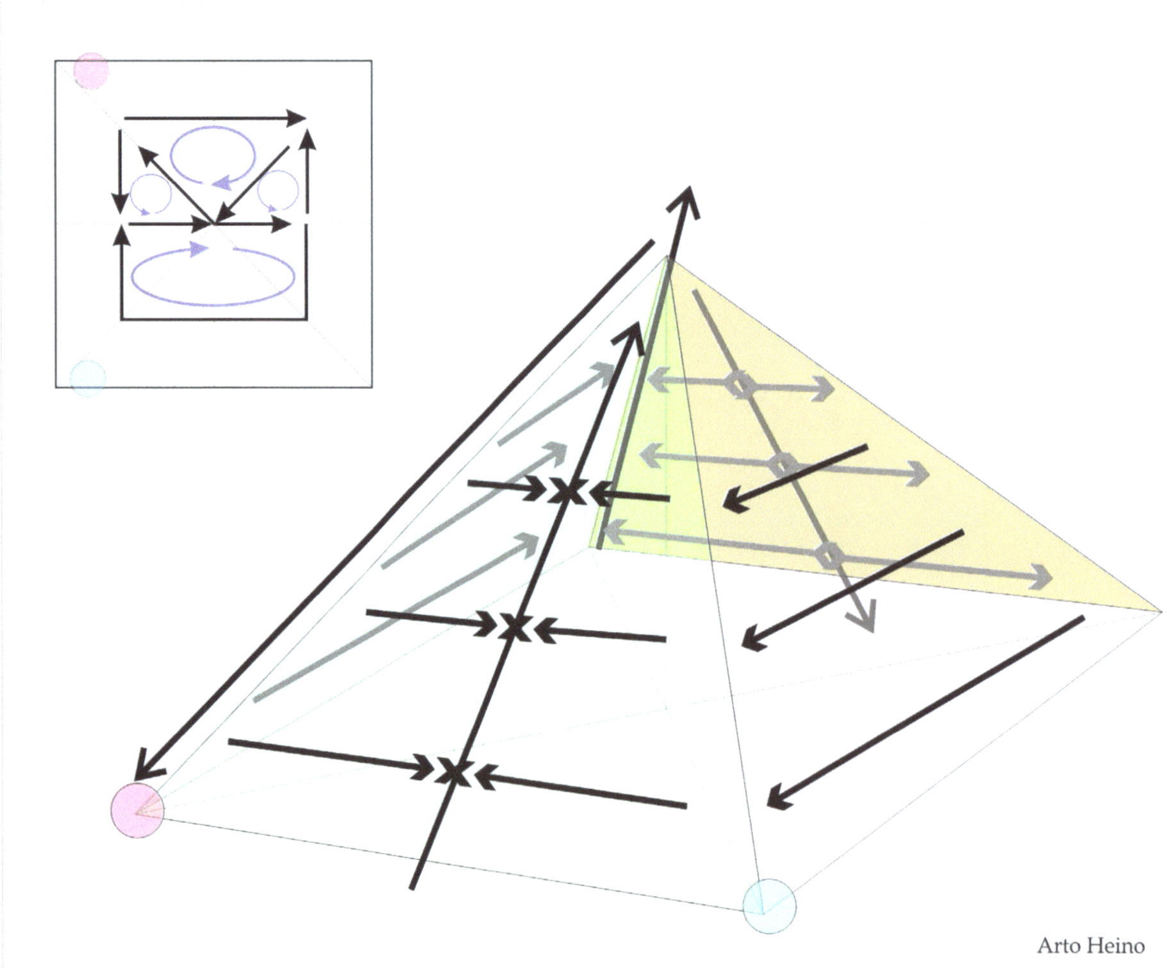

Nested Square 15

by Arto Juhani Heino 1996

	Sqrs	Ring Value	Square
Ring1	1	113	1
Ring2	8	904	3
Ring3	16	1808	5
Ring4	24	2712	7
Ring5	32	3616	9
Ring 6	40	4520	11
Ring 7	48	5424	13
Ring 8	56	6328	15
Total	**225**	**25425**	**64**

Pair 226

204	1	2	3	4	5	6	7	205	206	207	208	209	210	218
198	179	29	30	31	32	33	34	180	181	182	183	184	191	28
199	174	158	53	54	55	56	57	159	160	161	162	168	52	27
200	175	154	141	73	74	75	76	142	143	144	149	72	51	26
201	176	155	138	128	89	90	91	129	130	134	88	71	50	25
202	177	156	139	126	119	101	102	120	123	100	87	70	49	24
203	178	157	140	127	118	114	109	116	108	99	86	69	48	23
211	185	163	145	131	121	115	113	111	105	95	81	63	41	15
14	40	62	80	94	104	110	117	112	122	132	146	164	186	212
13	39	61	79	93	103	125	124	106	107	133	147	165	187	213
12	38	60	78	92	137	136	135	97	96	98	148	166	188	214
11	37	59	77	153	152	151	150	84	83	82	85	167	189	215
10	36	58	173	172	171	170	169	67	66	65	64	68	190	216
9	35	197	196	195	194	193	192	46	45	44	43	42	47	217
8	225	224	223	222	221	220	219	21	20	19	18	17	16	22

1695 1695 1695 1695 1695 1695 1695 1695 1695 1695 1695 1695 1695 1695 1695 1695 1695

1 2 3 4 5 6 7
14 13 12 11 10 9 8
15
16 17 18 19 20 21
22
28 27 26 25 24 23
29 30 31 32 33 34
40 39 38 37 36 35
41
42 43 44 45 46
47
52 51 50 49 48
53 54 55 56 57
62 61 60 59 58
63
64 65 66 67
68
72 71 70 69
73 74 75 76
80 79 78 77
81
82 83 84
85
88 87 86
89 90 91
94 93 92
95
96 97
98
100 99
101 102
104 103
105
106
107
108
109
110
111
112
113

28

1 2 3 4 5 6 7 8 9 10 11 12 13 14 15 16 17 18 19 20 21 22 23 24 25
225 224 223 222 221 220 219 218 217 216 215 214 213 212 211 210 209 208 207 206 205 204 203 202 201

24

29 30 31 32 33 34 35 36 37 38 39 40 41 42 43 44 45 46 47 48 49 50 51 52
197 196 195 194 193 192 191 190 189 188 187 186 185 184 183 182 181 180 179 178 177 176 175 174

20

53 54 55 56 57 58 59 60 61 62 63 64 65 66 67 68 69 70 71 72
173 172 171 170 169 168 167 166 165 164 163 162 161 160 159 158 157 156 155 154

16

73 74 75 76 77 78 79 80 81 82 83 84 85 86 87 88
153 152 151 150 149 148 147 146 145 144 143 142 141 140 139 138

12

89 90 91 92 93 94 95 96 97 98 99 100
137 136 135 134 133 132 131 130 129 128 127 126

8

101 102 103 104 105 106 107 108
125 124 123 122 121 120 119 118

4

109 110 111 112
117 116 115 114

1

113

Sqr Number	3	5	7	9	11	13	15
Line Value	339	565	791	1017	1243	1469	1695
Total Value	1017	2825	5537	9153	13673	19097	25425
Division of Center	9	25	49	81	121	169	225
Corner Value	452	452	452	452	452	452	452
Pair Value	226	226	226	226	226	226	226

Properties of 1695

x	$1695^{1/x}$	x/1695	1695/x	1695*x	z	1695/z
1	1695	0.00059	1695	1695	6	282.5
2	41.170378	0.00118	847.5	3390	12	141.25
3	11.92312	0.00177	565	5085	18	94.16667
4	6.4164147	0.00236	423.75	6780	24	70.625
5	4.4241917	0.00295	339	8475	30	56.5
6	3.4529871	0.00354	282.5	10170	36	47.08333
7	2.8927437	0.00413	242.14286	11865	42	40.35714
8	2.5330643	0.00472	211.875	13560	48	35.3125
9	2.2845288	0.00531	188.33333	15255	54	31.38889
10	2.1033763	0.0059	169.5	16950	60	28.25
11	1.9658976	0.00649	154.09091	18645	66	25.68182
12	1.8582215	0.00708	141.25	20340	72	23.54167
13	1.7717305	0.00767	130.38462	22035	84	20.17857
14	1.7008068	0.00826	121.07143	23730	96	17.65625
15	1.6416402	0.00885	113	25425	108	15.69444

10.4 - Coded Square 3

The last of the series on the magic square of three has led me to the understanding of re-sequenced events that can occur in nature, sometimes modifying the base code to improve the living flow or obstruct it and create a cancer. The number sequences on the right hand side are simple descriptions that will create the shapes using the sequential square 3 numbers on the left. The group below has the same number sequences but are now ordered according to a magic square arrangement. Note how some show amazing symmetry and most fail to coordinate harmony.

1 2 3
4 5 6
7 8 9

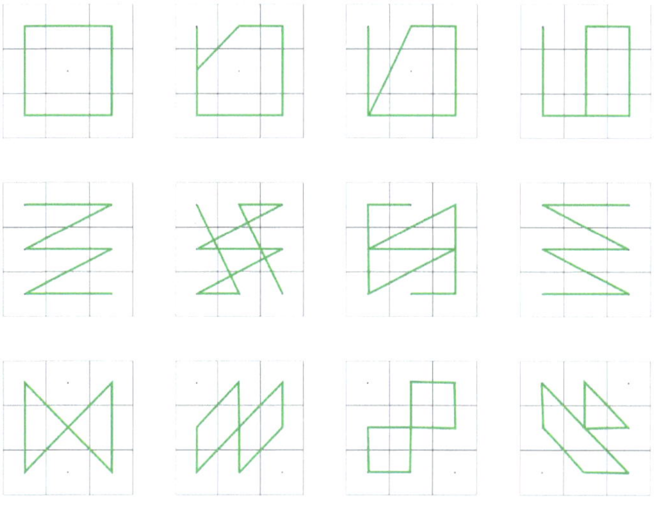

1 4 7 8 9 6 3 2 1
1 4 7 8 9 6 3 2 4
1 4 7 8 9 6 3 2 5
1 4 7 8 9 6 3 2 6

1 2 3 4 5 6 7 8 9
1 8 7 6 5 4 3 2 9
2 1 7 6 5 4 3 9 8
3 2 1 6 5 4 9 8 7

1 5 3 6 9 5 7 4 1
2 5 8 6 3 5 7 4 2
3 6 5 4 7 8 5 2 3
5 2 6 5 9 8 4 1 5

6 1 8
3 5 7
4 9 2

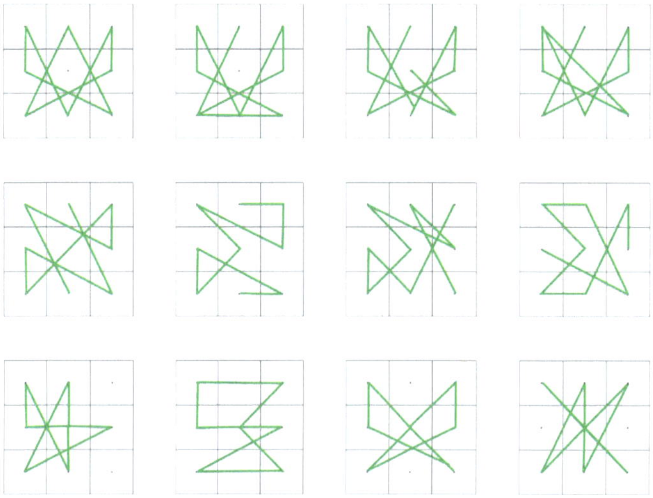

1 4 7 8 9 6 3 2 1
1 4 7 8 9 6 3 2 4
1 4 7 8 9 6 3 2 5
1 4 7 8 9 6 3 2 6

1 2 3 4 5 6 7 8 9
1 8 7 6 5 4 3 2 9
2 1 7 6 5 4 3 9 8
3 2 1 6 5 4 9 8 7

1 5 3 6 9 5 7 4 1
2 5 8 6 3 5 7 4 2
3 6 5 4 7 8 5 2 3
5 2 6 5 9 8 4 1 5

Chapter Eleven

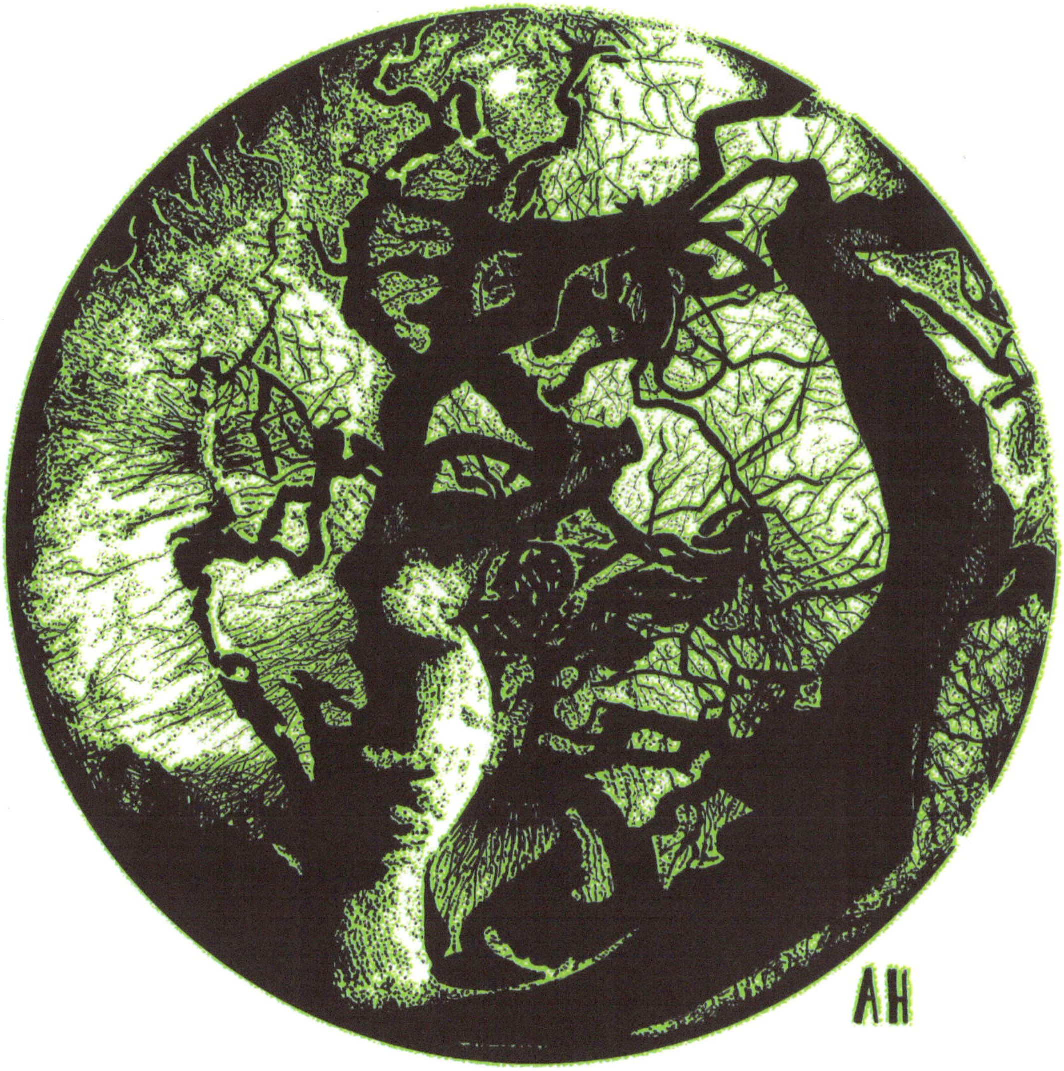

Earth

11.1 Quantum Solar System

PLANETARY ORBITAL EMPIRICAL VALUES AND QUANTUM EQUIVALENTS

	Empirical Semi Major Diam Millions KM	Quantum Semi Major Diam Millions KM	Empirical Perigee Millions KM	Empirical Apogee Millions KM	Quantum Perigee Millions KM	Quantum Apogee Millions KM	Empirical Eccentricity	Quantum Eccentricity
Mercury	57.9091008793	56.8846880000	46.0008696863	69.8173320723	44.5852960000	69.1840800000	0.2056366100	0.2162162162
Venus	108.2072844246	108.0767520000	107.4753714348	108.9391974143	105.9576000000	110.1959040000	0.0067639900	0.0196078431
Earth	149.5978976276	149.5872000000	147.0948809557	152.1009142995	147.0940800000	152.0803200000	0.0167316300	0.0166666667
Mars	227.9441350871	227.5387520000	206.6621058409	249.2261643334	206.0148160000	249.0626880000	0.0933651100	0.0945945946
Asteroids	549.5252000000	549.5252000000	520.8543200000	578.1960800000	520.8543200000	578.1960800000	0.0521739130	0.0521739130
Jupiter	778.2799587829	779.8894880000	740.5054405314	816.0544770344	740.0411200000	819.7378560000	0.0485359000	0.0510948905
Saturn	1427.3879082546	1422.1172000000	1348.1561033962	1506.6197131129	1345.2460000000	1498.9884000000	0.0555082500	0.0540540541
Uranus	2870.4808732433	2874.1102880000	2735.9776027734	3004.9841437132	2742.9721760000	3005.2484000000	0.0468574000	0.0456273764
Neptune	4498.3372899471	4497.6300320000	4458.0574235013	4538.6171563928	4456.6182080000	4538.6418560000	0.0089543900	0.0091185410
Pluto	5907.1502289883	5905.7442080000	4437.1418354870	7377.1586224895	4433.2244320000	7378.2639840000	0.2488523800	0.2493368700

A	a * a	C + G
B	b * b	C - G
C	2 * d * e	K - J
D	d * d	(G + F) / 2
E	e * e	G - D
F	a * b	D - E
G	d * d + e * e	D + E
H	2 * d * e + a * b	C + F
I	2 * d * e - a * b	C - F
J	d * b	K - C
K	d * a	D + X
L	(b * e * d * a)/6	(C * F) / 12
V	(d * e + d * e)-2*a*b	W - 2 * F
W	d * e + d * a	X + K
X	d * e	C / 2
Y	d * d + e * a	C + E
Z	e * e + a * d	E + K

	Empirical Time in Years	Quantum Time in Years	Empirical Time in Days	Quantum Time in Days
Mercury	0.2408420400	0.2344796425	87.9656283737	85.6418135916
Venus	0.6151723700	0.6140595633	224.6867867635	224.280343015
Earth	1.0000007000	0.9998934472	365.2422556694	365.2030824587
Mars	1.8808524000	1.8758371401	686.9662922808	685.1345087327
Asteroids	7.0403377200	7.0403377156	2571.4270295282	2571.4270279371
Jupiter	11.8663142100	11.9031435891	4334.0763346888	4347.5279707847
Saturn	29.4730508300	29.3099553144	10764.7960312509	10705.2266989477
Uranus	84.0512272500	84.2106882000	30699.0383432445	30757.2801795578
Neptune	164.8883975000	164.8495116623	60224.168079695	60209.9653385575
Pluto	248.1298978600	248.0413130687	90627.4601541821	90595.105267845

Data conversion by Arto Heino

Unit of Conversion	41552
Quantum Unit	D x 41552

QUANTUM ARITHMETIC VALUES FOR PLANETARY ELLIPSES

							Semi Major Diam	
	b	e	d	a	B	E	D	A
Mercury	29	8	37	45	841	64	**1369**	2025
Venus	50	1	51	52	2500	1	**2601**	2704
Earth	59	1	60	61	3481	1	**3600**	3721
Mars	67	7	74	81	4489	49	**5476**	6561
Asteroids	109	6	115	121	11881	36	**13225**	14641
Jupiter	130	7	137	144	16900	49	**18769**	20736
Saturn	175	10	185	195	30625	100	**34225**	38025
Uranus	251	12	263	275	63001	144	**69169**	75625
Neptune	326	3	329	332	106276	9	**108241**	110224
Pluto	283	94	377	471	80089	8836	**142129**	221841

	Distance Between Foci	Semi Latus Rectum	Cross Diagonal	Unit Area			Perigee	Apogee
	C	F	G	L	H	I	J	K
Mercury	592	1305	1433	64380	1897	713	**1073**	**1665**
Venus	102	2600	2602	22100	2702	2498	**2550**	**2652**
Earth	120	3599	3601	35990	3719	3479	**3540**	**3660**
Mars	1036	5427	5525	468531	6463	4391	**4958**	**5994**
Asteroids	1380	13189	13261	1516735	14569	11809	**12535**	**13915**
Jupiter	1918	18720	18818	2992080	20638	16802	**17810**	**19728**
Saturn	3700	34125	34325	10521875	37825	30425	**32375**	**36075**
Uranus	6312	69025	69313	36307150	75337	62713	**66013**	**72325**
Neptune	1974	108232	108250	17804164	110206	106258	**107254**	**109228**
Pluto	70876	133293	150965	787272889	204169	62417	**106691**	**177567**

Arto Heino

11.2 - Earth Constants

Using the gravity acceleration of the earth at sea level

First establish some constants

RC	= **0.523236546546474681** mtrs	Royal cubit based on QA
SC	= **0.44848846840686900** mtrs	Standard Cubits based on QA
C	= **299792458** mtrs sec	
CL	= C x pi/2	= **470912891.826213** mtrs per sec

So if a Royal Cubit has a hidden ratio then this is it

CLRC = CL / RC = **900000000** RC per sec

This is the Royal Cubit based on the velocity of light

G	= **6.67384 e -11**	text book Gravitational constant, CODATA 2010
m1	= **5.9726 e+24 kg**	text book mass of the Earth, WMAP data 2012
r0	= **6371009** mtrs	text book mean radius of the Earth. Wiki
cr0	= **40030230.1407** mtrs	derived circumference

g0 = G x m1 / r0^2

= **9.8202554113** mtrs per sec — gravity at sea level per Newton

then if

r1 = **6370961.6770147** — based on Light whole units of RC

and then

cr1 = **40029932.801623**

then

cq1	= r1/ RC	= **12175842**	whole number units
cq2	= r1/ SC	= **14205149**	
cqr	= cq2 / cq1	= **1.166666**	= 1 1/3 = 7/6

The difference of

rdf = **47.3229853017** = r0 - r1 does not relate to anything

crdf = **297.3390859365** = cr0 - cr1

w0 = crdf/rdf = 6.2832954 = pi x 2

w1 = rdf * pi = 148.66954297 = crdf / 2

If the pyramid of Giza base edge is

Ps1 = **230.22408044886** mtrs = 440 x RC

and the height

Ph1 = **146.50623301291** mtrs = 280 x RC

First Let us correct the sea level radius based on Giza

if

ph3 = Ph1/3 = **48.8354110043** mtrs

then the sea level Earth radius becomes

r3 = r1 + ph3 = **6371010.5124257** mtrs New average radius

then

g1 = **9.8202507488** new gravity acceleration at sea level

also

cr2 = **40030239.64335599** new circumference

Now the Great Pyramid is can be used as a measure of sea level change

then

Ph1 = 3 x (r1 – r3)

as you can see **Ph1** is the true size from a calculation over the whole Earth

Another problem is the Gravity constant, here is another way to derive the constant and get results

GC = 1 / (C x 50) = **6.67128190396304 e-11**

now let us go back to the sea level equation using some new data

g2 = GC x m1 / r1^2

= **9.8164866273** mtrs per sec the NEW gravity at sea level

Chapter Twelve

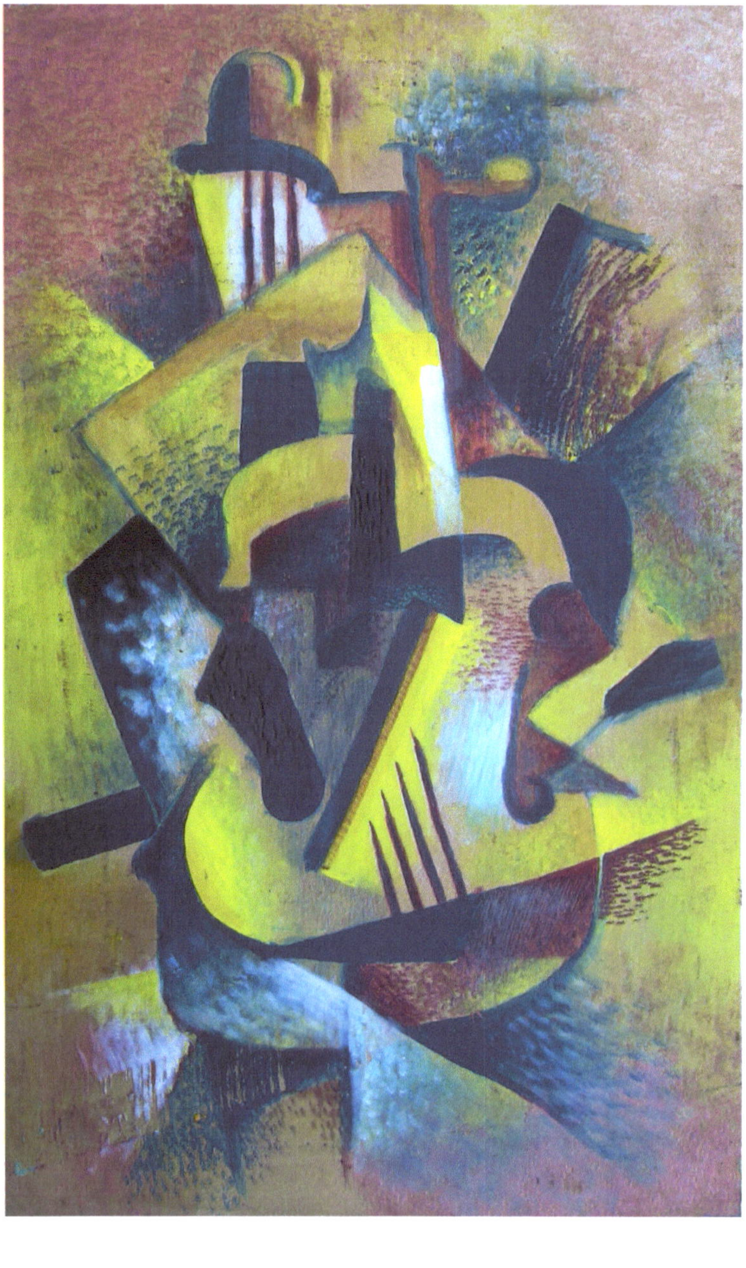

Inventions and Inventors

12.1 - Tesla Tower

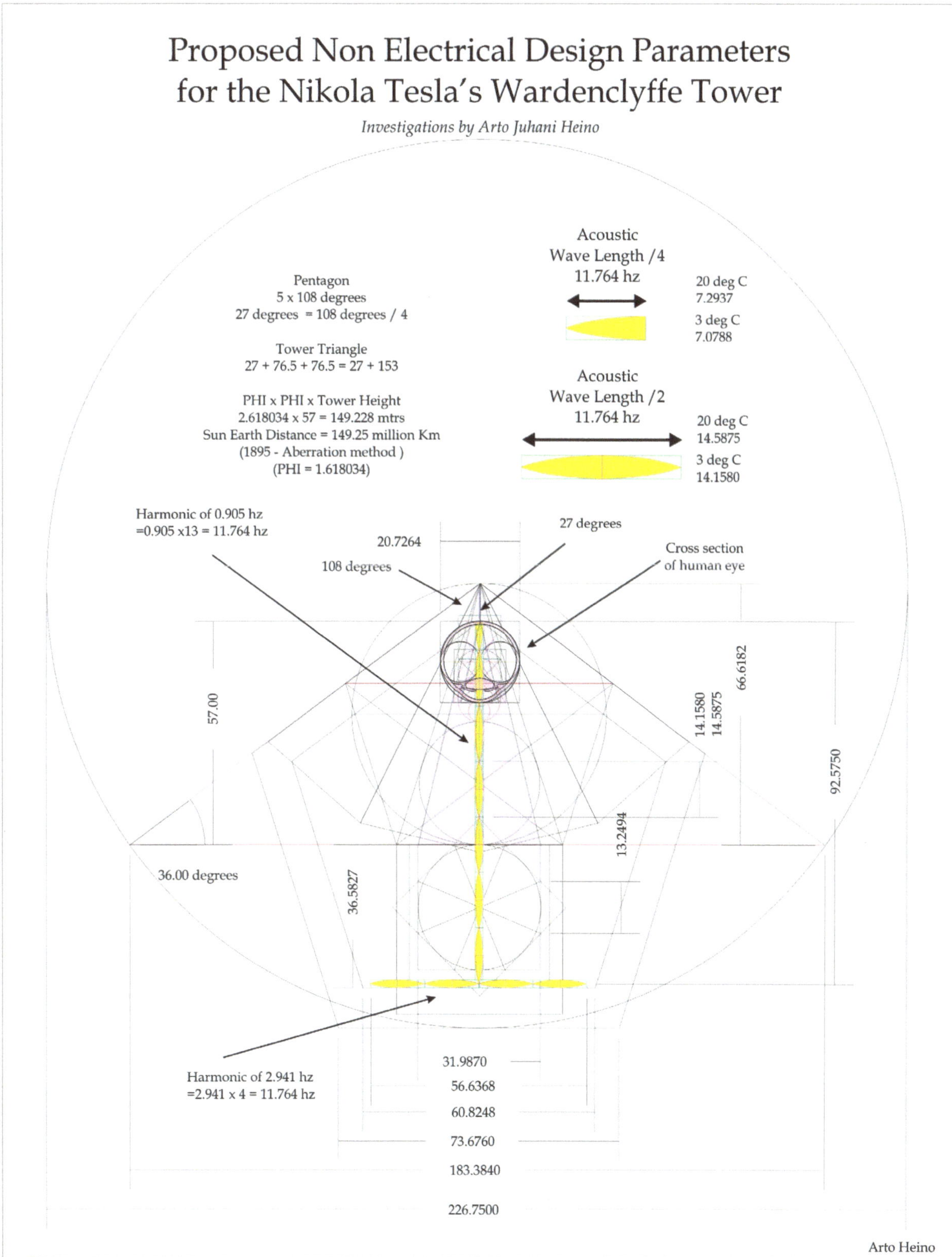

12.2 - Adams Motor

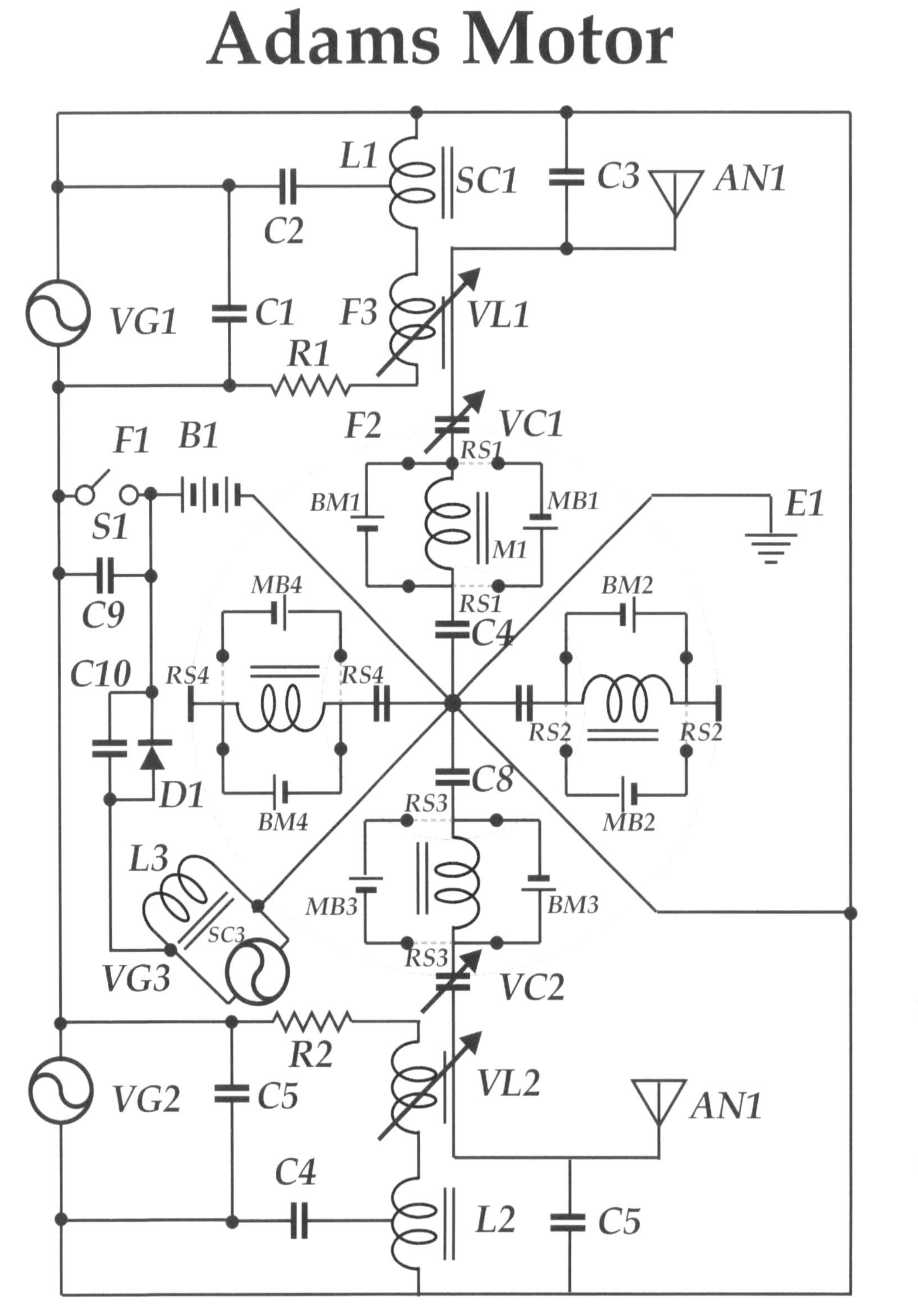

Arto Heino

The minimum Schematics and Parameters required to engineer the Adams Motor

Light Velocity = 299792458 m/s
Phi Ratio = 1.618034
Alnico Magnet = 0.15 Tesla or less
Core = Soft Iron, low remanence
Winding Resistance = 10 ohms approx

C1 Added Capacitance
C2 Inter-winding Capacitance
C3 Core to winding Capacitance
C4 Capacitance between central shaft to South Magnet end
C5 Added Capacitance
C6 Inter-winding Capacitance
C7 Core to winding Capacitance
C8 Capacitance between central shaft to South Magnet end
C9 Switch Capacitance
C10 Generator Pulse Capacitor
D1 Generator Charging Diode
VC1 Variable Capacitance between Stator core and North Magnet end
VC2 Variable Capacitance between Stator core and North Magnet end
L1 Static Inductance of Stator
L2 Static Inductance of Stator
L3 Static Inductance of Generator Coil
VL1 Variable Inductance of Stator
VL2 Variable Inductance of Stator
R1 Winding Resistance
R2 Winding Resistance
SC1 Stator core - material = Low magnetic remnant soft Iron
SC2 Stator core - material = Low magnetic remnant soft Iron
SC3 Generator core - material = Laminated silicon steel
VG1 Voltage generated AC

VG2 Voltage generated AC

VG3 Voltage generated AC

AN1 Core acting like Antenna

AN2 Core acting like Antenna

M1+BM1 Alnico Magnet Polarity NS

M1+MB1 Alnico Magnet Polarity SN

RS1 Polarity reversal - Resonant switch

M2+BM2 Alnico Magnet Polarity NS

M2+MB2 Alnico Magnet Polarity SN

RS2 Polarity reversal - Resonant switch

M3+BM3 Alnico Magnet Polarity NS

M3+MB3 Alnico Magnet Polarity SN

RS3 Polarity reversal - Resonant switch

M4+BM4 Alnico Magnet Polarity NS

M4_MB4 Alnico Magnet Polarity SN

RS4 Polarity reversal - Resonant switch

B1 Battery

E1 Earth

S1 Switch (Reed, Brush etc)

F1 Switch Frequency = RPM/60 x 4

F2 Parametric Capacitance Pump Frequency = F4 - F1, F4 + F1, F4 x 2

F3 Parametric Inductance Pump Frequency = F4 - F1, F4 + F1, F4 x 2

F4 Resonant Frequency of L1 and (C1+C2)

F5 Polarity reversal Frequency

F6 Generated voltage Frequency multiples of F1

T0 Time on/off = 1/F1

T1 On Time (S1) = T0/4

T2 Off Time = T1 x 3

GP1 Geometric position of on Time to Stator/Magnet alignment = 1.738 degrees = m =360/43/Phi x 2

GP2 Geometric position of off Time to Stator/Magnet alignment = 24.238 degrees = m + 90/4

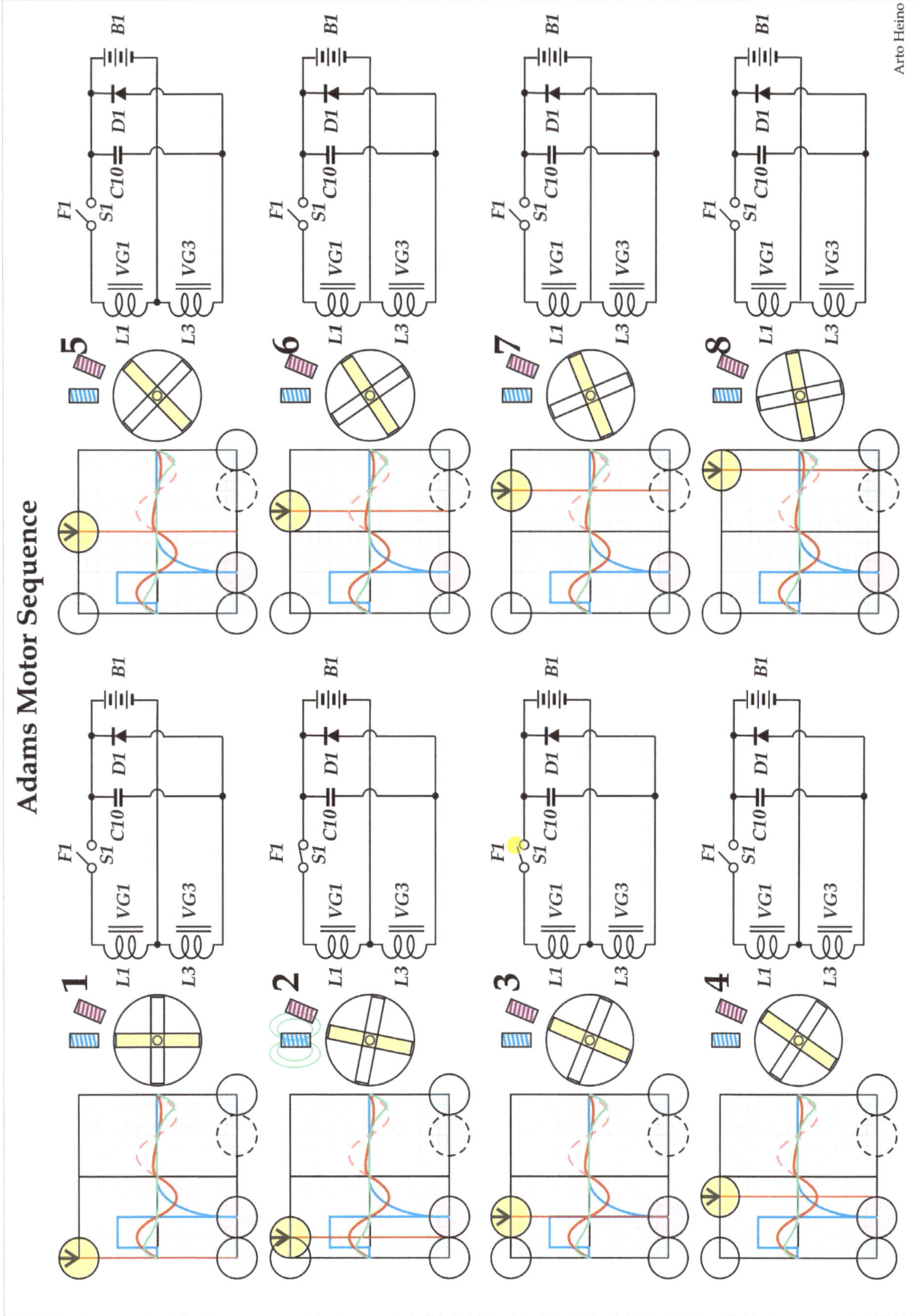
Adams Motor Sequence
1
2
3
4
5
6
7
8
L1
L3
VG1
VG3
F1
S1
C10
D1
B1
Arto Heino

The pursuit of an idea of the coupling of the great solar /earth system to a small rotating pulsed device was my initial concept that had haunted me since my childhood. To realize electricity was part of the universal mechanisms that was driving the massive bodies of the solar system, occurred to my young self through the lens of understanding electrostatic repulsion as was discovered by myself when making artistic observations with paper and plastic on my drawing table.

In 1986 I was designing some unique colour-music art (Synesthesia) and looking at ways to put this into electronics using leds, lasers, cathode ray tubes or computer programs, this led me to designing a colour organ that was fully integrated with audio correlation to the harmonic contents of the sound as well. I decided to apply for a patent and went to the patent office to do a search for similar devices.

During the search I listed a few names and previous attempts, which went all the way back to 1890's, while I was their I also looked for new developments in electric motors, this was the moment I found the Adams motor and noted the inventor was from New Zealand and he had patented his invention in 1969. I immediately concluded this simple design would be very efficient and unique as all the poles were of the same polarity, this was duly noted and I pressed on with other matters at the patent office. Amazingly enough I did end up the the Edison Electric offices where I explained the basic tenants and principles of this colour-organ to a few very attentive engineers. Later in that month I met with one of the patent holders for one of the colour-organs who just happen to be an artist as well and lived in Sydney. Before the month was out I naively wrote to the Sharp corporation and presented my findings and possible applications on my invention, they wrote back to me with a generous tone and kept me in mind for any future developments in this direction.

A brief description of my colour organ – Frequency correlated harmonic fourier separation to colour chroma addition translated to voltage steps controlling electron charge amplifiers creating a proportional light intensity and equated colour area and location in the visual field, the resultant beam or electric display (Tesla coils with or without gas enveloped top loads, closed electron tube/ phosphor display, optical wall/3D display, Holographic multi- projectors, laser projectors) would modulate with sensitive modes that equated to the overall mood and flowing structure of the music based on translating electronics. Some of these ideas have been used by others in recent years, due to the advent of readily available electronic component modules.

This chance study of the Adams motor at the patent office was originally prompted by a childhood dream of a free running motor that collected the energy from the universe to power it. This reoccurring dream was in the late 60's when I was building crystal radio's, high voltage coils, transistor circuits and single valve circuits. This was the time I listened to Short and Long wave radio with my home made aerials picking up whistlers and other radio noise that I found by pointing my wire dishes at different portions of the sky and at different times of the day and night. I was about 12 at the time when a lot of these ideas about how these things operated occurred to me.

In early 1990's I started compiling my various work on many subjects including motor design, when I was pleasantly surprised that Robert Adams published a ground breaking article in the Nexus magazine on his motor design. This was an exiting time as a lot of the backyard and professional engineers were collaborating and working on many new devices. At this time in Sydney there were Adams motor group meetings and we all were showing our current builds and exchanging information between members, the only problem was the new paradigm of over-unity engineering was not developed enough to handle the questions many members had. As always, I was the outsider who didn't put faith in the current techno-babble of electrical engineering, being self taught in these matters, I relied on basic tenants of creating the equations from the evidence, not forcing the build to comply with the equations. This is what I gleaned from most of the Scientists and Inventors of history mainly Leonardo Da Vinci, Nicola Tesla and Victor Schauberger. In a desperate attempt to help my fellow builders I met with Robert Adams at the Sydney Nexus conference, from that moment on I corresponded with Mr Adams a few times and sent him a few of my design sketches. I will include a copy of one of the letters I received from him, below.

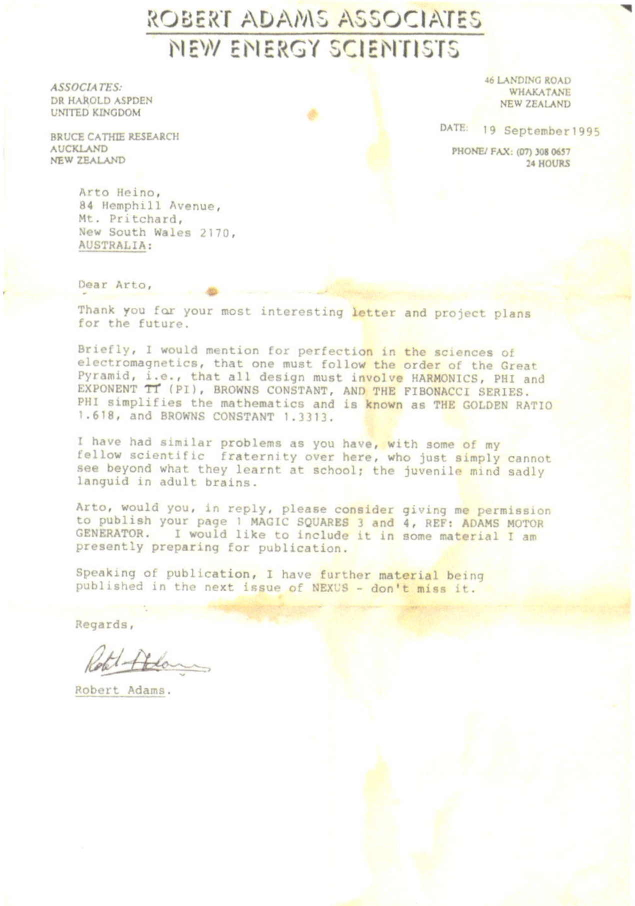

ROBERT ADAMS ASSOCIATES
NEW ENERGY SCIENTISTS

ASSOCIATES:
DR HAROLD ASPDEN
UNITED KINGDOM

BRUCE CATHIE RESEARCH
AUCKLAND
NEW ZEALAND

46 LANDING ROAD
WHAKATANE
NEW ZEALAND

DATE: 19 September 1995

PHONE/ FAX: (07) 308 0657
24 HOURS

Arto Heino,
84 Hemphill Avenue,
Mt. Pritchard,
New South Wales 2170,
AUSTRALIA:

Dear Arto,

Thank you for your most interesting letter and project plans for the future.

Briefly, I would mention for perfection in the sciences of electromagnetics, that one must follow the order of the Great Pyramid, i.e., that all design must involve HARMONICS, PHI and EXPONENT π (PI), BROWNS CONSTANT, AND THE FIBONACCI SERIES. PHI simplifies the mathematics and is known as THE GOLDEN RATIO 1.618, and BROWNS CONSTANT 1.3313.

I have had similar problems as you have, with some of my fellow scientific fraternity over here, who just simply cannot see beyond what they learnt at school; the juvenile mind sadly languid in adult brains.

Arto, would you, in reply, please consider giving me permission to publish your page 1 MAGIC SQUARES 3 and 4, REF: ADAMS MOTOR GENERATOR. I would like to include it in some material I am presently preparing for publication.

Speaking of publication, I have further material being published in the next issue of NEXUS - don't miss it.

Regards,

Robert Adams.

The testing process of my build of a number of Adams motor was arduous and drawn out over a least 2-3 years, I had to learn many new skills and I developed an intuitive insight into this simple motor that goes beyond equations and engineering. The great help came from the generous group I was involved with at the time, I was loaned a oscilloscope, that is when the next phase of my journey began. I tried to convey the results to my fellow researchers and found they were closed minded about developing new equations that included open systems, I met with many and explained the tenants of true energy as freely captured from the environment and conditions of its location, they turned to their university books and denounced anything that was not written in the science bibles. As I was not brainwashed in these matters, I could see their distress when I confronted their most hallowed beliefs of the system of modern science they were taught as sacrosanct, alas Einstein ruled. I am not a scientist or an engineer, I am an Artist from the school(mind set) of Leonardo da Vinci and Walter Russell, therefore I am a natural scientist and observer of phenomenon, please allow me at least this modesty for my years of study and practical application .

I moved away from trying to convince others of the errors hidden in the slavish conformity to the equations and I allowed the natural curiosity and time for them to realize and discover it in there own time, the greatest teacher is ones own experience. This has been the legacy of Robert Adams, a man who spent a lifetime serving the system only to confront it head-on with its own fallacies. Robert had been attacked on many occasions by the agents of a flawed science, only to be praised by those who have already awakened to the same dilemma such as the notable scientist Harold Aspen. Their written legacy still exists for other to learn from.

I had performed about 400 separate tests that gave me the understanding of the operations of the motor circuitry and to look for environmental correlations. I have seen many people approach the design of the Adams motor with a sledge hammer, with stronger magnets, faster speeds, larger disks, bigger coils etc, this has always been to no avail as the basic understanding has been missed. This motor will run in a resonant fashion, to itself and the environment. This is the key, do not underestimate what Robert Adams tried to convey with his descriptions and explanations. Look at the wave functions of the whole solar system, day/night, altitude, atmospherics, turrelian currents, temperature, EM fog and geographic location before you make conclusions about how this simple motor operates. Energy is everywhere, tapping into this resource takes patience, observation, curiosity, creativity and sound scientific principles. Do not be led into a blind alley of dogmatic procedures that avoid the hidden truth that we live in an OPEN system and energy is always available to be used through the correct gating mechanisms that we discover.

As shown here it is the schematic of the Adams motor design as I have come to understand, I have not included all the equations required to build such a motor/gen, just the basic schematic I have revealed.

12.3 - Sixto Ramos

When I first drew up the step by step motion of the Sixto Ramos Force Multiplier , I could see the joining shaft ends did not match the pivots, this was a revelation of an impulse of non-linear force was in action here. Here is a breakdown of what is really happening(refer diagrams), as you can see the velocity increases and decreases as the 2nd wheel turns. If you add a flywheel with one-way clutch bearings you will start to understand the incredible action he has accomplished here.

The upward arm is an resonant device that allows the impulse force to be added to the 2nd wheel at the time when the velocity decreases thus maintaining an increase in the rotational momentum of the external flywheel. Adding more wheels will multiply this action many fold. This arm could be placed at different parts of the system, as long as it is functional as a resonator.

I have not had much time to devote to building a prototype of Sixto's machine or do any of the equations that are required, I hope that the reader would take the baton. These diagrams should help in the design of a practical unit.

These points relate to the sequence drawings.

First - the joining shaft end on the second rotor did not match the pivots end, the arm length was alternatively moving off the end pivot.(2nd wheel)

Second - The rotation of the second rotor appeared even like the main rotor, this is not the way it is in reality, the 2nd rotor has a variable rotation.

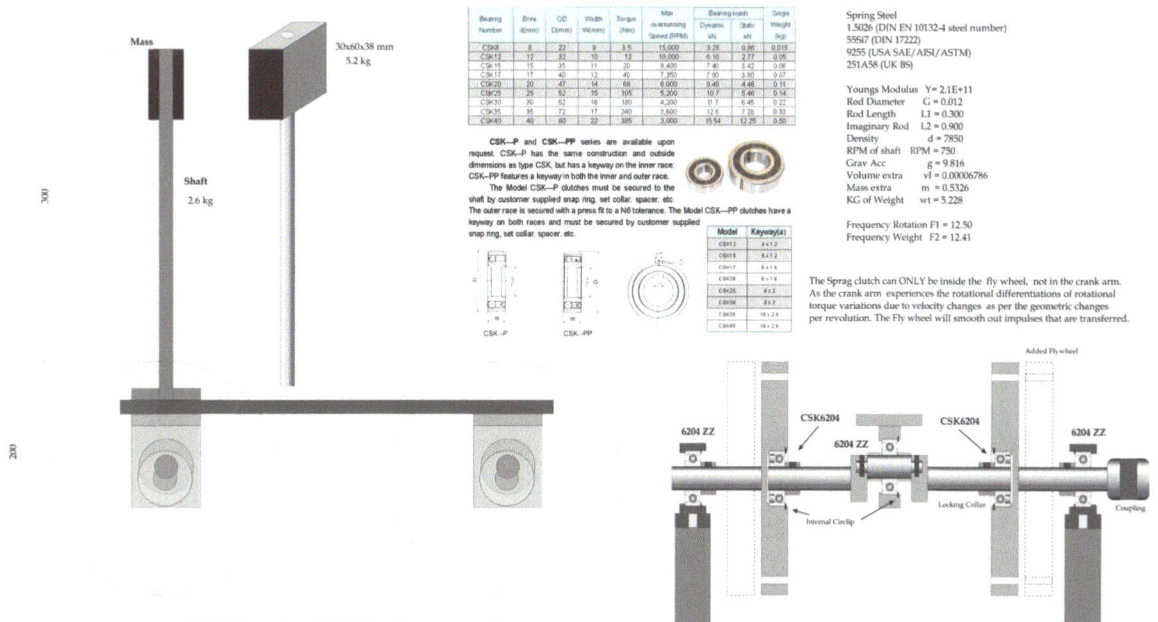

Bearing Number	Bore d(mm)	OD D(mm)	Width W(mm)	Torque (Nm)	Max overrunning Speed (RPM)	Bearing loads Dynamic kN	Static kN	Single Weight (kg)
CSK8	8	22	9	3.5	15,000	3.28	0.86	0.015
CSK12	12	32	10	12	10,000	6.10	2.77	0.05
CSK15	15	35	11	20	8,400	7.40	3.42	0.06
CSK17	17	40	12	40	7,350	7.90	3.80	0.07
CSK20	20	47	14	68	6,000	9.40	4.46	0.11
CSK25	25	52	15	105	5,200	10.7	5.46	0.14
CSK30	30	62	16	180	4,200	11.7	6.45	0.22
CSK35	35	72	17	240	3,600	12.6	7.28	0.33
CSK40	40	80	22	385	3,000	15.54	12.25	0.50

Model	Keyway(a)
CSK12	4x1.2
CSK15	5x1.2
CSK17	6x1.6
CSK20	6x1.6
CSK25	8x2
CSK30	8x2
CSK35	10x2.4
CSK40	10x2.4

Sixto Ramos Force Multiplier, design parameters

By Arto Heino

The flywheel with one-way clutch bearings action is such that when you have a retarding velocity, the bearing will act in free wheel mode and when you have a increasing velocity or an acceleration you will have a locked bearing which will allow the transfer of rotation to the flywheel.

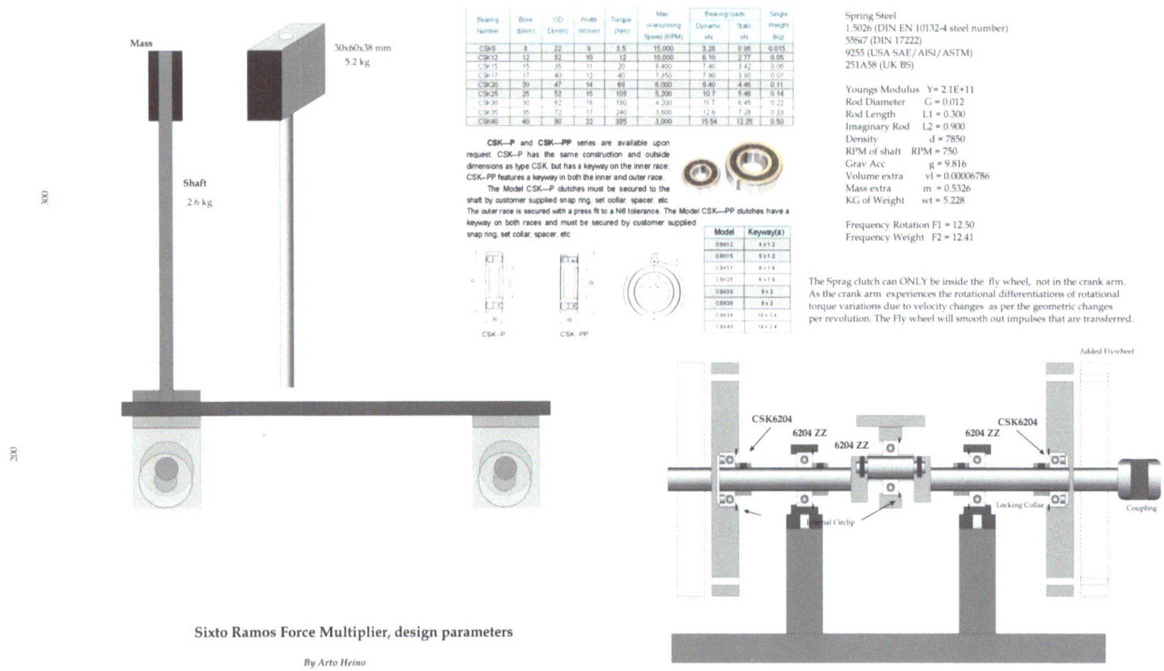

Sixto Ramos Force Multiplier, design parameters

By Arto Heino

I would hazard a guess that Sixto did not reveal all the parameters or hidden information, using a one-way clutch bearing seems to be the only way you can drive a generator without fluctuating AC voltages(frequency shifting). The flywheels are there to smooth-out the velocity changes and the one-way bearings are there to impart a impulse to the flywheel when the geometry changes.

It would be a clear and definite improvement on this machine utilizing the one-way bearings on the out put side only, maybe Sixto does not use them and maybe his offset is closer to the shaft, this sort of speculation is irrelevant,

On my diagrammatic model all you need to do is move the centers further apart and the offset becomes smaller, just a small device variation, the one-way bearings are good engineering practice as backlash will occur when the 2nd wheels velocity slows down. If you only rely on small shaft offsets without the clutch bearings, it will still cause the output to vibrate with large damping forces in the arms resonant action. Larger offsets with one-way bearings allow different modes of resonant operation.

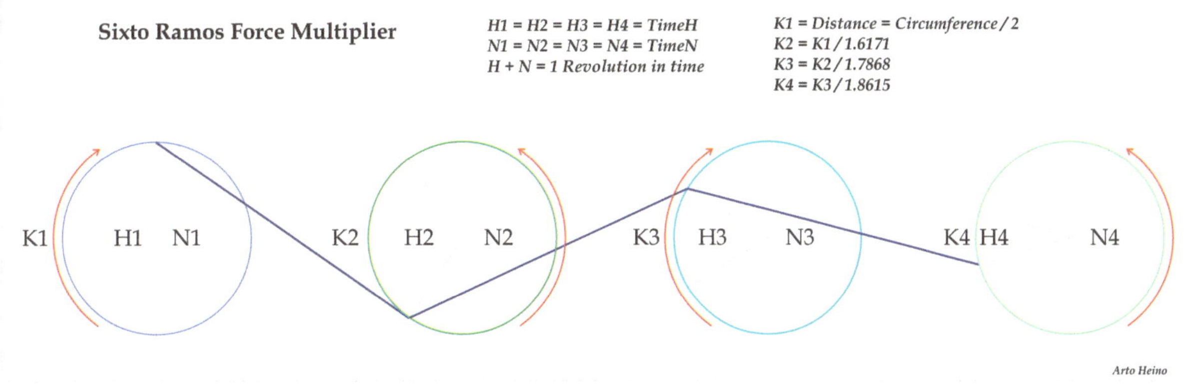

This machine is a transformer from a steady rotation to an impulse rotation, if you understand Tesla you will know that this machine is capable of much engineering if you do not lock down the system and allow loose coupling.

The real picture emerges when you try to understand HOW to couple the impulse forces and the resonant weight, this was the purpose of these diagrams, altruism is the key as I will give my researches freely so other can creatively construct the mathematical model, This machine that Sixto built is a crude model, that clearly works, nothing here is a mystery, only improvements and understanding must be imparted to those who will build it. I commend Sixto on his release of his idea to the world, he knows others will improve it and find out better ways make to it work,

"An idea that is not dangerous is unworthy of being called an idea at all"

Oscar Wilde

I would not build it as per my pictures, it is the exaggeration of the offset that helps you see the real situation. If you look at the impulse in the time domain, it is a 1/4 wave system. In the distance domain it is a impulse wave. There should be some effort to draw up the complete transformations and equivalence to the electrical equations, another possibility in the near future.

Over unity is not the goal, if there is any, then it will have to be engineered with a gravity system in secondary or tertiary connections. The useful energy will be a gain in the system at the time domain, increasing velocity without gearing and compressing force to a smaller time frame, translating geometrically. The advantage gain is only at the impulse 1/4 wave peak, just like a Tesla coil. The resonant arm is the real interest here, as the impulses are in true translation, making gravity and mechanical resonance add to the impulse velocity.

These diagrams below should clarify the understanding of the interesting geometry that Sixto has revealed in his machine. In my researches my Hyperbolic Impulse Drive, also shows the similarities in its internal geometry to the Ramos device.

"An idea that is developed and put into action is more important than an idea that exists only as an idea"

Buddha

Sixto Ramos Force Multiplier

34+55 =89 233/144=1.618055
55+89 =144
89+144=233

360/1.618034 = 222.4922344
360/1.620000 = 222.2222222
360/1.620357 = 222.1732618

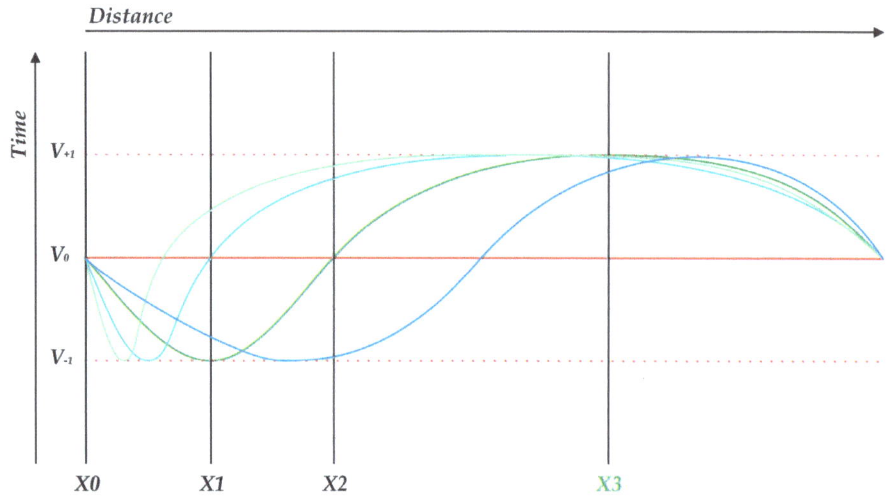

Y1= PI x D1 = 452.3893
Y2=Y1/360 x 111.0862 = 139.5850
Y3=Y1-Y2=312.7942

A	72 mm
A_1	72 mm
B	233 mm
B_1	233 mm
C	243.870867468831 mm
C_1	243.870867468831 mm

Arto Heino

The vibrating shaft should be made of spring steel, this results in less problems with breakage and will allow vibrations to be less damped. The situation with this machine is there no net gain in the basic design, it has only an impulse gain, not a continuous rotary gain. The impulse is lost when you have a rigid connection to the other flywheel, think of it more like a loose coupling in a resonant system such as a Tesla coil. Utilize a mechanical resonate formula for the vibrating arm, clear and usable information to improve the vibrational phases.

I can assure the builder will not have good results until a lot of the measures I have outlined have been implemented, it is an engineering reality of rotating machinery. Use of a spring steel arm, with a variable mass, adjusting its height and weight according to the resonant equation will also improve the action.

"Often the difference between a successful person and a failure is not one has better abilities or ideas, but the courage that one has to bet on one's ideas, to take a calculated risk - and to act." Andre Malraux

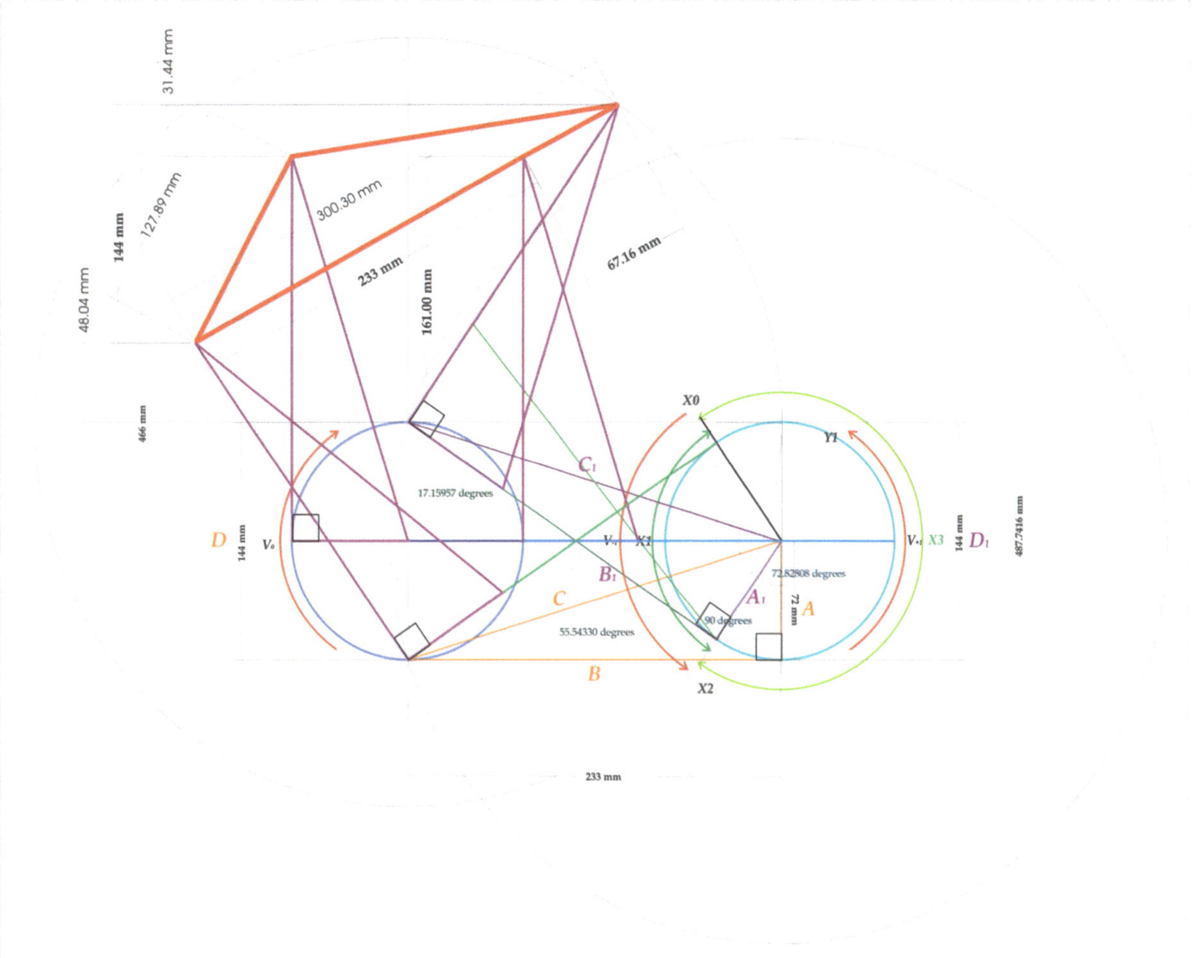

Arto Heino

The sprag clutches should be inside each output flywheel, it should lock with the same direction as the output rotation and freewheel in the opposite rotational direction. The output shaft should be part of the flywheel, which will be connected to the outer body of the clutch bearing the offset cranking output shaft should rotate inside the bearing, freely in the opposite direction and lock in the same direction as the output shaft. This arrangement will insure all impulses are transferred, and any retarding rotation will not brake the system but allow free movement. If this description is not clear, I will draw the arrangement if needed, if so, could you please give me some basic dimensions of your setup.

There is no magic engineering in the Ramos unit, many people TRYING to replicate, including good engineers, alas you cannot built it until you understand it, maybe there is something do not understand about how the forces operate in a design like this.

Ramos has on display is just what he wants you to see and know, as he is quoted as saying something like (from translation) "When I add/change one of the components in the production model it will create much more power"

To clarify about the action when you include one-way clutch bearings in the design like the first two drawings. It should lock with the same direction as the output rotation and freewheel in the opposite rotational direction. Driving the unit as a flywheel does, it cannot overrun in the same direction. It will have traction in the same direction and free runs in the opposite direction This means all retarding forces are allowed to run without creating traction against the rotation.

As I have 30 years of experience supplying bearings to engineers and solving their problems, this will never help the English language to convey actuality of counter rotating masses and gyroscopic spins that help alleviate frictional torques.

Gravity plays a very minor role in the Ramos device, it is all about coupling inertial impulses via geometric advantage with harmonic vibrational parametric energy growth. I have witnessed machines self destruct when all these forces are in resonance, the Ramos device uses this destructive potential of resonance to create energy increase.

This unit can be tuned, just like my piano, the sounding box is the flywheel, where energy can accumulate over time in accordance the the law of growth, exponential increases will occur. were it not for losses in this rotational system it would continue until destruction. This device in my conception is similar to to loosely coupled Tesla coil, I cannot speak for Sixto. As for free energy, this can only lead to idle speculation about trivial matters that can only distract our attention away from a unique mechanical assemblage.

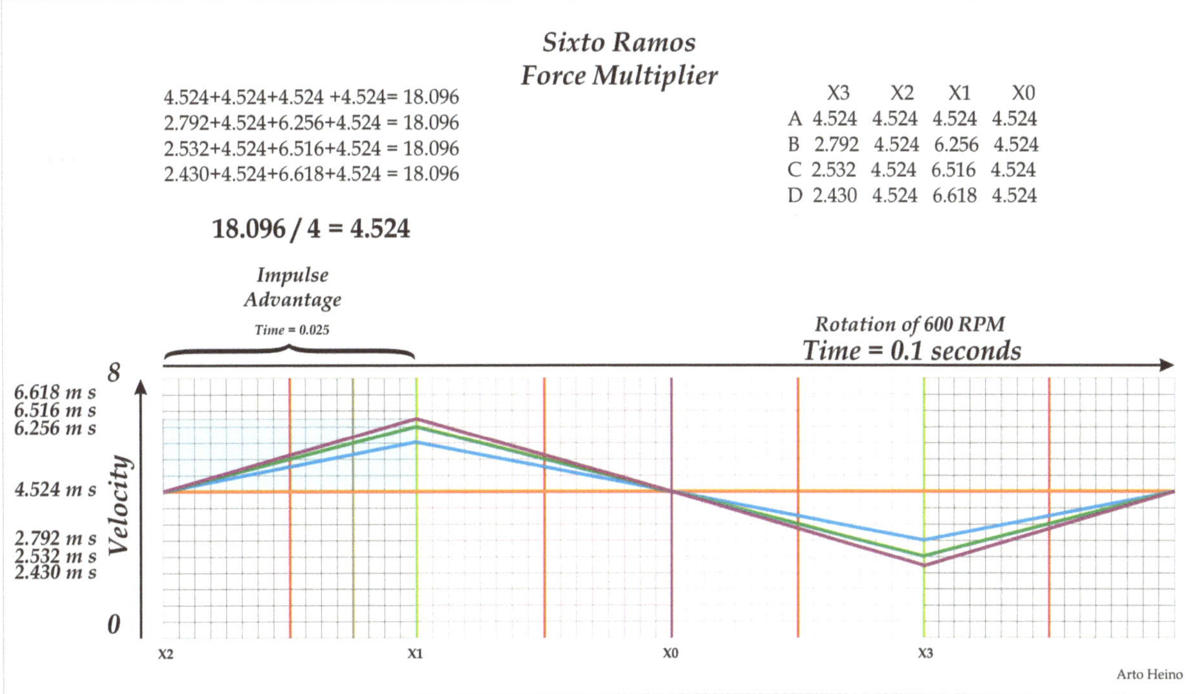

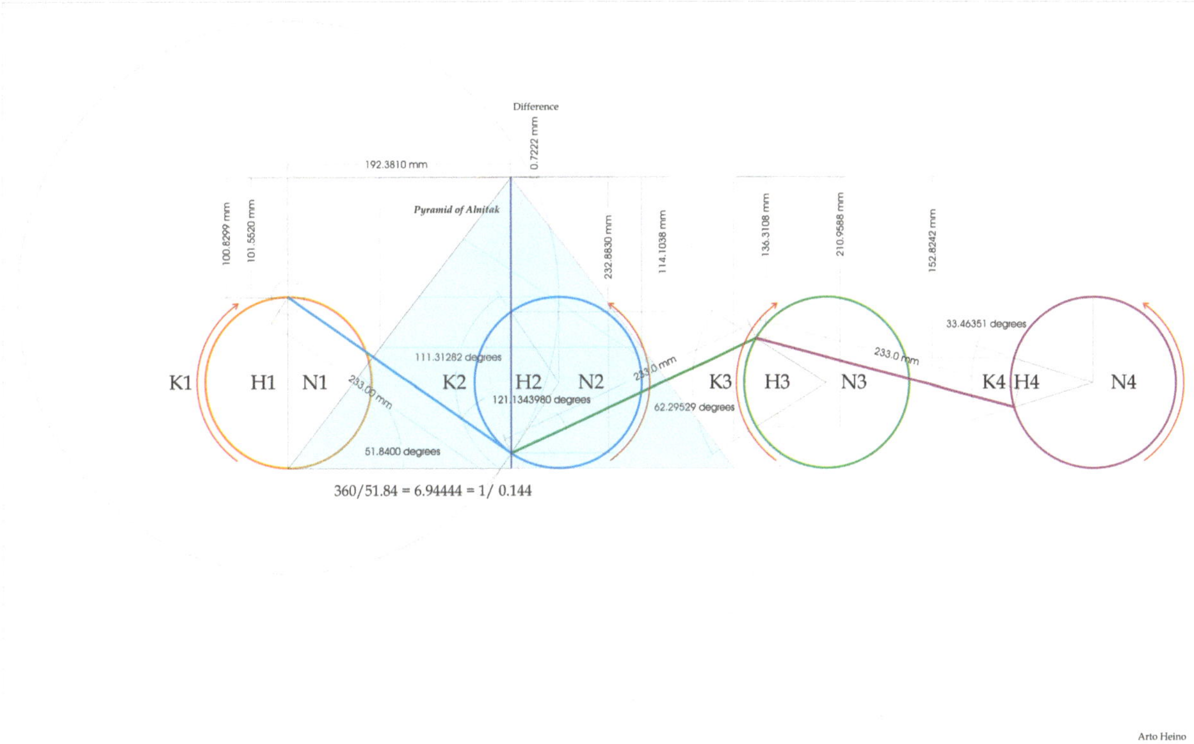

Fine Tuning the parameter of this machine, one point is the weighted arm does not have to be on one end, it should be in the center when you start to tune your arm and weight, resonance is what you are looking for in this combination. RPM, Weight , length of arm, thickness of arm, materials modulus, air resistance, all these factors should be finely tuned long before you worry about any thing else.

Moving the arm to different positions is part of the tuning process, small increments in all aspects of the combinations if you are doing this by trial and error, otherwise use a good dose of engineering equations to simplify your efforts. The old story of Tesla and Edison, where Tesla would work out the math and the problem is 95% solved and Edison would fiddle till the cows come home with 100 workers slogging with trial and error for weeks even months. There are times when there are no maths or equations, this is when you have great discoveries in Engineering and Science, the moment all the complexity gives way to a simple equation.

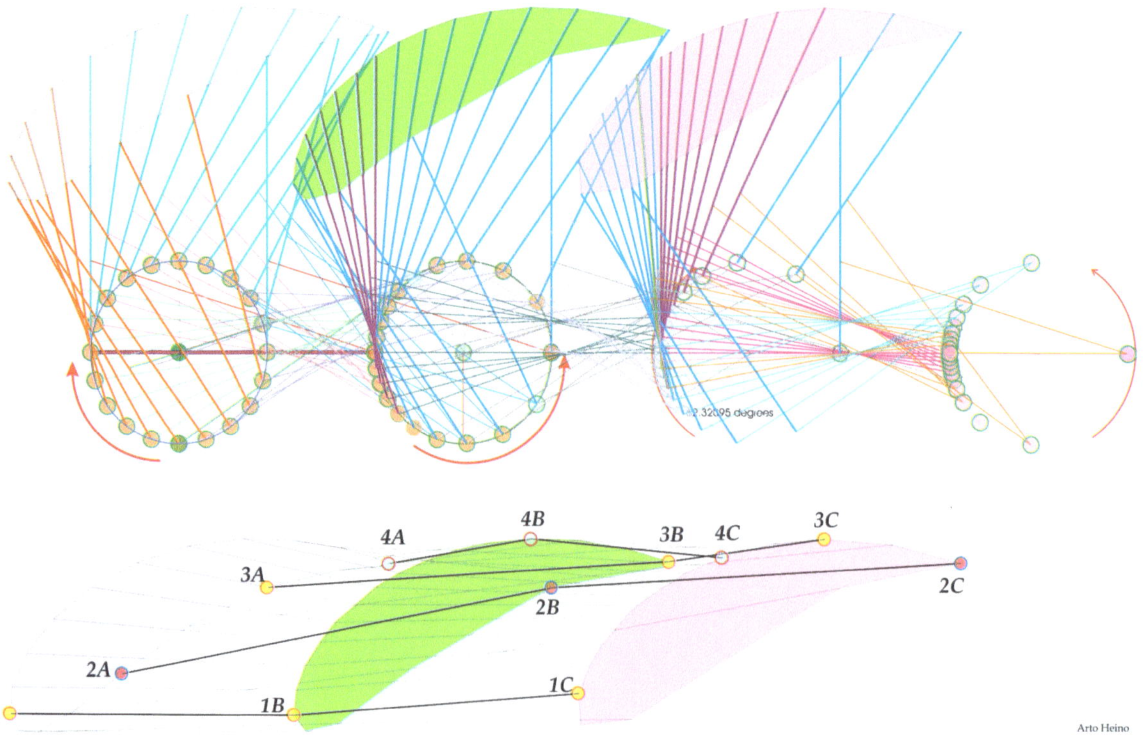

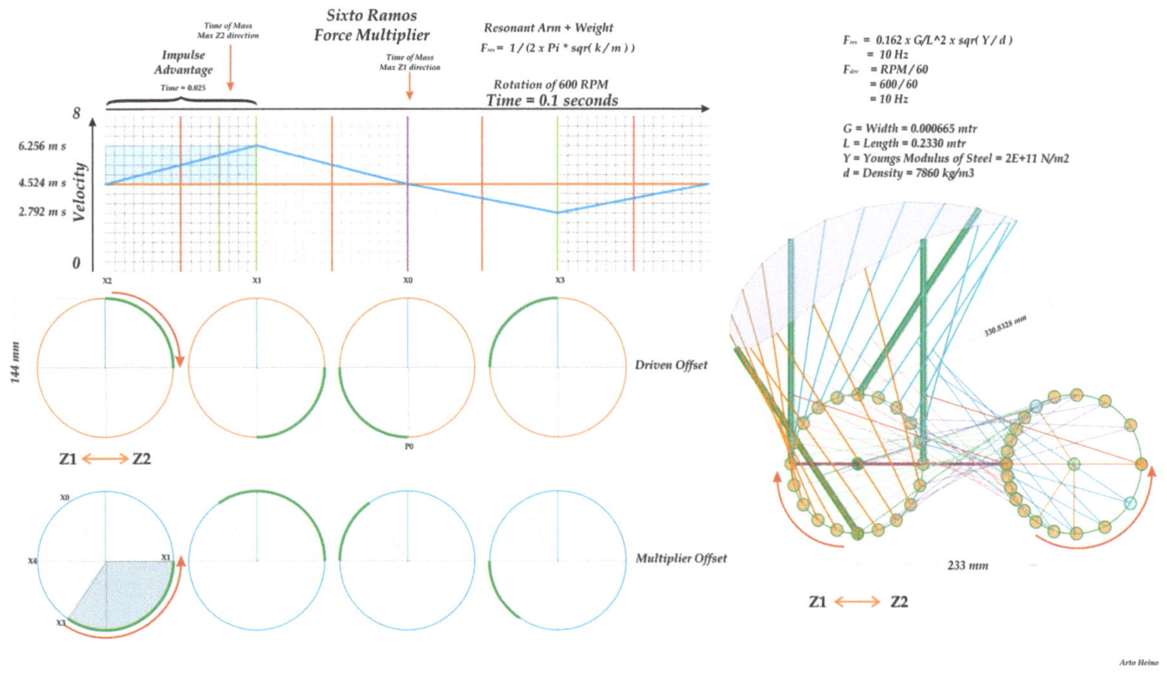

12.4 - Ucros Device

In the Ucros Drive drawing I have employed geometry which has multiple overlays to harmonic, pyramidic, Phi and Fibonacci relationships, this is so we can look for some correlations to ancient pictorial art. This type of multilayer geometric illustration has allowed me to expose these forms to open minded engineers, research like this can possibly unearth some engineering principles hidden within hieroglyphs and art from many ancient cultures(pre 3000 BC). If we can can see into the core of this type of sacred geometry and realize that it is more than pretty pictures that decorate the walls of ancient temples then we have found some parts of the hidden scientific language from antiquity. With the Ucros pivot you can easily see that it has a asymmetric velocity profile, a clue to a natural process or a combination of motions that have chaotic roots.

The UCROS Multiplier Drive by Carlos Ucros

(Corrected by weight pivot arm location)

Artwork by Arto Heino

Degrees rotation - clockwise

Motor	Generator
0	0
30	20.0000760
60	45.9001280
90	75.1556190
120	105.354936
150	137.818544
180	180.000000
210	239.449984
240	286.719519
270	312.699746
300	329.743987
330	344.299827

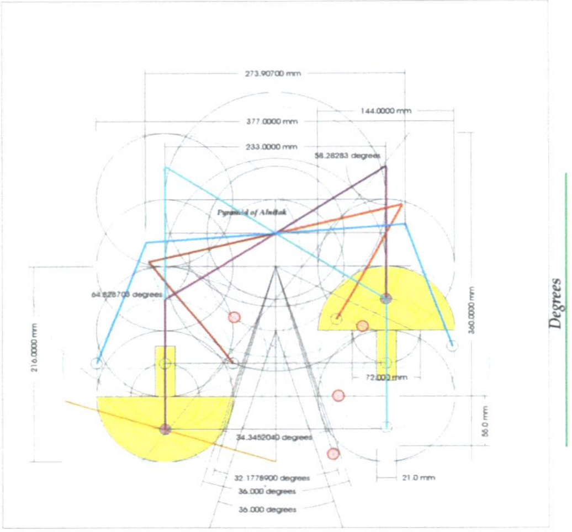

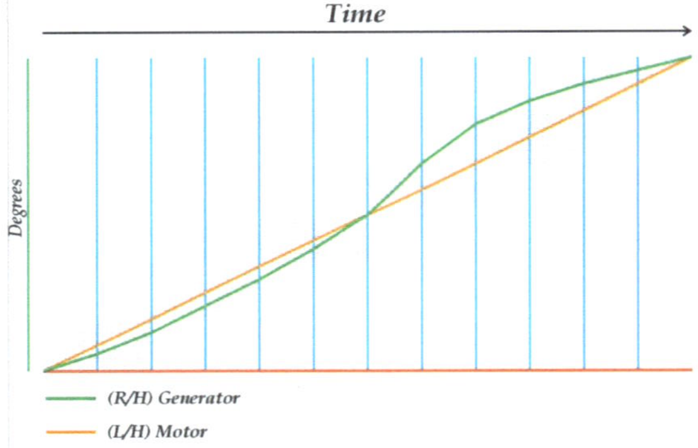

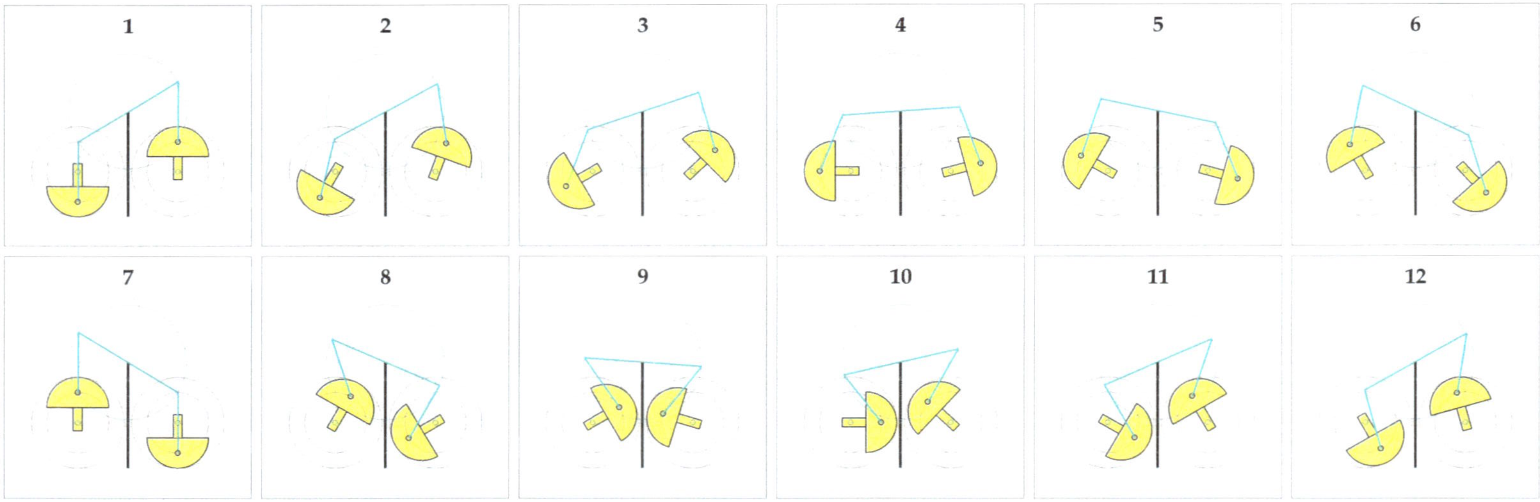

Arto Heino

12.5 - Tesla Receiving Circuit

Method of intensifying effects transmitted through natural media

When Nikola Tesla designed his transmitting and receiving circuits he needed to also design a way of testing his inventions so he could confirm the signals were really being sent as he predicted. In this design he had no diodes, no valves , no transistors, not even a coherer was needed. To grasp his inventiveness, the rotating commutator was his switching scheme, capacitors and inductors would shape the waves while commutators would allow some novel logic to be directed towards his goal of intercepting the nature of his transmitting apparatus.

In a previous chapter I showed how Tesla uses a commutator to create one of his early Tesla coils systems(Patent 583953). If we hope to utilize the natural media, either through magneto-dielectric ground conduction, using elevated radiators or other methods yet to be explored, the need to dump a lot of modern assumptions about the intervening space which for a better word is the Ether.

This medium is currently used as a free for all, highly directed, frequency partitioned, corporate sponsored, , high frequency radiation smog. The industry is full of corporate appointed experts that have no real understanding of the sacred nature of our living environment, they are just trolls for life damaging ignorance. If our vision had an extra couple of octaves added below what we can see, the reality of what is happening to our living system would become clear, you would be virtually blinded by our artificial bands of communications, not just mobile phones, blue tooth and radio but your computer and all of your personal technology would glow with radiations covering a few octaves.

The disappointment Tesla would have in our current use of wasteful Hertzian radiations would not only be his but many others who have not been indoctrinated via Einstein to a science devoid of reality and based on mathematical models that are nothing more than a patchwork quilt made from sterile assumptions by compromised career theoreticians steeped in metaphysical symbolism.

To recover what has been lost we must give our creative spirit a boost by allowing to revert back to more primitive methodologies so we can gain fresh insights to a lost form of electrical engineering that was present about 120 years ago, where the Ether was considered a medium rooted in reality. The circuit that Tesla designed is best described by his Patent descriptions in is own words.(extracts only)

Refer to Patents

US Patent 685953 , 5th November 1901

US Patent 695955 , 5th November 1901

US Patent 787412 , 18th April 1905

CA Patent 142352 , 17th April 1906

This arrangement consists of a cylinder A of insulating material, which is moved at a uniform rate of speed by clockwork or other suitable motive power and is provided with two metal shafts B and B1, upon which bear brushes a and a1, which are connected, respectively, in the manner shown to the terminal plates P and P1, above referred to.

From the rings B and B1 extend narrow metallic segments s and s1, which by the rotation of the cylinder A are brought alternately into contact with double brushes b and b1, carried by and in contact with conducting-holders h and h1, which are adjustable longitudinally in the metallic supports D and D1, as shown. The latter are connected to the terminals T and T1 of a condenser C, and it should be understood that they are capable of angular displacement, as ordinary brush-supports.

The object of using two brushes, as b and b1, in each of the holders h and h1 is to vary at will the duration of the electric contact of the plates P and P1 with the terminals T and T1, to which is connected a receiving-circuit including a receiver R and a device d of the kind above referred to, which performs the duty of closing the receiving-circuit at predetermined intervals of time and discharging the stored energy through the receiver.

In the present case this device consists of a cylinder d, made partly of conducting and partly of insulating material e and e1, respectively, which is rotated at the desired rate of speed by any suitable means. The conducting part e is in good electrical connection with the shaft S and is provided with tapering segments f and f1 upon which slides a brush k, supported on a conducting-rod l, capable of longitudinal adjustment in a metallic support m.

Another brush n is arranged to bear upon the shaft S, and it will be seen that whenever one of the segments f comes in contact with the brush k the circuit, including the receiver R, is completed and the condenser discharged through the same. By an adjustment of the speed of rotation of the cylinder d and a displacement of the brush k along the cylinder the circuit may be made to open and close in as rapid succession and remain open or closed during such intervals of time as may be desired.

The plates P and P1, through which the electrifications are conveyed to the brushes a and a1, may be at a considerable distance from each other and both in the ground or both in the air, or one in the ground and the other in the air, preferably at some height, or they may he connected to conductors extending to some distance or to the terminals of any kind of apparatus supplying electrical energy which is obtained from the energy of the impulses or disturbances transmitted from a distance through the natural media.

Tesla Receiving Circuit

Nikola Tesla

METHOD OF INTENSIFYING AND UTILIZING EFFECTS TRANSMITTED THOUGH THE NATURAL MEDIA
Patent 685953 , 5th November 1901

US Patent 685953 , 5th November 1901
US Patent 695955 , 5th November 1901
US Patent 787412 , 18th April 1905
CA Patent 142352 , 17th April 1906

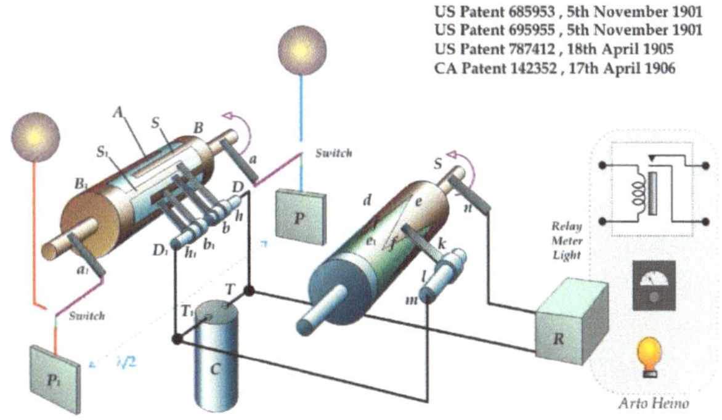

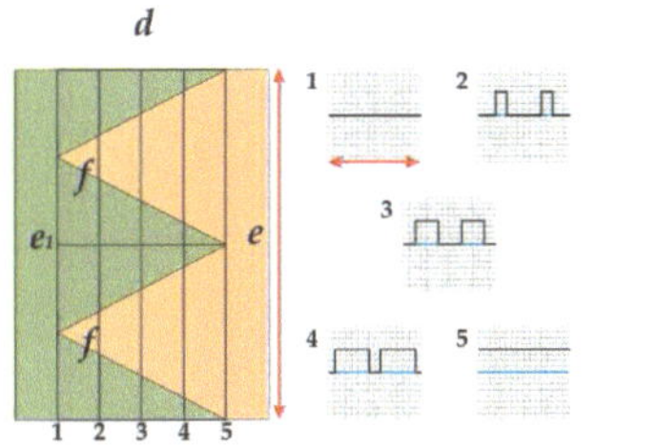

Location of brush k Wave forms at different positions

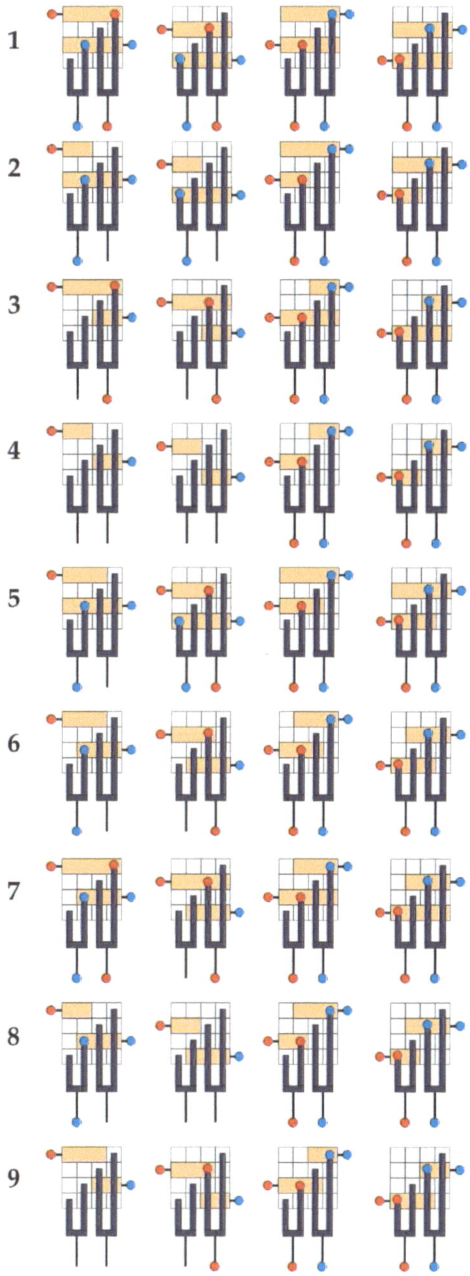

Brush Position Polarity Switching

1	+1	-1	+1	-1	-1	+1	-1	+1
2	+1	0	+1	0	-1	+1	-1	+1
3	0	-1	0	-1	-1	+1	-1	+1
4	0	0	0	0	-1	+1	-1	+1
5	+1	0	+1	-1	-1	+1	-1	+1
6	+1	0	0	-1	-1	+1	-1	+1
7	+1	-1	0	-1	-1	+1	-1	+1
8	+1	0	0	0	-1	+1	-1	+1
9	0	0	0	-1	-1	+1	-1	+1

<table>
<tr><td>4 0 0
0 0
-1 +1
-1 +1</td><td>9 0 0
0 -1
-1 +1
-1 +1</td><td>8 +1 0
0 0
-1 +1
-1 +1</td></tr>
<tr><td>3 0 -1
0 -1
-1 +1
-1 +1</td><td>2 +1 0
+1 0
-1 +1
-1 +1</td><td>6 +1 0
0 -1
-1 +1
-1 +1</td></tr>
<tr><td>5 +1 0
+1 -1
-1 +1
-1 +1</td><td>7 +1 -1
0 -1
-1 +1
-1 +1</td><td>1 +1 -1
+1 -1
-1 +1
-1 +1</td></tr>
</table>

Arto Heino

So I hope this gives clarity with fresh drawings to one of Nikola Teslas favorite receiving circuits, the one that gave him impetuous to carry out bolder experiments and to him verify the World Wireless System as a possibility not just a dream.

12.6 - Flux-gate Magnetometer

To explain the working of the flux-gate magnetometer is straight forward and simple to understand, the diagram F shows four windings and diagram G shows three windings. The two rods could be any material that would give you good design results, in these examples the rods could be two pieces of ferrite rod such as those to used for radio, say 100mmx8mm, these should be available from any local electronic shop.

To truly investigate this device you should wind 8 with a Right-hand Helix and 8 with a Left-hand Helix. Wind the rods with about 300 turns of 30-40 gauge wire, this should give you values that you can engineer with chip amplifiers and digital components, this way you can log data and analyze the results and make some basic conclusions about the preferred combinations that suite the need of your design. Each one of these arrangements has a useful topology when combined with other 2 or 3 point elements used in circuits. One of the most important requirements is the careful matching of the two coils(A,B), this must be within 0.1% tolerance to give truly accurate results. These two coils(A,B) can also have a central wire attached to the bottom connection which is crucial in keeping the system balanced.

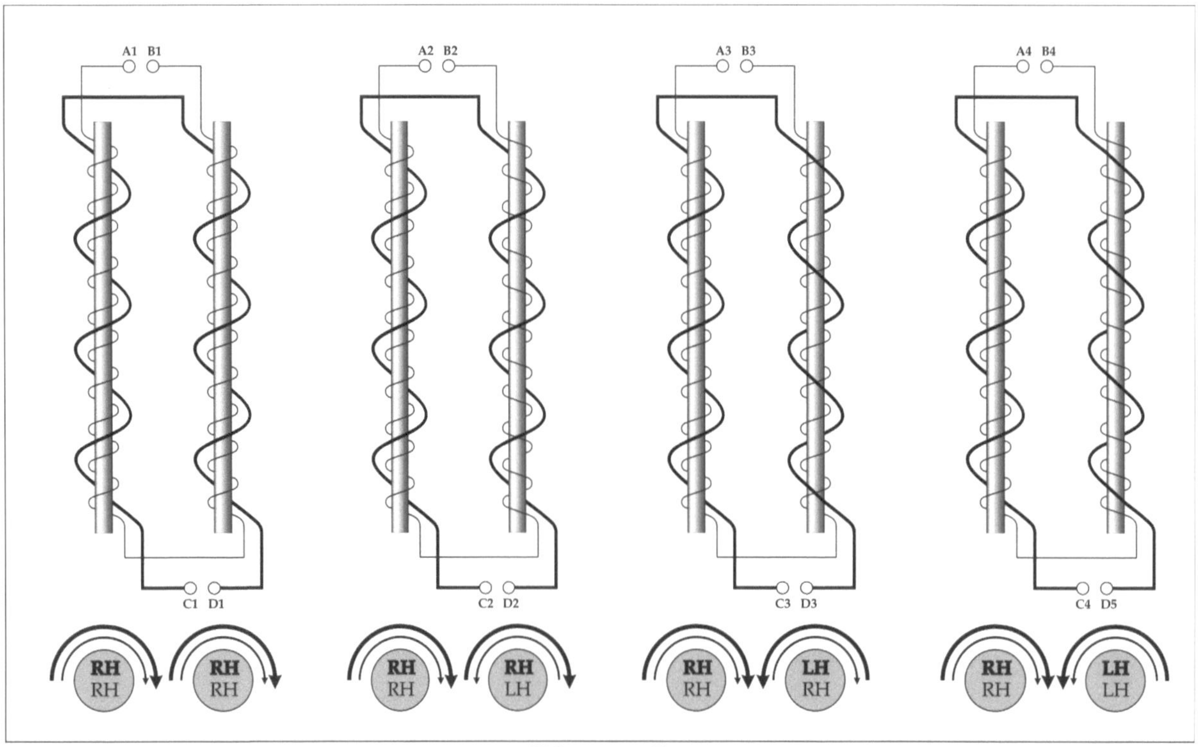

Diagram F

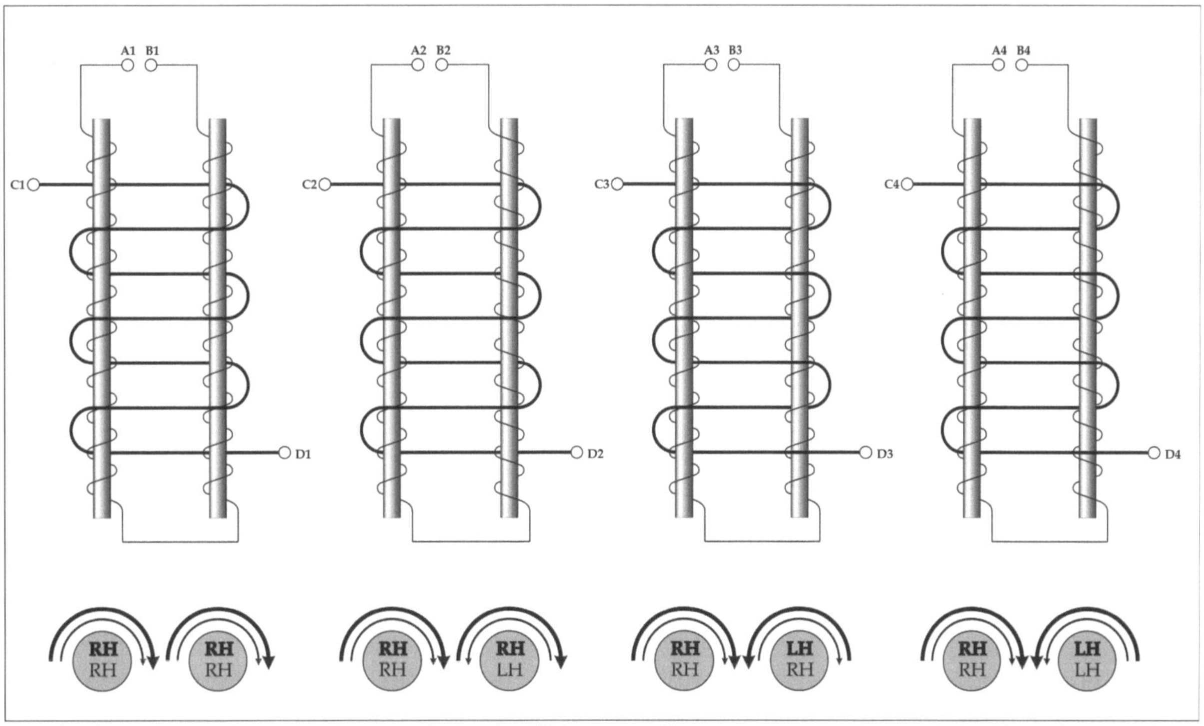

Diagram G

The Flux-gate Magnetometer uses the first arrangement(A1,B1,C1,D1) on each of the diagrams. The two thinner windings(A,B) are sounded by the thicker sense winding(C,D) which detects the excitation if there is a difference of flux between the two separate coils(A,B). The excitation should be at the limit of flux saturation of the inductors, this will make it very sensitive to any variation in the surrounding magnetic field, this is the key to its operation. This is when any stray the magnetic field will affect the coils and the tuning will go out of balance, showing a voltage across the sense winding(C,D)

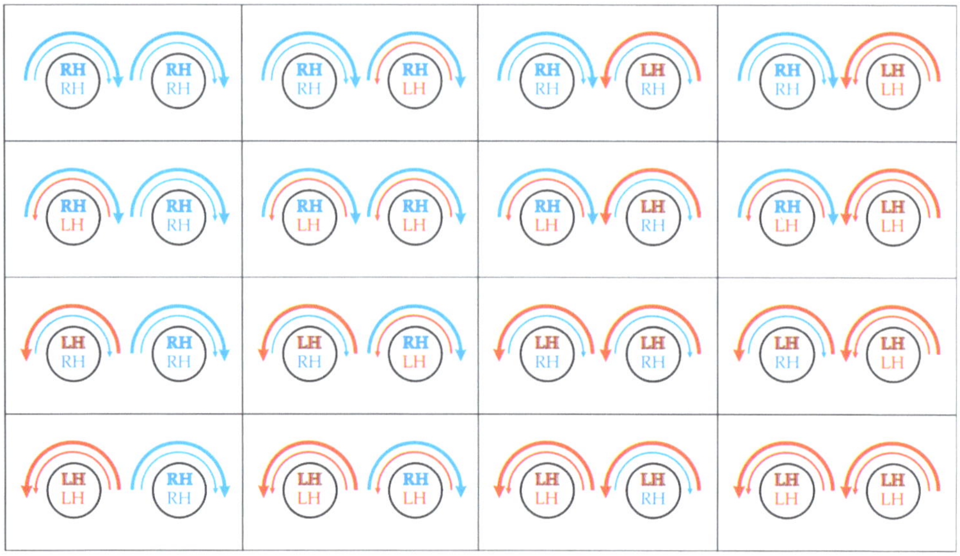

The other arrangements have not been fully explored as sometimes the needs the designer have not arisen, so as you can see from 32 different possibilities only 2 have been used. Some of the are redundant reflections and mirrors of each other, but as a complete system each of these could be switched in and out as required and can serve to enhance or retard voltages and currents.

The interesting aspect of this investigation is the understanding of how we can expand our experiences by simple rearrangements of basic elements. Lester Hendershot is one of those individuals who was inspired by these ideas and developed his own device based on this sort of knowledge. The possible key to Lester and his device could be the exciter frequency, when it is in the region of 60 hz then the coils will pickup any mains source and register a voltage across the pickup coil, this anomaly would have to be verified and accounted for if you choose this frequency for the exciter coils, generally you would choose a frequency that is in prime relation to the mains to avoid this sort of reading.

BIBLIOGRAPHY

The New Penguin Dictionary of Electronics	- E.C. Young	- Penguin Books
Electronic Pocket Reference	- Edward Pasahow	- McGraw Hill
Applied Electricity	- H Cotton	- Cleaver Press, (1960)
Radio Astronomy	- F. Graham Smith	- Penquin Books,
Inventions,Researches and Writings of Nikola Tesla	- Thomas Commerford Martin	- Angriff Press (1894)
Harmonic 695	- Bruce Cathie & Peter Temm	- Sphere Books
Stonehenge Decoded	- Gerald S. Hawkins	- Fontana/Collins
Radiant Energy	- Dr. Thomas Henry Moray	- Research Institute
Fingerprints of the Gods	- Graham Hancock	- Manadarine Paperbacks
Worlds in Collision	- Immanuel Velikovsky	- Sphere Books(1950)
Ages in Chaos	- Immanuel Velikovsky	- Sphere Books
City of Revelation	- John Michell	- Abacus (1973)
The Great Pyramid	- Tom Valentine	- Pantha UK
Prelude to Synchronous Harmonics	- Iver B. Iverson	- Ballard Printing and Publishing
Quantum Arithmetic I II III	- Iver B Iverson	- ITAM, Oregon
Pythagoras and the Quantum World	- Ben Iverson	- ITAM, Oregon
EGYPT: "Image of Heaven"	- Willem H. Zitman	- Frontier Publishing
The Standard Electrical Dictionary	- T. O'Connor Sloane	- Crosby Lockwood and Son, (1902)
Velikovsky Reconsidered	- Editors of Pensee	- Sphere Books
The Sacred Symbols of Mu	- James Churchwood	- Rider and Company Publishers
The Theoretic Arithmetic of the Pythagoreans	- Thomas Taylor (1816)	- Samuel Weiser Inc
The Thirteen Books of Euclid's Elements Vol1,2,3	- Sir Thomas L. Heath (1906)	- Dover Publications Inc (1925)
Colorado Springs Notes	- Nikola Tesla (1900)	- BN Publishing
Symbolic Representations of the Generalized Electric Wave	- Eric Dollard	- Boarderlands
Vacuum Tube Oscillators	- William A Edson	- John Wiley & Sons (1953)

Resources: Chapter 8

Silent Weapons for Silent Wars	2005, Kent Daniel Bentkowski
Radio recordings	William Cooper

The Secret Teachings of All Ages	Manly P. Hall
The Committee of 300	Dr John Coleman
Common Law	Robert Mendard
Through Its Enemies Eyes- Interview	Micheal Scheuer
Interview	Lord Christopher Monckton
The Zionist Roots Of Freemasonry	Micheal Tsarion
Lecture	Harry Beckhough
Zeitgeist	Peter Joseph
Interview	John Perkins
Resource-Based Economy	Jacques Fresco
Kieser Report	Max Kieser
Comedy Film Clips	George Carlin
Totally Absolutely Without Conflict	Jiddu. Krishnamurti
Secrets of the Federal Reserve	Eustace Mullins

Resources: Chapter 9

The Bible

Ancient Egypt.	-Lorna Oaks& Lucia Gahlin	
The Ancient Egyptians,	- Gardener Wilkinson	
Their Life and Customs, Volume Two		
The Great Year in Greek,	- B. L. Waerden	
Persian and Hindu astronomy		
The Dawning	- Terry MacKinnell	
Khemer	- Jean-Pierre Lacroix & Robert Bywater	
The Orion Mystery	- Robert Bauval and Adrian Gilbert	
Aritarchus of Samos	- Sir Thomas Heath	
Text Book on Practical Astronomy	- George L Hosmer	
Spherical and Practical Astronomy	- Dascom Greene	
The Papyrus of Ani,	- Translated by E.A. Wallis Budge	
(the Egyptian book of the dead)		
The Mayan Factor	- Jose Arguelles	
Hidden Records	- Wayne Herschel and Birgitt Lederer	
Fingerprints of the Gods	- Graham Hancock	
Canon and Proportions in Egyptian Art	- Erik Iversen	
The Great Year and the BRI website	- Walter Cruttenden	
John Anthony West	- Various Articles and Videos	
The Temple in Man	- R.A.Schwaller de Lubicz	- Inner Traditions Int.

The Land Of Osiris	- Stephen S. Mehler	- Adventures Unlimited Press
An article Published in Annales du Service Archaeologydes Antiquites de L'Egypte Volume 26, 1926, pages 197 – 202	- Battiscombe Gunn	- Printed by the Institute of France, Oriental Le Caire, France

Pictures:

Artwork and Diagrams x 215		Arto J Heino, Pen and Ink, Corel, GeoGebra	
Chapter paintings		Arto J Heino, Acrylic, Oil, Digital, Water color	
1	Pursuit in Warm	2	Structures in Movement 2
3	Russlan	4	Cellophane Suburbs
5	Blues at Glebe Pub	6	Modern Life
7	Still Life with Keys	8	Neighborhood
9	Dendera Wall Picture	10	Cubist Guitar Improvisation
11	Trees in Time (1967)	12	Still Life with Cubist Guitar
Front Cover, back Cover		Arto J Heino, Painting	
Charts and Tables		Arto J Heino, Open Office Calc	
Lester Hendershot		Hendershot Archives	
The Last Supper		Leonardo Da Vinci, Santa Maria delle Grazie, Italy	

Pictures Chapter 9

Narmer Pallet Pictures	Misha Pletinsky
Eta Leprois	S. M. Lawler
Crab Nebulae	Hubble, Wikipedia
M42, Orion Nebula, The Rosette Nebula,	
Zeta Tauri, Lambda Orionis,	
Cone Nebula	NASA, ESA
Barnards Loop	Hunter Wilson
NGC2301	Fausta nonnatuttua
Pleiades	Tim Easterday
Sumerian Tablet	Bettmann/CORBIS

Help and encouragement:

Robert Adams	Terry MacKinnell	Laird Scranton
Walter Cruttenden	Dale Pond	David G Dawson, Robert Malenica

A special Acknowledgment to my family for putting up with my creative temperament..

and to Shane Kunoh, for his unswerving help with suppressed information over many years

INDEX

AC, alternating current 87 109 111 121 326 332
acceleration 22 59
acoustic 17 35 59
acoustic impedance 35
acoustic inductance 35
acoustic resistance 36
action 73 332
Adams, Robert 43 94 119 120 124 125 128 324 328 329 330
adiabatic 59
aerial 148 160
air 32 133 135 342
Airy, Sir George 163
Aker 256
Alexandrian Library 40 79 293
algorithm 80 155
Allen, Thomas Gaskell 45
Alnitak 273 277 278 280 282
alpha 90 300
aluminium 102 235
Alzheimer 235
ambient 106
amperage 32 95
amplification 128
amplifier 79 90
ancient 92 155 169 243 248 254 256 261 264 267 277 293 300 339
Angkor 169 262
ante-chamber 18
antenna 147 149 150 326
apex 287
Aquarius 79 197 245 261
arc 158
archaeologist 18 39
Archeogeometry 268
Archimedes 155 174
archimedian 194
Aries 197 243 254
arm 44 64 65 166 331 338
art 114 158 160 195 200 217 228 229 244 312
artist 94 160 196 200 201 228 229 243 249 259 261 276 328
Artone 19 206 208 211
Aspen, Harold 330
astable 146
astrological 249
astronomy 17 155 244 249
Atkinson, Bo 116
atmospheric 37 47
atomic 72 73 74 161 167
Atum 274
augmented 208
Aztec 17 261
Ba 249
backlash 22
balance 119
Bank 221
basket weave 115
Bat 249
batteries 68 130
battery 108 109 111 326
Bauval, Robert 244
beta 90
biaxial 116
bible 197 254 255 258 260 291
bifilar 116 117 118 119 120 121 122 123 124 126 129
bifurcation 71
Bilderberger 221
binary 247
binomial 308
blocking 104 105 129
Boltzmann 160
Bonetti, Stefano 43
boost 108
Boskovic, Ruder Josip 160
boundary 72 73 75 80
Bourke 22
bronze 203 246
Brookesmith, Peter 160
Brownian 91
brush 40 110 111 112 342 343
bugle 204
bulk modulus 22 37
buzzer 104
Cadman, John 278
Callan, Nickolas 115
Cancer 197 243 260
cancer 235 236
capacitance 75 85 91 117 125 126 325
capacitor 102 103 109 121 132 133 135 136 137 139 140 147 150
Capricorn 197 243
Carter, Howard 202
Cathie, Bruce 154 155 158 162 167
cavity 17 18 46 273 274
Cayce, Edgar 249
Celsius 36
centroid 284 287 288
CFR 221
chaos 94
chaotic 94 127 339
charge 43 136
chemistry 17
chlorine 235
choke 89
chord 206 213 214
chromatic 205 206 208 209 210
CIA 221 231
circle 177 191 194 274 287 288 291
circuit 94 106 109 111 114 115 126 127 129 149 328 342 343 344
circular 115 294 312
circumcenter 284
circumcircle 287
circumference 187 320 321
civilisation 171 248
Cladorhiza 160
Cockcroft, John 110 137 139
code 154 162 163 166 169 274
coefficient 114 309

coil	27 28 79 91 92 94 102 103 104 111 112 114 116 117 119 120 121 122 124 125 126 127 128 148 150 328 330 344 346
Colorado Springs	82
commutator	111 113 141 341
compression	59 333
computer	80 226
condensation	32 36
condenser	111 112 342
conductor	70 72 139
conic	177
conical	115
connections	82 125
constant	136 157 170 280 320 321
Constantinesco, George	20 55
contra	59 61 62 63 64
controller	111
converter	108
Cook, Daniel	116
Cooper, Leon	278
copper	74 95
corporate	68 90 139 172 200 341
corporation	69 106 221 230
cosmic	90
cosmology	171
Costeley, Guillaume	210
coupling	114 127
CRMRVO	61 62 63 64
crown	116
Cruttenden, Walter	246
crystal	328
Crystallography	167
Cubist	200
cubit	155 158 251 255 267 268 269 272 275 320
current	71 86 110 119
cylinder	59 133 342
d'Arsonval, Jacques	106
Da Vinci, Leonardo	196 197 255 329
damper	58 332 335
DC, direct current	40 106 107 108 109 121
De Salinas, Francisco	210
decahedron	166
decan	249 260 261 300
decay	90
degree	37 93 174 175 181 182 183 184 259 276 291 326 340
Delivery Head	46 48
Delivery Valve	46
delta	79
Dendera	247
density	22 32 35 36 37 302 303 312
DHS	221
diameter	75 91 174
diatonic	156 205 213
dictator	221
didactic	169
dielectric	29 75 116 127
differential	20 86
dimension	170 176
diminished	206 208
diode	56 75 79 111 113 137 139 147 325
distribution	69
diverus	194
dolerite	18
Dollard, Eric	32 72 167 224
drive	64 333
Drive Pipe	46 49
Duau	256
Dunn, Christopher	18 167 279
duple	156
Dynamic Viscosity	35 36
dynamo	40
earth	25 32 91 92 94 149 150 163 166 292 318 320 321 326 328
earthquake	169
east	91 118
Easter	197 266
ecliptic	257
Edison, Thomas Alva	338
Edwards, I.E.S	279
efficiency	48 119
Egypt	116 169 248 249 255 259 261 266 267
Egyptian	17 34 79 244 245 249 250 254 255 256 258 261 262 268 274 289 295 297 298
Einstein, Albert	14 160 161 330 341
electrical	78 80 111 115 150 167 323 329 342
electricity	68 69 72 80
electromagnetic	149
electromagnetism	40
electron	72 328
electronic	68 74 106 114 214 328 344
electrostatic	43 127
element	72 73 83 157 172
elite	69 256
ellipse	170 177 179 180 269 289 319
Eltanin	160
energy	14 43 90 94 106 109 111 112 115 117 118 127 151 161 170 230 301 303 308 328 330 336 342
engineer	69 80 83 114 116 117 160 161 162 200 276 280 312 344
engineered	119 124 127
engineering	117 118 155 329 333 335 336 338
environment	93 94 127 128 147 150 172
equilateral	169 177 192
equilibrium	72
equinox	249 258
eternity	79
ether	14 69 72 73 75 78 162 170 341
Euclid	155 170

Everglades 163
Faraday, Michael 40 69 92 93 115
Farnsworth, Philo 116
fascist 69 160
FBI 221
Feigenbaum 280
Fibonacci 74 83 84 85 86 170 174 199 255 279 339
filter 85
finger 155 158 205 269 272 274
fire 79
Fitzgerald 162
flat 115
Florida 163
fluorine 235
Fluorosis 235
flux 93 95 96 344 345
Flux-gate Magnetometer 344 345
flywheel 59 64 331 335 336
force 14 44 65 91 308 331 332 337 339
fractal 104
free energy 43 69 106
frequency 59 90 91 93 95 96 106 108 109 110 113 114 117 136 145 170 300 301 303 326 328 341 346
front 60
fulcrum 64
Furlong, David 279
fusor 116
Futurists 200
Gailey, Ken 116
galvanometer 86
gamma 90
gas 18 20 36 47 59
gas constant 36 47
Gaussian 177
gear 64 69
gearbox 20
Gemini 197 243 248 251
General Electric 69
generator 22 43 91 92 106 325
geomatra, *geomatria* 155 156 162 164 165 167 291
geometric 46 72 90 115 153 155 169 177 213 268 326
geometry 17 74 80 116 118 124 170 172 173 174 176 178 181 194 197 268 277 283 314 333 339
germanium 74
Gilbert, Adrian 244
Giza, *Ghiza* 40 163 164 165 166 244 279 321
Gobekli Tepe 169 266
golden 197 199 246 259 260
golden age 79 244 255
Golden mean 255
Grand Gallery 18 19
gravitational 14
gravity 22 36 40 47 170 303 305 306 308 312 320 321 333 336
Great Pyramid 17 20 164 274 277 278 321
Great Year 243 246 292
Greek 17 34 197 244 245 246 255 261 293
Greenwich 163 164
grid 158 160 162 164 165 166
ground 147
growth 74
guitar 208 211
Gunn, Battiscombe 270
Hancock, Graham 268 279
harmonic 32 93 109 127 154 155 159 160 161 163 203 208 210 276 339
harmony 125 156 157 206 214 287
Hatch, Larry 160
heat 79 127
Heaviside, Oliver 69
helical 245 246
helix 116
Helmholtz Resonator 19 35
Helmholtz, Herman von 116
Hendershot, Lester 90 91 92 94 97 98 99 100 101 116 346
henry 146
Henry, Joseph 115
Hermite, Charles 284
Herschel, Wayne 244 249 250
hexagon 189
hexaphone 205
Hexartom 19 206 208 214 215 219
Hierakonopolis 254
hieroglyphs 79 255 260 261 339
Hilbert, David 160
Hipparchus 245
Hooper, William 116
Horus 197 258 262 274 293
House, Edward Mandell 221
Hubbard, Alfred 90 116
Hunkler, Tim G. 279
hydraulic 69 146
hydrogen 18
hyperbolic 64 333
Hyrdo-Automat 45
IC, impulse current 107 108 109 121
idle 58 111
Illuminate 221
impedance 35 85 203
impulse 64 65 111 331 333 335 337
in-circle, incircle 187 287 294
Inca 261
incenter 284
inch 158 160 168 203 274
Indian 45
indoctrination 69 72
inductance 27 28 91 114 117 125 126 325
induction 111
inductor 114 121 345
inertia 59 170
inertial 58 94
infinity 116 128 129 131 132 154

insulated 111 342
insulator 72 74 139
interval 157 208 217
inverse 75
inverter 108
iron 246 325
isosceles 169 177 178 181 182 183 184 192 193
Iverson, Ben 169 178
Iverson, Erik 255
Johnson, Charles William 160
joules 108 109 125
Keely, Worrell 22
kelvin 36
Khemetian 17 40 244 247 249 250 254 255 268 274 275 289 293
Khemit 248 275 276
Khmer 262
kinetic 161
Kings Chamber 17 18 273 274
kite 193
Klein, Felix 160
knowledge 45 80 157 160 248 261 346
Kronecker, Leopold 284
Laboy, Samuel 268 279
latitude 92 93 163 165 166
Lavrentiev, Oleg 116
layout 114
Lehner, Mark 279
Leo 197 243 259
lever 44
Libra 197 243 261
library 80
light 32 72 75 129 154 155 159 161 267 291 320 325
liquids 20
longitude 162 163 164 166
longitudinal 20 29 32
loop 118 122 124
Lorentz 162
M42 250 251
machine 39 40 45
Machu Pichu 155 172
MacKinnell, Terry 244
magic square 122 123 125 127 166 194 300 301 302 303 305 311 312 313 314 316
magnet 92 325 326 330
magnetic 59 61 62 63 64 87 91 92 93 115 121 345
magnetism 127
Magneto-Dielectric 32 341
magnify 117
major 205 206 207 208 213
mandela 116
Mandelbaum, Joel 210
mankind 68
Marconi 69
mass 14 47 161 301 302 303 312 336
mathematical 69
matter 14
Maxwell, James 116 177
Mayan 169 261
mechanical 53 56 69 79 110 336
media 222 342
medium 20
Mehler, Stephen 244
meridian 163
Meskhenet 258
metal 72
metaphysical 69 90 341
meter 158 163 267 275
methodology 69 292 293
microwave 114
mile 158 159 160
minor 206 207 208
minute 158
mobius 115
momentum 161 331
monument 39 155 171
Moray, Henry T. 72 90
motor 22 40 43 94 106 111 119 122 124 128 327 328 329 330
movement 13 170
MSA 221
Mu 79
multi-filar 117 126
multi-vibrator 24 94 145 146
multiplier 137 139 141 162 331 337 339 340
Munck, Carl 162 163 164 167 169
muror 206 207 208
music 155 156 202
mutual 114
mythology 196 249 250 256
Narmer 249 250 251 254 255 258 262
natural 80 127 155 157 167 169 204 280 339 342
nature 94 104 171 291 306 341
nautical 159
negative 65 111
network 125
New York Times 161
Newton, Isaac 14 320
Nexus 329
NGC2238 250
Nineveh 157
non-inductive 116
nonagon 190
north 163 166
novtuple 157
NPN 78 145 146
O'Leary, William 116
OC, oscillating current 107 108 109 121
octagon 190
octave 205 208
octuple 157
Oersted, Hans Christian 115
ohm 136 325
Ol Jergen Bryn Apex 278
One-way 56 65 331
opposite 128
organism 127

Orion 197 244 248 250 251 254 258 260 279
orthocenter 284
oscillate 90
oscillating 72
oscillator 53 59 61 62 63 64 94 104 105 129
Osiris 197 259 274
Ostrakon 268
over pressure 24 35 36 37
oxygen 18 19
Panther Mound 163
parallel 83 85 87 91 94 129 140 145 146 148 149 194
parametric 22 59 91 94 128
Parker 187
particle 72 90
particle velocity 35 37
parvatic 208 209 210 211 217
pascal 23 310
patent CA142352 342
patent US1597664 45
patent US381970 71
patent US568176 141 142
patent US568177 142
patent US568178 143
patent US568179 144
patent US568180 144
patent US583953 110 341
patent US685953 341
patent US695955 341
patent US787412 341
pattern 125 127 205 206 314
pendulum 40
pentagon 160 189 194 296
pentangle 194
periodic 184
Petrie, William Flinders 18
Pharaoh 254
phase 22 170
phi 14 91 159 255 274 279 300 325
PHI 14 84 155 157 159 168 169 272 275 277 280 281 283 291
Philo 186
photon 90
physicists 69 72
physics 167 171 225
pi 153 154 157 164 165 168 169 187 272 274 275 277 279 280 281 282 283 294
piano 157 336
Picasso, Pablo 200
piezo 18
pipe 23 24 35 36 37 45 47 48 219
Pisces 197 243 245 246 254 260 261
piston 22 59
pivot 59 65 331
Plank, Max 160 161
plate 90 111 112
Pleaides 251 258
polarity 69 326 343
polarized 137
poloid 116
poloidal 115
polyhedral 115
polynomial 308
Pond, Dale 178
positive 111
power 20 87 91 95 108 111 221 293 328 336 342
pressure 14 22 23 32 35 36 37 170
primary 95 96 111 112 166
prime 155 163 164 167 168 274 275 284 291 346
pulse 43 111 122 124 128
Puma-Punka 40 155
pump 43 45 91 93 110 326
pyramid 163 244 247 268 273 277 278 279 282 284 293
Pythagoras 155 293
Pythagorean 170 176 177 217
quadrangle 183 193 194
quadrifilar 117 126
quadrilateral 185
quadruple 156
quantum 17 40 155 169 170 176 177 178 179 180 181 182 183 184 192 193 194 213 214 225 255 268 272 273 274 275 276 277 278 279 283 284 287 288 291 294 297 318 319
Queens Chamber 19 20
Quibell, J.E. 249
quintuple 156
Ra 34 254
radial 166
radiation 32 127 147 148 170 341
radiator 90
radio 32 69 87 90 227 328
radioactive 90
radium 90 91
radius 73 187 294 321
Ram Pump 45 48
Ramanujan, Srinivasa 284
Ramos, Sixto 331 333 334 336 337 339
random 104 127
ratio 155 159 169 187 222 279 282 296
ray 90
reactance 117
rectified 110
reflector 72
relativity 225
relaxation 59 104
reluctance 128
Renaissance 196 200 246 255 261
repulsion 59 61 62 63 64
resistance 95 96 136 325
resistor 79 83 84 121
resolution 44
resonance 59 106 128 337
resonant 117 273 330 332 333

resonator 17 18 19 25 85 86 91
resurrection 79
retarding 59 65 336
rhombus 193
rings 115
Rodin, Marco 116
Roman 17 40 255 260 261
Romanian 20
root 176 177 269 293 339
rotate 64 306 336 342
rotating 59 69 335
rotation 58 65 115 331 337
rotational 59 61 62 63 64 194 336
rotor 128
royal cubit 31 34 158 251 255 267 274 320
Royal Observatory 163
Royal Raymond Rife 72
rubber 74
Ruhmkorff, Heinrich 116
Sagittarius 187 243
Sakkara, *Saqqara* 155 268
salt 91
scale 205 206 209 211 214
scalene 192
Schauberger, Victor 72 160 224 329
schematic 80 97 98 99 100 101 106 113 139 330
school 223
Schumann Cavity 32
Schwaller de Lubicz 167 266
science 90 94 117 155 225 248 312 341
scientific 69 104 330 339
scientist 72 90 115 116 160 162 330
Scorpio 197 243 260
scribe 80
sea 36 37 47 48 320 321
second 24 154 158 163 267
secondary 96 111
Sef 256
segments 111 205 312 313 342
Sekhmet 244
Semi-conductors 74 77 139
septuple 157
sequence 84 153 314 326
series 83 111 140 145 148 176
sesquiquialter 155
sesquitertian 155
sextuple 157
shaft 18 20 336
side 60
silicon 141
silver 203 204 234 246
Sirius 246 251 258
Smith , Don 106 107 109
Smith, Henry J.S. 284
Smith, Wilber 116
Sol 246
solenoid 115 117 118 123 129
solids 20
Solstice 258
sonics 20
sound velocity 36 37 38
source 46
space 14 72 115
spherical 115 161
Sphinx 163
spin 91
spiral 27 115
Spitzer, Lyman 116
sprag clutch 56
spring 59 79 335
square 82 188 193
stages 49 50
star 115 243 247 250 260 262 295 297
starship 116
stator 119 126 325
Steinmetz 69 72
Stellartator 116
Stravinsky, Igor 200
stress 69 72 73 74
Sumer 169 267
Sumerian 40 262
Sun 247
superpaticulars 155
Supply Head 46
surface area 48 176
switch 87 110 111 139 140 141 325 326
symbol 79 80 87 258 260 268 341
tank 45 47 48
Taurus 197 243 248 249 251 259
Technetium 184
teeth 57
temperature 36
tempered 214
terahertz 90
Tesla coil 136 141 142 143 144 166 328 333 335 336
Tesla Turbine 52
Tesla, Nikola 20 25 26 27 28 29 30 32 40 41 42 69 70 71 72 82 90 106 110 111 113 114 116 117 119 120 125 141 160 161 162 167 224 323 329 338 341 343
tetragon 193 194
Theodosius 260
theoretical 69 72 74 90 158 225
theory 167 169
theta 34
Thoth 197 274
time 14 154 326 333 337 339 342
Tokamak 116
Toltec 40 261
tonic 156
top 60
toroidal 115 116 217
Torque converter 55
torus 70

transceiver 29 30
transformer 94 95 111 118 121 333
transient 72
transistor 78 111 113 328
transmit 25
trapezoid 193 194
triad 205
triangle 188 192 284 285 287 294 310
trifilar 126
trinity 79
triple 156 177
trumpet 203 204
tube 38 66 67 113 216
Tutankhamen 202 203 204
Ucros, Carlos 346
unifilar 126
unity 170
Uranium 90
vacuum 22 61 62 63
valve 18 46 49 90
vector 305 306
Vedic 17
velocity 22 23 31 32 33 35 43 72 75 93 154 155 159 161 170 203 267 303 325 333 339
vibrate 94 332 335
vibrator 87 93 94
Virgo 197 243
voltage 32 93 94 95 110 113 117 119 139 141 149 325 328 345 346
volts 117 129 133 136 147
volume 28 35 49 74 75 95 176 302 312
volumetric velocity 32 35 37
vortex 66 67
Walter Russell 72 160
Walton, Ernest 110 137 139
war 230
Wardenclyffe 32 69 323
Washington 163
Waste Valve 46
water 20 22 45 47 79 132 235 261
water-hammer 22 24
watt 108 129
wax 90
west 91 92 118 166
Weston Fisher 22
wheel 64 65 332
Wiegand, John 116
Wilhelm Reich 72 224
Wilson, Woodrow 221
winding 116 117 124 129 325 344 345
wire 27 70 74 87 93 95 102 103 118 125 126
Woolhouse, Wesley 210
Yuga 244 246 260 261
Yukteswar, Sri 244
Zep Tepi 244 259
Zodiac 197 245 247 257

DATES

12000 254
9100 259
5466 250
4000 262
3100 249
3000 244 247 262 339
1333 202
727 244
702 BC 260
1 AD 244
10 260
200 260
391 260
500 260 262
721 260
727 260
800 260
900 262
1054 250
1433 245 246
1577 210
1588 210
1700 261
1800 177
1820 115
1824 262
1831 115
1835 210
1836 115
1851 116 163
1859 284
1863 284
1869 116
1871 116
1873 116
1884 163
1890 116 328
1891 116
1894 116
1895 139
1897 249
1901 341
1905 341
1906 342
1913 20 221 284
1917 69
1918 104 116
1920 102 116 137
1921 221
1922 202
1925 268
1926 45
1946 116
1950 116 178
1952 116
1961 210
1964 116
1965 284
1968 116
1969 328
1980 154
1986 116 328
1987 244
1988 277
1990 116 128 162 172 178
1991 116
1992 116
1993 116
1994 116
1996 124 205 214
1998 206
1999 122 124
2004 206
2007 116
2009 214
2012 115
2145 261

FREQUENCIES

1093.75 hz 108
120 hz 91 108
15.362 hz 92
1.59099 hz 109
1.76Mhz 19
2 hz 19
216000 hz 166
236000 hz 91
2.4766 hz 19
270.80685 hz 109
3500 hz 108
35000 hz 108
43 hz 91
47 hz 19
5.45 jz 91
60 hz 91 96 108 346
62.76 hz 92
68 hz 93
7000 hz 108
70000 hz 108
70710.687 hz 109
7.85Hz 18
82 hz 92
8246 hz 91
98994.94 hz 108

www.ingramcontent.com/pod-product-compliance
Lightning Source LLC
LaVergne TN
LVHW070945210926
848482LV00002B/101
9781876406035